Springer Series in
CHEMICAL PHYSICS 69

Springer
Berlin
Heidelberg
New York
Hong Kong
London
Milan
Paris
Tokyo

Springer Series in
CHEMICAL PHYSICS

The purpose of this series is to provide comprehensive up-to-date monographs in both well established disciplines and emerging research areas within the broad fields of chemical physics and physical chemistry. The books deal with both fundamental science and applications, and may have either a theoretical or an experimental emphasis. They are aimed primarily at researchers and graduate students in chemical physics and related fields.

Series homepage – http://www.springer.de/phys/books/chemical-physics/

Volumes 1–62 are listed at the end of the book

Igor S. Osad'ko

Selective Spectroscopy of Single Molecules

With 87 Figures

Springer

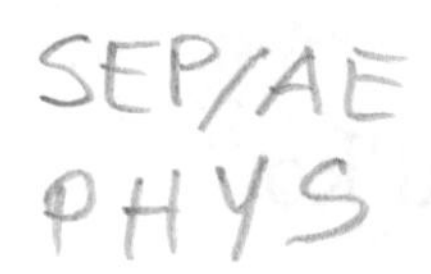

Professor Igor S. Osad'ko
Lebedev Physical Institute of RAS
Department of Luminescence
Leninsky Pr., 53
119991 Moscow, Russia

Series Editors:

Professor F.P. Schäfer
Max-Planck-Institut für Biophysikalische Chemie
37077 Göttingen-Nikolausberg, Germany

Professor J.P. Toennies
Max-Planck-Institut für Strömungsforschung
Bunsenstrasse 10
37073 Göttingen, Germany

Professor W. Zinth
Universität München,
Institut für Medizinische Optik
Öttingerstr. 67
80538 München, Germany

ISSN 0172-6218

ISBN 3-540-44101-8 Springer-Verlag Berlin Heidelberg New York

Library of Congress Cataloging-in-Publication Data applied for.

Die Deutsche Bibliothek – CIP-Einheitsaufnahme

Osad'ko, Igor' S.:
Selective spectroscopy of single molecules/Igor S. Osad'ko. –
Berlin; Heidelberg; New York; Hong Kong; London; Milan; Paris; Tokyo:
Springer, 2002
(Springer series in chemical physics; 69)
(Physics and astronomy online library)
ISBN 3-540-44101-8

Springer-Verlag Berlin Heidelberg New York
a member of BertelsmannSpringer Science+Business Media GmbH

http://www.springer.de

Typesetting: Camera-ready copy by the author
Cover concept: eStudio Calamar Steinen
Cover production: *design & production* GmbH, Heidelberg

Printed on acid-free paper SPIN: 10867292 57/3141 - 5 4 3 2 1 0

Preface

The main topic in this book is the low temperature selective spectroscopy of impurity centers. It deals not only with the well developed methods of selective spectroscopy of molecular ensembles such as spectral hole-burning, fluorescence line-narrowing and femtosecond photon echoes, but also new methods of nanospectroscopy such as single-molecule spectroscopy. Both theory and experimental data are discussed. Theory united with experimental data enables one to organize the experimental facts. This organization is of great importance, especially for young researchers and students, because it can provide better insight into the nature of the rather complicated dynamics of guest molecules in polymers and glasses, where our understanding is still somewhat lacking.

The book mainly addresses young researchers and graduate students studying modern methods of spectroscopy of solid solutions. For this reason, all formulas commonly used in practice are derived in this book. The reader will also find new approaches and formulas which have been used by the author, but are not well known to the majority researchers. When the derivation of a formula seems rather complicated, it is included in appendix. The reader is then free to look the derivation up if he or she is interested. Otherwise, the book can be read without reference to these sections. The book includes many examples where theory helps us to interpret experimental data of various types, for instance, the data on line- and hole-broadening.

The theory presented in the book includes my original results. These were obtained in collaboration with my students S.N. Gladenkova, A.U. Jalmukhambetov, S.A. Kulagin, M.A. Mikhailov, A.A. Shtygashev, S.L. Soldatov, L.B. Yershova, and N.N. Zaitsev. I thank them and Dr. L.A. Bykovskaya for their help, criticism and collaboration. I am also grateful to Prof. Michel Orrit for valuable discussions when I started research on single molecules in his group at Bordeaux University. Some problems presented in the book have already been discussed in my lectures for graduate students at the Moscow State Pedagogical University and for Soros teachers in Russian schools. This experience was used while writing the book. I thank Prof. M.D. Galanin for comments on the Russian version of the book issued by Nauka Publishing House and graduate student Michel Savin who prepared the LaTeX version of this book.

Moscow, July 2002 *I.S. Osad'ko*

Contents

Part II. Phonons and Tunneling Excitations

Part III. Spectroscopy of a Single Impurity Center

Part IV. Optical Band Shape Theory for Impurity Centers

Part V. Methods of Selective Spectroscopy

Part VI. Transient Coherent Phenomena in Solids

Part VII. Low Temperature Spectral Diffusion in Polymers and Glasses

Introduction

Quite recently researchers could only dream of investigating a single molecule or atom. However, recent impressive progress in nanomicroscopy has enabled scientists to realize this dream. First of all, it was understood that the Coulomb part of the oscillating electromagnetic field enables one to increase the space resolution of an optical device by a considerable factor. This Coulomb field, the so-called near field, is used in scanning near-field optical microscopy (SNOM). SNOM is able to overcome the optical limit connected with the light wavelength and to reach space resolutions of a few nanometers [1].

However, the most impressive progress in space resolution is connected with the invention of scanning tunneling microscopy (STM) [2]. In STM, light as the source of information is replaced by a tunneling current. STM provides detailed images even of single atoms and single molecules lying on a surface. The space resolution of STM exceeds that of any device using light excitation. It seemed for a while that light had lost the competition with tunneling electrons as a source of information about nano-objects. However, single-molecule spectroscopy (SMS) with selective laser excitation refutes such a conclusion. This is because selective SMS allows us to research single molecules both on surfaces and in bulk. STM cannot be used to research single molecules in the bulk.

SMS emerged about ten years ago [3,4], puzzling researchers at the time because they had to deal with a wide variety of fluctuations in each spectroscopic experiment with single molecules. Jumping optical lines, blinking fluorescence and the like are typical phenomena in SMS. Quantum mechanics is not able to calculate the appearance of fluctuations in the time scale because it only gives prescriptions for calculating probabilities. How can we reconcile the fluctuating character of the data measured in an experiment with a single molecule with the fact that quantum mechanics delivers only probabilities? This is the problem discussed in detail in this book.

As a rule, the spectroscopy of polyatomic organic molecules deals with molecules dissolved in solid or liquid matrices. Under such conditions, the electronic excitation of the molecules always interacts with intermolecular vibrations, i.e., phonons. Less than forty years ago, this interaction was considered detrimental because it was assumed to be responsible for the low resolution of molecular spectral bands. Since this interaction cannot be removed, it was commonly accepted that spectroscopy of polyatomic molecules was destined to be a low-resolution spectroscopy.

The first sound argument against this point of view appeared after studies performed by the Shpol'skii group [5, 6], which showed that the optical spectra of polyatomic molecules of various classes dissolved in normal alkanes, now called the Shpol'skii matrices, already exhibit a well-resolved vibronic structure at 77 K and reveal a phonon structure at 4.2 K. As was established later, the advantage of Shpol'skii matrices over other nonpolar matrices is the absence of a large inhomogeneous broadening of the optical spectrum, which strongly deteriorates the resolution of the optical spectra in other matrices. The electron–phonon and vibronic interactions do not impair the degree of resolution of the optical bands, and now these interactions can become the subject of study, i.e., they are treated as useful rather than adverse, whereas the inhomogeneous broadening was declared detrimental.

The inhomogeneous broadening caused by a spectral distribution of the electronic frequencies of molecules found in various local environments dominates in amorphous media such as glasses, alcohols, polymers, etc. In order to remove it, one can use selective excitation [7, 8] or, better, laser excitation [9, 10] of fluorescence. A monochromatic laser light efficiently excites only those impurity centers whose zero-phonon lines (ZPLs) are resonant with its frequency, thereby performing a frequency selection of the molecules. A similar frequency selection is performed in hole-burning spectroscopy, which has gained wide acceptance after the discovery of persistent spectral holes in the optical spectra of many polyatomic molecules [11, 12]. At present, fluorescence line-narrowing and spectral hole-burning are the main methods of selective spectroscopy for solid solutions. This spectroscopy is based on the selection of molecules from an inhomogeneous ensemble over their ZPL frequencies.

Would it be possible to reduce the number of impurity centers selected by the laser light to unity? This question seemed reasonable for selective spectroscopy of solids. It arose naturally in experiments aiming to detect single atoms in beams [13, 14], in radio-frequency traps [15], on solid surfaces [16], and in liquid flows [17, 18]. The concept of single-molecule spectroscopy was already ripe by the early 1990s [19–22]. Following the papers [3] and [4], in which an optical line of a single pentacene molecule in a *p*-terphenyl crystal was measured, the number of papers on single-molecule spectroscopy (SMS) in polymers, glasses, and polycrystals began to grow rapidly.

This growth testifies that selective SMS possesses attractive features and offers a variety of applications. By detecting the probability that a molecule resides in the triplet state, a magnetic resonance of one spin [23, 24] and the Hahn echo on the triplet state of a single molecule [25] have been observed, the vibronic spectrum of a single molecule has been measured [26], and some interesting data on the local dynamics of impurity centers have been obtained [27–29]. The detection of single-molecule fluorescence is now the most promising tool for the direct observation of single protein molecules [30–32], single light harvesting systems [33], and fluorescence resonance energy transfer in single biomolecules [34]. A review of the first papers on SMS is presented in [35].

In the author's opinion, the main achievement of the first papers on selective SMS is that they have demonstrated an unexpectedly wide variety of local conditions in which the individual molecules dissolved in amorphous media may find themselves. As a result, the study of the local dynamics of a molecule interacting with its environment, i.e., of an impurity center, has become one of the main trends in selective SMS. SMS has revealed the fascinating variety of local dynamics in amorphous media. An impurity molecule is used in SMS mainly as a probe for studying its environment. A photon emitted by a molecule provides information on the local dynamics of the given impurity center. For this reason, SMS is a particularly efficient means for studying impurity centers in amorphous media.

The first experimental observations in SMS were not treated theoretically. However, the theory brings more insight into the experimental facts obtained via SMS methods. The theory of impurity centers developed in detail over the last four decades is, in fact, the theory of single molecules which interact with phonons and tunneling degrees of freedom. What is the current state of affairs in the field of single-molecule theory? To what extent can the dynamical theory of impurity centers in crystals developed over the last four decades be used in SMS? All these topics are discussed in the present book.

SMS is, in fact, the spectroscopy of a guest-molecule-doped amorphous solid. However, in contrast to single crystals, amorphous solids which serve as matrices for polyatomic molecules have additional tunneling degrees of freedom. These manifest themselves, for example, in hole-burning spectroscopy and yield unusual temperature and temporal effects, such as spectral diffusion. Therefore, the dynamical theory of impurity centers in crystals must be generalized in order to take into account low frequency tunneling excitations of amorphous solids. This generalization has been achieved by the author over the last decade and the theory has recently been adjusted to the specific problems of SMS as well. This theory and its relation to other theories for impurity centers is discussed in detail in the book.

Part I

Single Atom in Transverse Electromagnetic Field

1. Quantum Principles of Two-Level Atomic Spectroscopy

Until very recently a single atom interacting with an electromagnetic field could only be a topic of theoretical consideration because experimentally only ensembles of atoms were accessible. Atomic spectra and the probabilities of transitions in atoms had been successfully studied in atomic ensembles for a long time. Quantum mechanics enables one to calculate both spectra and probabilities. However, quantum mechanics predicts that any transition happens at a random moment of time. In accordance with this prediction a single atom irradiated by light will make random jumps between its ground and excited electronic states. However, these jumps cannot be observed if we detect emission from the whole atomic ensemble, because a statistical average hides the random character of the processes. Therefore they cannot be observed in ensembles. This is only one example of new aspects which emerge when we move from atomic ensembles to a single atom. In this book we intend to discuss not only problems of probabilistic and spectral calculations, but also various random characteristics of atoms and molecules, the so-called quantum trajectories, which are typical in the spectroscopy of single atoms or molecules. In order to prepare for the discussion of quantum trajectories, we have to discuss the basic principles underlying the spectroscopy of a single two-level atom interacting with a transverse electromagnetic field.

1.1 Hamiltonian of an Electron–Photon System

Any dynamical system is completely determined if the Lagrangian of the system is known. An electron in an electromagnetic field is described by the following Lagrangian [36]:

$$L = -mc^2\sqrt{1-\frac{u^2}{c^2}} - e\varphi + \frac{e}{c}\boldsymbol{u}\cdot\boldsymbol{A}\,, \tag{1.1}$$

where m, e and $\boldsymbol{u}$ are the mass, charge and velocity of an electron and φ and $\boldsymbol{A}$ are the scalar and vector potential of the electromagnetic field, respectively.

The momentum of a particle in a system which includes an electromagnetic field is a derivative of the Lagrangian with respect to the generalized velocity $\boldsymbol{u}$ of the particle [37]:

$$P_i = \frac{\partial L}{\partial u_i} = \frac{m u_i}{\sqrt{1-u^2/c^2}} + \frac{e}{c} A_i = p_i + \frac{e}{c} A_i \, . \tag{1.2}$$

Here p_i is a component of the particle mechanical momentum for the case when the electromagnetic field is absent.

Since the density of the Lagrangian is a function of coordinates $\boldsymbol{r}$ and velocities $\boldsymbol{u}$, it is a relativistic invariant. Therefore it can describe particles moving with velocities which are near to the velocity of light c. If the initial values of the coordinates and velocities are known, the relativistic Lagrangian determines the system completely.

In parallel to such an approach, based on the Lagrangian, there is another approach to any dynamical system. This is based on the use of the Hamiltonian of the system. Neither the Hamiltonian nor its density are relativistic invariants. However, the velocities of atomic electrons satisfy the strong inequality $u/c \ll 1$ and therefore atomic electrons can be studied in the non-relativistic limit. The Hamiltonian has an advantage over the Lagrangian in describing quantum systems such as atoms, because use of the Hamiltonian facilitates the procedure of transforming classical dynamical variables to the relevant operators, i.e., transforming the classical system to a quantum system. Therefore the dynamics of an atom is described with the help of its quantum Hamiltonian.

The Hamiltonian is a function of the generalized coordinates $\boldsymbol{r}$ and the generalized momenta $\boldsymbol{P}$. These coordinates and momenta are called conjugate dynamical variables of the system. The substitution of the generalized velocity $\boldsymbol{u}$ by the generalized momentum $\boldsymbol{P}$ can be carried out with the help of the Legendre transformation. This transformation is often employed in thermodynamics to transform one function of the thermodynamic state to another. In our case, the Legendre transformation is given by

$$H(\boldsymbol{r}, \boldsymbol{P}) = \boldsymbol{u} \cdot \boldsymbol{P} - L(\boldsymbol{r}, \boldsymbol{u}) \, . \tag{1.3}$$

It is easy to prove that the full differential of the Hamiltonian $H(\boldsymbol{r}, \boldsymbol{P})$ is indeed a function of the independent variables $\boldsymbol{r}$ and $\boldsymbol{P}$. Making use of (1.1), (1.2) and (1.3) and assuming that $u/c << 1$, one easily finds that

$$H(\boldsymbol{r}, \boldsymbol{P}) = \frac{mc^2}{\sqrt{1-u^2/c^2}} + e\varphi \approx mc^2 + \frac{mu^2}{2} + e\varphi \, . \tag{1.4}$$

Taking into account the relation $\boldsymbol{u} = \boldsymbol{p}/m$, which is correct in the non-relativistic case, and omitting the rest mass energy of the electron mc^2, we arrive at the following equation for the Hamiltonian of a non-relativistic electron in an electromagnetic field:

$$H(\boldsymbol{r}, \boldsymbol{P}) = \frac{\left(\boldsymbol{P} - \frac{e}{c}\boldsymbol{A}\right)^2}{2m} + V(\boldsymbol{r}) \, , \tag{1.5}$$

where $V(\boldsymbol{r}) = e\varphi$ is the Coulomb electronic energy. In accordance with the general principles of quantization, we must replace the generalized momentum $\boldsymbol{P}$ by the operator $-\mathrm{i}\hbar\nabla$. After this substitution the classical Hamiltonian described by (1.5) is transformed into the quantum Hamiltonian of the atomic electron in an external electromagnetic field.

1.2 Transverse Electromagnetic Field as a Photon Gas

Approaches based on the Lagrangian or Hamiltonian for the dynamical system were first developed for mechanical systems. However, it was understood later that these approaches catch fundamental features of any dynamical system and can also be applied to non-mechanical systems, such as the electromagnetic field. The Lagrangian of the transverse electromagnetic field is given by [38]

$$L = \frac{1}{8\pi}\int\limits_V \left[\frac{\dot{\boldsymbol{A}}^2}{c^2} - (\mathrm{rot}\,\boldsymbol{A})^2\right]\mathrm{d}V\,. \tag{1.6}$$

The electric and magnetic vectors are

$$\boldsymbol{E} = -\dot{\boldsymbol{A}}/c\,, \quad \boldsymbol{B} = \mathrm{rot}\,\boldsymbol{A}\,. \tag{1.7}$$

By Fourier transforming the vector potential, we express it in terms of plane, standing monochromatic waves,

$$\boldsymbol{A}(\boldsymbol{r},t) = \sum_{\boldsymbol{k}} \boldsymbol{e}_{\boldsymbol{k}} A_{\boldsymbol{k}}(t)\cos\Psi_{\boldsymbol{k}}\,, \tag{1.8}$$

where $\Psi_{\boldsymbol{k}} = \boldsymbol{k}\cdot\boldsymbol{r} + \alpha_{\boldsymbol{k}}$ is a space phase and $\boldsymbol{e}_{\boldsymbol{k}}$ is a unit polarization vector. Using (1.7) we find the following expressions for the electric and magnetic fields:

$$\boldsymbol{E}(\boldsymbol{r},t) = -\sum_{\boldsymbol{k}} \boldsymbol{e}_{\boldsymbol{k}} \frac{\mathrm{d}A_{\boldsymbol{k}}}{c\mathrm{d}t}\cos\Psi_{\boldsymbol{k}} = \sum_{\boldsymbol{k}} \boldsymbol{E}_{\boldsymbol{k}}(t)\cos\Psi_{\boldsymbol{k}}\,, \tag{1.9}$$

$$\boldsymbol{B}(\boldsymbol{r},t) = \sum_{\boldsymbol{k}} (\boldsymbol{e}_{\boldsymbol{k}}\times\boldsymbol{k})A_{\boldsymbol{k}}\sin\Psi_{\boldsymbol{k}} = \sum_{\boldsymbol{k}} \boldsymbol{B}_{\boldsymbol{k}}(t)\sin\Psi_{\boldsymbol{k}}\,. \tag{1.10}$$

We can now write the Lagrangian of the electromagnetic field in the form

$$L = \int\limits_V \frac{E^2 - B^2}{8\pi}\mathrm{d}V = V\frac{\langle E^2\rangle - \langle B^2\rangle}{8\pi}\,. \tag{1.11}$$

Here $\langle E^2\rangle$ and $\langle B^2\rangle$ are averaged values calculated in accordance with the following rule:

$$\langle a \rangle = \frac{1}{V} \int\limits_V a \, \mathrm{d}V \, . \tag{1.12}$$

Since $\langle \sin \Psi_{\boldsymbol{k}} \sin \Psi_{\boldsymbol{k}'} \rangle = \langle \cos \Psi_{\boldsymbol{k}} \cos \Psi_{\boldsymbol{k}'} \rangle = (1/2)\delta_{\boldsymbol{k}\boldsymbol{k}'}$, where $\delta_{\boldsymbol{k}\boldsymbol{k}'}$ is the Kronecker symbol, we easily find

$$L = \sum_{\boldsymbol{k}} \frac{E_{\boldsymbol{k}}^2(t) - B_{\boldsymbol{k}}^2(t)}{16\pi} V = \sum_{\boldsymbol{k}} L_{\boldsymbol{k}} \, . \tag{1.13}$$

The Lagrangian is the sum of the Lagrangians of individual modes. By taking into account (1.9) and (1.10) and the relation $k = \omega_k / c$ between wave vector and frequency, we can write the Lagrangian of the individual mode as

$$L_{\boldsymbol{k}} = \frac{V}{16\pi c^2} \left(\dot{A}_{\boldsymbol{k}}^2 - \omega_{\boldsymbol{k}}^2 A_{\boldsymbol{k}}^2 \right) \, . \tag{1.14}$$

Dynamical variables $A_{\boldsymbol{k}}$ and $\dot{A}_{\boldsymbol{k}}$ can be considered as the generalized coordinate and the generalized velocity of the kth dynamical system.

In order to quantize the system, we must find its Hamiltonian. Using the expression

$$p_{\boldsymbol{k}} = \frac{\mathrm{d}L_{\boldsymbol{k}}}{\mathrm{d}\dot{A}_{\boldsymbol{k}}} = \frac{V}{8\pi c^2} \dot{A}_{\boldsymbol{k}} \tag{1.15}$$

for the conjugate momentum, we carry out the Legendre transformation

$$H_{\boldsymbol{k}} = p_{\boldsymbol{k}} \dot{A}_{\boldsymbol{k}} - L_{\boldsymbol{k}} = \frac{V}{16\pi c^2} \left(\dot{A}_{\boldsymbol{k}}^2 + \omega_{\boldsymbol{k}}^2 A_{\boldsymbol{k}}^2 \right) \, . \tag{1.16}$$

This expression for the Hamiltonian can be transformed to the form

$$H_{\boldsymbol{k}} = \frac{\hbar \omega_{\boldsymbol{k}}}{2} [P_{\boldsymbol{k}}^2 + Q_{\boldsymbol{k}}^2] \, , \tag{1.17}$$

where

$$P_{\boldsymbol{k}} = \sqrt{\frac{V}{c^2 \hbar \omega_{\boldsymbol{k}} 8\pi}} \dot{A}_{\boldsymbol{k}} \, , \quad Q_{\boldsymbol{k}} = \sqrt{\frac{V \omega_{\boldsymbol{k}}}{c^2 \hbar 8\pi}} A_{\boldsymbol{k}} \tag{1.18}$$

are the dimensionless generalized momentum and the generalized coordinate of the kth dynamical system describing the kth mode of the electromagnetic field. The Planck constant has been introduced into the classical Hamiltonian in a formal manner. The justification for this will be made after the system has been quantized.

The classical Hamiltonian of the kth mode described by (1.17) looks like the classical Hamiltonian of an oscillating particle with mass. In the case of an oscillator with mass, (1.18) would be different. However, this circumstance does not influence the fact that both an oscillator with mass and the kth electromagnetic mode are described by the same Hamiltonian, given by (1.17).

This means that both dynamical systems behave like an oscillator. For an oscillator with mass, vibrations correspond to the transformation of potential energy, which depends quadratically on the generalized coordinate, to kinetic energy, which depends quadratically on the generalized momentum. The energy of the transverse electromagnetic field can take either electric or magnetic form. The periodic mutual transformation of electric energy to magnetic energy resembles the mutual transformation of potential and kinetic energy in an oscillator with mass. Therefore the variable electric and magnetic fields play the roles of the generalized momentum and coordinate of the dynamical system, respectively. The generalized momentum and mechanical momentum coincide in mechanical systems. However, they differ considerably in an electromagnetic field. Indeed the density of the momentum of the electromagnetic field is proportional to the Poynting vector, which includes both electric and magnetic fields.

Quantization of the system involves replacing the generalized coordinate q and the generalized momentum p of the system by the operators $\hat{q}$ and $\hat{p}$ with the commutation relation $\hat{q}\hat{p}-\hat{p}\hat{q}=\mathrm{i}\hbar$. After this quantization, conjugate variables q and p cannot both be measured exactly. The commutation relation for the dimensionless generalized coordinate and momentum is given by

$$[Q,P]=\mathrm{i}\,. \tag{1.19}$$

It is found from this relation that if $Q=Q_{\boldsymbol{k}}$, then $P=-\mathrm{i}/\mathrm{d}Q_{\boldsymbol{k}}$. After replacing the classical generalized coordinate and momentum by their operators, the quantum Hamiltonian takes the form

$$H_{\boldsymbol{k}}=\frac{\hbar\omega_{\boldsymbol{k}}}{2}\left(-\frac{\mathrm{d}^2}{\mathrm{d}Q_{\boldsymbol{k}}^2}+Q_{\boldsymbol{k}}^2\right)\,. \tag{1.20}$$

Let us find the eigenfunctions and eigenvalues of this Hamiltonian. We omit the index $\boldsymbol{k}$. By introducing new operators

$$a=\frac{1}{\sqrt{2}}\left(\frac{\mathrm{d}}{\mathrm{d}Q}+Q\right)\,,\quad a^+=\frac{1}{\sqrt{2}}\left(-\frac{\mathrm{d}}{\mathrm{d}Q}+Q\right) \tag{1.21}$$

and taking into account that their commutator is $[a,a^+]=1$, we can rewrite (1.20) as

$$H=\hbar\omega(a^+a+1/2)\,. \tag{1.22}$$

With the help of the relation

$$[a,(a^+)^n]=n(a^+)^{n-1}\,, \tag{1.23}$$

we easily find that the solution of the Schrödinger equation $H|n\rangle=E_n|n\rangle$ is

$$|n\rangle=\frac{(a^+)^n}{\sqrt{n!}}|0\rangle\,, \tag{1.24}$$

$$E_n = \hbar\omega(n + 1/2) , \tag{1.25}$$

where $n = 0, 1, 2, \ldots$. These are the normalized eigenfunctions and eigenvalues of the Hamiltonian described by (1.22). The functions in (1.24) are the harmonic oscillator functions. The function $|0\rangle$ describes the ground state of the electromagnetic field, the so-called zero-point vibrations. The energy $E_0 = \hbar\omega/2$ is the energy of the zero-point vibrations.

In accordance with (1.25), the energy of the electromagnetic field consists of energy quanta. The quantum of the electromagnetic field with energy $\hbar\omega$ is called a photon. The functions defined by (1.24) with $n \geq 1$ describe an electromagnetic field consisting of n photons. Using (1.23), we find that

$$a|n\rangle = \sqrt{n}|n-1\rangle , \quad a^+|n\rangle = \sqrt{n+1}|n+1\rangle . \tag{1.26}$$

The operators a and a^+ decrease and increase the photon number by one unit, respectively. They are therefore called photon annihilation and creation operators. The following equations can be derived for the operators corresponding to the generalized coordinate and momentum:

$$Q|n\rangle = \sqrt{\frac{n}{2}}|n-1\rangle + \sqrt{\frac{n+1}{2}}|n+1\rangle , \tag{1.27}$$

$$P|n\rangle = -\mathrm{i}\frac{\mathrm{d}}{\mathrm{d}Q}|n\rangle = -\mathrm{i}\left(\sqrt{\frac{n}{2}}|n-1\rangle - \sqrt{\frac{n+1}{2}}|n+1\rangle\right) . \tag{1.28}$$

Up to now we have only considered a single mode of the electromagnetic field. However, the Hamiltonian of the transverse electromagnetic field is the sum over all modes,

$$H_\perp = \sum_k \hbar\omega_k(a_k^+ a_k + 1/2) , \tag{1.29}$$

where $k = \boldsymbol{k}j$ and j numbers the two possible independent polarizations of the transverse electromagnetic field. The energy of the transverse electromagnetic field is the sum over all photons,

$$E = \sum_k \hbar\omega_k(n_k + 1/2) . \tag{1.30}$$

Since the modes of the electromagnetic field are independent, the eigenfunction of the electromagnetic field consisting of several photons is a product:

$$|n_1, n_2, \ldots, n_s, \ldots\rangle = \prod_k |n_k\rangle . \tag{1.31}$$

By introducing a vector $\boldsymbol{n} = (n_1, n_2, \ldots)$ and a vector $\boldsymbol{\omega} = (\omega_1, \omega_2, \ldots)$ with an infinite number of components, we can rewrite (1.30) and (1.31) in the more concise form

$$E_{\boldsymbol{n}} = \hbar \boldsymbol{n} \cdot \boldsymbol{\omega} + E_0 \, , \quad \prod_k |n_k\rangle = |\boldsymbol{n}\rangle \, . \tag{1.32}$$

After quantizing the electromagnetic field, the electric and magnetic fields become operators. Therefore, using (1.18) we can rewrite (1.8), (1.9) and (1.10) as

$$\hat{\boldsymbol{A}}(\boldsymbol{r},t) = \sum_{\boldsymbol{k}} \boldsymbol{e}_{\boldsymbol{k}} \frac{c}{\omega_{\boldsymbol{k}}} \sqrt{\frac{\hbar\omega_{\boldsymbol{k}} 8\pi}{V}} \hat{Q}_{\boldsymbol{k}} \cos \Psi_{\boldsymbol{k}} \, , \tag{1.33}$$

$$\hat{\boldsymbol{E}}(\boldsymbol{r},t) = -\sum_{\boldsymbol{k}} \boldsymbol{e}_{\boldsymbol{k}} \sqrt{\frac{\hbar\omega_{\boldsymbol{k}} 8\pi}{V}} \hat{P}_{\boldsymbol{k}} \cos \Psi_{\boldsymbol{k}} \, , \tag{1.34}$$

$$\hat{\boldsymbol{B}}(\boldsymbol{r},t) = \sum_{\boldsymbol{k}} (\boldsymbol{e}_{\boldsymbol{k}} \times \boldsymbol{s}) \sqrt{\frac{\hbar\omega_{\boldsymbol{k}} 8\pi}{V}} \hat{Q}_{\boldsymbol{k}} \sin \Psi_{\boldsymbol{k}} \, . \tag{1.35}$$

The field vectors are marked by the operator sign because they are expressed via the operators for the generalized coordinate $\hat{Q}_{\boldsymbol{k}}$ and momentum $\hat{P}_{\boldsymbol{k}}$, and we use the relation $\boldsymbol{k} = \boldsymbol{s}\omega_{\boldsymbol{k}}/c$, where $\boldsymbol{s}$ is a unit vector.

1.3 Electron–Photon Interaction. Rabi Frequency

Equation (1.5) for the Hamiltonian describes an atom in an external electromagnetic field. If we add (1.29) for the Hamiltonian of the transverse electromagnetic field, we arrive at the full Hamiltonian for the electron–photon system:

$$H = \frac{\left(\boldsymbol{P} - \frac{e}{c}\boldsymbol{A}\right)^2}{2\mu} + V(\mathbf{r}) + \sum_{\boldsymbol{k}} \hbar\omega_{\boldsymbol{k}} (a_{\boldsymbol{k}}^{+} a_{\boldsymbol{k}} + 1/2) \, , \tag{1.36}$$

where $\boldsymbol{P} = -\mathrm{i}\hbar\nabla$. The first term can be expressed as the sum of three terms. Therefore (1.36) can be written as

$$H = H_{\mathrm{a}} + H_{\mathrm{ph}} + \Lambda \, , \tag{1.37}$$

where

$$H_{\mathrm{a}} = \frac{\boldsymbol{P}^2}{2\mu} + V(\boldsymbol{r}) \quad \text{and} \quad H_{\mathrm{ph}} = \sum_{\boldsymbol{k}} \hbar\omega_{\boldsymbol{k}} (a_{\boldsymbol{k}}^{+} a_{\boldsymbol{k}} + 1/2) \tag{1.38}$$

describe the atom and the transverse electromagnetic field, respectively, and the operator

$$\Lambda = -\frac{e}{\mu c}\boldsymbol{P}\cdot\boldsymbol{A} + \frac{e^2}{2\mu c^2}\boldsymbol{A}^2 = \Lambda_1 + \Lambda_2 \tag{1.39}$$

determines their interaction.

The first term Λ_1 depends on the mechanical momentum of the atomic electron. It describes the interaction of the moving electron with the electromagnetic field. The second term Λ_2 does not depend on the electronic dynamical variables. Therefore this term cannot change the quantum state of the electron. The operator Λ_2 contributes to light scattering. However, we may neglect this term when considering light absorption and emission because these processes are accompanied by a change of electronic quantum state. Allowing for this fact, we will take $\Lambda = \Lambda_1$ when considering light absorption and emission.

Substituting (1.33) for the vector potential in (1.39), we find the following expression for the electron–photon interaction:

$$\Lambda = \Lambda_1 = -\frac{e}{\mu c}\sum_{\boldsymbol{k}} \boldsymbol{P}\cdot\boldsymbol{A}_{\boldsymbol{k}} = \sum_{\boldsymbol{k}} \Lambda_{\boldsymbol{k}} , \tag{1.40}$$

where

$$\boldsymbol{A}_{\boldsymbol{k}} = \boldsymbol{e}_{\boldsymbol{k}}\frac{c}{\omega_{\boldsymbol{k}}}\sqrt{\frac{\hbar\omega_{\boldsymbol{k}}8\pi}{V}}\hat{Q}_{\boldsymbol{k}} . \tag{1.41}$$

Here we assume that the atom occupies the point of space where $\cos\Psi_{\boldsymbol{k}} = 1$. The operator $\boldsymbol{A}_{\boldsymbol{k}}$ acts on the kth electromagnetic mode. Hence, allowing for (1.27), we can write

$$\boldsymbol{A}_{\boldsymbol{k}}|n_{\boldsymbol{k}}\rangle = \boldsymbol{e}_{\boldsymbol{k}}\frac{c}{\omega_{\boldsymbol{k}}}\sqrt{\frac{\hbar\omega_{\boldsymbol{k}}4\pi}{V}}\left(\sqrt{n_{\boldsymbol{k}}}|n_{\boldsymbol{k}}-1\rangle + \sqrt{n_{\boldsymbol{k}}+1}|n_{\boldsymbol{k}}+1\rangle\right) . \tag{1.42}$$

Let us now consider the action of the operator P on the dynamical variable of the atomic electron. Calculation of the commutation relation yields

$$[H_{\mathrm{a}}, \boldsymbol{r}] = \frac{1}{2\mu}[\boldsymbol{P}^2, \boldsymbol{r}] = -\mathrm{i}\frac{\hbar}{\mu}\boldsymbol{P} . \tag{1.43}$$

Calculation of the matrix elements of the left and right hand sides of (1.43), with the help of the eigenfunctions of the atomic Hamiltonian, yields the following relation for the matrix elements of the electron coordinate and mechanical momentum:

$$\boldsymbol{P}_{lm} = \mathrm{i}\mu\,\omega_{lm}\,\boldsymbol{r}_{lm} . \tag{1.44}$$

Here $\omega_{lm} = (E_l - E_m)/\hbar$ is the Bohr resonant frequency of the atom. With the help of (1.43), we find the relation

$$\frac{e}{\mu}\boldsymbol{P}|m\rangle = \sum_l \frac{e}{\mu}\boldsymbol{P}_{lm}|l\rangle = \mathrm{i}\sum_l \omega_{lm}\boldsymbol{d}_{lm}|l\rangle , \tag{1.45}$$

where $\boldsymbol{d}_{lm} = e\boldsymbol{r}_{lm}$ is the matrix element of the electronic dipole moment, which describes the action of the momentum operator on the mth electronic eigenfunction. It is obvious that the product $|m\rangle|n_k\rangle$ is an eigenfunction of the Hamiltonian $H_\mathrm{a} + H_\mathrm{ph}$.

Using (1.42) and (1.45), we find how the operator $\Lambda_{\boldsymbol{k}}$ acts on the function $|m\rangle|n_{\boldsymbol{k}}\rangle$:

$$\Lambda_{\boldsymbol{k}}|m\rangle|n_{\boldsymbol{k}}\rangle = \sum_l \Lambda_{lm}(\boldsymbol{k})|l\rangle\left[\sqrt{n_{\boldsymbol{k}}}|n_{\boldsymbol{k}} - 1\rangle + \sqrt{n_{\boldsymbol{k}} + 1}|n_{\boldsymbol{k}} + 1\rangle\right] , \tag{1.46}$$

where

$$\Lambda_{lm}(\boldsymbol{k}) = -\mathrm{i}\frac{\omega_{lm}}{\omega_k}\sqrt{\frac{4\pi\hbar\omega_k}{V}}\boldsymbol{e}_{\boldsymbol{k}}\cdot\boldsymbol{d}_{lm} . \tag{1.47}$$

The electromagnetic field can be considered as a classical field when considering light-induced optical transitions. Let us find a relation between the matrix element described by (1.47) and the matrix element with the classical field. This is possible if we can find a relation between the amplitude of the classical electromagnetic field on the one hand and the number of photons on the other hand. Because of the independence of electromagnetic modes, we can find this relation for each mode. The moving electromagnetic wave is described by the following electric vector:

$$\boldsymbol{E}_{\boldsymbol{k}}(\boldsymbol{r}, t) = \boldsymbol{E}\cos(\boldsymbol{k}\cdot\boldsymbol{r} - \omega_k t) . \tag{1.48}$$

The standing wave is a sum of two waves moving in opposite directions. Therefore the electric vector in the standing wave is given by

$$\boldsymbol{E}(\boldsymbol{r}, t) = \boldsymbol{E}_{\boldsymbol{k}}(\boldsymbol{r}, t) + \boldsymbol{E}_{-\boldsymbol{k}}(\boldsymbol{r}, t) = 2\boldsymbol{E}\cos\boldsymbol{k}\cdot\boldsymbol{r}\cos\omega_k t . \tag{1.49}$$

The electric energy in the standing electromagnetic wave oscillates. At times satisfying the equation $\omega_k t = \pi(2n+1)/2$, where $n = 0, 1, 2, \ldots$, the electric energy is transformed completely into magnetic energy. At the instant $t = 0$ when the energy has purely electric form, we can write

$$\int_V \frac{\boldsymbol{E}^2(\boldsymbol{r}, 0)}{8\pi}\mathrm{d}V = \frac{(2\boldsymbol{E})^2}{16\pi}V = \hbar\omega_k(n_{\boldsymbol{k}} + 1/2) . \tag{1.50}$$

This equation allows us to find the following relation between the amplitude of the classical electric vector and the number of photons:

$$\boldsymbol{E} = \boldsymbol{e}_{\boldsymbol{k}}\sqrt{\frac{4\pi\hbar\omega_k}{V}(n_{\boldsymbol{k}} + 1/2)} \approx \boldsymbol{e}_{\boldsymbol{k}}\sqrt{\frac{4\pi\hbar\omega_k}{V}n_{\boldsymbol{k}}} . \tag{1.51}$$

Using this relation with (1.46) and (1.47), we can find the following relation for the matrix element of the interaction with a classical electromagnetic field:

$$\begin{aligned} \Lambda &= \frac{\Lambda_{10}(\boldsymbol{k})}{\hbar}\sqrt{n_{\boldsymbol{k}}} = -\mathrm{i}\frac{\boldsymbol{d}_{01}\cdot\boldsymbol{E}_{\boldsymbol{k}}}{\hbar} = -\mathrm{i}\chi \,, \\ \Lambda^* &= \frac{\Lambda_{01}(\boldsymbol{k})}{\hbar}\sqrt{n_{\boldsymbol{k}}+1} \approx \mathrm{i}\frac{\boldsymbol{d}_{10}\cdot\boldsymbol{E}_{\boldsymbol{k}}}{\hbar} = \mathrm{i}\chi \,. \end{aligned} \tag{1.52}$$

In deriving (1.52), we took $\omega_{10}/\omega_k = -\omega_{01}/\omega_k = 1$. These equations express the matrix elements of the atom–field interaction via the amplitude of the classical electric field and the matrix element of the dipole moment. The real value denoted by χ is called the Rabi frequency. It will be used frequently right through the book.

1.4 Equations for Transition Amplitudes

Spectroscopy is the study of quantum transitions. Each quantum transition is characterized by its probability. The probability of finding an atom or molecule in a definite quantum state changes with time. The probability of the quantum transition is a quadratic function of the relevant transition amplitude. In this section, the general equations for these transition amplitudes will be derived.

Let H be the Hamiltonian for some physical system, for example, atom + electromagnetic field, and $|m\rangle$ a function describing an initial state of the system. This state will evolve with time in accordance with the Schrödinger equation. Its general solution is

$$|m,t\rangle = \mathrm{e}^{-\mathrm{i}tH/\hbar}|m\rangle \,. \tag{1.53}$$

The new state $|m,t\rangle$ is a superposition of various states $|m'\rangle$. Let $|l\rangle$ be one of these states. According to the general principles of quantum mechanics, the probability of finding the system in the state $|l\rangle$ equals $|\langle l|m,t\rangle|^2 = W_{l\leftarrow m}$. The relevant transition amplitude will be denoted

$$G_{lm}(t) = -\mathrm{i}\langle l|m,t\rangle = -\mathrm{i}\langle l|\mathrm{e}^{-\mathrm{i}tH/\hbar}|m\rangle \,. \tag{1.54}$$

It is obvious that $|G_{lm}|^2 = W_{lm}$. If $|m\rangle$ is an eigenfunction of the Hamiltonian H, i.e., it satisfies an equation $H|m\rangle = E_m|m\rangle$, we find

$$G_{lm} = -\mathrm{i}\mathrm{e}^{-\mathrm{i}tE_m/\hbar}\langle l|m\rangle = -\mathrm{i}\mathrm{e}^{-\mathrm{i}tE_m/\hbar}\delta_{lm} \,, \tag{1.55}$$

where δ_{lm} is the Kronecker symbol, equal to unity when $l = m$ and zero in other cases. Therefore the probability of finding the system in any state except the initial state m equals zero. This means there are no transitions in the case considered. This is not surprising because $|m\rangle$ is a stationary state of the system. However, if $|m\rangle$ is not an eigenstate of the Hamiltonian H the transition to the state $|l\rangle$ is possible because $|G_{lm}|^2 = W_{lm}$ for $l \neq m$. Let us now find an equation for the transition amplitude G_{lm}.

Let the initial state $|m\rangle$ be an eigenstate of the Hamiltonian H_0. Mathematically this means that

$$H_0|m\rangle = E_m|m\rangle \ . \tag{1.56}$$

This Hamiltonian can describe, for instance, a system consisting of an atom that is not interacting with the electromagnetic field. The Hamiltonian H_0 is a part of the Hamiltonian H of the full system. The set of eigenfunctions of each Hamiltonian is complete and the functions are orthogonal so that

$$\langle l|m\rangle = \delta_{lm} \ , \quad \sum_m |m\rangle\langle m| = 1 \ . \tag{1.57}$$

Differentiating both sides of (1.54) with respect to time gives

$$\dot{G}_{lm}(t) = -(\mathrm{i}/\hbar)\sum_s \langle l|H|s\rangle G_{sm}(t) \ . \tag{1.58}$$

Here we have used the completeness of the set of functions. By writing the Hamiltonian H of the system in the form $H = H_0 + V$, we can transform (1.58) to

$$\dot{G}_{lm}(t) = -\mathrm{i}\omega_l G_{lm}(t) - (\mathrm{i}/\hbar)\sum_s V_{ls}G_{sm}(t) \ , \tag{1.59}$$

where $\omega_l = E_l/\hbar$. This is the infinite set of coupled equations for transition amplitudes. It can be solved if the matrix elements V_{ls} of the perturbation are known.

In practice the set of linear differential equations (1.59) for all realistic physical systems is solved with the help of the Laplace transformation. This transformation allows us to convert equations of differential type into algebraic equations. If initial values of all the amplitudes are known at $t = 0$, (1.59) enables us to predict the temporal behavior of all the amplitudes at $t > 0$. It does not matter what values the amplitudes had for $t < 0$. Therefore, without loss of generality, we may set all amplitudes equal zero at $t < 0$. Then the Laplace transformation is given by

$$(G(t))_\omega = G(\omega) = \int_0^\infty G(t)\mathrm{e}^{\mathrm{i}t(\omega+\mathrm{i}0)}\,\mathrm{d}t \ . \tag{1.60}$$

The vanishing positive term i0 in the exponent ensures disappearance of the amplitude for $t < 0$. Indeed, due to this term the Laplace component $G(\omega)$ is an analytical function in the upper half-plain of the complex variable ω. Therefore, in accordance with the rule for integrating functions with definite analytical properties, we find the following inverse transformation:

$$G(t) = \int_{-\infty}^{\infty} G(\Omega)\mathrm{e}^{-\mathrm{i}\Omega t}\frac{\mathrm{d}\Omega}{2\pi} = \begin{cases} 0 \ , & t < 0 \ , \\ G(t) \ , & t > 0 \ . \end{cases} \tag{1.61}$$

Equations (1.60) and (1.61) determine the Laplace transformation and the inverse transformation for the amplitudes.

Let us apply (1.60) to a damping exponential function. We obtain

$$\int_0^\infty e^{i(\omega-\omega_0+i\gamma)t}\,dt = \frac{i}{\omega-\omega_0+i\gamma} = i\frac{\omega-\omega_0}{(\omega-\omega_0)^2+\gamma^2} + \frac{\gamma}{(\omega-\omega_0)^2+\gamma^2}\,. \tag{1.62}$$

If $\gamma \to 0$, the real term approaches the Dirac delta function, i.e.,

$$\lim_{\gamma\to+0}\frac{\gamma}{(\omega-\omega_0)^2+\gamma^2} = \pi\delta(\omega-\omega_0)\,. \tag{1.63}$$

The inverse transformation to (1.62) yields

$$\int_{-\infty}^{\infty}\frac{d\Omega}{2\pi}\frac{e^{-i\Omega t}}{\Omega-\omega_0+i\gamma} = -ie^{-i(\omega_0-i\gamma)t}\,. \tag{1.64}$$

Applying (1.60) to the time derivative, we find

$$(\dot{G}(t))_\omega = -G(t=0) - i(\omega+i0)G(\omega). \tag{1.65}$$

Let us carry out the Laplace transformation of both sides of (1.59). Then taking into account (1.65), we can transform the differential equation (1.59) to the following set of algebraic equations for the Laplace components of the transition amplitudes:

$$G_{lm}(\omega) = \frac{iG_{lm}(0)}{\omega-\omega_l+i0} + \frac{1}{\omega-\omega_l+i0}\sum_s \frac{V_{ls}}{\hbar}G_{sm}(\omega)\,, \tag{1.66}$$

where $iG_{lm}(0) = \delta_{lm}$. The solution of these algebraic equations is an easier task.

1.5 Density Matrix of the System. Relation to Transition Amplitudes

As already discussed in Sect. 1.4, the temporal evolution of the system is determined by the evolution operator which acts on the function $|m\rangle$ describing an initial state of the system,

$$|m,t\rangle = e^{-itH/\hbar}|m\rangle\,, \tag{1.67}$$

where H is the Hamiltonian of the system. An arbitrary dynamical variable F which can be measured in an experiment is determined by

$$F_m(t) = \langle m,t|\hat{F}|m,t\rangle \; . \tag{1.68}$$

Taking into account that

$$\langle m,t| = \langle m|\mathrm{e}^{\mathrm{i}tH/\hbar} \; , \tag{1.69}$$

we may write (1.68) in the form

$$F_m(t) = \langle m|\mathrm{e}^{\mathrm{i}tH/\hbar}\hat{F}\mathrm{e}^{-\mathrm{i}tH/\hbar}|m\rangle = \sum_{l,l'} G_{l'm}(t)G^*_{lm}(t)F_{ll'} \; . \tag{1.70}$$

The average value of the dynamical variable F depends on the quantum index m of the initial state of the system. The function

$$\rho_{ls}(t) = G_{lm}(t)G^*_{sm}(t) \tag{1.71}$$

is called the density matrix of the system. It certainly depends on the initial state of the system. Keeping this fact in mind, we will henceforth omit the index of the initial state in the expression for the density matrix. Using the density matrix we can express the average value of the dynamical variable described by (1.70) as follows:

$$F(t) = \sum_{s,l} \rho_{ls}F_{sl} = \mathrm{Tr}\left(\hat{\rho}(t)\hat{F}\right) \; . \tag{1.72}$$

Here we introduce a density operator which depends on time. Its matrix elements $\rho_{ls}(t) = \langle l|\hat{\rho}(t)|s\rangle$ relate to the transition amplitudes in accordance with (1.71). It is obvious that

$$\mathrm{Tr}\left(\hat{\rho}(t)\right) = \sum_s \rho_{ss}(t) = \sum_s |G_{sm}(t)|^2 = 1 \; . \tag{1.73}$$

We may claim that the diagonal element ρ_{ss} of the density matrix determines the probability of finding our quantum system in the s quantum state. If the density matrix is known, we can determine both the probability of finding the system in various quantum states and the average value of the dynamical variable.

Let us find equations for the elements of the density matrix. By differentiating both parts of (1.71) with respect to time, we find that

$$\dot{\rho}_{ls}(t) = \dot{G}_{lm}G^*_{sm}(t) + G_{lm}\dot{G}^*_{sm}(t) \; . \tag{1.74}$$

With the help of (1.58) for the transition amplitudes and the equation conjugate to it, we can express the time derivative via the right hand side of these equations. Substituting the time derivative in (1.74) by the relevant expression on the right hand side, we find that

$$\dot{\rho}_{ls}(t) = -\frac{\mathrm{i}}{h}\sum_p \left[H_{lp}\rho_{ps}(t) - \rho_{lp}(t)H_{ps}\right] \; . \tag{1.75}$$

This equation for the matrix elements can be transformed to the equation for the density operator,

$$\dot{\hat{\rho}} = -\frac{\mathrm{i}}{\hbar}[H, \hat{\rho}(t)] , \tag{1.76}$$

where $[\dots]$ is the commutator of the operators.

The Hamiltonian of the system can be written as a sum $H = H_0 + V$, where H_0 is the Hamiltonian of that part of the system for which the eigenfunctions $|l\rangle$ and eigenvalues E_l are known. By inserting $H = H_0 + V$ into (1.75) and using the eigenfunctions $|l\rangle$, $|p\rangle$ and $|s\rangle$ of the Hamiltonian H_0 in this equation, we arrive at the following equations for the density matrix:

$$\dot{\rho}_{ls}(t) = -\mathrm{i}\omega_{ls}\rho_{ls}(t) - \frac{\mathrm{i}}{\hbar}\sum_p [V_{lp}\rho_{ps} - \rho_{lp}(t)V_{ps}] . \tag{1.77}$$

Here $\omega_{ls} = (E_l - E_s)/\hbar$, where E_l and E_s are eigenvalues of the Hamiltonian H_0.

2. Two-Photon Start–Stop Correlator

2.1 Photon Counting in the Start–Stop Regime

Let us consider a situation when a two-level atom is irradiated by continuous wave (CW) monochromatic laser light. Such an atom can absorb and consequently emit only one photon. In accordance with the principles of quantum mechanics, the instant of time at which a photon is emitted or absorbed is not determined. Therefore the atom will jump at random times between the ground and excited electronic states.

A train of photons emitted at random time intervals by a single atom irradiated by CW laser light is shown in Fig. 2.1. These photons will produce

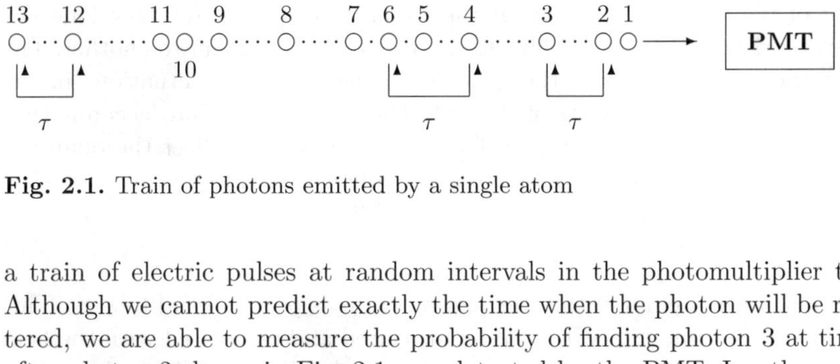

Fig. 2.1. Train of photons emitted by a single atom

a train of electric pulses at random intervals in the photomultiplier tube. Although we cannot predict exactly the time when the photon will be registered, we are able to measure the probability of finding photon 3 at time τ after photon 2 shown in Fig. 2.1 was detected by the PMT. In other words we are able to measure the probability of finding a photon pair with a delay τ between the two photons.

It is obvious that there are two types of photon pairs, as shown in Fig. 2.1. For instance, we can count in the so-called start–stop regime, taking into account only those pairs consisting of consecutively emitted photons like pairs (2,3) and (12,13). The count rate $s(t)$ of such pairs is called the start–stop correlator. However, there is another possibility, namely, to count all pairs of photons with delay τ including additional pairs like the pair (4,6) with one intermediate photon 5. The count rate of all pairs with an arbitrary number of intermediate photons is called the full two-photon correlator. In this section we consider measurements in the start–stop regime. The count rate of photon pairs with a definite delay τ is proportional to the probability of finding this type of pair and it can be calculated quantum mechanically.

Let W_0, W_1 and $W_{\boldsymbol{k}}$ be the probabilities of finding a two-level atom in the ground electronic state, in the excited electronic state and again in the ground electronic state after spontaneous emission of a photon with wave vector $\boldsymbol{k}$. If we take the registration time of photon 2 shown in Fig. 2.1 as $t = 0$, we find that $W_0(0) = 1$, $W_1(0) = W_{\boldsymbol{k}}(0) = 0$. Due to laser excitation these probabilities will change. The sum

$$\sum_{\boldsymbol{k}} W_{\boldsymbol{k}}(\tau) = S(\tau) \tag{2.1}$$

is the probability of finding that a photon is spontaneously emitted at time τ. $S(\tau)$ increases with time and approaches unity, because $S(\infty) = 1$.

We are interested in the probability of finding an emitted photon in a time interval $(\tau, \tau + \mathrm{d}\tau)$. The probability is determined by

$$\mathrm{d}S(\tau) = \dot{S}(\tau)\,\mathrm{d}\tau\,. \tag{2.2}$$

It is obvious that the number $N(t, \tau)$ of photon pairs with time delay lying in the interval $(\tau, \tau + d\tau)$ is proportional to $\mathrm{d}S(\tau)$. The number $N(t, \tau)$ as a function of the delay τ reflects the quantum dynamics of the two-level atom interacting with a transverse electromagnetic field. It also depends on the time resolution of the setup, the quantum efficiency of the PMT, and the time t of photon pair counting. If the counting time t is not very long the number of detected events will fluctuate from one experiment to another. The longer the counting time t, the smaller the ratio of the fluctuations in the number $N(t, \tau)$ to the value of $N(t, \tau)$. Therefore taking into account that $N(t, \tau) \propto \mathrm{d}S(\tau)$, and omitting all factors which do not reflect the quantum dynamics of the system, we may write the following relation:

$$\lim_{t\to\infty} \frac{N(t, \tau)}{t} \propto \frac{\mathrm{d}}{\mathrm{d}\tau} S(\tau) = s(\tau)\,. \tag{2.3}$$

The count rate of pairs in the start–stop regime is proportional to $s(\tau)$. This function of the time interval τ can be measured and calculated with the help of quantum mechanics. It depends on the probabilities of light absorption and emission. In the next three sections we consider these probabilities in detail.

2.2 Spontaneous Fluorescence. Temporal Evolution of Fluorescence Line Shape

We start our considerations with the simplest task of finding the probability of spontaneous emission. In this case the system Hamiltonian, the perturbation operator and zero-order basic eigenfunction are given by

$$\begin{aligned} H_0 &= H_a + H_{\text{ph}}\,, \qquad V = \Lambda\,, \\ |m\rangle &= |0\rangle|\boldsymbol{n}\rangle\,, \qquad |1\rangle|\boldsymbol{n}\rangle\,, \end{aligned} \tag{2.4}$$

where $|0\rangle$ and $|1\rangle$ are eigenfunctions of the atomic Hamiltonian H_{a} relating to the eigenvalues 0 and E. The state $|\boldsymbol{n}\rangle$ is an eigenfunction of the Hamiltonian H_{ph} of the transverse electromagnetic field relating to the energy $\hbar\boldsymbol{n}\cdot\boldsymbol{\omega}+E_0$.

Initially, the atom is in the excited state with unit probability. This quantum state is described by the function $|1\rangle|0\rangle \equiv |1,0\rangle$ and energy $E+E_0$. However, this quantum state is not a stationary state of the system because of the electron–photon interaction Λ. Due to this interaction, the atomic electron emits a photon with energy $\hbar\omega_{\boldsymbol{k}}$ and the atom jumps into the ground electronic state, i.e., the system moves into the quantum state $|0\rangle|1_{\boldsymbol{k}}\rangle \equiv |0,\boldsymbol{k}\rangle$ with energy $\hbar\omega_{\boldsymbol{k}}+E_0$, as shown in Fig. 2.2. However, the new state is not a stationary state either. To begin with, the inverse transition to the initial state, i.e., the atomic excitation accompanied by the absorption of photon $\boldsymbol{k}$, is now possible. Secondly, simultaneously with atomic excitation, the electron–photon operator Λ can create a state with photon $\boldsymbol{k}'$. The latter transition to the state $|1,\boldsymbol{k},\boldsymbol{k}'\rangle$ with energy $E+\hbar\omega_{\boldsymbol{k}}+\hbar\omega_{\boldsymbol{k}'}+E_0$ is a transition of the virtual type because the final energy differs from the initial energy by the energy of two light quanta. For the chosen initial state, this quantum state cannot actually be reached. However, this virtual transitions will contribute to the value of the transition amplitude. Fortunately, its contribution is proportional to a small ratio $\Lambda_{10}(\boldsymbol{k})/(\hbar\omega_{\boldsymbol{k}}+\hbar\omega_{\boldsymbol{k}'})$. Therefore we may neglect virtual processes which do not conserve the full number of excitations. In this approximation, which we call the resonant approximation, we neglect matrix elements in the electron–photon interaction operator which do not conserve the full number of excitations. In the literature this approximation is sometimes called the rotating wave approximation.

$$|1,0\rangle \longleftrightarrow |0,\boldsymbol{k}\rangle \longleftrightarrow |1,\boldsymbol{k},\boldsymbol{k}'\rangle \longleftrightarrow |0,\boldsymbol{k},\boldsymbol{k}',\boldsymbol{k}''\rangle \longleftrightarrow \ldots$$

Fig. 2.2. States of the system connected by the electron–photon interaction

Using the resonant approximation, we need only take into account two states, described by the function $|1\rangle|0\rangle$ with energy $E+E_0$ and the function $|0\rangle|1_{\boldsymbol{k}}\rangle \equiv |0\rangle|\boldsymbol{k}\rangle$ with energy $\hbar\omega_{\boldsymbol{k}}+E_0$, in the infinite chain of coupled quantum states shown in Fig. 2.2. For simplicity, we denote them as 1 and $\boldsymbol{k}$, respectively. Then (1.66) takes the form

$$\begin{aligned} G_1(\omega) &= \frac{1}{\omega+\mathrm{i}0} + \frac{1}{\omega+\mathrm{i}0}\sum_{\boldsymbol{k}} \lambda_{\boldsymbol{k}} G_{\boldsymbol{k}}(\omega)\,, \\ G_{\boldsymbol{k}}(\omega) &= \frac{\lambda^*_{\boldsymbol{k}}}{\omega-\Delta_{\boldsymbol{k}}+\mathrm{i}0} G_1(\omega)\,, \end{aligned} \tag{2.5}$$

where the resonant frequency $\Omega = E/\hbar$ is included in the variable ω, the difference is described by $\Delta_{\boldsymbol{k}} = \omega_{\boldsymbol{k}} - \Omega$, and the matrix elements of the electron–photon interaction are given by

$$\lambda_{\boldsymbol{k}} = \frac{\Lambda_{10}(\boldsymbol{k})}{\hbar} = -\mathrm{i}\Omega\sqrt{\frac{4\pi}{\hbar\omega_{\boldsymbol{k}}V}}\boldsymbol{e}_{\boldsymbol{k}}\cdot\boldsymbol{d} \,. \tag{2.6}$$

The index of the initial state is omitted in the transition amplitudes. The expression for the amplitude G_1 can be found from (2.5):

$$G_1(\omega) = \frac{1}{\omega - \Delta(\omega) + \mathrm{i}\dfrac{\gamma(\omega)}{2}} \,, \tag{2.7}$$

where

$$\Delta(\omega) = \sum_{\boldsymbol{k}} \frac{|\lambda_{\boldsymbol{k}}|^2}{\omega - \Delta_{\boldsymbol{k}}} \,, \qquad \gamma(\omega) = 2\pi \sum_{\boldsymbol{k}} |\lambda_{\boldsymbol{k}}|^2 \delta(\omega - \Delta_{\boldsymbol{k}}) \,. \tag{2.8}$$

The summation over $\boldsymbol{k}$ runs up to infinity. Therefore the frequency shift $\Delta = \infty$. This is the so-called ultraviolet divergence of quantum electrodynamics. Quantum electrodynamics is a renormalizable theory [39]. This means that the divergences in the theory can be removed by substituting the 'bare' electronic charge and mass by their 'dressed' values. After renormalization the frequency shift is described by a finite value and it can be included in the frequency ω. This means that formally we may set $\Delta(\omega) = 0$ in (2.8). The expression for $\gamma(\omega)$ is calculated in Appendix A. The result is given by

$$\gamma(\omega) = \frac{4d^2\Omega^2(\Omega + \omega)}{3c^3\hbar} \,. \tag{2.9}$$

This function of the frequency ω is almost constant over a frequency scale of the order of γ. Therefore we may take the function $\gamma(\omega)$ at the resonance point $\omega = 0$:

$$\gamma(0) = \gamma = \frac{1}{T_1} = \frac{4d^2}{3\hbar}\left(\frac{\Omega}{c}\right)^3 \,. \tag{2.10}$$

This constant determines the rate of fluorescence. Indeed, by taking into account our comments concerning the frequency shift Δ, we may rewrite (2.7) as

$$G_1(\omega) = \frac{1}{\omega + \mathrm{i}\gamma/2} \,. \tag{2.11}$$

Carrying out the inverse Laplace transformation with the help of (1.64), we find

$$G_1(t) = -\mathrm{i}\,\mathrm{e}^{-\gamma t/2} . \tag{2.12}$$

Hence, the probability of finding the two-level atom in the excited electronic state decreases exponentially:

$$W_1(t) = |G_1(t)|^2 = \mathrm{e}^{-t\gamma} = \mathrm{e}^{-t/T_1} . \tag{2.13}$$

The time T_1 is called the fluorescence lifetime or the energy relaxation time.

Let us estimate the value of T_1 for dipolar optical transitions. Allowing for $d \approx ea$, $e^2/a \approx \hbar\Omega$, where a is the Bohr radius, we can express the dipolar moment d in terms of the electronic charge and the energy of the light quantum, viz., $d = e^3/\hbar\Omega$. Substituting this value for the dipolar moment in (2.10), we find

$$\gamma = \frac{1}{T_1} \approx \left(\frac{e^2}{\hbar c}\right)^3 \Omega \approx \left(\frac{1}{137}\right)^3 10^{15}\,\mathrm{s}^{-1} \approx \frac{1}{2} 10^9\,\mathrm{s}^{-1} ,$$

i.e., the fluorescent lifetime is of the order of a few nanoseconds.

The probability of light emission can be found with the help of the transition amplitude $G_{\boldsymbol{k}}$. Using the second equation of (2.5) with (2.11), we find the following expression for the Laplace component of the transition amplitude:

$$G_{\boldsymbol{k}}(\omega) = \frac{\lambda_{\boldsymbol{k}}^*}{\Delta_{\boldsymbol{k}} + \mathrm{i}\gamma/2}\left(\frac{1}{\omega - \Delta_{\boldsymbol{k}} + \mathrm{i}0} - \frac{1}{\omega + \mathrm{i}\gamma/2}\right) . \tag{2.14}$$

With the help of this equation and (1.64), we find the time dependent transition amplitude to be

$$G_{\boldsymbol{k}}(t) = \frac{-\mathrm{i}\lambda_{\boldsymbol{k}}^*}{\Delta_{\boldsymbol{k}} + \mathrm{i}\gamma/2}\left(\mathrm{e}^{-\mathrm{i}\Delta_{\boldsymbol{k}} t} - \mathrm{e}^{-\gamma t/2}\right) . \tag{2.15}$$

Using this expression, it is shown in Appendix B that the probability of photon emission is given by

$$W_{\mathrm{ph}}(t) = \sum_{\boldsymbol{k}} W_{\boldsymbol{k}}(t) = \sum_{\boldsymbol{k}} |G_{\boldsymbol{k}}(t)|^2 = \int_{-\infty}^{\infty} \frac{\mathrm{d}W_{\mathrm{ph}}(t)}{\mathrm{d}\Delta}\,\mathrm{d}\Delta , \tag{2.16}$$

where

$$I_{\mathrm{fl}}(\Delta, t) = \frac{\mathrm{d}W_{\mathrm{ph}}(t)}{\mathrm{d}\Delta} = \frac{\gamma/2\pi}{\Delta^2 + (\gamma/2)^2}\left(1 + \mathrm{e}^{-t\gamma} - 2\mathrm{e}^{-\gamma t/2}\cos\Delta t\right) . \tag{2.17}$$

Here $\Delta = \Omega - \omega$, where ω is the frequency of the emitted photon. The function I_{fl} is the probability density for finding a photon with frequency ω. This function describes the fluorescence line shape. Allowing for the equation

$$\int_{-\infty}^{\infty} \cos\Delta t \frac{\gamma/2\pi}{\Delta^2 + (\gamma/2)^2}\,\mathrm{d}\Delta = \mathrm{e}^{-\gamma t/2} , \tag{2.18}$$

we can rewrite (2.16) in the form

$$W_{\mathrm{ph}}(t) = 1 - \mathrm{e}^{-t\gamma} \,. \tag{2.19}$$

This means that probability of finding an emitted photon increases exponentially towards unity. Therefore the conservation law for probabilities is satisfied:

$$W_1(t) + W_{\mathrm{ph}}(t) = 1 \,. \tag{2.20}$$

Since W_{ph} is also the probability of finding the atom in the ground electronic state, the last relation means that the probability of finding the two-level atom either in the excited or in the ground state equals one. The probability density for finding the emitted photon at time t is described by

$$\dot{W}_{\mathrm{ph}}(t) = \gamma \exp(-\gamma t) \,. \tag{2.21}$$

The fluorescence line shape is described by (2.17). At long times, when $\gamma t \gg 1$, we may neglect the exponential functions in (2.17). In the long time limit, the fluorescence line is Lorentzian with full-width at half-maximum (FWHM) γ. It equals the inverse fluorescence lifetime $1/T_1$. This is a well known result.

At short times, when $\gamma t \ll 1$, the fluorescence line shape depends on time. In this case the line shape can differ considerably from the Lorentzian and can even be described by a function with more than one maximum. Figure 2.3 shows examples of this type. The figure demonstrates some features which might be called anomalies because they are not observed in ordinary experiments where fluorescence from an atomic ensemble is measured. However, they are not anomalies from the point of view of quantum mechanics.

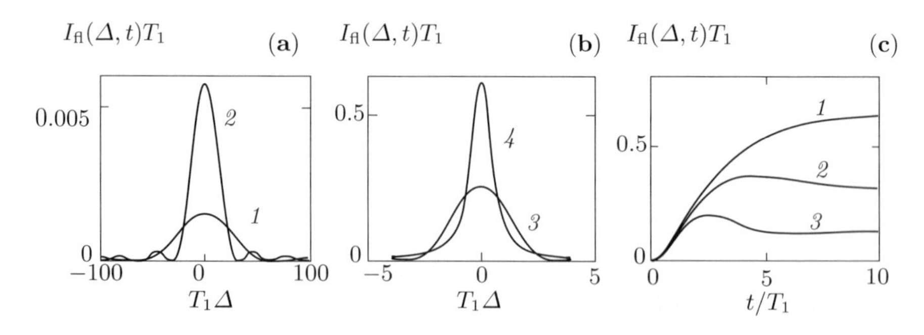

Fig. 2.3. Fluorescence line shape dependence on time delay between times of photon absorption and emission for (**a**) $t/T_1 = 0.1$ (curve 1) and 0.2 (curve 2), (**b**) $t/T_1 = 2$ (curve 3) and 10 (curve 4), and (**c**) dependence of fluorescence intensity on time at $\Delta = 0$ (curve 1), 0.5 (curve 2) and 1 (curve 3)

In accordance with Fig. 2.3a, the FWHM of the fluorescence line can be much more than γ. This is in agreement with the Heisenberg uncertainty relation. However, the line shape described by curve 2 is a function with several maxima. Curve 3 in Fig. 2.3c demonstrates another anomalous effect – non-monotonic temporal behavior of the probability of finding a photon if its frequency ω differs considerably from the atomic Bohr frequency Ω. It is very difficult indeed to observe such anomalies in fluorescence line shapes at short times in ensembles. However, it will be shown later that such anomalies can be observed in two-photon correlators of single molecules.

2.3 Optical Absorption Line. Equations for Transition Amplitudes

When the first photon from a photon pair is detected by the PMT, this means that an atom occurs in the ground electronic state with unit probability. Such an initial state of the system consisting of an atom and a laser field is described by the following function and energy:

$$|m\rangle = |0\rangle|n\rangle , \qquad E_m \equiv E_n = \hbar\omega n + E_0 , \tag{2.22}$$

where n is the number of photons in the laser field exciting an atom in the ground electronic state $|0\rangle$. The ground state atomic energy is set to zero.

This initial atomic state is not a stationary state because of the presence of the laser field. Due to the electron–photon interaction Λ, the atom absorbs one photon from the laser mode and the system makes a transition to the quantum state $|1\rangle|n-1\rangle$ with energy $E+\hbar\omega_0(n-1)+E_0$, where E is the energy of the excited atom. However, this quantum state is not the stationary state either. Due to the electron–photon interaction, two transitions are possible from this state. Firstly, light-induced emission of one photon of the laser mode is possible with an atomic jump to the ground electronic state, i.e., returning it to the initial state. Secondly, spontaneous emission of a photon with wave vector $\boldsymbol{k}$ and transition to a quantum state $|0\rangle|n-1, \boldsymbol{k}\rangle$ with energy $\hbar\omega_0(n-1) + \hbar\omega_{\boldsymbol{k}} + E_0$ is possible. If we use the resonant approximation the infinite set of possible quantum states are coupled as shown in Fig. 2.4. Here we have used a modified notation for the functions discussed above.

$$\left|\begin{matrix}0\\ n\end{matrix}\right\rangle \longleftrightarrow \left|\begin{matrix}1\\ n-1\end{matrix}\right\rangle \longleftrightarrow \left|\begin{matrix}0\\ n-1, \boldsymbol{k}\end{matrix}\right\rangle \longleftrightarrow \left|\begin{matrix}1\\ n-2, \boldsymbol{k}\end{matrix}\right\rangle \longleftrightarrow \dots$$

Fig. 2.4. Quantum states of the system consisting of a two-level atom and a transverse electromagnetic field coupled by the operator Λ in the resonant approximation

In accordance with Fig. 2.4, we have an infinite set of coupled quantum states, even if we use the resonant approximation. Therefore the chain of coupled equations for the transition amplitudes will consist of an infinite set of equations. It is worth noting that the resonant approximation allows us to take into account only two coupled equations when considering spontaneous fluorescence. In the case of light absorption, (1.66) takes the following form:

$$\begin{aligned} G_n^0 &= \frac{1}{\omega + \mathrm{i}0} + \frac{1}{\omega + \mathrm{i}0} \Lambda^* G_{n-1}^1 , \\ G_{n-1}^1 &= \frac{1}{\omega - \Delta + \mathrm{i}0} \left[\Lambda G_n^0 + \sum_{\boldsymbol{k}} \lambda_{\boldsymbol{k}} G_{n-1\boldsymbol{k}}^0 \right] , \\ G_{n-1\boldsymbol{k}}^0 &= \frac{1}{\omega - \Delta_{\boldsymbol{k}} + \mathrm{i}0} \left[\lambda_{\boldsymbol{k}}^* G_{n-1}^1 + \Lambda^{*\prime} G_{n-2\boldsymbol{k}}^1 \right] , \\ G_{n-2\boldsymbol{k}}^1 &= \frac{1}{\omega - \Delta - \Delta_{\boldsymbol{k}} + \mathrm{i}0} \left[\Lambda' G_{n-1\boldsymbol{k}}^0 + \sum_{\boldsymbol{k}'} \lambda_{\boldsymbol{k}'} G_{n-2\boldsymbol{k}\boldsymbol{k}'}^0 \right] , \quad \text{etc.} \end{aligned} \tag{2.23}$$

Here we omit everywhere the index for the initial quantum state and use the notation

$$\Delta = \Omega - \omega_0 \, , \quad \Delta_{\boldsymbol{k}} = \omega_{\boldsymbol{k}} - \omega_0 \, , \quad \Lambda = \lambda_0 \sqrt{n} \, , \quad \Lambda' = \lambda_0 \sqrt{n-1} \, , \tag{2.24}$$

where

$$\lambda_0 = -\mathrm{i}\Omega \sqrt{\frac{4\pi}{\hbar \omega_0 V}} \boldsymbol{e}_0 \cdot \boldsymbol{d} \tag{2.25}$$

is deduced from (2.6) by assuming index 0 for the laser mode.

The solution of (2.23) can be expressed as a continued fraction, e.g.,

$$G_{n-1}^1 = \cfrac{\Lambda G_n^0}{\omega - \Delta - \sum\limits_{\boldsymbol{k}} \cfrac{|\lambda_{\boldsymbol{k}}|^2}{\omega - \Delta_{\boldsymbol{k}} - \cfrac{|\Lambda'|^2}{\omega - \Delta - \Delta_{\boldsymbol{k}} - \sum\limits_{\boldsymbol{k}'} |\lambda_{\boldsymbol{k}'}|^2 \cdots}}} . \tag{2.26}$$

The matrix elements $\lambda_{\boldsymbol{k}}$ are small, not involving the number of photons. They determine the strength of the electromagnetic interaction, which is weak and can be treated as a small perturbation [39]. In contrast, the matrix elements $\Lambda, \Lambda', \ldots$ are large because they include $\sqrt{n}$, where n is the large number of photons in the laser mode. In the continued fraction (2.26), the large and small matrix elements alternate. It is obvious that the influence of the large matrix element Λ' will be considerably decreased by the small matrix element $\lambda_{\boldsymbol{k}}$. Therefore we may set $\Lambda' = 0$. After this approximation is made, the first three equations in the set described by (2.23) become independent of the others:

$$
\begin{aligned}
G_n^0 &= \frac{1}{\omega + \mathrm{i}0} + \frac{1}{\omega + \mathrm{i}0} \Lambda^* G_{n-1}^1 \,, \\
G_{n-1}^1 &= \frac{1}{\omega - \Delta + \mathrm{i}0} \left[\Lambda G_n^0 + \sum_{\boldsymbol{k}} \lambda_{\boldsymbol{k}} G_{n-1\boldsymbol{k}}^0 \right] \,, \\
G_{n-1\boldsymbol{k}}^0 &= \frac{1}{\omega - \Delta_{\boldsymbol{k}} + \mathrm{i}0} \lambda_{\boldsymbol{k}}^* G_{n-1}^1 \,.
\end{aligned}
\tag{2.27}
$$

Using the third equation for the amplitude $G_{n-1\boldsymbol{k}}^0$, we find the following important relation:

$$
\sum_{\boldsymbol{k}} \lambda_{\boldsymbol{k}} G_{n-1\boldsymbol{k}}^0 = \sum_{\boldsymbol{k}} \frac{|\lambda_{\boldsymbol{k}}|^2}{\omega - \Delta_{\boldsymbol{k}} + \mathrm{i}0} G_{n-1}^1 = -\mathrm{i}\frac{\gamma}{2} G_{n-1}^1 \,. \tag{2.28}
$$

Here we have taken into account the comments after (2.8) concerning the real part of the function described by the sum over $\boldsymbol{k}$. Equation (2.28) enables one to substitute the second amplitude G_{n-1}^1 for the third amplitude $G_{n-1\boldsymbol{k}}^0$. This approximation plays a decisive role in uncoupling the coupled equations in the infinite set (2.23). Using (2.28) we can write these equations in the form

$$
\begin{aligned}
G_n^0 &= \frac{1}{\omega + \mathrm{i}0} + \frac{1}{\omega + \mathrm{i}0} \Lambda^* G_{n-1}^1 \,, \\
G_{n-1}^1 &= \frac{\Lambda G_n^0}{\omega - \Delta + \mathrm{i}\gamma/2} \,.
\end{aligned}
\tag{2.29}
$$

A solution of these equations is given by

$$
\begin{aligned}
G_n^0 &= \frac{1}{\omega - \dfrac{|\Lambda|^2}{\omega - \Delta + \mathrm{i}\gamma/2}} \,, \qquad G_{n-1}^1 = \frac{\Lambda}{\omega - \Delta + \mathrm{i}\gamma/2} G_n^0 \,, \\
G_{n-1\boldsymbol{k}}^0 &= \frac{\lambda_{\boldsymbol{k}}^*}{\omega - \Delta_{\boldsymbol{k}} + \mathrm{i}0} G_{n-1}^1 \,.
\end{aligned}
\tag{2.30}
$$

These equations enable one to find the temporal behavior of the transition amplitudes with the help of the inverse Laplace transformation. This will be carried out in the next section.

The three equations (2.27) include all the transition amplitudes playing an important role when we work in the start–stop regime. Indeed the amplitudes allow us to find the probabilities $W_0 = |G_n^0(t)|^2$, $W_1 = |G_{n-1}^1(t)|^2$, $W_{\boldsymbol{k}} = |G_{n-1\boldsymbol{k}}^0(t)|^2$ discussed in Sect. 2.1. By setting $\Lambda' = 0$, we switch off the laser after spontaneous emission of photon $\boldsymbol{k}$. Under these physical conditions the sum of all the written probabilities must be one. Equation (2.27) enables one to derive this conservation law.

In order to prove this statement we should find equations for the time dependent amplitudes which match the three equations (2.27). They are given by

$$\begin{aligned}
\dot{G}^0_n(t) &= -\mathrm{i}\Lambda^* G^1_{n-1} \,, \\
\dot{G}^1_{n-1}(t) &= -\mathrm{i}\Delta G^1_{n-1}(t) - \mathrm{i}\left[\Lambda G^0_n(t) + \sum_{\boldsymbol{k}} \lambda_{\boldsymbol{k}} G^0_{n-1\boldsymbol{k}}(t)\right] \,, \\
\dot{G}^0_{n-1\boldsymbol{k}}(t) &= -\mathrm{i}\Delta_{\boldsymbol{k}} G^0_{n-1\boldsymbol{k}}(t) - \mathrm{i}\lambda^*_{\boldsymbol{k}} G^1_{n-1} \,.
\end{aligned} \tag{2.31}$$

Using the Laplace transformation described by (1.60) and (1.65), we can easily verify that (2.31) exactly match (2.27). The equations for the conjugate amplitudes needed to calculate the relevant probabilities can be obtained from (2.31) by the operation of conjugation. The derivative of the probability is expressed via transition amplitudes as follows:

$$\frac{\mathrm{d}}{\mathrm{d}t} |G(t)|^2 = \dot{G}G^* + G\dot{G}^* \,. \tag{2.32}$$

Then, using the (2.31) and their conjugated counterparts, we can express the time derivative of the amplitudes in terms of the amplitudes themselves. After that we can find expressions for the time derivative for all three probabilities. The following conservation law is proven in Appendix C:

$$\frac{\mathrm{d}}{\mathrm{d}t}\left[|G^0_n(t)|^2 + |G^1_{n-1}(t)|^2 + \sum_{\boldsymbol{k}} |G^0_{n-1\boldsymbol{k}}(t)|^2\right] = 0 \,. \tag{2.33}$$

Since $|G^0_n(0)|^2 = W_0(0) = 1$ and $|G^1_{n-1}|^2 = W_1(0) = |G^0_{n-1\boldsymbol{k}}(0)|^2 = W_{\boldsymbol{k}}(0) = 0$, we find that

$$W_0(t) + W_1(t) + \sum_{\boldsymbol{k}} W_{\boldsymbol{k}}(t) = 1 \,. \tag{2.34}$$

2.4 Temporal Evolution of Probabilities

Let us introduce a simplified notation for the amplitudes: $G^0_n = G_0$, $G^1_{n-1} = G_1$ and $G_{n-1\boldsymbol{k}} = G_{\boldsymbol{k}}$. We need to use (2.30) to calculate the temporal behavior of only two amplitudes G_0 and G_1, because the temporal behavior of the probability $\sum_{\boldsymbol{k}} W_{\boldsymbol{k}} = S(t)$ can be found from the conservation law (2.34), viz.,

$$S(t) = 1 - |G_0(t)|^2 - |G_1(t)|^2 \,, \tag{2.35}$$

proved in the previous section.

Equation (2.30) for the Laplace component of these two amplitudes can be written

$$G_0(\omega) = \frac{\omega - \Delta + \mathrm{i}\gamma/2}{\omega(\omega - \Delta + \mathrm{i}\gamma/2) - |\Lambda|^2} \,, \quad G_1(\omega) = \frac{\Lambda}{\omega(\omega - \Delta + \mathrm{i}\gamma/2) - |\Lambda|^2} \,. \tag{2.36}$$

The polynomial in the denominator is given by

$$\omega^2 - (\Delta - \gamma/2)\omega - |\Lambda|^2 = (\omega - \omega_1)(\omega - \omega_2) . \tag{2.37}$$

It can be expressed via its roots

$$\omega_{1,2} = b \pm R , \tag{2.38}$$

where

$$b = \frac{\Delta}{2} - \mathrm{i}\frac{\gamma}{4} , \quad R = \sqrt{b^2 + |\Lambda|^2} . \tag{2.39}$$

Using (2.37), we can transform the formulas (2.36) to the form

$$G_0(\omega) = \frac{1}{\omega_1 - \omega_2}\left(\frac{\omega_1}{\omega - \omega_2} - \frac{\omega_2}{\omega - \omega_1}\right) , \tag{2.40}$$

$$G_1(\omega) = \frac{\Lambda}{\omega_1 - \omega_2}\left(\frac{1}{\omega - \omega_1} - \frac{1}{\omega - \omega_2}\right) . \tag{2.41}$$

The inverse Laplace transform of these amplitudes to the time dependent functions is easily carried out with the help of (1.64). The result is given by

$$G_0(t) = \frac{-\mathrm{i}}{\omega_1 - \omega_2}\left(\omega_1 \mathrm{e}^{-\mathrm{i}\omega_2 t} - \omega_2 \mathrm{e}^{-\mathrm{i}\omega_1 t}\right) , \tag{2.42}$$

$$G_1(t) = \frac{\mathrm{i}\Lambda}{\omega_1 - \omega_2}\left(\mathrm{e}^{-\mathrm{i}\omega_2 t} - e^{-\mathrm{i}\omega_1 t}\right) . \tag{2.43}$$

The complex values $\omega_{1,2}$ are determined by (2.38).

The formulas for the amplitudes are simplified for the resonant case, i.e., at $\Delta = 0$. In this particular case, (2.43) takes the form

$$G_0(t) = -\mathrm{i}\mathrm{e}^{-ibt}\left(\cos Rt - \mathrm{i}\frac{b}{R}\sin Rt\right) , \tag{2.44}$$

$$G_1(t) = \mathrm{e}^{-ibt}\frac{\Lambda}{R}\sin Rt , \tag{2.45}$$

where

$$b = -\mathrm{i}\frac{\gamma}{4} , \qquad R = \sqrt{|\Lambda|^2 - \frac{\gamma^2}{16}} . \tag{2.46}$$

It is obvious that the type of temporal behavior displayed by the amplitudes depends strongly on the intensity of the laser light, which is determined by the number of photons n in the matrix elements Λ.

At weak laser intensity we find $|\Lambda|^2 < \gamma^2/16$ and therefore $R = \mathrm{i}|R|$, where $|R|$ is a positive quantity. In this case, inserting (2.44) and (2.45) into the expressions for the probabilities,

$$W_0 = |G_0|^2 = \mathrm{e}^{-\gamma t/2} \left(\cosh^2 |R|t + \frac{\gamma^2}{16|R|^2} \sinh^2 |R|t \right) , \tag{2.47}$$

$$W_1 = |G_1|^2 = \mathrm{e}^{-\gamma t/2} \frac{|\Lambda|^2}{|R|^2} \sinh^2 |R|t . \tag{2.48}$$

Here W_0 and W_1 are the probabilities of finding an atom in the ground and excited electronic states. The temporal behavior of these probabilities is shown in Fig. 2.5. The probability $S(t)$ was calculated with the help of (2.35).

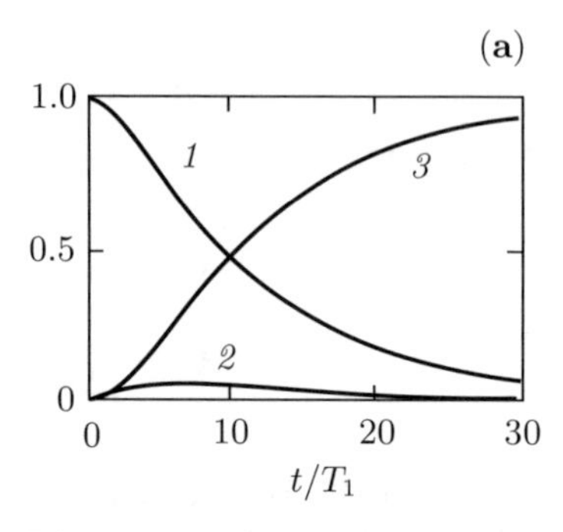

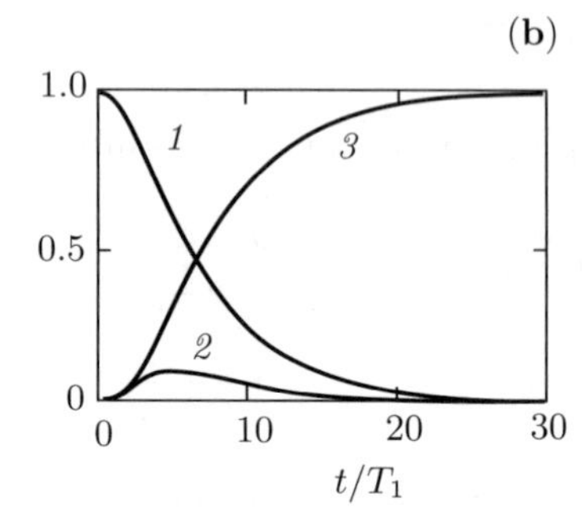

Fig. 2.5. Temporal behavior of the probabilities W_0 (curve 1), W_1 (curve 2) and $\sum_k |G_k(t)|^2 = S(t)$ (curve 3) in a weak electromagnetic field with $\Delta = 0$ and (**a**) $\Lambda/\gamma = 0.15$, (**b**) $\Lambda/\gamma = 0.2$

When the laser intensity is increased, the rates of change of the probabilities also increase.

At strong laser intensity, we find $|\Lambda|^2 > \gamma^2/16$ and therefore R is a positive quantity. In this case, using (2.44) and (2.45), we find the following expressions for the probabilities:

$$W_0 = \mathrm{e}^{-\gamma t/2} \left(\cos Rt + \frac{\gamma}{16R^2} \sin Rt \right)^2 , \tag{2.49}$$

$$W_1 = \mathrm{e}^{-\gamma t/2} \frac{|\Lambda|^2}{R^2} \sin^2 Rt . \tag{2.50}$$

The temporal behavior of all three probabilities at strong pumping is shown in Fig. 2.6. Oscillations appear at strong pumping. The frequency of the oscillations increases as the laser intensity increases.

Let us find the value of the critical laser intensity I_c which separates the domains of weak and strong intensity. This can be found from the equation $R = 0$, which yields

$$|\Lambda|^2 = |\lambda_0|^2 n = \frac{\gamma^2}{16} . \tag{2.51}$$

Substituting the right hand side of (2.10) for one of the factors γ and using (2.25) for λ_0, we obtain

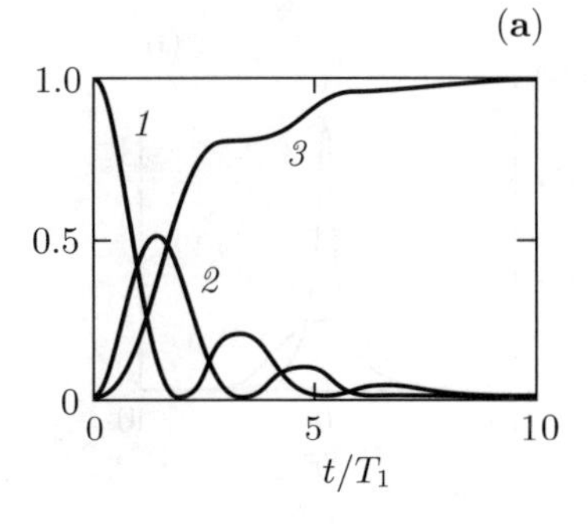

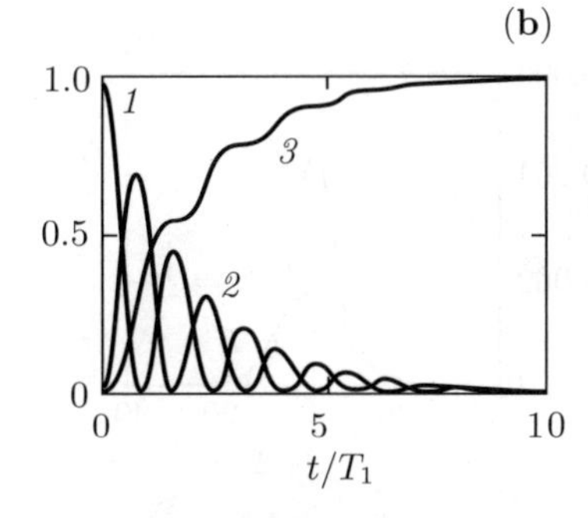

Fig. 2.6. Temporal behavior of the probabilities W_0 (curve 1), W_1 (curve 2) and $\sum_{\boldsymbol{k}} |G_{\boldsymbol{k}}(t)|^2 = S(t)$ (curve 3) in a strong electromagnetic field, with $\Delta = 0$ and (**a**) $\Lambda/\gamma = 1$, (**b**) $\Lambda/\gamma = 2$

$$\frac{n}{V} = \frac{\gamma}{48\pi} \frac{\Omega^2}{c^3} , \tag{2.52}$$

where we have taken $\omega = \Omega$ and $(\boldsymbol{e} \cdot \boldsymbol{d})^2 = d^2$. The volume V occupied by the laser radiation equals Sct where S is the area of the laser spot on the sample, c is the light speed and t is the duration of the laser pulse. Substituting Sct for V in (2.52), we find the following expression for the critical intensity $I_\mathrm{c} = n/St$:

$$I_\mathrm{c} = \frac{\gamma}{48\pi} \left(\frac{\Omega}{c}\right)^2 = \frac{\pi}{12} \frac{\gamma}{\lambda^2} , \tag{2.53}$$

where γ and λ are the FWHM and wavelength of the optical line, respectively. Obviously the critical intensity I_c determines the number of photons arriving per second per unit area of the target. Taking $\gamma = 10^8\ \mathrm{s}^{-1}$, $\lambda = 500$ nm, we find $I_\mathrm{c} \approx 10^{16}$ ph/cm^{2}s. This laser intensity corresponds to 3 mW cm^{-2}. At this laser intensity the rate of light-induced optical transitions reaches the value of the spontaneous emission rate. The simple equation (2.53) is also applicable in the more complicated case when the FWHM is two or three orders of magnitude larger than $1/T_1$, because of the electron–phonon interaction which has not yet been taken into account here.

2.5 Temporal Evolution of Absorption Line Shape

The concept of optical line in common use today refers to the optical line of an ensemble of identical atoms or molecules. The optical line shape of a two-level atom is described by a Lorentzian whose FWHM equals the inverse fluorescence lifetime $1/T_1$ of the excited electronic state.

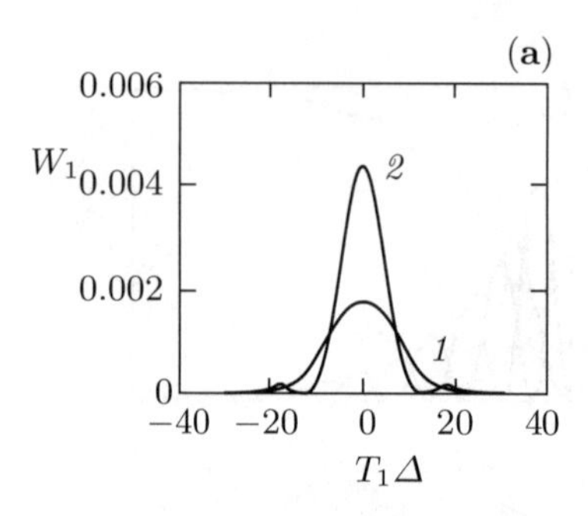

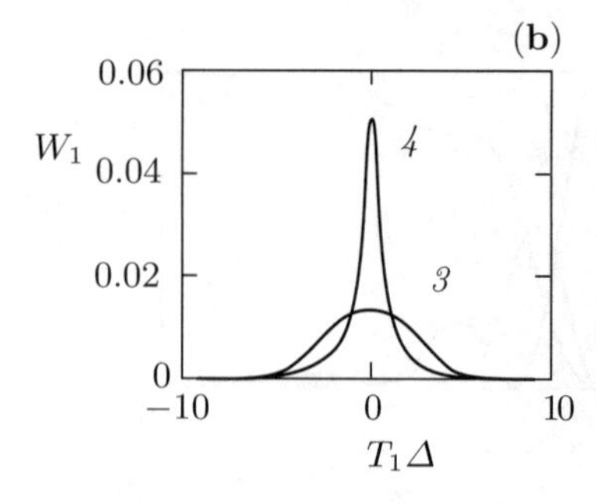

Fig. 2.7. (a,b) Temporal evolution of an atomic absorption line in a weak electromagnetic field. $\Lambda T_1 = 0.15$, $t/T_1 = 0.3$ (curve 1), 0.5 (curve 2), 1 (curve 3) and 10 (curve 4)

By considering the fluorescence line of a single atom in Sect. 2.2, we found that the fluorescence line shape of a single atom depends on time. At short times satisfying the inequality $t/T_1 \ll 1$, the fluorescence line is wider than $1/T_1$ and has a complicated shape. In the long time limit, when $t/T_1 \gg 1$ the optical line becomes narrower and its shape approaches the Lorentzian whose FWHM equals $1/T_1$. The large FWHM of a fluorescence line in the short time limit agrees with the Heisenberg uncertainty relation.

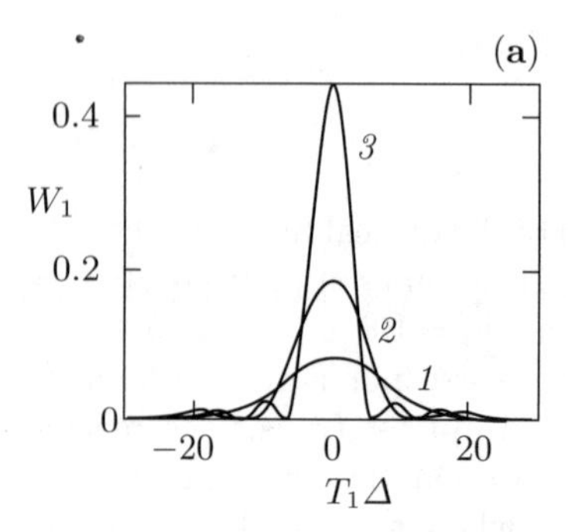

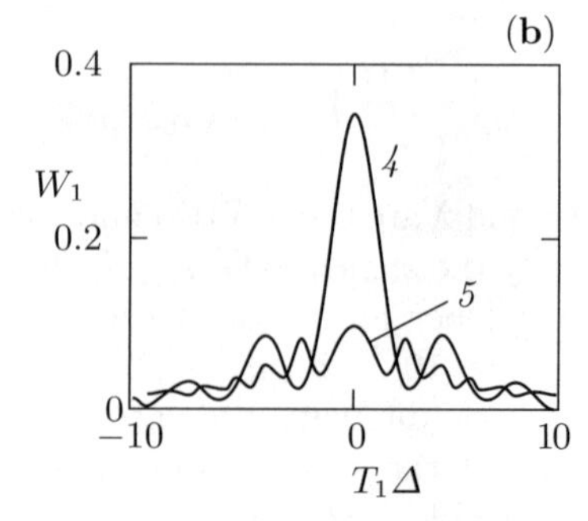

Fig. 2.8. (a,b) Temporal evolution of atomic absorption line in a strong electromagnetic field. $\Lambda T_1 = 1$, $t/T_1 = 0.3$ (curve 1), 0.5 (curve 2), 1 (curve 3), 2 (curve 4) and 5 (curve 5)

The absorption line shape also depends on time. However, the shape depends additionally on the intensity of the exciting laser light. The absorption line shape can be analyzed using

$$W_1(t) = |G_1(t)|^2 = \left| \frac{\mathrm{i}\Lambda}{\omega_1 - \omega_2} (\mathrm{e}^{-\mathrm{i}\omega_2 t} - \mathrm{e}^{-\mathrm{i}\omega_1 t}) \right|^2 \tag{2.54}$$

for the probability of finding the two-level atom in the excited state. Here ω_1 and ω_2 are the complex frequencies determined by (2.38) and (2.39).

Figure 2.7 shows the result of the calculation with (2.54) for weak intensity of the exciting light.

It is clear that the temporal behavior of the absorption line in the short time limit is similar to the behavior of the fluorescence line: the line becomes narrower with time and its shape approaches a Lorentzian whose FWHM equals $1/T_1$.

If we increase the laser intensity so that the probability of light-induced transitions exceeds the probability of spontaneous transitions, we obtain the picture shown in Fig. 2.8.

Oscillations in the frequency scale are observed even in the long time limit. They are of the same nature as the well known Rabi oscillations observed in the probability of finding an atom in the excited state.

2.6 Two-Photon Start–Stop Correlator

All the probabilities considered in the preceding sections refer to a single atom irradiated by CW laser light. Since this atom is in the ground electronic state at $t = 0$, we can use these probabilities to find the start–stop correlator. The count rate of pairs consisting of two consecutively emitted photons is described by (2.3). In order to find this rate we have to calculate $s(t) = \dot{S}(t)$. This time derivative can be found with the help of the conservation law in the form described by (2.33). However, a more convenient expression exists in practice. It can be derived with the help of the third equation of (2.31). Using this and the conjugate equation we can easily find the following relation for the two-photon start–stop correlator:

$$
\begin{aligned}
s(t) &= \frac{\mathrm{d}}{\mathrm{d}t} \sum_{\boldsymbol{k}} |G^0_{n-1\boldsymbol{k}}|^2 \\
&= -\mathrm{i} \sum_{\boldsymbol{k}} \left(\lambda^*_{\boldsymbol{k}} G^{*0}_{n-1\boldsymbol{k}} G^1_{n-1} - \lambda_{\boldsymbol{k}} G^0_{n-1\boldsymbol{k}} G^{*1}_{n-1} \right) \\
&= \left[-\mathrm{i} \left(\frac{\mathrm{i}\gamma}{2} \right) + \mathrm{i} \left(\frac{-\mathrm{i}\gamma}{2} \right) \right] |G^1_{n-1}|^2 = \frac{W_1(t)}{T_1} \, .
\end{aligned}
\tag{2.55}
$$

In deriving this equation for the two-photon correlator, we used (2.32) and (2.28). Although (2.28) is written for the Laplace components of the amplitudes, it is obvious that this equation is also correct for the time dependent amplitudes. Allowing for (2.55), we can conclude that curves 2 which illustrate the temporal behavior of the probability W_1 in Figs. 2.5 and 2.6 in fact describe the dependence of the start–stop correlator on the delay t between photons.

The probability W_1 can be calculated using the equations for the transition amplitudes, as has been done in the preceding sections. Unfortunately, the electron–phonon interaction cannot be taken into consideration when we use the equations for the transition amplitudes. However, there is another way to carry out the calculation. It is based on the equations for the density matrix of the system. This means of calculation is preferable because it allows us to consider systems with an electron–phonon interaction.

However, if we try to apply the general (1.77) for the density matrix to the system consisting of a two-level atom and a transverse electromagnetic field, we face the problem of solving an infinite set of coupled equations. It is not easy to find approximations allowing us to uncouple these equations for the density matrix. However, the problem of uncoupling has already been solved for the equations governing the transition amplitudes. Therefore we will derive equations for the density matrix by using the uncoupled equations for the transition amplitudes.

Let us consider (2.29) for the Laplace components of the transition amplitudes. By taking the inverse Laplace transformation, we obtain the following equations for the time dependent amplitudes and the amplitudes conjugate to them:

$$\begin{aligned} \dot{G}_0 &= -\mathrm{i}\Lambda^* G_1 \,, \qquad \dot{G}_0^* = \mathrm{i}\Lambda G_1^* \,, \\ \dot{G}_1 &= -\mathrm{i}\Delta G_1 - \mathrm{i}\left[\Lambda G_0 - \frac{\mathrm{i}}{2T_1} G_1\right] \,, \\ \dot{G}_1^* &= \mathrm{i}\Delta G_1^* + \mathrm{i}\left[\Lambda^* G_0^* + \frac{\mathrm{i}}{2T_1} G_1^*\right] \,. \end{aligned} \tag{2.56}$$

Using the two complex amplitudes we can create the following four elements of the density matrix:

$$\begin{aligned} |G_0(t)|^2 &= W_0 \,, \qquad & |G_1(t)|^2 &= W_1 \,, \\ G_0(t)G_1^*(t) &= W_{01} \,, \qquad & G_1(t)G_0^*(t) &= W_{10} \,. \end{aligned} \tag{2.57}$$

By differentiating these matrix elements with respect to time and substituting the right hand sides of (2.56) for the time derivatives of the amplitudes, we find the following equations for the elements of the density matrix:

$$\begin{aligned} \dot{W}_{10} &= -\mathrm{i}\left(\Delta - \frac{\mathrm{i}}{2T_1}\right) W_{10} - \chi(W_0 - W_1) \,, \\ \dot{W}_{01} &= \mathrm{i}\left(\Delta + \frac{\mathrm{i}}{2T_1}\right) W_{01} - \chi(W_0 - W_1) \,, \\ \dot{W}_1 &= -\chi(W_{10} + W_{01}) - \frac{W_1}{T_1} \,, \\ \dot{W}_0 &= \chi(W_{10} + W_{01}) \,, \end{aligned} \tag{2.58}$$

where

$$\chi = \mathrm{i}\Lambda = \Omega\sqrt{\frac{4\pi}{\hbar\omega_0}\frac{n_0}{V}}\boldsymbol{e}_0{\cdot}\boldsymbol{d} \approx \frac{\boldsymbol{d}\cdot\boldsymbol{E}}{\hbar} \tag{2.59}$$

is the Rabi frequency and $\boldsymbol{E}$ is the electric vector of the laser electromagnetic field. Equations (2.56) enable one to calculate the start–stop correlator, and to derive the expression $s(t) = W_1/T_1$ for it. Indeed, summing the third and fourth equations yields

$$\dot{W}_0 + \dot{W}_1 = -\frac{W_1}{T_1} \,. \tag{2.60}$$

However, the conservation law for the probabilities in the form described by (2.33) can be rewritten in the form

$$\dot{W}_0 + \dot{W}_1 = -\frac{\mathrm{d}}{\mathrm{d}t} \sum_{\boldsymbol{k}} W_{\boldsymbol{k}} = -s(t) \,. \tag{2.61}$$

Comparing the two equations, we arrive at the result $s(t) = W_1/T_1$.

It will be shown further that two-photon correlators are a powerful tool for the study of relaxation processes in polymers and glasses with the help of photons emitted by a single guest molecule. However, curves 2 in Figs. 2.5 and 2.6 for the probability W_1, which display the start–stop correlator, show that it approaches zero on a time scale of the order of T_1. Due to this fact the start–stop correlator cannot be used to study single-atom dynamics on a longer time scale than T_1. Therefore, another type of two-photon correlator is used in practice.

3. Full Two-Photon Correlator

3.1 Counting All Photon Pairs

In the start–stop regime, only those pairs consisting of consecutively emitted photons are counted. However, the probability of a photon being emitted by an excited atom approaches zero on the time scale of the fluorescence lifetime T_1. Therefore the start–stop correlator approaches zero on this short time scale and it cannot be used to study the slow relaxation existing in molecular systems.

However, there is another way to count photon pairs. To understand this approach, we turn back to Fig. 2.1. Here we see three photon pairs (2,3), (4,6) and (12,13) with the same time delay τ between photons in the pair. In contrast to the pairs (2,3) and (12,13), the pair (4,6) includes one intermediate photon 5. It is obvious that the number of such intermediate photons is arbitrary. If we count pairs without intermediate photons, we work in the start–stop regime. However, other photon-counting regimes are possible. For instance, we can count all pairs with the definite delay τ, independently of how many intermediate photons were emitted during time τ. The count rate in this regime is called the full two-photon correlator and we will denote it by $p(\tau)$ [40].

In the last chapter, we showed that the start–stop correlator is determined by the equation

$$S(t) = \frac{W_1(t)}{T_1}, \tag{3.1}$$

where W_1 is a probability which can be found from (2.58). We assume that the full two-photon correlator $p(t)$ can be written in a similar form, viz.,

$$p(t) = \frac{\rho_{11}(t)}{T_1}, \tag{3.2}$$

where $\rho_{11}(t)$ is an unknown probability. Our task is to find equations governing this probability.

We start with the simplest case, when the time delay between two emitted photons exceeds the relaxation time of all possible processes. In this case the second photon 'forgets' about the first photon. This means there is no correlation between the two photons. The probability of measuring such a photon pair equals the probability of measuring the second photon at $\tau = \infty$ under CW laser excitation. However, an equilibrium between excited and unexcited atoms exists under CW laser excitation, and the intensity of the emission is proportional to the number $N_1(\infty) = N\rho_{11}(\infty)$ of excited atoms. Here N is the total number of atoms irradiated by light and $\rho_{11}(\infty)$ is the probability of finding an excited atom. It is obvious that the rate of photon emission is given by

$$p(\infty) = \frac{\rho_{11}(\infty)}{T_1} . \tag{3.3}$$

For a two-level atom excited by CW laser light, the probability $\rho_{11}(\infty)$ can be found as the stationary solution of the ordinary balance equations or the optical Bloch equations. We can suppose that (3.3) is correct also at short time delays τ. If this suggestion is correct we can replace $\rho_{11}(t)$ in (3.2) by the relevant probability found from the optical Bloch equations at arbitrary time. The conclusion reached here on an intuitive level will be proven mathematically in the following sections.

3.2 Infinite Set of Equations for Transition Amplitudes. Discussion of Main Approximations

It is obvious that the numerator in (3.2) must take into account the probability of finding a photon pair with an arbitrary number of intermediate photons. This probability depends on quantum states of the system consisting of the atom and the electromagnetic field with an arbitrary number of emitted photons. There are an infinite number of such states. Therefore the density matrix for the system satisfies an infinite set of equations which can only be solved with certain approximations. It will be shown that summation over the photon pairs with various numbers of intermediate photons provides the transition from the infinite set of equations for the full density matrix to four optical Bloch equations. However, discussion of these approximations is more convenient if we use equations for the transition amplitudes. We must therefore turn back to the infinite set of equations for the transition amplitudes derived in previous chapters.

Equations (2.31) are the first three equations from the infinite chain of coupled equations. This chain of equations for the transition amplitudes can be found with the help of the general (1.66) and appears as follows:

$$
\begin{aligned}
\dot{G}^0 &= -\mathrm{i}\Lambda^* G^1 \,, \\
\dot{G}^1 &= -\mathrm{i}\Delta G^1 - \mathrm{i}\left[\Lambda G^0 + \sum_{\boldsymbol{k}} \lambda_{\boldsymbol{k}} G^0_{\boldsymbol{k}}\right] \,, \\
\dot{G}^0_{\boldsymbol{k}} &= -\mathrm{i}\Delta_{\boldsymbol{k}} G^0_{\boldsymbol{k}} - \mathrm{i}\left[\lambda^*_{\boldsymbol{k}} G^1 + \Lambda^{*\prime} G^1_{\boldsymbol{k}}\right] \,, \\
\dot{G}^1_{\boldsymbol{k}} &= -\mathrm{i}(\Delta + \Delta_{\boldsymbol{k}}) G^1_{\boldsymbol{k}} - \mathrm{i}\left[\Lambda' G^0_{\boldsymbol{k}} + \sum_{\boldsymbol{k}'} \lambda_{\boldsymbol{k}'} G^0_{\boldsymbol{k}\boldsymbol{k}'} + \lambda_{\boldsymbol{k}}\sqrt{2}\underline{G^0_{2\boldsymbol{k}}}\right] \,, \\
\underline{\dot{G}^0_{2\boldsymbol{k}}} &= \ldots \,, \\
&\vdots \\
\dot{G}^0_{\boldsymbol{k}\boldsymbol{k}'} &= -\mathrm{i}(\Delta_{\boldsymbol{k}} + \Delta_{\boldsymbol{k}'}) G^0_{\boldsymbol{k}\boldsymbol{k}'} - \mathrm{i}\left[\lambda^*_{\boldsymbol{k}} G^1_{\boldsymbol{k}'} + \lambda^*_{\boldsymbol{k}'} G^1_{\boldsymbol{k}} + \Lambda^{*\prime\prime} G^1_{\boldsymbol{k}\boldsymbol{k}'}\right] \,, \\
\dot{G}^1_{\boldsymbol{k}\boldsymbol{k}'} &= -\mathrm{i}(\Delta + \Delta_{\boldsymbol{k}} + \Delta_{\boldsymbol{k}'}) G^1_{\boldsymbol{k}\boldsymbol{k}'} \\
&\quad -\mathrm{i}\left[\Lambda'' G^0_{\boldsymbol{k}\boldsymbol{k}'} + \sum_{\boldsymbol{k}''} \lambda_{\boldsymbol{k}''} G^0_{\boldsymbol{k}\boldsymbol{k}'\boldsymbol{k}''} + \lambda_{\boldsymbol{k}}\sqrt{2}\underline{G^0_{2\boldsymbol{k}\boldsymbol{k}'}} + \lambda_{\boldsymbol{k}'}\sqrt{2}\underline{G^0_{\boldsymbol{k}2\boldsymbol{k}'}}\right] \,, \\
&\vdots
\end{aligned}
\tag{3.4}
$$

Here we use a simplified notation by omitting indices n, $n-1$, $n-2, \ldots$ of the exciting laser mode. This infinite set of equations takes into account all quantum states shown in Fig. 2.4. We make three approximations.

First Approximation. We neglect those quantum states in which wave vectors $\boldsymbol{k}$, $\boldsymbol{k}'$, $\boldsymbol{k}''$, ... coincide with each other. In other words we neglect the states with $\boldsymbol{k} = \boldsymbol{k}'$, $\boldsymbol{k}' = \boldsymbol{k}''$, and so on. Since $\boldsymbol{k}$ is a continuous variable, the statistical weight of the states we neglect is zero. Therefore we can omit all underlined terms and equations in the set (3.4).

Second Approximation. This approximation concerns the matrix elements of the electron–photon interaction $\Lambda = \sqrt{n}\lambda_0$, $\Lambda' = \sqrt{n-1}\lambda_0$, $\Lambda'' = \sqrt{n-2}\lambda_0$, and so on, which include the number n of photons in the laser mode. Because this number is large, we can assume $\Lambda = \Lambda' = \Lambda'' = \ldots$.

Third Approximation. This can be written as follows:

$$
\sum_{\boldsymbol{k}} \lambda_{\boldsymbol{k}} G^0_{\boldsymbol{k}} = -\frac{\mathrm{i}}{2T_1} G^1 \,, \qquad \sum_{\boldsymbol{k}'} \lambda_{\boldsymbol{k}'} G^0_{\boldsymbol{k}\boldsymbol{k}'} = -\frac{\mathrm{i}}{2T_1} G^1_{\boldsymbol{k}} \,,
$$
$$
\sum_{\boldsymbol{k}''} \lambda_{\boldsymbol{k}''} G^0_{\boldsymbol{k}\boldsymbol{k}'\boldsymbol{k}''} = -\frac{\mathrm{i}}{2T_1} G^1_{\boldsymbol{k}\boldsymbol{k}'} \,, \quad \text{etc.} \tag{3.5}
$$

We have already used the first of these relations when carrying out the transition from (2.27) to (2.29). The evidence for (2.28) describing the first equation is based on neglecting the real part of the sum over $\boldsymbol{k}$ in (2.28). This real part describes a frequency shift and it can be included in the frequency ω. The same holds for the second, the third, and all the other relations in (3.5). Although these relations are proven for the frequency-dependent components, we can carry out the inverse Laplace transformation of the time-dependent amplitudes on both sides of these relations. Therefore (3.5) can be applied both to frequency- and time-dependent amplitudes. The third approximation allows us to transform the infinite set of coupled equations (3.4) to the following infinite set of uncoupled pairs of equations:

$$
\begin{aligned}
\dot{G}^0 &= -\mathrm{i}\Lambda^* G^1 \, , \\
\dot{G}^1 &= -\mathrm{i}\Delta G^1 - \mathrm{i}\left[\Lambda G^0 - \frac{\mathrm{i}}{2T_1} G^1\right] , \\
\dot{G}^0_{\boldsymbol{k}} &= -\mathrm{i}\Delta_{\boldsymbol{k}} G^0_{\boldsymbol{k}} - \mathrm{i}\left[\lambda^*_{\boldsymbol{k}} G^1 + \Lambda^* G^1_{\boldsymbol{k}}\right] , \\
\dot{G}^1_{\boldsymbol{k}} &= -\mathrm{i}(\Delta + \Delta_{\boldsymbol{k}}) G^1_{\boldsymbol{k}} - \mathrm{i}\left[\Lambda G^0_{\boldsymbol{k}} - \frac{\mathrm{i}}{2T_1} G^1_{\boldsymbol{k}}\right] , \\
\dot{G}^0_{\boldsymbol{k}\boldsymbol{k}'} &= -\mathrm{i}(\Delta_{\boldsymbol{k}} + \Delta_{\boldsymbol{k}'}) G^0_{\boldsymbol{k}\boldsymbol{k}'} - \mathrm{i}\left[\lambda^*_{\boldsymbol{k}} G^1_{\boldsymbol{k}'} + \lambda^*_{\boldsymbol{k}'} G^1_{\boldsymbol{k}} + \Lambda^* G^1_{\boldsymbol{k}\boldsymbol{k}'}\right] , \\
\dot{G}^1_{\boldsymbol{k}\boldsymbol{k}'} &= -\mathrm{i}(\Delta + \Delta_{\boldsymbol{k}} + \Delta_{\boldsymbol{k}'}) G^1_{\boldsymbol{k}\boldsymbol{k}'} - \mathrm{i}\left[\Lambda G^0_{\boldsymbol{k}\boldsymbol{k}'} - \frac{\mathrm{i}}{2T_1} G^1_{\boldsymbol{k}\boldsymbol{k}'}\right] , \\
&\vdots
\end{aligned}
\tag{3.6}
$$

We can then solve each pair of equations in turn.

The first pair of equations describes the absorption of one laser photon and spontaneous emission of one photon in the presence of the laser field. This situation corresponds to the start–stop regime, considered in detail in previous chapters. The second pair of equations describes the temporal evolution in the presence of one extra photon $\boldsymbol{k}$. The third pair of equations describes the temporal evolution in the presence of two extra photons $\boldsymbol{k}$ and $\boldsymbol{k}'$, and so on. Equations (3.6) can be described by the infinite set of diagrams shown in Fig. 3.1. By considering the temporal evolution of the amplitudes in the start–stop regime, we took into account only the first pair of equations in (3.6). This pair is described by the first diagram. In the next section we consider all pairs of equations and carry out the summation shown in Fig. 3.1.

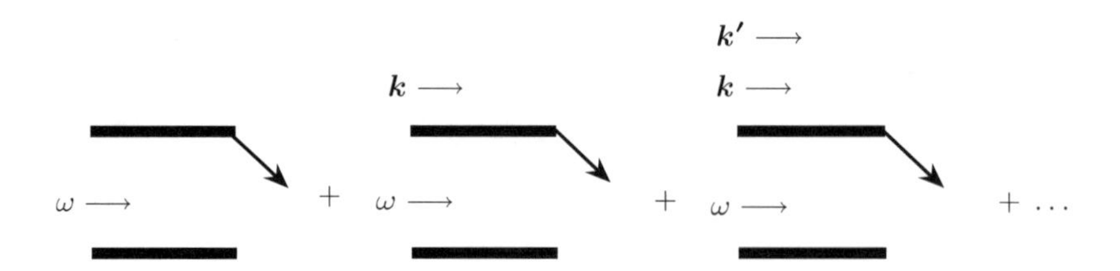

Fig. 3.1. Summation of quantum states with various numbers of intermediate photons, carried out in the next section

3.3 Equations Governing the Full Two-Photon Correlator

By using the first pair of (3.6) and equations conjugate to them, we have derived the four equations (2.58) which govern the probability W_1, i.e., the start–stop correlator. Now we are ready to derive equations governing the full two-photon correlator. For this purpose we consider the second pair of (3.6). Using these equations for the transition amplitudes and the equations conjugate to them, we find the following four equations for the corresponding elements of the full density matrix:

$$\begin{aligned}
\frac{\mathrm{d}}{\mathrm{d}t}\sum_{\boldsymbol{k}} G^1_{\boldsymbol{k}} G^{*0}_{\boldsymbol{k}} &= -\mathrm{i}\left(\Delta - \frac{\mathrm{i}}{2T_1}\right)\sum_{\boldsymbol{k}} G^1_{\boldsymbol{k}} G^{*0}_{\boldsymbol{k}} \\
&\quad -\chi\left(\sum_{\boldsymbol{k}} |G^0_{\boldsymbol{k}}|^2 - \sum_{\boldsymbol{k}} |G^1_{\boldsymbol{k}}|^2\right) + \underline{\mathrm{i}\sum_{\boldsymbol{k}} \lambda_{\boldsymbol{k}} G^1_{\boldsymbol{k}} G^{*1}}\,, \\
\frac{\mathrm{d}}{\mathrm{d}t}\sum_{\boldsymbol{k}} G^0_{\boldsymbol{k}} G^{*1}_{\boldsymbol{k}} &= \left(\frac{\mathrm{d}}{\mathrm{d}t}\sum_{\boldsymbol{k}} G^1_{\boldsymbol{k}} G^{*0}_{\boldsymbol{k}}\right)^*, \\
\frac{\mathrm{d}}{\mathrm{d}t}\sum_{\boldsymbol{k}} |G^1_{\boldsymbol{k}}|^2 &= -\chi\sum_{\boldsymbol{k}}(G^1_{\boldsymbol{k}} G^{*0}_{\boldsymbol{k}} + G^0_{\boldsymbol{k}} G^{*1}_{\boldsymbol{k}}) - \frac{1}{T_1}\sum_{\boldsymbol{k}} |G^1_{\boldsymbol{k}}|^2\,, \\
\frac{\mathrm{d}}{\mathrm{d}t}\sum_{\boldsymbol{k}} |G^0_{\boldsymbol{k}}|^2 &= \chi\sum_{\boldsymbol{k}}(G^1_{\boldsymbol{k}} G^{*0}_{\boldsymbol{k}} + G^0_{\boldsymbol{k}} G^{*1}_{\boldsymbol{k}}) + \frac{1}{T_1}|G^1|^2\,.
\end{aligned} \tag{3.7}$$

Here we have used the approximations described by (3.5). The summation over $\boldsymbol{k}$ in these equations takes into account all possible wave vectors of the intermediate photon.

Fourth Approximation. This concerns the term underlined in the first equation of (3.7). It is based on the following equations:

$$\begin{aligned}
&\sum_{\boldsymbol{k}} \lambda_{\boldsymbol{k}} G^1_{\boldsymbol{k}} \approx \frac{\mathrm{i}\gamma}{2}\frac{\Lambda}{\Omega} G^1\,, \qquad \sum_{\boldsymbol{k}'} \lambda_{\boldsymbol{k}'} G^1_{\boldsymbol{k}\boldsymbol{k}'} \approx \frac{\mathrm{i}\gamma}{2}\frac{\Lambda}{\Omega} G^1_{\boldsymbol{k}}\,, \\
&\qquad \sum_{\boldsymbol{k}''} \lambda_{\boldsymbol{k}''} G^1_{\boldsymbol{k}\boldsymbol{k}'\boldsymbol{k}''} \approx \frac{\mathrm{i}\gamma}{2}\frac{\Lambda}{\Omega} G^1_{\boldsymbol{k}\boldsymbol{k}'}\,, \quad \text{etc.}\,,
\end{aligned} \tag{3.8}$$

derived in Appendix D. Allowing for $\Lambda/\Omega \ll 1$, we may neglect the underlined term in the first equation of (3.7). Similar terms can be dropped in the equations for the density matrix elements derived with the help of the third, fourth and other pairs of (3.6). The fourth approximation consists in neglecting the terms which include expressions from the left hand sides of (3.8).

By using the third pair of (3.6) for the transition amplitudes and taking into account all four approximations, we find the following equations:

$$
\begin{aligned}
\frac{\mathrm{d}}{\mathrm{d}t}\frac{1}{2!}\sum_{\boldsymbol{k}\boldsymbol{k}'} G^1_{\boldsymbol{k}\boldsymbol{k}'}G^{*0}_{\boldsymbol{k}\boldsymbol{k}'} &= -\mathrm{i}\left(\Delta - \frac{\mathrm{i}}{2T_1}\right)\frac{1}{2!}\sum_{\boldsymbol{k}\boldsymbol{k}'} G^1_{\boldsymbol{k}\boldsymbol{k}'}G^{*0}_{\boldsymbol{k}\boldsymbol{k}'} \\
&\quad -\chi\left(\frac{1}{2!}\sum_{\boldsymbol{k}\boldsymbol{k}'}|G^0_{\boldsymbol{k}\boldsymbol{k}'}|^2 - \frac{1}{2!}\sum_{\boldsymbol{k}\boldsymbol{k}'}|G^1_{\boldsymbol{k}\boldsymbol{k}'}|^2\right), \\
\frac{\mathrm{d}}{\mathrm{d}t}\frac{1}{2!}\sum_{\boldsymbol{k}\boldsymbol{k}'} G^0_{\boldsymbol{k}\boldsymbol{k}'}G^{*1}_{\boldsymbol{k}\boldsymbol{k}'} &= \left(\frac{\mathrm{d}}{\mathrm{d}t}\frac{1}{2!}\sum_{\boldsymbol{k}\boldsymbol{k}'} G^1_{\boldsymbol{k}\boldsymbol{k}'}G^{*0}_{\boldsymbol{k}\boldsymbol{k}'}\right)^*, \\
\frac{\mathrm{d}}{\mathrm{d}t}\frac{1}{2!}\sum_{\boldsymbol{k}\boldsymbol{k}'} |G^1_{\boldsymbol{k}\boldsymbol{k}'}|^2 &= -\chi\frac{1}{2!}\sum_{\boldsymbol{k}\boldsymbol{k}'}(G^1_{\boldsymbol{k}\boldsymbol{k}'}G^{*0}_{\boldsymbol{k}\boldsymbol{k}'} + G^0_{\boldsymbol{k}\boldsymbol{k}'}G^{*1}_{\boldsymbol{k}\boldsymbol{k}'}) \\
&\quad -\frac{1}{T_1}\frac{1}{2!}\sum_{\boldsymbol{k}\boldsymbol{k}'}|G^1_{\boldsymbol{k}\boldsymbol{k}'}|^2, \\
\frac{\mathrm{d}}{\mathrm{d}t}\frac{1}{2!}\sum_{\boldsymbol{k}\boldsymbol{k}'} |G^0_{\boldsymbol{k}\boldsymbol{k}'}|^2 &= \chi\frac{1}{2!}\sum_{\boldsymbol{k}\boldsymbol{k}'}(G^1_{\boldsymbol{k}\boldsymbol{k}'}G^{*0}_{\boldsymbol{k}\boldsymbol{k}'} + G^0_{\boldsymbol{k}\boldsymbol{k}'}G^{*1}_{\boldsymbol{k}\boldsymbol{k}'}) + \frac{1}{T_1}\sum_{\boldsymbol{k}}|G^1_{\boldsymbol{k}}|^2 .
\end{aligned}
\tag{3.9}
$$

These equations describe the situation where two intermediate photons $\boldsymbol{k}$ and $\boldsymbol{k}'$ are emitted. Double summation takes into account all possible quantum states of two intermediate photons. The rules for deriving equations taking into account an arbitrary number of intermediate photons are now clear.

Let us introduce the following infinite series:

$$
\begin{aligned}
\rho_{00} &= W_0 + \sum_{\boldsymbol{k}}|G^0_{\boldsymbol{k}}|^2 + \frac{1}{2!}\sum_{\boldsymbol{k}\boldsymbol{k}'}|G^0_{\boldsymbol{k}\boldsymbol{k}'}|^2 + \cdots, \\
\rho_{11} &= W_1 + \sum_{\boldsymbol{k}}|G^1_{\boldsymbol{k}}|^2 + \frac{1}{2!}\sum_{\boldsymbol{k}\boldsymbol{k}'}|G^1_{\boldsymbol{k}\boldsymbol{k}'}|^2 + \cdots, \\
\rho_{10} &= W_{10} + \sum_{\boldsymbol{k}}G^1_{\boldsymbol{k}}G^{*0}_{\boldsymbol{k}} + \frac{1}{2!}\sum_{\boldsymbol{k}\boldsymbol{k}'}G^1_{\boldsymbol{k}\boldsymbol{k}'}G^{*0}_{\boldsymbol{k}\boldsymbol{k}'} + \cdots, \\
\rho_{01} &= W_{01} + \sum_{\boldsymbol{k}}G^0_{\boldsymbol{k}}G^{*1}_{\boldsymbol{k}} + \frac{1}{2!}\sum_{\boldsymbol{k}\boldsymbol{k}'}G^0_{\boldsymbol{k}\boldsymbol{k}'}G^{*1}_{\boldsymbol{k}\boldsymbol{k}'} + \cdots .
\end{aligned}
\tag{3.10}
$$

Each new matrix element ρ_{00}, ρ_{11}, ρ_{10}, ρ_{01} is the trace of the corresponding element of the full density matrix of the system consisting of a two-level atom and a transverse electromagnetic field with respect to the indices of the spontaneously emitted photons. Hence the new matrix elements do not depend on the indices of the emitted photons. They depend only on atomic indices 0 and 1 describing two states of the atom. However, the atom is only a subsystem of the full system, which also includes the electromagnetic field. Such a transition from the matrix elements of the density matrix of the full system to matrix elements of a subsystem is called the reduction of the density matrix. With the help of the new matrix elements, we are able to describe the temporal behavior of any dynamical variable relating to the two-level atom in an external electromagnetic field. For instance, the average value of the atomic dipolar moment induced by the external electromagnetic field is given by

$$\overline{\boldsymbol{d}(t)} = \mathrm{Tr}\left(\hat{\rho}_\alpha(t)\hat{\boldsymbol{d}}\right) = \rho_{00}\boldsymbol{d}_{00} + \rho_{01}\boldsymbol{d}_{10} + \rho_{10}\boldsymbol{d}_{01} + \rho_{11}\boldsymbol{d}_{11} \,, \tag{3.11}$$

where elements ρ_{00}, ρ_{11}, ρ_{10}, ρ_{01} of the electronic density matrix are described by (3.10). Since any free atom has no dipolar moment in the stationary state, we should set $\boldsymbol{d}_{00} = \boldsymbol{d}_{11} = 0$. An atomic dipolar moment can only emerge in an external electromagnetic field when $\rho_{01} \neq 0$, $\rho_{10} \neq 0$.

The equations governing matrix elements ρ_{00}, ρ_{11}, ρ_{10}, ρ_{01} can be found by summing (2.58), (3.8), (3.9), and so on, which describe the temporal evolution of the density matrices relating to the situation without intermediate photons, with one intermediate photon, and so on. Such a summation, shown also in Fig. 3.1, is a direct way to find the full two-photon correlator. We carry out a summation of all the first equations from the sets (2.58), (3.7), (3.9) and so on. Then we sum all the second, all the third and all the fourth equations from these sets. After such a summation we arrive at the following four equations:

$$\begin{aligned}
\dot{\rho}_{10} &= -\mathrm{i}\left(\Delta - \frac{\mathrm{i}}{2T_1}\right)\rho_{10} - \chi(\rho_{00} - \rho_{11}) \,, \\
\dot{\rho}_{01} &= \mathrm{i}\left(\Delta + \frac{\mathrm{i}}{2T_1}\right)\rho_{01} - \chi(\rho_{00} - \rho_{11}) \,, \\
\dot{\rho}_{11} &= -\chi(\rho_{10} + \rho_{01}) - \frac{\rho_{11}}{T_1} \,, \\
\dot{\rho}_{00} &= \chi(\rho_{10} + \rho_{01}) + \frac{\rho_{11}}{T_1} \,.
\end{aligned} \tag{3.12}$$

These equations are similar to the optical Bloch equations. Since we did not allow for any additional dephasing processes except spontaneous emission, the equations in (3.12) include the dephasing rate $1/2T_1$ that is the lowest limit of the true dephasing rate $1/T_2$ occurring in the true optical Bloch equations [41]. It will be shown further that the electron–phonon interaction results in additional dephasing processes. When the electron–phonon interaction is taken into account, the rate constant in the first two equations of (3.12) is replaced by the rate

$$\frac{1}{T_2} = \frac{\gamma(T)}{2} + \frac{1}{2T_1} \,,$$

where $\gamma(T)/2$ is a contribution to the dephasing rate from the electron–phonon interaction. The contribution from the electron–phonon interaction depends on temperature.

Let us now find an expression for the full two-photon correlator. This correlator can be found with the help of the summation shown in Fig. 3.1. Mathematically, an expression for the full two-photon correlator is given by

$$
\begin{aligned}
p(t) &= \frac{\mathrm{d}}{\mathrm{d}t}\left[\sum_{\boldsymbol{k}} |G^0_{\boldsymbol{k}}|^2 + \frac{1}{2!}\sum_{\boldsymbol{k}\boldsymbol{k}'} |G^0_{\boldsymbol{k}\boldsymbol{k}'}|^2 + \frac{1}{3!}\sum_{\boldsymbol{k}\boldsymbol{k}'\boldsymbol{k}''} |G^0_{\boldsymbol{k}\boldsymbol{k}'\boldsymbol{k}''}|^2 + \cdots\right] \\
&= \frac{1}{T_1}\left[|G^1|^2 + \sum_{\boldsymbol{k}} |G^1_{\boldsymbol{k}}|^2 + \frac{1}{2!}\sum_{\boldsymbol{k}\boldsymbol{k}'} |G^1_{\boldsymbol{k}\boldsymbol{k}'}|^2 + \cdots\right] \\
&= \frac{\rho_{11}(t)}{T_1} \, . \qquad (3.13)
\end{aligned}
$$

Here we have taken into account (3.5). Equation (3.13) tells us that the excited state population found from the optical Bloch equations permits us to calculate the full two-photon correlator.

3.4 Relating the Full and Start–Stop Two-Photon Correlators

We now seek the relation between the full and start–stop correlators determined by the equations $p(t) = \rho_{11}(t)/T_1$ and $s(t) = W_1(t)/T_1$, where the probabilities $\rho_{11}(t)$ and $W_1(t)$ can be found from (3.12) and (2.58), respectively. These relations look simpler when expressed in terms of the Laplace components of both probabilities.

Full Two-Photon Correlator. This correlator is described by the equation $p(t) = \rho_{11}(t)/T_1$, where $\rho_{11}(t)$ can be found from (3.12). Using (1.60) and (1.65) for the Laplace transformation of both the function and its time derivative, we can transform (3.12) for time-dependent components to the following equations for Laplace components:

$$
\begin{aligned}
&[\mathrm{i}\omega - \mathrm{i}(\Delta - \mathrm{i}\Gamma)]\,\rho_{10} = \chi(\rho_{00} - \rho_{11}) \, , \\
&[\mathrm{i}\omega + \mathrm{i}(\Delta + \mathrm{i}\Gamma)]\,\rho_{01} = \chi(\rho_{00} - \rho_{11}) \, , \\
&\mathrm{i}\left(\mathrm{i}\omega - \frac{1}{T_1}\right)\rho_{11} = \chi(\rho_{10} + \rho_{01}) \, , \\
&\mathrm{i}\omega\rho_{00} + \frac{1}{T_1}\rho_{11} = -\chi(\rho_{10} + \rho_{01}) - \rho_{00}(0) \, ,
\end{aligned}
\qquad (3.14)
$$

where $\Gamma = 1/2T_1$. Expressing the off-diagonal elements in terms of the diagonal elements, we find the following equations for the diagonal matrix elements:

$$
\begin{aligned}
&\left(\mathrm{i}\omega - \frac{1}{T_1} - k\right)\rho_{11} + k\rho_{00} = 0 \, , \\
&\left(\frac{1}{T_1} + k\right)\rho_{11} + (\mathrm{i}\omega - k)\rho_{00} = -1 \, ,
\end{aligned}
\qquad (3.15)
$$

where

$$
k(\omega) = \mathrm{i}2\chi^2 \frac{\omega + \mathrm{i}\Gamma}{(\omega + \mathrm{i}\Gamma)^2 - \Delta^2} \, . \qquad (3.16)
$$

By solving (3.15), we find the following equation for the Laplace component of the full two-photon correlator:

$$p(\omega) = \frac{\rho_{11}(\omega)}{T_1} = \frac{k}{T_1 D_p} , \tag{3.17}$$

where the determinant of (3.15) is given by

$$D_p = \mathrm{i}\omega \left(\mathrm{i}\omega - \frac{1}{T_1} - 2k \right) . \tag{3.18}$$

Start–Stop Correlator. This correlator is described by the equation $s(t) = W_1(t)/T_1$, where $W_1(t)$ can be found from (2.58). By carrying out the Laplace transformation of both sides of (2.58), we find the following equations for the Laplace components:

$$\begin{aligned}
&[\mathrm{i}\omega - \mathrm{i}(\Delta - \mathrm{i}\Gamma)]\, W_{10} = \chi(W_0 - W_1) , \\
&[\mathrm{i}\omega + \mathrm{i}(\Delta + \mathrm{i}\Gamma)]\, W_{01} = \chi(W_0 - W_1) , \\
&\left(\mathrm{i}\omega - \frac{1}{T_1} \right) W_1 = \chi(W_{10} + W_{01}) , \\
&\mathrm{i}\omega W_0 = -\chi(W_{10} + W_{01}) - W_0(0) .
\end{aligned} \tag{3.19}$$

We can express the off-diagonal elements in terms of the diagonal elements, giving

$$\left(\mathrm{i}\omega - \frac{1}{T_1} - k \right) W_1 + kW_0 = 0 , \qquad kW_1 + (\mathrm{i}\omega - k)W_0 = -1 . \tag{3.20}$$

The solution is

$$s(\omega) = \frac{k}{T_1 D_s} , \tag{3.21}$$

where

$$D_s = D_p + \frac{k}{T_1} . \tag{3.22}$$

Using (3.17), (3.18), (3.21) and (3.22), we find the following relation between the Laplace components of the two correlators:

$$p(\omega) = s(\omega) + s(\omega)p(\omega) . \tag{3.23}$$

It is shown in Appendix E that

$$\int_{-\infty}^{\infty} s(\omega)p(\omega)\mathrm{e}^{-\mathrm{i}t\omega} \frac{\mathrm{d}\omega}{2\pi} = \int_0^t s(t-x)p(x)\,\mathrm{d}x . \tag{3.24}$$

Using this result, equation (3.23) for the Laplace components can be transformed to the relation between time-dependent components of the two-photon correlators:

$$p(t) = s(t) + \int_0^t s(t-x)p(x)\,\mathrm{d}x \,. \tag{3.25}$$

This integral equation can be solved by means of an iteration procedure. The solution is given by

$$\begin{aligned} p(t) &= s(t) \\ &+ \sum_{n=1}^{\infty} \int_0^t \mathrm{d}x_1 \int_0^{x_1} \mathrm{d}x_2 \ldots \int_0^{x_{n-1}} \mathrm{d}x_n \, s(t-x_1)s(x_1-x_2)s(x_2-x_3)\ldots s(x_n) \,. \end{aligned} \tag{3.26}$$

We see that the full two-photon correlator is an infinite series with terms which include only start–stop correlators. The terms of the series have a simple physical meaning. The first term $s(t)$ relates to the start–stop counting regime. The second term of the series is an integrated bilinear function of the start–stop correlator. It determines the counting rate of pairs with one intermediate photon, like the pair (4,6) with intermediate photon 5, shown in Fig. 2.1. The nth term in the sum describes the count rate of pairs with n intermediate photons.

3.5 Frequency and Time Dependence of Full and Start–Stop Two-Photon Correlators

The full two-photon correlator is a function of the time delay t between photons and the frequency ω of the excitation. Inserting (3.16) into (3.17) yields the following expression for the Laplace component of the full two-photon correlator:

$$p(\omega) = \frac{2\mathrm{i}\chi^2}{T_1}\frac{\omega + \mathrm{i}\Gamma}{Q_p(\omega)} \,, \tag{3.27}$$

where

$$\begin{aligned} Q_p(\omega) &= \omega\left[4\chi^2(\omega+\mathrm{i}\Gamma) - \left(\omega + \frac{\mathrm{i}}{T_1}\right)\left[(\omega+\mathrm{i}\Gamma)^2 - \Delta^2\right]\right] \\ &= (\omega-\omega_1)(\omega-\omega_2)(\omega-\omega_3)(\omega-\omega_4) \,, \end{aligned} \tag{3.28}$$

where ω_j are the roots of the equation $Q_p(\omega) = 0$. These four roots are the poles of the function $p(\omega)$. Therefore it can be written as the sum of terms with these poles:

$$p(\omega) = \sum_{j=1}^{4} \frac{p_j}{\omega - \omega_j} , \tag{3.29}$$

where

$$p_j = [(\omega - \omega_j)p(\omega)]_{\omega=\omega_j} \tag{3.30}$$

is the residue at the pole ω_j. The time-dependent correlator can easily be found if we apply (1.64) to each resonant term of (3.29):

$$p(t) = -\mathrm{i} \sum_{j=1}^{4} p_j \mathrm{e}^{-\mathrm{i}\omega_j t} . \tag{3.31}$$

One conclusion can be drawn with the help of this general expression. Due to the conservation of total probability for the atomic population, expressed by the equation $\dot{\rho}_{00} + \dot{\rho}_{11} = 0$, one root of the optical Bloch equations, say ω_1, equals zero. The other roots will have finite negative imaginary parts. Therefore the full two-photon correlator approaches p_1 when time tends to infinity. This temporal behavior of the full two-photon correlator at large t is in contrast to the temporal behavior of the start–stop correlator.

An expression for the Laplace component of the start–stop correlator can be found by inserting (3.16) into (3.21). The result is given by

$$s(\omega) = \frac{2\mathrm{i}\chi^2}{T_1} \frac{\omega + \mathrm{i}\Gamma}{Q_s(\omega)} , \tag{3.32}$$

where

$$\begin{aligned} Q_s(\omega) &= \left(\omega + \frac{\mathrm{i}}{2T_1}\right) 4\chi^2(\omega + \mathrm{i}\Gamma) - \omega \left(\omega + \frac{\mathrm{i}}{T_1}\right) \left[(\omega + \mathrm{i}\Gamma)^2 - \Delta^2\right] \\ &= (\omega - \nu_1)(\omega - \nu_2)(\omega - \nu_3)(\omega - \nu_4) , \end{aligned} \tag{3.33}$$

where ν_j are the roots of the equation $Q_s(\omega) = 0$. The start–stop correlator can be expressed in terms of these roots as

$$s(t) = -\mathrm{i} \sum_{j=1}^{4} s_j \mathrm{e}^{-\mathrm{i}\nu_j t} , \tag{3.34}$$

where

$$s_j = [(\omega - \nu_j)s(\omega)]_{\omega=\nu_j} . \tag{3.35}$$

Equations (3.26), (3.31) and (3.34), (3.35) can be used to calculate the dependence of both types of correlator on the delay t, the laser detuning Δ, and the excitation intensity χ^2. The numerical calculations for two-photon correlators can be carried out at arbitrary values of the rate constants Γ and

$1/T_1$. However, we can find simple analytical expressions for both correlators in two special cases, when $\Delta = 0$ or $\Gamma = 1/T_1 = \gamma$.

The dependence of the correlator $p(t)$ on the frequency of the excitation can be used to find the temporal evolution of the optical absorption line. We consider this temporal evolution for the case when $\Gamma = 1/T_1 = \gamma$. In this special case the equation $Q_p(\omega) = 0$ is easily solved to give the roots

$$\omega_1 = -\mathrm{i}0\,, \quad \omega_2 = -\mathrm{i}\gamma\,, \quad \omega_3 = -\mathrm{i}\gamma + Q\,, \quad \omega_4 = -\mathrm{i}\gamma - Q\,, \tag{3.36}$$

where

$$Q = \sqrt{\Delta^2 + 4\chi^2}\,. \tag{3.37}$$

Inserting these equations into (3.29), we find the following expression for the Laplace component of the full correlator:

$$\begin{aligned} p(\omega) = \frac{2\mathrm{i}\chi^2\gamma}{Q^2+\gamma^2}\bigg[\frac{1}{\omega+\mathrm{i}0} &- \frac{1}{2}\left(1-\mathrm{i}\frac{\gamma}{Q}\right)\frac{1}{\omega+Q+\mathrm{i}\gamma} \\ &- \frac{1}{2}\left(1+\mathrm{i}\frac{\gamma}{Q}\right)\frac{1}{\omega-Q+\mathrm{i}\gamma}\bigg]\,. \end{aligned} \tag{3.38}$$

Carrying out the inverse transformation of (3.38) to the time-dependent function, the exponential functions with these poles appear in accordance with (1.64). The result of the inverse transformation is given by

$$p(t) = \frac{2\chi^2\gamma}{Q^2+\gamma^2}\left[1 - \mathrm{e}^{-t\gamma}\left(\cos Qt + \frac{\gamma}{Q}\sin Qt\right)\right]\,. \tag{3.39}$$

At large t this correlator, in contrast to the start–stop correlator, approaches a finite function

$$p(\Delta) = \frac{2\chi^2\gamma}{\Delta^2 + 4\chi^2 + \gamma^2}\,. \tag{3.40}$$

It describes an absorption line with FWHM

$$\Delta\omega_{1/2} = 2\sqrt{4\chi^2+\gamma^2}\,, \tag{3.41}$$

which depends on the intensity of the laser light.

The two-photon correlator enables one to study the absorption line shape at short time intervals. Here we are seeing a temporal evolution in the absorption line shape. This effect is shown in Fig. 3.2. Curves 1, 2 and 3 relate to various time delays t. They demonstrate the temporal transformation of the absorption line. At short times, the line is very broad and has a rather complicated shape with few maxima. As time goes by, the line becomes narrower and its shape approaches a Lorentzian. In moving from Fig. 3.2a to d, we see how the laser intensity influences the absorption line shape.

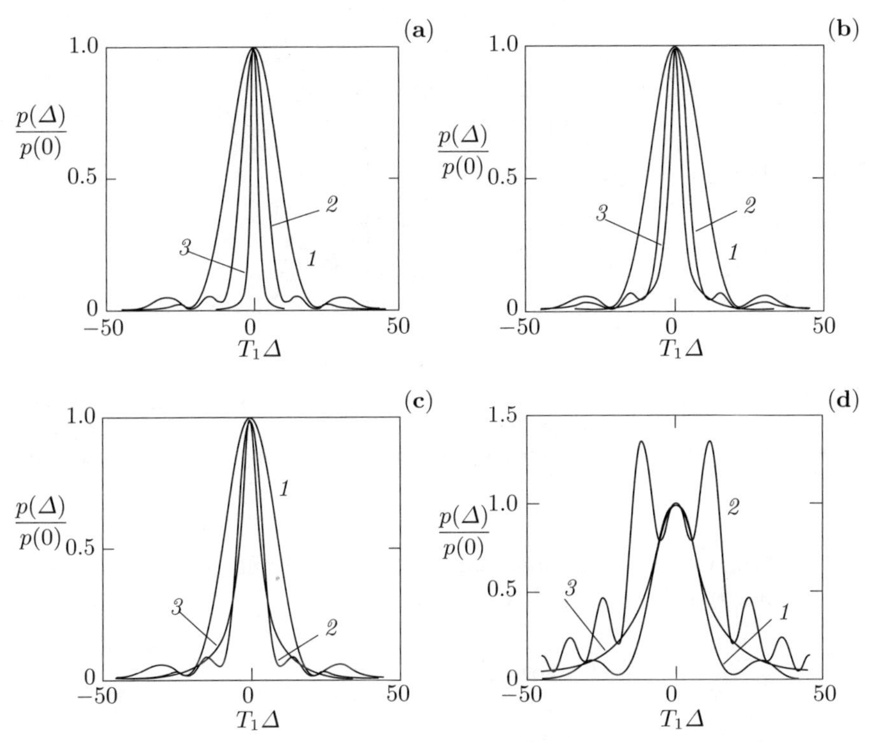

Fig. 3.2. Dependence of full two-photon correlator on frequency detuning Δ at $t/T_1 = 0.3$ (curve 1), 0.6 (curve 2) and 5 (curve 3), and at various values of the Rabi frequency χ, viz., (**a**) $\chi T_1 = 0.1$, (**b**) $\chi T_1 = 1$, (**c**) $\chi T_1 = 2$, and (**d**) $\chi T_1 = 5$

Let us now consider the temporal dependence of the full two-photon correlator and compare it with the temporal dependence of the start–stop correlator. We will consider the special case of the resonance when $\Delta = 0$. Taking into account that $\Gamma = 1/2T_1 = \gamma/2$, we find the following set of roots for the equations $Q_s(\nu) = 0$ and $Q_p(\omega) = 0$:

$$\nu_{1,2} = -\mathrm{i}\frac{\gamma}{2}\,, \quad \nu_{3,4} = -\mathrm{i}\frac{\gamma}{2} \pm R_s\,, \quad R_s = \sqrt{4\chi^2 - \frac{\gamma^2}{4}}\,,$$
$$\omega_1 = -\mathrm{i}0\,, \quad \omega_2 = -\mathrm{i}\frac{\gamma}{2}\,, \quad \omega_{3,4} = -\mathrm{i}\frac{3\gamma}{2} \pm R_p\,, \quad R_p = \sqrt{4\chi^2 - \frac{\gamma^2}{16}}\,. \tag{3.42}$$

If the intensity χ increases, the functions R_s and R_p become real. The values of the intensity χ at which R_s and R_p become real are different. This means that the critical intensity separating the domain of low and high intensities differs considerably for the full and start–stop correlators. This results from the difference in the relaxation matrices of the sets (3.14) and (3.19).

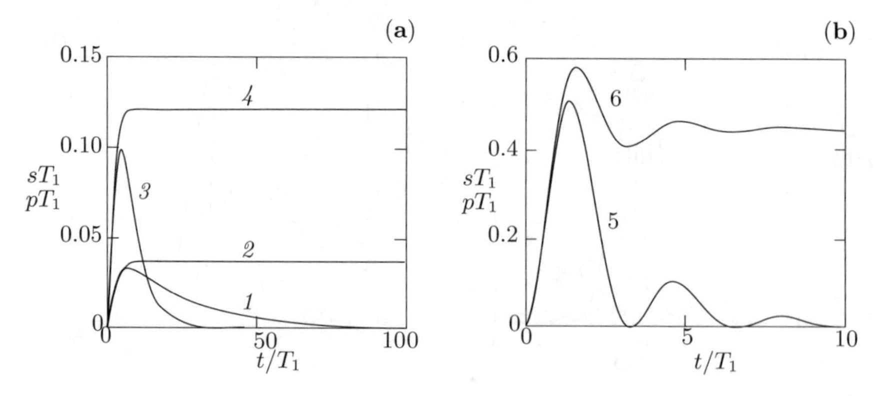

Fig. 3.3. Dependence of two-photon correlators $s(t)$ (curves 1, 3, and 5) and $p(t)$ (curves 2, 4, and 6) on the time delay of the photon pair. (**a**) Weak field $\chi T_1 = 0.2$ (curves 1 and 2) and $\chi T_1 = 0.3$ (curves 3 and 4). (**b**) Strong field $\chi T_1 = 1$ (curves 5 and 6)

Using the roots ν_j determined by (3.42), we can find the start–stop correlator. Inserting these roots into (3.34), we find

$$s(t) = \frac{2\chi^2\gamma}{R_s^2}\, \mathrm{e}^{-\gamma t/2}\,(1 - \cos R_s t)\ . \tag{3.43}$$

This formula coincides exactly with (2.50) found earlier using the transition amplitudes, if we take into account that $R_s = 2R$ and $|\Lambda|^2 = \chi^2$.

The roots ω_j determined by (3.42) allow us to find the full two-photon correlator. Inserting these roots into (3.30), we find

$$p(t) = \frac{2\chi^2\gamma}{R_p^2 + \gamma_p^2}\left[1 - \mathrm{e}^{-\gamma_p t}\left(\cos R_p t + \frac{\gamma_p}{R_p}\sin R_p t\right)\right]\ , \tag{3.44}$$

where $\gamma_p = 3\gamma/2$. This formula resembles (3.39), found earlier for another special case. The correlators $s(t)$ and $p(t)$ determined by (3.43) and (3.44) are shown in Fig. 3.3.

The stronger the intensity of the laser light, the faster the growth of the correlators. The decrease of the start-stop correlator at large times is also determined by the laser intensity. Figure 3.3 shows clearly that the start–stop correlator cannot be used to study the slow dynamics inherent in guest molecules embedded in a polymer or glass.

Part II

Phonons and Tunneling Excitations

4. Adiabatic Interaction

In the last three chapters we considered the main principles of light emission and absorption by a single atom. Of course, any sample contains a large number of atoms, and we have not yet discussed how a system consisting of a single atom can be prepared. If we deal with atoms in the vapor phase, the problem of preparing a trap for a single atom can be solved, although it is not easy to realize from a technical standpoint. However, research with complex organic molecules in the vapor phase is an insoluble problem, because many of these molecules will be destroyed in the course of evaporation. Therefore, complex organic molecules are usually studied in a condensed phase. Matrix isolation of complex molecules in a transparent, low temperature solution is the main method in single-molecule spectroscopy [3, 35].

In such systems, we face the problem of light absorption by a single impurity center. The optical electron in such a center interacts not only with photons of the external laser field, but also with phonons from the solid matrix. Due to this electron–phonon interaction, electronic excitation of the impurity center is accompanied by creation and annihilation of phonons. Cooling the solid sample can suppress only the annihilation of phonons, whilst phonon creation persists even at zero temperature. The theory discussed in Part I did not account for the existence of phonons and the electron–phonon interaction. A generalized theory allowing for the electron–phonon interaction will be discussed in Part III. However, phonons are not the only possible type of low frequency excitation in solid matrices. In amorphous solids the situation is more complex because there are also the so-called tunneling excitations in the matrix. In this and the next two chapters, we discuss various types of low frequency excitation existing in amorphous matrices.

4.1 Theoretical Description of Interacting Electrons and Nuclei

Optical spectroscopy deals with systems comprising electrons and nuclei interacting via Coulomb forces. The main principles for a theoretical description of electrons and nuclei bound by Coulomb forces were set out by Born and Oppenheimer at the very inception of quantum mechanics [42–45]. Although these principles were derived for electrons and nuclei in a molecule, they apply equally to any system consisting of interacting electrons and nuclei, including impurity molecules interacting with the molecules in a solid solvent.

In their original work [42], Born and Oppenheimer used a perturbation theory in which wave functions and energies are expressed as a power series in a parameter $(m/M)^{1/4}$, m and M being the masses of the electron and the nucleus, respectively. Although this parameter is much smaller than unity, even for the molecule H_2, such an approach has one serious shortcoming. The problem is that it assumes that all analytical expressions in the power series are of order unity. However, this is far from being true, for instance, when the electronic states are degenerate.

Born later developed a new method which was free from the above-mentioned shortcoming [43]. This method is now known as the Born–Oppenheimer method. It is based on the idea that the interaction of electrons with their own nucleus is stronger than their interaction with other nuclei, or the interaction of the nuclei with one another. This assumption corresponds to the fact that the energies of electronic excitations are of the order of 10^4 cm^{-1} and energies of vibronic excitations and phonons are one or two orders of magnitude smaller. This Born–Oppenheimer method will be used throughout the book.

Let r and R be the coordinates of the electrons and nuclei. Then the Hamiltonian of the system can be written as

$$H(r,R) = H_0(r,R) + T(R) = H_0(r,R) - \frac{\hbar^2}{2}\sum_n \frac{\mathrm{d}^2}{\mathrm{d}R_n^2} . \tag{4.1}$$

Here the square root of the nuclear mass is included in the nuclear coordinate R. The operator H_0 describes electrons moving around the fixed nuclei. The nuclear coordinates R are parameters, but not dynamic variables. In fact, the eigenfunctions and eigenvalues of the Hamiltonian H_0 depend on the parameter R so that

$$H_0(r,R)\varphi^f(r,R) = E^f(R)\varphi^f(r,R) . \tag{4.2}$$

Any eigenfunction $\Psi(r,R)$ of the Hamiltonian $H(r,R)$ can be expanded in terms of the eigenfunctions $\varphi^f(r,R)$ of the electronic Hamiltonian H_0,

$$\Psi(r,R) = \sum_f \Phi^f(R)\varphi^f(r,R) . \tag{4.3}$$

Let us insert this function $\Psi(r,R)$ into the stationary Schrödinger equation with Hamiltonian $H(r,R)$ and multiply both sides by an electronic function $\varphi^f(r,R)$. After integrating over electronic coordinates, we find the set of equations

$$\left[T(R) + U^f(R) - E\right]\Phi^f(R) + \sum_{f'} \hat{U}^{ff'}(R)\Phi^{f'}(R) = 0 \tag{4.4}$$

for the unknown function $\Phi^f(R)$, where

$$U^f(R) = E^f(R) + \hat{U}^{ff}(R) , \tag{4.5}$$

$$\hat{U}^{ff'}(R) = -\sum_n \int \mathrm{d}r\, \varphi^f(r,R) \left[\frac{\mathrm{d}}{\mathrm{d}R_n} \varphi^{f'}(r,R) \frac{\mathrm{d}}{\mathrm{d}R_n} + \frac{1}{2} \frac{\mathrm{d}^2}{\mathrm{d}R_n^2} \varphi^{f'}(r,R) \right] .$$

The operator $U^{ff'}(R)$ is called the nonadiabatic operator. It involves all electronic states, so that the mathematical task becomes rather complicated. Fortunately, we can neglect this operator if the distance between electronic levels is much larger than the values of vibrational quanta. In this approximation, which is called the adiabatic approximation, the function for the whole system is given by a product,

$$\Psi(r,R) = \varphi^f(r,R)\Phi^f(R) , \tag{4.6}$$

where the electronic function is determined by (4.2) and the function $\Phi^f(R)$ can be found from the equation

$$\left[T(R) + U^f(R) - E\right] \Phi^f(R) = 0 . \tag{4.7}$$

Equation (4.6) describes the function for the whole system in the adiabatic approximation and the Hamiltonian

$$H^f(R) = T(R) + U^f(R) = \int \varphi^f(r,R) H(r,R) \varphi^f(r,R)\, \mathrm{d}r \tag{4.8}$$

is called the adiabatic Hamiltonian of the system. As usual one neglects the derivative in the equation for $U^f(R)$. Then this function of the nuclear coordinates describes the so-called adiabatic potential of the electron–nuclei system in the fth electronic quantum state.

4.2 Franck–Condon and Herzberg–Teller Interactions

The adiabatic function described by (4.6) takes into account that part of the full electron–nuclei interaction which is called the adiabatic interaction. The adiabatic interaction describes the mutual influences of the electronic and nuclear dynamics.

The influence of the electronic dynamics on the nuclear manifests itself through the dependence of the adiabatic potential $U^f(R)$ on the electronic index f. In the harmonic approximation, the adiabatic potential is a quadratic function of the nuclear coordinates R, i.e.,

$$U^f(R) = \left(\boldsymbol{R} + \boldsymbol{a}^f\right) \frac{U^f}{2} \left(\boldsymbol{R} + \boldsymbol{a}^f\right) , \tag{4.9}$$

where a^f and R are multidimensional vectors with components a_n^f and R_n which describe the equilibrium positions of the nuclei and their displacements from those equilibrium positions, respectively, and U^f is a force matrix. It is obvious that the adiabatic Hamiltonian is given by

$$H^f(R) = T(R) + \left(\boldsymbol{R} + \boldsymbol{a}^f\right) \frac{U^f}{2} \left(\boldsymbol{R} + \boldsymbol{a}^f\right) . \tag{4.10}$$

The adiabatic potential $U^f(R)$ is in fact a multidimensional surface called the Franck–Condon surface [44,45]. The difference between the two adiabatic Hamiltonians

$$H^{\mathrm{e}} - H^{\mathrm{g}} = \boldsymbol{a}\frac{U^{\mathrm{e}}}{2}\boldsymbol{a} + \boldsymbol{a}U\boldsymbol{R} + \boldsymbol{R}\frac{W}{2}\boldsymbol{R} , \tag{4.11}$$

where indices g and e relate to the ground state and some excited electronic state, is called the Franck–Condon interaction (FC interaction). Here $\boldsymbol{a}^{\mathrm{g}} = 0$ and

$$\boldsymbol{a} = \boldsymbol{a}^{\mathrm{e}} , \quad W = U^{\mathrm{e}} - U^{\mathrm{g}} \tag{4.12}$$

describe a shift in the equilibrium positions and a change in the force matrix upon electronic excitation of the electron–nuclei system. These are parameters of the FC interaction. The FC interaction consists of two terms which are linear and quadratic functions of the nuclear dynamic variable R. The linear FC interaction mainly determines intensities of vibronic optical lines and phonon sidebands in the optical spectra of impurity centers. The quadratic FC interaction results in line-broadening. The role of both types of FC interaction will be discussed in detail in Part IV.

The FC interaction can be very large for ions because of the static polarization they generate in their vicinity. The polarization is accompanied by large shifts in the nuclear equilibrium positions, which are the parameters of the linear FC interaction. Such a situation is typical, for instance, for F centers in NaCl and other ionic crystals. Optical bands of these centers are very broad. They result mainly from electron–phonon transitions.

However, the FC interaction is only a part of the whole adiabatic interaction because nuclear vibrations also influence electronic states. This influence manifests itself through the dependence of the electronic wave function $\varphi^f(r, R)$ on the vibration coordinates R. The nuclear vibrations will influence matrix elements of the electronic dipolar moment. Therefore the matrix element of the operator $\boldsymbol{d}(r)\cdot E$ that describes a dipolar interaction with light will be modulated by vibrations:

$$\int \varphi^{\mathrm{e}}(r, R)\boldsymbol{d}(r)\cdot\boldsymbol{E}\varphi^{g}(r, R)\,\mathrm{d}r = M^{eg}(R) . \tag{4.13}$$

This modulation-type interaction is referred to as the Herzberg–Teller interaction (HT interaction). The function $M^{eg}(R)$ can be expanded as a power series in the dynamical variable R. The coefficients of this series, i.e., derivatives of the function $M^{eg}(R)$ with respect to R, are parameters of the HT interaction. In fact, intensities of vibronic and electron–phonon transitions will be proportional to the square of such derivatives.

The HT interaction describes the influence of other electronic transitions on the transition $e \leftarrow g$. This influence emerges as a result of nuclear vibrations. In order to demonstrate this fact, we rewrite (4.2) in the form

$$\left[T(r) + V(r,R) - E^f(R)\right] \varphi^f(r,R) = 0 \,, \tag{4.14}$$

where $T(r)$ is the electronic kinetic energy operator. Let equilibrium positions of the vector R in the fth electronic state be denoted by 0. Then, using conventional perturbation theory, the following expressions for the eigenvalues and eigenfunctions of (4.14) are found:

$$\begin{aligned} E^f(R) &= E^f(0) + V^{ff}(R) \,, \\ \varphi^f(r,R) &= \varphi^f(r,0) + \sum_{f'} \frac{V^{f'f}(R)}{E^f(0) - E^{f'}(0)} \varphi^{f'}(r,0) \,, \end{aligned} \tag{4.15}$$

where

$$V^{f'f} = \int \varphi^{f'}(r,R) V(r,R) \varphi^f(r,R) \, \mathrm{d}r \tag{4.16}$$

is a matrix element of the electron–nuclei interaction. This matrix element mixes various electronic states.

As a rule the HT interaction is weaker than the FC interaction. Despite this fact, the HT interaction plays a decisive role in some cases. For instance, it is this HT interaction which determines the value of the cross-section for non-resonant Raman scattering. In addition, it is due to the HT interaction that molecules of high symmetry whose electronic transitions are forbidden do have vibronic optical lines. The closer the electronic levels are to one another, the larger the HT interaction, as can be seen from (4.15).

5. Natural Vibrations of Solids

5.1 Acoustic and Optical Phonons

In accordance with the Born–Oppenheimer theory, the dynamics of nuclei is determined by an adiabatic Hamiltonian $H^f(R)$. Let us consider the adiabatic potential for the ground electronic state. The equilibrium positions in the ground state are assumed to be zero. By omitting the electronic index g we can write the adiabatic potential in the harmonic approximation as

$$U(\boldsymbol{R}) = \boldsymbol{R}\frac{U}{2}\boldsymbol{R} = \frac{1}{2}\sum_{nm} R_n U_{nm} R_m \ . \tag{5.1}$$

If R_n denotes a coordinate of an atom, it can only be a coordinate of translational type. However, if R_n denotes a molecular coordinate, it can be of both translational and angular type.

Let us consider first a solid consisting of atoms. Then R_n is a coordinate of the translational type. It is obvious that if we translate all atoms in space by a vector $\boldsymbol{b}$, the potential energy of the solid is not changed, i.e.,

$$U(\boldsymbol{R}+\boldsymbol{b}) - U(\boldsymbol{R}) = 0 \ . \tag{5.2}$$

This equation must be true at arbitrary values of vectors $\boldsymbol{R}$ and $\boldsymbol{b}$. This is possible if

$$\sum_m U_{mn} = \sum_m U_{nm} = 0 \ . \tag{5.3}$$

Hence the elements of the force matrix are not independent and every element can be expressed in terms of the sum of the remaining elements. For instance,

$$U_{nn} = -\sum_{m(\neq n)} U_{nm} = -\sum_{m(\neq n)} U_{mn} \ . \tag{5.4}$$

The potential energy can be rewritten so that (5.3) will be fulfilled automatically. Taking into account the fact that permutation of indices in the force matrices does not change the potential energy in (5.1), we may write

$U_{nm} = U_{mn}$. With the help of this symmetry condition, we can transform the potential energy to the form

$$U(\boldsymbol{R}) = \frac{1}{2}\sum_{nm} U_{nm}(R_n - R_m)^2 \,. \tag{5.5}$$

This expression guarantees that (5.2) will be fulfilled.

In the case when (5.1) describes the potential energy of a solid consisting of molecules with definite orientation in space, (5.2) will not be fulfilled because the potential energy is changed if all molecules are rotated through the same angle. In this case the potential energy cannot be described by (5.5). The position of every molecule in space is described by three coordinates of translational type and three coordinates of angular type. The adiabatic potential of such a solid is given by

$$H(\boldsymbol{R}) = \frac{\hat{P}^2}{2} + \boldsymbol{R}\frac{U}{2}\boldsymbol{R} = \sum_{\boldsymbol{n}j} \frac{P_{\boldsymbol{n}j}^2}{2} + \sum_{\boldsymbol{n}i,\boldsymbol{m}j} R_{\boldsymbol{n}i}\frac{U_{\boldsymbol{n}i\,\boldsymbol{m}j}}{2}R_{\boldsymbol{m}j} \,. \tag{5.6}$$

Here $\boldsymbol{n}$ and $\boldsymbol{m}$ are vectors defining positions of molecules in space. The indices j, i =1,2,3 and j, i = 4,5,6 relate to translational and angular degrees of freedom. Mass coefficients are included in coordinates R and mechanical momenta P.

It is obvious that vibration of one molecule cannot be a natural mode. Intermolecular interactions cause such a vibration to pass to other molecules. Therefore natural modes R_q are linear combinations of molecular displacements $R_{\boldsymbol{n}j}$. Since natural modes do not interact with each other, the Hamiltonian of a solid is given by $H = \sum_q H_q$, where

$$H_q = \frac{1}{2}(P_q^2 + \omega_q^2 R_q^2) \tag{5.7}$$

is the Hamiltonian of a natural mode.

Molecular displacements $R_{\boldsymbol{n}j}$ can be represented as components of a multidimensional vector. The transition to natural modes in the vibrational Hamiltonian can be considered as a rotation in the multidimensional space. This rotation is described by linear transformations

$$R_{\boldsymbol{n}i} = \sum_q u(\boldsymbol{n}i, q)R_q \,, \quad P_{\boldsymbol{n}i} = \sum_q u(\boldsymbol{n}i, q)P_q \,. \tag{5.8}$$

These transformations must conserve the length of the multidimensional vector R. Therefore the real matrix u should be of orthogonal type. This means that matrix elements must satisfy

$$\sum_{\boldsymbol{n}i} u(q, \boldsymbol{n}i)u(\boldsymbol{n}i, q) = \delta_{qq'} \,, \quad \sum_q u(\boldsymbol{n}i, q)u(q, \boldsymbol{m}j) = \delta_{\boldsymbol{n}\boldsymbol{m}}\delta_{ij} \,. \tag{5.9}$$

Since any orthogonal transformation conserves the lengths of all multidimensional vectors we can write

$$P^2 = \sum_{ni} P_{ni}^2 = \sum_q P_q^2 \,. \tag{5.10}$$

Allowing for (5.9), we can transform the vibrational Hamiltonian to the view described by (5.7), if coefficients u satisfy the set of equations

$$\sum_{mj} U_{ni\,mj} u(mj, q) = \omega_q^2 u(ni, q) \,. \tag{5.11}$$

Here ω_q^2 are roots of the determinant of the force matrix. They determine the frequencies of the natural modes, i.e., phonon frequencies. In accordance with the general principles of quantum mechanics, we must assume that natural coordinates and conjugate mechanical momenta are operators. Then, introducing a dimensionless coordinate Q_q, we find

$$R_q = \sqrt{\frac{\hbar}{\omega_q}} Q_q \,, \quad P_q = -\mathrm{i}\hbar \frac{\mathrm{d}}{\mathrm{d}R_q} = -\mathrm{i}\sqrt{\hbar\omega_q} \frac{\mathrm{d}}{\mathrm{d}Q_q} \,. \tag{5.12}$$

Using (5.9)–(5.12), we can easily transform the adiabatic Hamiltonian described by (5.6) to the form

$$H = \sum_q \frac{\hbar\omega_q}{2} \left(-\frac{\mathrm{d}^2}{\mathrm{d}Q_q^2} + Q_q^2 \right) \,. \tag{5.13}$$

A Hamiltonian of this type has already been examined in Sect. 1.2. All discussions and equations flowing from (1.20) can be repeated, substituting the word 'phonon' for the word 'photon'. Introducing phonon creation and annihilation operators for each q,

$$b_q = \frac{1}{\sqrt{2}} \left(\frac{\mathrm{d}}{\mathrm{d}Q_q} + Q_q \right) \,, \quad b_q^+ = \frac{1}{\sqrt{2}} \left(-\frac{\mathrm{d}}{\mathrm{d}Q_q} + Q_q \right) \,, \tag{5.14}$$

we can express the adiabatic Hamiltonian as a sum of phonon Hamiltonians, viz.,

$$H = \sum_q \frac{\hbar\omega_q}{2} \left(b_q^+ b_q + \frac{1}{2} \right) \,. \tag{5.15}$$

Using (5.8), (5.12) and (5.14), we find the following expressions for coordinates and mechanical momenta:

$$\begin{aligned} R_{ni} &= \sum_q u(ni, q) \sqrt{\frac{\hbar}{2\omega_q}} (b_q + b_q^+) \,, \\ P_{ni} &= -\mathrm{i}\hbar \frac{\mathrm{d}}{\mathrm{d}R_{ni}} = -\mathrm{i} \sum_q u(ni, q) \sqrt{\frac{\hbar\omega_q}{2}} (b_q - b_q^+) \,. \end{aligned} \tag{5.16}$$

The transition to natural coordinates and mechanical momenta carried out above is valid not only for an ordered solid system such as a crystal, but also for disordered systems like polymers and glasses. However, the matrix U_{nimj} relating to a crystal depends only on the distance $\boldsymbol{n}-\boldsymbol{m}$ between molecules, i.e.,

$$U_{nimj} = U_{i,j}(\boldsymbol{n}-\boldsymbol{m}) \;, \tag{5.17}$$

because of the translational invariance of the crystal. In such a homogeneous medium, the natural modes are plane waves, i.e.,

$$u(\boldsymbol{n}i,q) = \frac{e_s^i(\boldsymbol{q})}{\sqrt{N}}\mathrm{e}^{-\mathrm{i}\boldsymbol{q}\cdot\boldsymbol{n}}, \quad u(q,\boldsymbol{n}i) = \frac{e_s^{*i}(\boldsymbol{q})}{\sqrt{N}}\mathrm{e}^{\mathrm{i}\boldsymbol{q}\cdot\boldsymbol{n}} \;, \tag{5.18}$$

where $\boldsymbol{q}$ is a wave vector and s is the number of phonon branches. Using

$$\frac{1}{N}\sum_{\boldsymbol{n}}\mathrm{e}^{\mathrm{i}(\boldsymbol{q}-\boldsymbol{q}')\cdot\boldsymbol{n}} = \delta_{qq'} \;, \quad \frac{1}{N}\sum_{\boldsymbol{q}}\mathrm{e}^{\mathrm{i}(\boldsymbol{n}-\boldsymbol{m})\cdot\boldsymbol{q}} = \delta_{\boldsymbol{nm}} \;, \tag{5.19}$$

we can transform (5.9) to the following equations for the polarization vectors:

$$\sum_{i=1}^{6} e_s^{*i}(\boldsymbol{q})e_{s'}^i(\boldsymbol{q}) = \delta_{ss'} \;, \quad \sum_{s=1}^{6} e_s^{*i}(\boldsymbol{q})e_s^j(\boldsymbol{q}) = \delta_{ij} \;. \tag{5.20}$$

In a homogeneous medium, (5.11) is transformed to a set of six equations for the polarization vectors, namely

$$\sum_{j=1}^{6} U_{ij}(\boldsymbol{q})e_s^j(\boldsymbol{q}) = \omega_s^2(\boldsymbol{q})e_s^i(\boldsymbol{q}) \;, \tag{5.21}$$

where

$$U_{ij}(\boldsymbol{q}) = \sum_{\boldsymbol{n}} U_{ij}(\boldsymbol{n})\mathrm{e}^{\mathrm{i}\boldsymbol{q}\cdot\boldsymbol{n}} \;. \tag{5.22}$$

The explicit form of this matrix is determined by the symmetries of the crystal lattice.

The matrix $U_{ij}(q)$ allows us to analyze the types of mode that can exist in a solid. Let us examine a crystal consisting of atoms. Every atom has three degrees of freedom, i.e., i and j =1,2,3. Equation (5.3) is valid for any monatomic solid solution. For a crystal it takes the form

$$\sum_{\boldsymbol{n}} U_{ij}(\boldsymbol{n}) = U_{ij}(\boldsymbol{q}=0) = 0 \;. \tag{5.23}$$

A non-zero solution of (5.21) at $q \to 0$ can only exist if $\omega_s(q) \to 0$. Modes whose frequencies approach zero when their wavelengths $2\pi/q$ are increased are called acoustic modes because the conventional sound in a solid is described by long wavelength vibrations of the acoustic type.

However, (5.3) is not satisfied for coordinates of an orientational type, in a molecular crystal. Therefore, the frequencies of the natural vibrations relating to orientational displacements of molecules do not approach zero when the wavelength is increased. Such modes are called modes of optical type. Therefore $\omega_s(0) \neq 0$ for optical phonons.

The conclusion concerning the existence of acoustic and optical modes is also correct for a medium with a disordered disposition of molecules, as occurs in mixed crystals and amorphous solids. However, in contrast to pure crystals, such a medium has modes of localized type. These modes are discussed in the next section.

5.2 Localized Phonon Modes

Any natural vibration in a homogeneous medium is described by a wave. The shorter the wavelength, the larger the wave number and frequency of a phonon. Therefore natural vibrations of any homogeneous condensed medium are vibrations of a delocalized type, because many molecules take part in each natural vibration. An impurity molecule can differ from host molecules by its mass or by an interaction with host molecules. Such a guest molecule perturbs the lattice of a solid solvent and hence also the homogeneity of space. Under such perturbations, the natural modes of such a solid are changed and new modes of a localized type appear.

Let us find a mathematical quantity that could characterize the amplitude of molecular vibration in a qualitative manner when this molecule takes part in a natural vibration of a solid with a definite frequency. Since frequencies of an optical type are higher than frequencies of an acoustic type, we may neglect optical modes when examining the low frequency modes of the solid. Let $x_{\boldsymbol{n}i}$ be three translational displacements of the $\boldsymbol{n}$th molecule. They are expressed in terms of the former displacements $R_{\boldsymbol{n}i}$ by

$$x_{\boldsymbol{n}i} = \frac{R_{\boldsymbol{n}i}}{\sqrt{M}} , \tag{5.24}$$

where M is the mass of the host molecules. Let one molecule be characterized by $\boldsymbol{n} = 0$ and the nearest neighbor molecule by $\boldsymbol{n} = 1$. Then with the help of (5.24) and (5.16), one can find the following expression for the relative displacement:

$$x_{0i} - x_{1i} = \sum_q \sqrt{\frac{\hbar}{M\omega_q}} \left[u(0i, q) - u(1i, q)\right] \frac{b_q + b_q^+}{\sqrt{2}} . \tag{5.25}$$

The mean square of this displacement is described by

$$\langle 0|(x_{0i} - x_{1i})^2|0\rangle = \sum_q \frac{\hbar}{2M\omega_q}\left[u(0i,q) - u(1i,q)\right]^2 = \int_0^\infty \Gamma_{\text{ph}}(\omega)\,\mathrm{d}\omega\,, \tag{5.26}$$

where

$$\begin{aligned} \Gamma_{\text{ph}}(\omega) &= \frac{\hbar}{M\omega} g(\omega)\,, \\ g(\omega) &= \frac{1}{2}\sum_q \left[u(0i,q) - u(1i,q)\right]^2 \delta(\omega - \omega_q)\,. \end{aligned} \tag{5.27}$$

The distance between two molecules vibrates with the frequency ω of a natural mode with a definite amplitude. The function $\Gamma_{\text{ph}}(\omega)$ describes the square of this amplitude for a natural mode with frequency ω. The function $\Gamma_{\text{ph}}(\omega)$ shows, in fact, how the distance is involved in a vibration with definite frequency. This function is a product of the square of the zero-point vibration amplitude $\hbar/M\omega$ and the function $g(\omega)$ that determines the probability of finding the distance involved in the vibration with frequency ω. Such an interpretation of the function $g(\omega)$ is confirmed by the equation

$$\int_0^\infty g(\omega)\,\mathrm{d}\omega = 1\,. \tag{5.28}$$

This is easily proved using (5.9). If the function $g(\omega)$ equals zero at $\omega = \omega_0$, this means that molecules 0 and 1 do not vibrate with frequency ω_0. However, the inverse situation can emerge in a solid with a guest molecule: the function $g(\omega)$ has a small value everywhere except in a narrow region around $\omega = \omega_0$. In fact the function $g(\omega)$ looks like a narrow peak with a resonant frequency ω_0. This means that molecules 0 and 1 vibrate mainly with the frequency ω_0 and do not vibrate with other frequencies. Hence, this very function $g(\omega)$ is able to inform us about vibrational modes localized in space.

Let a host molecule in the zeroth node be substituted by a guest molecule. Neglecting a possible difference in masses, one accounts for the fact that an interaction between the guest and host molecules differs from the interaction existing between host molecules. The adiabatic potential of such a system is given by

$$H = H_0 + V\,, \tag{5.29}$$

where the Hamiltonian

$$H_0 = T(R) + \frac{1}{2}\sum_{nm} U_{nm}(R_n - R_m)^2 \tag{5.30}$$

describes a homogeneous solid and the operator

$$V = \Delta U(R_0 - R_1)^2 \tag{5.31}$$

describes the perturbation of the solid due to the existence of a guest molecule in the zeroth node. Inserting (5.16) into the operator V instead of R_0 and R_1, we obtain the expression

$$V = \frac{\Delta U}{2}(b + b^+)^2 \,, \tag{5.32}$$

where

$$b = \sum_q l_q b_q \,, \quad l_q = \sqrt{\frac{1}{\omega_q}}\big[u(0i, q) - u(1i, q)\big] \,. \tag{5.33}$$

Our task is to find the type of change that can be caused by the guest molecule in vibrations of the medium if, prior to the substitution, the solid was characterized by the Hamiltonian H_0. The task will be solved in two steps. At the first stage, we retain two terms that conserve the total number of excitations in the operator of the perturbation, i.e., the operator in this approximation takes the form

$$V' = \frac{\Delta U}{2}(bb^+ + b^+b) \,. \tag{5.34}$$

At the second stage, the task will be completed by using the total operator of the perturbation. The influence of the simplified interaction V' on the phonon modes can be taken into account much more simply than the influence of the total interaction. However, modes of localized type emerge even when we restrict ourselves to the simplified interaction V'.

The changes in the vibrational system caused by the simplified perturbation V' can be found with the help of the equations for the probability amplitudes derived in Part I. In fact, the probability amplitude

$$D_q^0(t) = -\mathrm{i}\langle 0|b_q \mathrm{e}^{-\mathrm{i}H_0 t/\hbar} b_q^+|0\rangle \tag{5.35}$$

describes the temporal evolution of the phonon q in a crystal lattice. Here $|0\rangle$ is the ground state of the system described by the Hamiltonian (5.30). This Hamiltonian can be transformed to (5.15). The amplitude is easily found and the result is given by

$$D_q^0(t) = -\mathrm{i}\mathrm{e}^{-\mathrm{i}\omega_q t} \,. \tag{5.36}$$

The amplitude oscillates with the phonon frequency and its modulus does not depend on time, being unity. This means that the natural mode of a perfect crystal lives infinitely long in the harmonic approximation. With the help of (1.60), we find the following expression for the Laplace amplitude:

$$D_q^0(\omega) = \frac{1}{\omega - \omega_q + \mathrm{i}0} \,. \tag{5.37}$$

Let us now examine the temporal behavior of the same mode in the same medium with a guest molecule in the zeroth node. The Hamiltonian of this system is $H = H_0 + V'$. Therefore the amplitude is described by the expression

$$D_{q'q}(t) = -\mathrm{i}\langle 0|b_{q'}\mathrm{e}^{-\mathrm{i}Ht/\hbar}b_q^+|0\rangle \,. \tag{5.38}$$

In order to find an equation that determines the temporal evolution of the amplitude, we can use (1.65) derived in Part I. Since the operator V' conserves the number of phonons, the equation will only include indices of one-phonon states, i.e., it looks like

$$D_{q'q}(\omega) = \delta_{q'q}D^0_{q'}(\omega) + D^0_{q'}(\omega)\sum_{q''}\frac{V'_{q'q''}}{\hbar}D_{q''q}(\omega) \,. \tag{5.39}$$

Using the Bose commutation relations, the operator V' can be transformed to the form

$$V' = \Delta U\, b^+b + \mathrm{Const.} \tag{5.40}$$

The constant can be added to the energy of zero-point vibrations. Because this energy does not manifest itself in the final equations, we may omit the constant in (5.40) and hence the matrix element of the perturbation V' is given by

$$\langle q''|V'/\hbar|q'\rangle = \frac{V'_{q''q'}}{\hbar} = \Delta U\, l_{q''}l_{q'} \,. \tag{5.41}$$

Inserting this equation into (5.39), we find the relation

$$D(\omega) = \frac{D_0(\omega)}{1 - \Delta U\, D_0(\omega)} \,, \tag{5.42}$$

where

$$D_0(\omega) = \sum_q l_q^2 D_q^0(\omega) \,, \quad D(\omega) = \sum_{qq'} l_{q'}D_{q'q}(\omega)l_q \,. \tag{5.43}$$

Equation (5.42) is a solution to the task discussed above. The relation expresses the phonon amplitude D for a phonon in the medium with a guest molecule via the phonon amplitude D_0 for a phonon in the medium unperturbed by the guest molecule. In fact, the following expression for the Laplace amplitude can be found using (5.37):

$$D_0(\omega) = \sum_q \frac{l_q^2}{\omega - \omega_q + \mathrm{i}0} = \int_0^\infty \frac{\mathrm{d}\nu}{\pi}\,\frac{\Gamma_0(\nu)}{\omega - \nu + \mathrm{i}0} = \Delta_0(\omega) - \mathrm{i}\Gamma_0(\omega) \,, \tag{5.44}$$

where

$$\Gamma_0(\nu) = \pi \frac{g_0(\nu)}{\nu}, \quad \Delta_0(\omega) = \mathrm{P} \int_0^\infty \frac{\mathrm{d}\nu}{\pi} \frac{\Gamma_0(\nu)}{\omega - \nu}. \tag{5.45}$$

Here the function $g_0(\nu)$ for unperturbed phonons is described by (5.27). The letter P before the integral indicates the principal value of the integral.

The function $D(\omega)$ is an analytical function in the upper half plane of the complex variable ω. Therefore the function $D(\omega)$ can be expressed via its own imaginary part $\Gamma(\omega)$ as follows:

$$D(\omega) = \int_{-\infty}^{\infty} \frac{\mathrm{d}\nu}{\pi} \frac{\Gamma(\nu)}{\omega - \nu + \mathrm{i}0} = \Delta(\omega) - \mathrm{i}\Gamma(\omega). \tag{5.46}$$

A similar formula has already been derived for the function $D_0(\omega)$. Using (5.42), we find the relation

$$1 + \Delta U\, D(\omega) = \frac{1}{1 - \Delta U\, D_0(\omega)}. \tag{5.47}$$

By calculating the imaginary part of this equation, we find the following relation between the imaginary parts of perturbed and unperturbed Laplace amplitudes:

$$\Gamma(\omega) = \frac{\Gamma_0(\omega)}{[1 - \Delta U\, \Delta_0(\omega)]^2 + [\Delta U\, \Gamma_0(\omega)]^2}. \tag{5.48}$$

The functions $\Gamma(\omega)$ and $\Gamma_0(\omega)$ equal zero at negative values of the frequency ω. Both functions have dimensions of inverse frequency. Their physical meaning is similar to that of the function $\Gamma_{\mathrm{ph}}(\omega)$ discussed above. Hence, they characterize the amplitude of oscillations in the distance between molecules 0 and 1 in a perfect crystal and the crystal perturbed by a guest molecule. Therefore we can determine how the amplitude is changed after a guest molecule is introduced. This means the task can be solved even using the simplified expression for the perturbation operator.

Let us take the next step and find a relation similar to (5.48), using the total perturbation operator described by (5.32). In this case we may still use the equations for the transition amplitudes. Since the interaction described by (5.32) does not conserve the total number of excitations, we arrive at an infinite set of coupled equations. Therefore the problem of finding a solution becomes too complicated and we must use a more effective method to solve the problem. This is discussed in detail in Appendix J. The result is as follows: equation (5.48) happens to be correct even when we use the total perturbation. However, we must now replace the amplitudes $D(\omega)$ and $D_0(\omega)$ by the functions

$$D_0(\omega) = \int_{-\infty}^{\infty} \frac{\mathrm{d}\nu}{\pi} \Gamma_0(\nu) \left(\frac{1}{\omega - \nu + \mathrm{i}0} - \frac{1}{\omega + \nu - \mathrm{i}0} \right), \tag{5.49}$$

$$D(\omega) = \int_{-\infty}^{\infty} \frac{\mathrm{d}\nu}{\pi} \Gamma(\nu) \left(\frac{1}{\omega - \nu + \mathrm{i}0} - \frac{1}{\omega + \nu + \mathrm{i}0} \right) . \quad (5.50)$$

The new Laplace amplitudes are not analytical functions, either in the upper or in the lower half plane of the complex variable ω. They can be found by means of the Laplace transformation of the so-called causal phonon Green functions. A special mathematical technique for calculating the causal Green functions can be borrowed from quantum field theory and is used in Appendix J, where the Laplace transform of the causal phonon Green functions is found.

It is interesting that the function $\Gamma_0(\omega)$ in (5.49) is described by (5.45) and that the relation between the functions $\Gamma(\omega)$ and $\Gamma_0(\omega)$ is true even in the case when the total perturbation operator is used. We only have to take into account the fact that the function $\Delta_0(\omega)$ is now described by the expression

$$\Delta_0(\omega) = \mathrm{P} \int_0^{\infty} \frac{\mathrm{d}\nu}{\pi} \Gamma_0(\nu) \frac{2\nu}{\omega^2 - \nu^2} . \quad (5.51)$$

It is easy to see that this function is the real part of the new phonon Green function.

The unitarity conditions given by (5.9) and (5.11) for finding the coefficients $u(\boldsymbol{n}j, q)$ are correct for a force matrix of arbitrary type. In our case, when the force matrix $U_{\boldsymbol{n}i,\boldsymbol{m}j}$ relates to a crystal with an impurity center, we come back to

$$\Gamma(\omega) = \frac{g(\omega)}{\omega} , \quad g(\omega) = \frac{1}{2} \sum_q \left[u(0i, q) - u(1i, q) \right]^2 \delta(\omega - \omega_q) , \quad (5.52)$$

and to the condition

$$\int_0^{\infty} g(\omega) \, \mathrm{d}\omega = 1 , \quad (5.53)$$

which can be deduced from (5.52). The type of force matrix determines the appearance of the coefficients $u(\boldsymbol{n}j, q)$. With the help of (5.48), we find that

$$g(\omega) = \frac{g_0(\omega)}{[1 - \Delta U \, \Delta_0(\omega)]^2 + [\pi \Delta U \, g_0(\omega)/\omega]^2} . \quad (5.54)$$

The function $g(\omega)$ describes the amplitude of vibrations with frequency ω in the distance between the guest molecule 0 and the host molecule 1. Equations (5.52) and (5.54) relate to the same function $g(\omega)$. Therefore in accordance with (5.53), an increase in one part of the function must accompanied by a decrease in the rest of the function. For instance, the appearance of a large peak at frequency ω_0 in the function $g(\omega)$ means that the 0–1 bond will oscillate effectively with frequency ω_0 and that oscillations of the bond with other frequencies are suppressed.

This general property of the function $g(\omega)$ can be demonstrated with the help of a model. Let us first find the spectral function $\Gamma_0(\omega)$ for the perfect crystal. By inserting (5.18) into (5.27), we find

$$g_0(\omega) = \frac{1}{2N}\sum_{\boldsymbol{q}s} e^{*i}_{\ s}(\boldsymbol{q})e^i_s(\boldsymbol{q})|1 - \mathrm{e}^{-\mathrm{i}\boldsymbol{q}\cdot\boldsymbol{a}}|^2\delta\big(\omega - \omega_s(\boldsymbol{q})\big) , \tag{5.55}$$

where $\boldsymbol{a}$ is the vector connecting the guest molecule with the nearest host molecule. In a homogeneous isotropic medium, $\omega_s(\boldsymbol{q}) = v\boldsymbol{q}$, where v is the sound velocity. For an isotropic medium, (5.55) takes the form

$$g_0(\omega) = \frac{1}{N}\sum_{\boldsymbol{q}}(1 - \cos\boldsymbol{q}\cdot\boldsymbol{a})\delta\big(\omega - \omega(\boldsymbol{q})\big) . \tag{5.56}$$

In order to simplify the calculation of this sum, the coefficient in front of the delta function can be replaced by its expression for small q, i.e., $1-\cos\boldsymbol{q}\cdot\boldsymbol{a} = A\omega^2(\boldsymbol{q})$. Using the density of states of the Debye model, we arrive at the expression

$$g_0(\omega) = A\omega^2\frac{1}{N}\sum_{\boldsymbol{q}}\delta\big(\omega - \omega(\boldsymbol{q})\big) = A\omega^2 3\frac{\omega^2}{\omega_\mathrm{D}^3} , \quad 0 \le \omega \le \omega_\mathrm{D} , \tag{5.57}$$

where ω_D is the bounding frequency of the acoustic phonons. The coefficient A can be found from the condition given by (5.28). Then the formulas characterizing oscillations of the 0–1 bond in a crystal are given by

$$g_0(\omega) = 5\frac{\omega^4}{\omega_\mathrm{D}^5} , \quad \Gamma_0(\omega) = 5\pi\frac{\omega^3}{\omega_\mathrm{D}^5} , \quad 0 \le \omega \le \omega_\mathrm{D} . \tag{5.58}$$

Using this equation for the function $\Gamma_0(\omega)$ together with (5.51), we find

$$\Delta_0(\omega) = \frac{10}{\omega_\mathrm{D}^2}\left[-\left(\frac{1}{3} + y^2\right) + \frac{y^3}{2}\ln\frac{y+1}{|y-1|}\right] , \quad y = \frac{\omega}{\omega_\mathrm{D}} . \tag{5.59}$$

Let us now inspect the function $\Gamma(\omega)$ describing a crystal with a guest molecule. Inserting (5.58) and (5.59) into (5.48) and carrying out the calculations, we obtain the results plotted in Fig. 5.1. The results are easily understood if we take into account the fact that equation

$$1 - \Delta U\,\Delta_0(\omega) = 0 \tag{5.60}$$

can have roots. Graphical solution of the equation is shown in Fig. 5.2. The equation has two roots at $\Delta U < 0$. Both roots lie in the phonon band. The

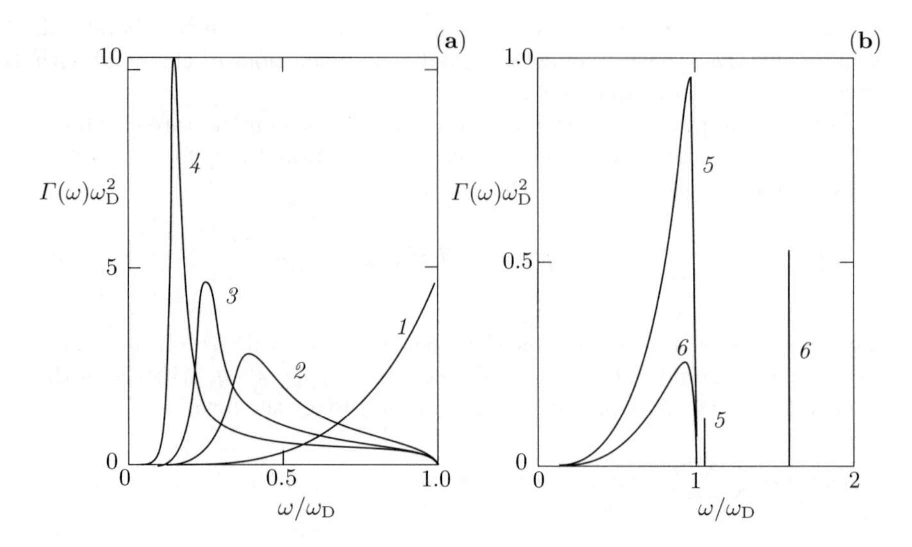

Fig. 5.1. Appearance of pseudo-localized (**a**) and localized (**b**) phonon modes in a crystal perturbed by an impurity molecule with $\Delta U/\omega_D = 0$ (curve 1), -0.2 (curve 2), -0.25 (curve 3), -0.28 (curve 4), 0.15 (curve 5), and 0.3 (curve 6)

high frequency root weakly influences the shape of the function $\Gamma(\omega)$ because the function $\Gamma_0(\omega)$ has a large value at this root. The situation is different with the low frequency root ω_0. The value $\Gamma_0(\omega_0)$ is small and therefore the function $\Gamma(\omega)$ has a sharp peak at $\omega = \omega_0$. The peak shows that the 0–1 bond takes part mainly in vibrations with frequencies close to ω_0. In this case a localized mode exists in the solid.

If $\Delta U = 0$, the 0–1 bond can oscillate with various frequencies. This is illustrated by curve 1 in Fig. 5.1a. If $\Delta U < 0$, the force between the guest and a host molecule decreases compared to its natural value, and a peak with frequency ω_0 appears in the low frequency range of the function $\Gamma(\omega)$. In parallel with the appearance of this peak, oscillations with other frequencies are suppressed in accordance with the integral relation (5.53). The appearance

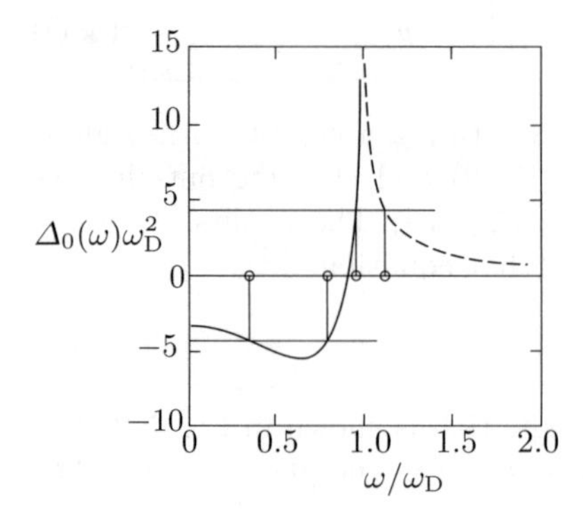

Fig. 5.2. Solution of (5.60) at $\Delta U/\omega_D = -0.25$ and 0.25

of the peak means that a localized mode emerges in the solid. The lower the frequency of the localized mode, the narrower the peak.

We face another situation when the 0–1 bond becomes stronger, i.e., $\Delta U > 0$. In this case (5.60) has two roots and one of them lies beyond the phonon band. The larger the perturbation ΔU, the further the root lies from the phonon band. This describes a high frequency localized mode. The further this frequency lies from the phonon band, the weaker the oscillations of the 0–1 bond with the frequencies of the phonon band. This situation is clearly illustrated in Fig. 5.1b.

6. Tunneling Systems in Solids

6.1 Tunneling Degrees of Freedom in Complex Molecules and Amorphous Solids

A system consisting of N nuclei has six degrees of freedom which describe three translations and three rotations of the whole system, and $3N-6$ internal degrees of freedom which relate to displacements of all nuclei. By choosing a potential that was quadratic in nuclear displacements, we assumed that the nuclei carried out small vibrations near their equilibrium positions. However, we have not yet discussed how many equilibrium positions can exist in a solid. One typically assumes that the number of equilibrium positions equals the number of vibrational degrees of freedom. However, a simple example demonstrates that such an assumption may be unjustified.

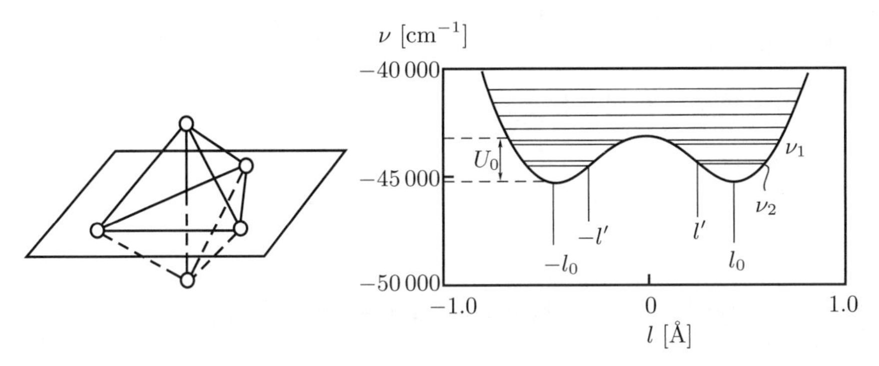

Fig. 6.1. NH_3 molecule and its adiabatic potential along the coordinate l, which is normal to the plane shown [46]

Figure 6.1 shows an NH_3 molecule. The N atom can occur both above and below the horizontal plane. This means that its potential as a function of the coordinate l connecting these two possible positions of the N atom has two minima. In making the transition from the upper equilibrium position to the lower one, the N atom overcomes a potential barrier separating these two minima. If the molecule cannot borrow a portion of the energy somewhere, this transition can only be of the tunneling type. However, if the molecule has such a source of energy, it can acquire the amount of energy needed to overcome the barrier in a classical way, i.e., it can pass over the barrier. This is a transition of Arrhenius type. The existence of degrees of freedom of tunneling type has to be taken into account when considering adiabatic potentials.

These tunneling degrees of freedom exist not only in complex molecules, but also in amorphous solids like polymers and glasses. They are responsible for the unusual low-temperature behavior of various physical properties of polymers and glasses and therefore deserve a more detailed discussion.

The multitude of nuclear coordinates R can be divided into two subsets: q and x. Then the adiabatic potential is a function of the coordinates q and x. If we move along any coordinate q, the potential energy is described by a double-well potential. However, the potential with respect to any coordinate x is a function with a single minimum. The adiabatic potential of such a system can be expressed in the form

$$U(R) = V(q) + U(q,x) \,. \tag{6.1}$$

Here the potential $V(q)$ is of double-well type and the remaining part $U(q,x)$ of the whole potential is described by many single-well potentials. It is obvious that the division of the whole potential into such parts can be carried out in various ways. However, this circumstance will not influence the final result, just as choosing a different set of eigenfunctions will not influence the final result of a calculation. Substituting the adiabatic potential (6.1) into (4.7), we arrive at the equation

$$\big[T(q) + V(q) + T(x) + U(q,x) - E\big]\Phi(q,x) = 0 \,. \tag{6.2}$$

Suppose we know the solution of the Schrödinger equation with the double-well potential, i.e.,

$$[T(q) + V(q) - E_l]\,\psi_l(q) = 0 \,. \tag{6.3}$$

The full function $\Phi(q,x)$ can be expanded in a series consisting of eigenfunctions of (6.3):

$$\Phi(q,x) = \sum_l \varphi^l(x)\psi_l(q) \,. \tag{6.4}$$

Inserting this function into (6.2), multiplying by a function $\psi_l(q)$ and carrying out the integration over coordinate q, we arrive at the equation

$$\big[T(x) + U_{ll}(x) + E_l - E\big]\varphi^l(x) + \sum_{l'} U_{ll'}(x)\varphi^{l'}(x) = 0 \,, \tag{6.5}$$

where

$$U_{ll'}(x) = \int \psi_l(q) U(q,x) \psi_{l'}(q)\, \mathrm{d}q \,. \tag{6.6}$$

In deriving (6.5) from (6.2), we did not use any approximations. It is obvious that $U_{ll'}(x)$ plays the role of an adiabatic potential for the vibrational system. It depends on an index l of the quantum state of the double-well system. The Hamiltonian

$$H_l(x) = T(x) + U_{ll}(x) \tag{6.7}$$

can be called the adiabatic Hamiltonian of a vibrational system. In fact, (6.5) is similar to (4.4), which describes an electron–phonon system. The index l of the quantum state of the double-well system plays the same role as the electronic index f in (4.4). By neglecting anharmonisms in the vibrational system, we may write the adiabatic potential in the form

$$H_l(x) = T(x) + (x + a^l)\frac{U^l}{2}(x + a^l) \,. \tag{6.8}$$

Hence, the equilibrium positions and the force matrix depend on the quantum state l of the double-well system.

The matrix elements $U_{ll'}(x)$ determine probabilities of quantum transitions between quantum states of the double-well system. In fact there are no such transitions at $U_{ll'}(x) = 0$. In this approximation, wave functions of the system consisting of the double-well system and harmonic oscillators are given by a product

$$\Phi(q,x) = \psi_l(q)\varphi_n^l(x) \,, \tag{6.9}$$

where the functions $\psi_l(q)$ can be found from (6.3), and the functions $\varphi_n^l(x)$ are solutions of

$$\left[H_l(x) - E_n^l\right] \varphi_n^l(x) = 0 \,. \tag{6.10}$$

The eigenfunctions and eigenvalues of this equation are the functions and the energies of a harmonic oscillator.

Solutions of (6.3) with the double-well potential can only be found numerically. The double-well potential can be described by various functions. One of the most convenient approaches is to approximate to it by trigonometric functions. For instance, the function

$$V(q) = V_0(\cos kq - \xi \cos 2kq + \eta \sin kq) \tag{6.11}$$

describes a periodic double-well potential. If q is an angular coordinate, this periodic character of the potential is a natural property. If q is not an angular coordinate, such a periodic character of the potential is an unphysical property. However, this is of no importance if the barrier V_0 separating two wells is high enough. In this case, the variable q can be considered within

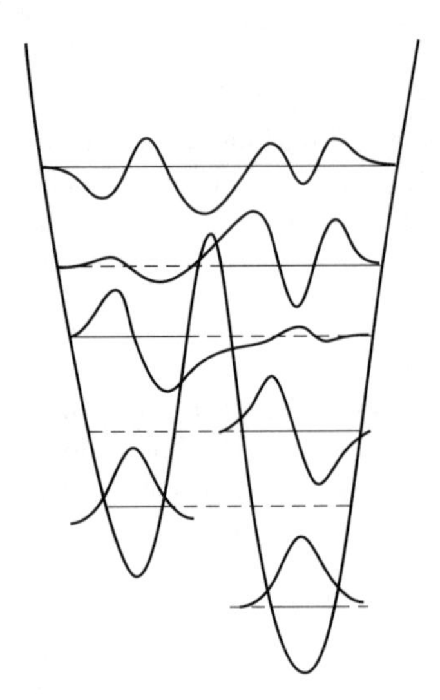

Fig. 6.2. Adiabatic potential calculated using (6.11) at $2V_0\mu/(hk)^2 = 1\,200$, $\xi = 0.37$ and $\eta = 0.02$, and eigenvalues and eigenfunctions of the Schrödinger equation with this potential [47,48]

the interval $0 < kq < 2\pi$. Then the parameter ξ determines the height of the barrier separating the two wells. The parameter η determines the asymmetry of the two minima.

A double-well potential $V(x)$ and related energies E_l and wave functions $\psi_l(x)$ calculated with the help of the adiabatic potential described by (6.11) are shown in Fig. 6.2. The distances between energies correspond to those calculated. The wave functions of two of the lowest levels recall the ground state functions of a harmonic oscillator. However, the functions for the upper levels differ considerably from those of a harmonic oscillator. This is particularly clear in the region of the barrier.

6.2 Rate Equations for Tunneling Systems

The double-well systems discussed above will henceforth be called tunneling systems. The wave functions of the lower states of the tunneling system shown in Fig. 6.2 are localized in different wells. These states are stationary states, according to the way they have been determined, implying that there are no transitions between them. However, in realistic systems there are transitions of intra-well and inter-well type. The physical reason for these transitions is an interaction between tunneling systems and phonons. Vibrations modulate the barrier between wells and therefore transitions between quantum states of the double-well system become possible. In our theoretical approach, the above-mentioned modulation results from the functions $U_{ll'}(x)$ determined by (6.6). In this section we derive equations for the probabilities of finding a double-well system in a quantum state l and a phonon subsystem in a quantum state n. The Hamiltonian of such a system is given by

$$H(q,x) = T(q) + V(q) + T(x) + U(q,x) = H(q) + T(x) + U(q,x) \,, \tag{6.12}$$

and its matrix elements in terms of a basis $|nl\rangle = \Phi(q,x) = \psi_l(q)\varphi_n^l(x)$ are given by

$$\langle nl|H|nl\rangle = E_l + \sum_q \hbar\omega_q^l \left(n_q + \frac{1}{2}\right) \equiv E_{nl} \,, \quad \langle nl|H|n'l'\rangle = \hbar U_{nl,n'l'} \,. \tag{6.13}$$

By inserting these matrix elements into (1.76) for the density matrix, we arrive at the following set of equations:

$$\begin{aligned} \dot\rho_{nl,nl} &= -\mathrm{i}\sum_{n'l'} (U_{nl,n'l'}\rho_{n'l',nl} - \rho_{nl,n'l'}U_{n'l',nl}) \,, \\ \dot\rho_{nl,n'l'} &= -\mathrm{i}\omega_{nl,n'l'}\rho_{nl,n'l'} \\ &\quad - \mathrm{i}\sum_{n''l''} (U_{nl,n''l''}\rho_{n''l'',n'l'} - \rho_{nl,n''l''}U_{n''l'',n'l'}) \,, \end{aligned} \tag{6.14}$$

where $\omega_{nl,n'l'} = (E_{nl} - E_{n'l'})/\hbar$. We shall look for a solution to this set of equations for the first non-vanishing approximation in the perturbation U. Therefore, in the sum over $n''l''$, we set $n''l'' = n'l'$ in the first term and $n''l'' = nl$ in the second term. In this approximation the sum in the second equation disappears and we arrive at the following set of equations:

$$\begin{aligned} \dot\rho_{nl,nl} &= -\mathrm{i}\sum_{n'l'} (U_{nl,n'l'}\rho_{n'l',nl} - \rho_{nl,n'l'}U_{n'l',nl}) \,, \\ \dot\rho_{nl,n'l'} &= -\mathrm{i}\omega_{nl,n'l'}\rho_{nl,n'l'} + \mathrm{i}U_{nl,n'l'}(\rho_{nl,nl} - \rho_{n'l',n'l'}) \,. \end{aligned} \tag{6.15}$$

The off-diagonal elements $\rho_{nl,n'l'}$ depend on the difference $\omega_{nl,n'l'}$ between the frequencies of two quantum states. Therefore these elements supply us with information about the phase $\omega_{nl,n'l'}t$. If the off-diagonal elements approach zero, the phase information disappears. It will be shown later that phase relaxation, or so-called dephasing, is realized faster than relaxation of the diagonal elements of the density matrix, referred to as energy relaxation.

Hence (6.15) describes both fast dephasing and the slower energy relaxation. Because energy relaxation is what interests us here, we can assume that phase relaxation is already over when energy relaxation begins. In this approximation we can set $\mathrm{d}\rho_{nl,n'l'}/\mathrm{d}t = \mathrm{d}\rho_{n'l',nl}/\mathrm{d}t = 0$. Then from the second equation of (6.15), we find the following expression for the off-diagonal matrix elements:

$$\rho_{nl,n'l'}(t) = \frac{U_{nl,n'l'}}{\omega_{nl,n'l'} - \mathrm{i}0}[\rho_{nl,nl}(t) - \rho_{n'l',n'l'}(t)] \,. \tag{6.16}$$

During energy relaxation, the off-diagonal matrix elements follow the temporal behavior of the diagonal elements of the density matrix. The imaginary vanishing term in the denominator guarantees the true relations between the direct and inverse Laplace transformation.

Inserting (6.16) into the first equation in (6.15), one arrives at rate equations for the the diagonal matrix elements of the tunneling system + phonons:

$$\dot{\rho}_{nl,nl} = -2\pi \sum_{n'l'} U_{nl,n'l'} U_{n'l'} \delta(\omega_{n'l',nl}) \rho_{nl,nl} + 2\pi \sum_{n'l'} U_{nl,n'l'} U_{n'l'} \delta(\omega_{n'l',nl}) \rho_{n'l',n'l'} . \tag{6.17}$$

During energy relaxation in the tunneling system, phonons are in thermal equilibrium and are therefore described by the Boltzmann function. Hence the diagonal elements of the tunneling system + phonons are given by

$$\rho_{nl,nl}(t) = \rho_n^l(T) \rho_{ll}(t) = \frac{\mathrm{e}^{-E_n^l/kT}}{\sum_n \mathrm{e}^{-E_n^l/kT}} \rho_{ll}(t) , \tag{6.18}$$

where E_n^l is the energy of phonons when the tunneling system is in the quantum state l. By inserting (6.18) into (6.17) and carrying out a summation over phonon indices n, we arrive at the equation

$$\dot{\rho}_{ll}(t) = -\rho_{ll}(t) \sum_{l'} \gamma_{l'l} + \sum_{l'} \gamma_{ll'} \rho_{l'l'}(t) , \tag{6.19}$$

where

$$\gamma_{ll'} = 2\pi \sum_n \rho_n^{l'}(T) \sum_{n'} U_{nl,n'l'} U_{n'l',nl} \delta(\omega_{n'l',nl}) . \tag{6.20}$$

The coefficients $\gamma_{l'l}(T)$ have inverse time dimensions and they determine the probability of the quantum transition $l' \longleftarrow l$ per second. It is obvious that the diagonal matrix element $\rho_{ll}(t)$ determines the probability of finding the tunneling system in quantum state l, i.e., in the left or the right well (see Fig. 6.2).

The rate equations we have derived enable us to calculate the temporal behavior of the probabilities of finding the tunneling system in a definite quantum state. The coefficients $\gamma_{ll'}(T)$ depend on temperature. They determine the probabilities of intra- and inter-well transitions.

6.3 One-Phonon Transition Probabilities in Tunneling Systems

The matrix elements $U_{nl,n'l'}$ in the expression for the transition probability are given by

$$U_{nl,n'l'} = \int \varphi_n^l(x) U_{ll'}(x) \varphi_{n'}^{l'}(x) \, \mathrm{d}x \; , \tag{6.21}$$

where φ_n^l are solutions of (6.10) for a harmonic oscillator. The most general expression for the rate matrix $\gamma_{ll'}(T)$ will be found in the case when both displacements a^l of the equilibrium positions of the oscillators and the force matrix U^l of the vibration system depend on an index l of the tunneling system. In this section we neglect this dependence on the index l by setting $a^l = a$, $U^l = U$. Then $\varphi_n^l(x) = \varphi_n(x)$ and therefore values of the matrix elements described by (6.21) are determined solely by the x-dependence of the function $U_{ll'}(x)$. Let us expand this function as a power series in the dynamical variable x. By taking into account only the linear term in the expansion, we arrive at the expression

$$U_{ll'}(x) = \sum_s \frac{\mathrm{d}U_{ll'}}{\mathrm{d}x_s} = \int \psi_l(q) S(q) \psi_{l'}(q) \, \mathrm{d}q \sum_s u_s x_s = S_{ll'} \sum_s u_s x_s \; , \tag{6.22}$$

where u_s are components of a multidimensional vector u with modulus unity. Therefore the value of the matrix element is determined by the value of the constant $S_{ll'}$. Insert (6.22) into (6.21). The matrix elements which do not equal zero exist only for $n' = n+1$ and $n' = n-1$. Therefore the expression for the rate constant takes the form

$$\begin{aligned} \gamma_{l'l} &= 2\pi S_{l'l}^2 \sum_s u_s^2 \sum_{n_s} \mathrm{e}^{(F_s - \hbar\omega_s n_s)/kT} \\ &\times \left[\langle n_s | x_s | n_s + 1 \rangle^2 \delta(\omega_{l'l} + \omega_s) + \langle n_s | x_s | n_s - 1 \rangle^2 \delta(\omega_{l'l} - \omega_s) \right] \; , \end{aligned} \tag{6.23}$$

where s is the index of a phonon mode and F_s is the free energy of the phonon system. By calculating the matrix elements with the help of the harmonic oscillator functions and summing over n_s, we find the following expression for the rate constant:

$$\gamma_{l'l} = \pi S_{l'l}^2 \sum_s u_s^2 \Big[[n_s(T) + 1] \delta(\omega_{l'l} + \omega_s) + n_s(T) \delta(\omega_{l'l} - \omega_s) \Big] \; , \tag{6.24}$$

where $n_s(T) = [\exp(\hbar\omega_s/kT) - 1]^{-1}$ is the average number of photons with frequency ω_s at temperature T. Since (6.24) determines a transition probability with creation and annihilation of one photon, the probability equals zero if the difference between the levels involved in transitions exceeds the energy of one photon. The first term in the square brackets describes a transition from the upper to the lower level with creation of one photon. This term does not vanish when the temperature approaches zero. The second term in the square brackets relates to transitions from the lower to the upper level of the double-well potential with annihilation of one phonon. This term disappears at $T = 0$.

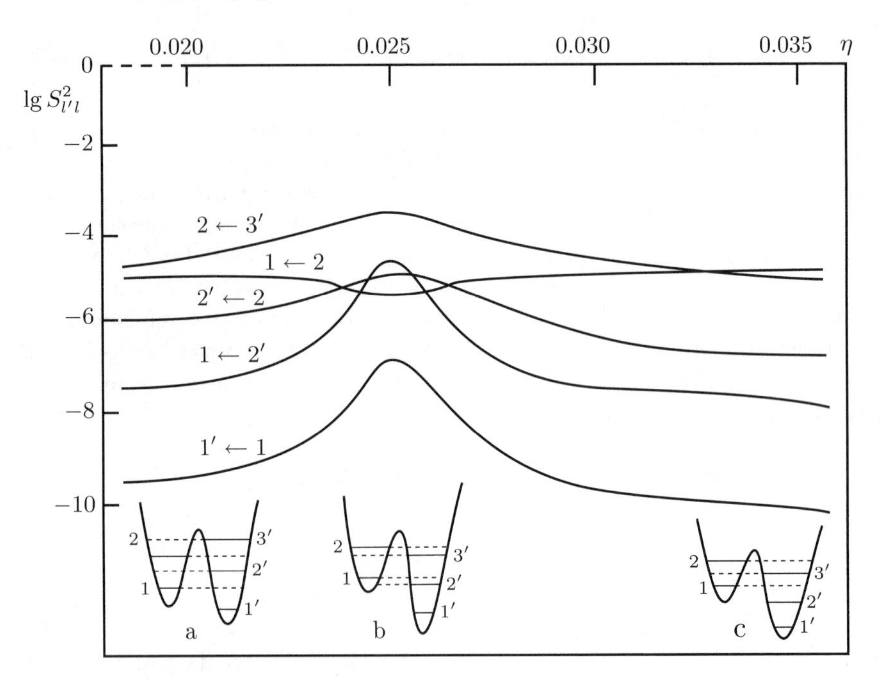

Fig. 6.3. Influence of resonance of the energy levels on the matrix element for intra-well and inter-well transitions [47, 48]. $V = 1200$ and $\xi = 0.37$ for all potentials. Asymmetry parameter (**a**) $\eta = 0.02$, (**b**) $\eta = 0.025$, (**c**) $\eta = 0.035$

The energy levels can relate to states localized either in one well or in different wells. In these cases we are talking about intra-well and inter-well transitions, respectively. The transition probability is determined by the constant $S^2_{ll'}$. The dependence of this matrix element on the indices of the energy levels is shown in Fig. 6.3. This matrix element is changed when the asymmetry parameter η is varied. The calculations were carried out in [48] with a potential determined by (6.11). Three potentials are placed directly under the relevant values of the asymmetry parameter η. The function $S(q)$ was chosen in the form $S_0 \sin kq$. The probabilities of intra-well transitions are one order of magnitude larger than those of inter-well transitions. In Fig. 6.3b there is a quasi-resonance between levels of states localized in different wells. It is obvious that probabilities of inter-well transitions are increased and probabilities of intra-well transitions are decreased in this case.

Let us now examine the temperature dependence of transitions in the double-well system. It is convenient to introduce a function

$$\varphi(\omega) = \sum_s u_s^2 \delta(\omega - \omega_s) . \tag{6.25}$$

If the coefficients u_s^2 equal $1/N$, where N is the number of degrees of freedom, then this function describes the density of states of the vibrational system.

For the Debye phonon model, this density equals $3\omega^2/\omega_D^3$. Although these coefficients are of the order of $1/N$, they do not exactly equal $1/N$, because they also determine an interaction of Herzberg–Teller type between a phonon mode and the double-well system. Therefore the function $\varphi(\omega)$ is called the weighted density of states. Using this function, an expression for the rate constant is given by

$$\gamma_{l'l} = \pi S_{l'l}^2 \Big[[n(\omega_{ll'}) + 1]\varphi(\omega_{ll'}) + n(\omega_{l'l})\varphi(\omega_{l'l}) \Big] , \tag{6.26}$$

where $n(\omega) = [\exp(\hbar\omega/kT) - 1]^{-1}$. In the high temperature limit when $kT/\hbar\omega_{ll'} \gg 1$, the temperature dependence of the rate constant is determined by

$$\gamma_{l'l} \approx \pi S_{l'l}^2 \frac{kT}{\hbar|\omega_{ll'}|} \varphi(|\omega_{ll'}|) . \tag{6.27}$$

In the low temperature limit when $kT/\hbar\omega_{ll'} \ll 1$, the rate constant is described by

$$\gamma_{l'l} \approx \begin{cases} \pi S_{l'l}^2 \varphi(|\omega_{ll'}|) , & E_{l'} < E_l , \\ \pi S_{l'l}^2 \varphi(|\omega_{ll'}|) \mathrm{e}^{-\hbar\omega_{l'l}/kT} , & E_{l'} > E_l . \end{cases} \tag{6.28}$$

For the transition with phonon creation, the constant does not depend on temperature. The transition with phonon annihilation has an Arrhenius-type temperature dependence.

By taking into account $n(\omega) + 1 = n(\omega)\exp(\hbar\omega/kT)$, one can easily find the following relation between the direct and inverse transitions:

$$\gamma_{l'l} = \mathrm{e}^{\hbar\omega_{ll'}/kT} \gamma_{ll'} . \tag{6.29}$$

This relation is general and it is true not only for one-phonon transitions. Probabilities ρ_{ll} in thermal equilibrium are easily found with the help of this relation. By setting derivatives in the left-hand side of (6.19) to zero and using (6.29), we find that

$$\sum_{l'} \gamma_{l'l} \left[\rho_{ll} - \mathrm{e}^{-\hbar\omega_{ll'}/kT} \rho_{l'l'} \right] = 0 . \tag{6.30}$$

The equation must be true for arbitrary values of $\gamma_{ll'}$. This is possible if

$$\mathrm{e}^{E_{l'}/kT} \rho_{l'l'} = \mathrm{e}^{E_l/kT} \rho_{ll} . \tag{6.31}$$

Since the sum of all the probabilities must be equal to one, we arrive at the well known distribution of probabilities in thermal equilibrium:

$$\rho_{ll}(T) = \frac{\mathrm{e}^{-E_l/kT}}{\sum_l \mathrm{e}^{-E_l/kT}} = \mathrm{e}^{(F_l - E_l)/kT} . \tag{6.32}$$

6.4 Kinetics of Tunneling Systems

The barrier separating wells means that the probabilities of inter-well transitions are much lower than the probabilities of intra-well transitions. Therefore thermal equilibrium in the quantum states localized in one well is established faster than thermal equilibrium between states localized in different wells. In other words the system has fast intra-well relaxation and slow inter-well relaxation.. This fact can be used to find simple formulas for describing the inter-well relaxation.

We denote the states whose wave functions are localized in the left and right wells by indices $l = a$ and $l = b$, respectively. Levels above the barrier can be excluded from consideration if the temperature is low enough. Then (6.19) can be rewritten in the form

$$\begin{aligned} \dot{\rho}_{aa} &= -\sum_{a'} \Gamma_{a'a}\rho_{aa} + \sum_{a'} \Gamma_{aa'}\rho_{a'a'} - \sum_{b} \gamma_{bu}\rho_{aa} + \sum_{b} \gamma_{ab}\rho_{bb}\,, \\ \dot{\rho}_{bb} &= -\sum_{b'} \Gamma_{b'b}\rho_{bb} + \sum_{b'} \Gamma_{bb'}\rho_{b'b'} - \sum_{a} \gamma_{ab}\rho_{bb} + \sum_{a} \gamma_{ba}\rho_{aa}\,. \end{aligned} \tag{6.33}$$

Here capital letters denote the rate constants of fast intra-well relaxation and small letters denote rate constants of slow inter-well relaxation. It is obvious that

$$\rho_1 = \sum_a \rho_{aa}\,, \quad \rho_2 = \sum_b \rho_{bb} \tag{6.34}$$

are the probabilities of finding a system in some quantum state localized either in the left or in the right well, respectively. The relaxation described by (6.33) has two stages: fast and slow. For the time of fast relaxation a thermal equilibrium between states localized in the same well is established and the probabilities thus take the form

$$\rho_{aa} = \rho_a(T)\rho_1(t)\,, \quad \rho_{bb} = \rho_b(T)\rho_2(t)\,, \tag{6.35}$$

where the probabilities, which depend on temperature, are probabilities of Boltzmann type and hence described by (6.32). Therefore the distribution of probabilities inside each well is of the Boltzmann type. They depend solely on the energy distance measured from the lowest level of the well and on temperature. However, the probabilities of finding the system in the left or in the right well are not in thermal equilibrium with each other.

Let us insert (6.35) into (6.33). Allowing for the relation described by (6.31) and taking into account (6.29), we find the relations

$$\begin{aligned} -\sum_{a'} \Gamma_{a'a}\rho_a(T) + \sum_{a'} \Gamma_{aa'}\rho_{a'}(T) &= 0\,, \\ -\sum_{b'} \Gamma_{b'b}\rho_b(T) + \sum_{b'} \Gamma_{bb'}\rho_{b'}(T) &= 0\,, \end{aligned} \tag{6.36}$$

i.e., all sums with capital letters in (6.33) disappear. By carrying out the summation over indices a and b in (6.33), we arrive at the following simple equations:

$$\begin{aligned} \dot{\rho}_1 &= -p_1\rho_1 + p_2\rho_2 , \\ \dot{\rho}_2 &= p_1\rho_1 - p_2\rho_2 , \end{aligned} \tag{6.37}$$

where

$$p_1 = \sum_{b,a} \gamma_{ba}\rho_a(T) , \quad p_2 = \sum_{a,b} \gamma_{a,b}\rho_b(T) . \tag{6.38}$$

These two equations determine the kinetics of inter-well relaxation. The kinetics of the multilevel system is reduced to the kinetics of the two-level system with effective rate constants described by (6.38). A solution of these equations is given by the functions

$$\begin{aligned} \rho_1(t) &= \frac{p_2}{p} + \left[\rho_1(0) - \frac{p_2}{p}\right] \mathrm{e}^{-pt} , \\ \rho_2(t) &= \frac{p_1}{p} + \left[\rho_2(0) - \frac{p_1}{p}\right] \mathrm{e}^{-pt} , \end{aligned} \tag{6.39}$$

where $p = p_1 + p_2$ and $\rho_{1,2}(0)$ are the initial probabilities. The rate of inter-well relaxation is determined by a constant p that depends on temperature. The temperature dependence can be calculated with the help of (6.38), if the double-well potential is definite.

Figure 6.4 shows the result of the temperature-dependence calculation for the partial rate constants and for the effective rate constants which are the sums of the partial rate constants. The weighted density of states $\varphi(\omega)$ was taken in the quasi-Debye form $\varphi(\omega) = \omega^3/\omega_{\mathrm{D}}^4$. It equals zero beyond the interval $0 < \omega < \omega_{\mathrm{D}}$. The matrix elements $S_{ll'}$ were calculated in the same way as those shown in Fig. 6.3. Calculations were carried out for the potentials placed under the relevant curves depicted in Fig. 6.4. The partial probabilities are given by

$$p_{ba}(T) = \gamma_{ba}(T)\rho_a(T) , \quad p_{ab}(T) = \gamma_{ab}(T)\rho_b(T) . \tag{6.40}$$

The sum of these partial probabilities, i.e., p_1, is shown by a dashed line. The shallow well is chosen as well 1. Therefore p_1 relates to the transition from the shallow well to the deep well. When the temperature approaches zero, all the partial probabilities except one approach zero as well. Therefore p_1 at $T = 0$ describes the probability of a pure tunneling transition through the barrier separating the wells.

When the temperature is raised, the main contribution to the probability p_1 results from the upper levels of the well, which are populated with the Boltzmann probabilities. The partial probabilities of the transitions from the

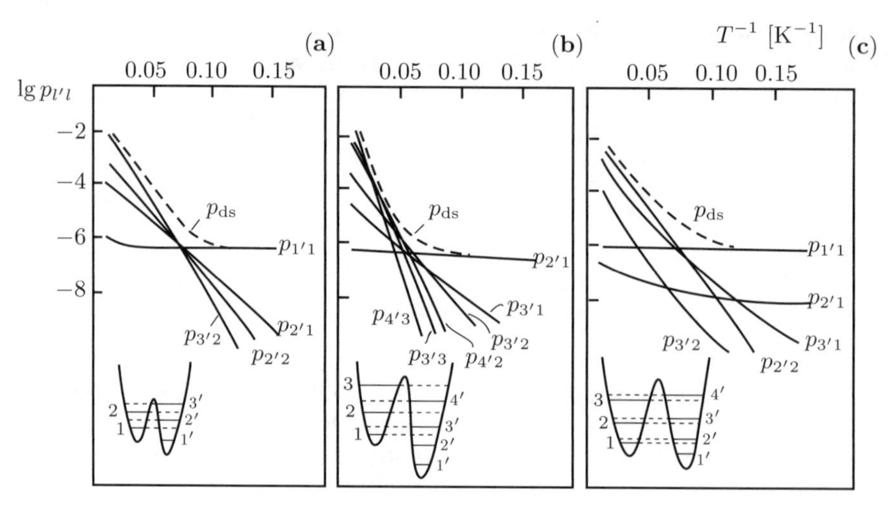

Fig. 6.4. (a–c) Temperature dependence of the partial probabilities of transitions from the levels of a shallow well to the levels of a deep well, and temperature dependence of the effective probability $p_{ds} = p_1$ [47,48]

upper levels of the well depend on temperature according to $\exp(-E/kT)$. They describe transitions along the classical path above the barrier. Such transitions are possible with annihilation of a phonon. The probability p_1 of the transition from the shallow well to the deep one can be approximated by the simple expression

$$p_1 = a + b\mathrm{e}^{-U/kT}\,, \tag{6.41}$$

where a and b are temperature-independent constants and U is an energy approximately equal to the height of the barrier. The slope of the curve p_{ds} must be proportional to the value of the parameter U. Figure 6.4 demonstrates that the interpretation of the energy U as the barrier height may turn out to be incorrect in a multilevel double-well system. In fact, the slope of the curve for p_{ds} calculated for the double-well potential shown in Fig. 6.4c is larger than that in Fig. 6.4a. This fact should be contrasted with the height of the barriers in these figures. The reason for this is a quasi-resonance existing between levels in the case shown in Fig. 6.4c. As has already been said in connection with Fig. 6.3, the quasi-resonance increases the probabilities of inter-well transitions.

6.5 Tunneling Systems in Polymers and Glasses

Thirty years ago it was believed that phonons comprised a single type of low frequency excitation both in crystals and in amorphous solids. Indeed, it was believed that the only difference between crystalline and amorphous solids was the existence of a huge number of localized modes in the latter, due to imperfections in the atomic or molecular lattice of such solids.

However, thirty years ago Zeller and Pohl [49] discovered that the low temperature behavior of the heat capacity of a glass cannot be explained if one assumes that phonons are the only type of low frequency excitation existing in a glass. The low temperature behavior of the heat capacity in a silicate glass exhibits a linear temperature dependence, in contrast to the well known T^3 dependence measured in many crystals, and was explained by the Debye model for phonons at the beginning of the last century. Since Zeller and Pohl's discovery served as a starting point for the new theory proposed in [50] and [51] for low frequency excitations in glasses, we shall examine their experiment in more detail.

In accordance with the Debye theory, the low temperature behavior of the heat capacity is determined by acoustic phonons. The heat capacity of a crystal is given by

$$C = \frac{\mathrm{d}E_{\mathrm{ph}}(T)}{\mathrm{d}T} . \tag{6.42}$$

Here $E_{\mathrm{ph}}(T)$ is the phonon energy at temperature T, given by

$$E_{\mathrm{ph}}(T) = \int_0^{E_{\mathrm{D}}} n(\varepsilon)\varepsilon\rho_{\mathrm{ph}}(\varepsilon)\,\mathrm{d}\varepsilon , \tag{6.43}$$

where ε is the phonon energy, $\rho_{\mathrm{ph}}(\varepsilon)$ is the density of phonon states, and $n(\varepsilon) = [\exp(\varepsilon/kT) - 1]^{-1}$ is the average number of phonons with energy ε at temperature T. The density of states in the Debye model is given by

$$\rho_{\mathrm{ph}}(\varepsilon) = 9n_{\mathrm{a}}\frac{\varepsilon^2}{E_{\mathrm{D}}^3} , \tag{6.44}$$

where n_{a} is the number of atoms per unit volume of the crystal and E_{D} is the bounding energy of acoustic phonons. By inserting the phonon density of states into the expression for the energy, we find the following expression for the phonon energy:

$$E_{\mathrm{ph}}(T) = \frac{(kT)^4}{E_{\mathrm{D}}^3}9n_{\mathrm{a}}I_{\mathrm{ph}}\left(\frac{E_{\mathrm{D}}}{kT}\right) , \tag{6.45}$$

where

$$I_{\mathrm{ph}}(z) = \int_0^z \frac{x^3\,\mathrm{d}x}{\mathrm{e}^x - 1} . \tag{6.46}$$

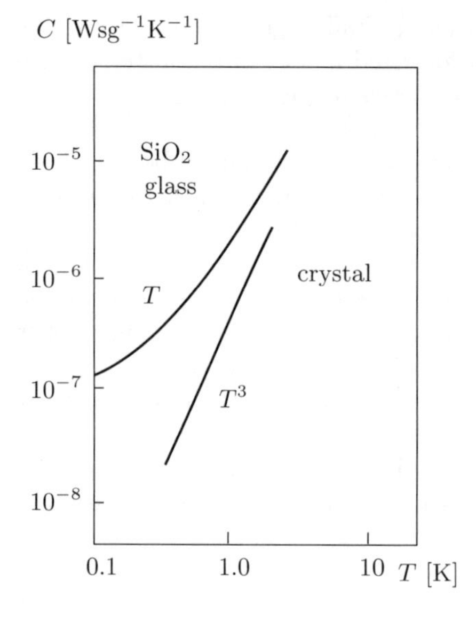

Fig. 6.5. Temperature dependence of specific heat in crystal and glass [49]

In the low temperature limit, this integral does not depend on T, and therefore the heat capacity is a cubic function of temperature:

$$C = 36kn_{\mathrm{a}} \left(\frac{kT}{E_{\mathrm{D}}} \right)^3 I_{\mathrm{ph}}(\infty) . \tag{6.47}$$

Figure 6.5 shows this cubic temperature dependence of the heat capacity in a crystal. However, the heat capacity in a silicate glass depends linearly on temperature in the low temperature region. What is the reason for this anomalous temperature dependence? The answer to this question can be found with the help of (6.42). If the energy of low frequency excitations in a glass were a quadratic function of temperature, the heat capacity could be a linear function of temperature. We can reach a desirable result for the energy if we assume that the density of states in the low frequency range is a constant. However, this is impossible for phonons. Then Andreson, Halperin, and Varma [50] and independently Phillips [51] assumed that additional degrees of freedom exist in glasses and that additional excitations relate to these degrees of freedom. A crucial assumption was that these new excitations should have a finite density of states in the low frequency range. We must therefore ask where these additional degrees of freedom might come from.

In order to answer this question let us turn back to the double-well potential depicted in Fig. 6.2. In the energy region lying below the top of the barrier, the potential looks like two independent parabolas. In this energy region we have in fact two types of oscillator with different equilibrium positions relating to the minima of the wells. Therefore, at low temperature, when the levels existing above the barrier do not play a conspicuous role, we may replace this double-well potential by two parabolas with different equilibrium positions. The phonon is a quantum of the excitation in one well. The tunneling transition between different wells is accompanied by creation or annihilation of a quantum with energy ε equal to the energy difference between the lowest levels in the two wells. The energy ε can be ascribed to a new particle called the tunnelon. This particle must be a quasi-particle of Fermi type because it exists only in two quantum states with quantum numbers 0 and 1. If the tunnelon density of states has finite value at $\varepsilon = 0$, the energy of all tunnelons can depend quadratically on temperature. The tunnelon contribution to the heat capacity will be proportional to the first power of the temperature.

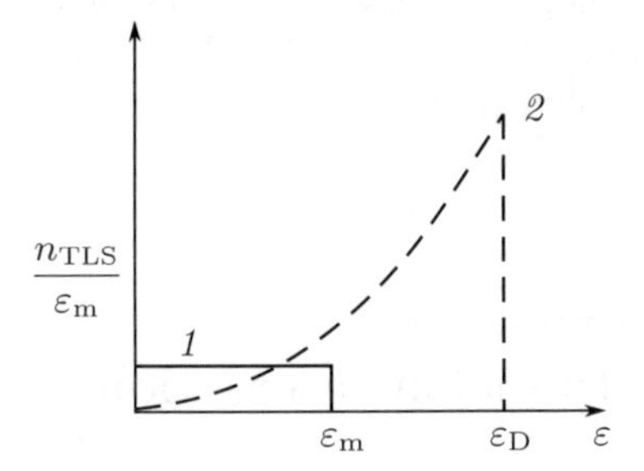

Fig. 6.6. Density of states of acoustic phonons (*dashed line*) and tunneling systems (*solid line*)

The density of phonon states in the Debye model and the density of tunnelon states are shown in Fig. 6.6. The tunnelons contribute to the heat capacity. Their contribution equals $\mathrm{d}E_{\mathrm{TLS}}/\mathrm{d}T$, where the tunnelon energy is determined by

$$E_{\mathrm{TLS}}(T) = \int_0^{E_{\mathrm{m}}} f(\varepsilon)\varepsilon g_{\mathrm{TLS}}(\varepsilon)\,\mathrm{d}\varepsilon\,. \tag{6.48}$$

Here $f(\varepsilon) = [\exp(\varepsilon/kT) + 1]^{-1}$ is the average number of tunnelons with energy ε at temperature T. By inserting the tunnelon density of states $g_{\mathrm{TLS}} = n_{\mathrm{TLS}}/E_{\mathrm{m}}$, we find

$$E_{\mathrm{TLS}}(T) = \frac{(kT)^2}{E_{\mathrm{m}}} n_{\mathrm{TLS}} I_{\mathrm{TLS}}\left(\frac{E_{\mathrm{m}}}{kT}\right), \tag{6.49}$$

where

$$I_{\mathrm{TLS}}(y) = \int_0^{y} \frac{x\,\mathrm{d}x}{\mathrm{e}^x - 1}\,. \tag{6.50}$$

This integral does not depend on temperature at low T.

Differentiating the tunnelon energy with respect to temperature and adding this result to the right hand side of (6.47), we find the following expression for the heat capacity of a glass:

$$C = 36kn_{\mathrm{a}} \left(\frac{kT}{E_{\mathrm{D}}}\right)^3 I_{\mathrm{ph}}(\infty) + 2kn_{\mathrm{TLS}} \left(\frac{kT}{E_{\mathrm{m}}}\right) I_{\mathrm{TLS}}(\infty) . \tag{6.51}$$

If the number n_{TLS} of tunneling systems per unit volume is large enough, the second term in (6.51) exceeds the first term at low temperature. Therefore the temperature dependence of the heat capacity will exhibit a conversion of the cubic dependence on T into a linear dependence at low temperature, as Fig. 6.5 shows. It turns out that these experimental data can be fitted by (6.51) if a silicate glass has a concentration of tunneling systems which corresponds to approximately one tunneling system per 10^4 atoms of glass. The assumption concerning the existence of tunneling systems in glasses has been confirmed in a huge number of experiments involving measurements of various physical characteristics of low temperature glasses [52]. Moreover, tunneling systems have been discovered in polymers and other amorphous solids.

6.6 Two-Level Systems (TLS). Tunnelons. Tunnelon–Phonon and Electron–Tunnelon Interaction

It was shown in Sect. 6.4 that once the intra-well energy relaxation is over, an inter-well relaxation begins. This inter-well relaxation is governed by (6.37). These two equations include effective probabilities of inter-well transitions. At low temperature, all levels of the double-well potentials apart from the lowest levels in each well will be depopulated. In fact only one lowest level from each well takes part in the inter-well relaxation at low temperature. Such a situation is realized in low temperature polymers and glasses.

In this case, (6.5) consists of two equations. In matrix form they are

$$\begin{pmatrix} H_2(x) + E_2 & U(x) \\ U(x) & H_1(x) + E_1 \end{pmatrix} \begin{pmatrix} \varphi^2(x) \\ \varphi^1(x) \end{pmatrix} = E \begin{pmatrix} \varphi^2(x) \\ \varphi^1(x) \end{pmatrix} , \tag{6.52}$$

where the adiabatic Hamiltonians are described by (6.8). Equation (6.52) is the Schrödinger equation for a complex system consisting of a phonon system and a tunneling system. By using the Pauli matrices

$$\sigma_x = \begin{pmatrix} 0 & 1 \\ 1 & 0 \end{pmatrix} , \qquad \sigma_y = \begin{pmatrix} 0 & -\mathrm{i} \\ \mathrm{i} & 0 \end{pmatrix} , \qquad \sigma_z \begin{pmatrix} 1 & 0 \\ 0 & -1 \end{pmatrix} , \tag{6.53}$$

we can write the Hamiltonian of such a system in the form

$$H = H(x) + E + \frac{\varepsilon + V(x)}{2} \sigma_z + U(x)\sigma_x . \tag{6.54}$$

The following notation has been used:

$$\begin{aligned} E &= \frac{E_2 + E_1}{2} , \quad H(x) = \frac{H_2(x) + H_1(x)}{2} , \\ \varepsilon &= E_2 - E_1 , \quad V(x) = H_2(x) - H_1(x) . \end{aligned} \tag{6.55}$$

The Pauli matrices can be expressed in terms of operators c^+ and c which satisfy the Fermi commutation relation

$$c^+ c + c c^+ = 1 . \tag{6.56}$$

The relation takes the form

$$c^+ = \frac{\sigma_x + \mathrm{i}\sigma_y}{2} , \quad c = \frac{\sigma_x - \mathrm{i}\sigma_y}{2} , \quad c^+ c - c c^+ = 2c^+ c - 1 = \sigma_z . \tag{6.57}$$

Substituting the Fermi-type operators c^+ and c for the Pauli matrices in the adiabatic Hamiltonian and setting $E_1 = 0$, we arrive at the Hamiltonian

$$H = H_{\mathrm{ph}} + H_{\mathrm{TLS}} + H_{\mathrm{TLS\text{–}ph}} , \tag{6.58}$$

where

$$\begin{aligned} &H_{\mathrm{ph}} = H_1(x) , \quad H_{\mathrm{TLS}} = \varepsilon c^+ c , \\ &H_{\mathrm{TLS\text{–}ph}} = V(x) c^+ c + U(x)(c^+ + c) \end{aligned} \tag{6.59}$$

are Hamiltonians for the phonons and the two-level system (TLS), and an operator representing the TLS–phonon interaction, respectively.

We introduce the following notation for the eigenvectors of the matrix σ_z:

$$\begin{pmatrix} 0 \\ 1 \end{pmatrix} = |0\rangle , \quad \begin{pmatrix} 1 \\ 0 \end{pmatrix} = |1\rangle . \tag{6.60}$$

Using the relation between the Fermi operators c^+ and c and the Pauli matrices, the following equations are found:

$$\begin{aligned} &c^+ |0\rangle = |1\rangle , \quad c|1\rangle = |0\rangle , \\ &c^+ |1\rangle = 0 , \quad c|0\rangle = 0 . \end{aligned} \tag{6.61}$$

According to these equations the operators c^+ and c can be called creation and annihilation operators for the TLS excitation. The latter is called a tunnelon. This stresses the fact that the creation of a tunnelon is accompanied by tunneling through the barrier. A tunnelon is a quasi-particle of Fermi type, in contrast to phonons, which are Bosons.

The operator $H_{\mathrm{TLS\text{–}ph}}$ describes the interaction of the TLS with phonons, i.e., a tunnelon–phonon interaction. The operator is a sum of two terms. One of them is a linear function of the tunnelon operators and the other is a quadratic function of these operators. The linear interaction results in tunneling transitions. This interaction determines a value for the tunnelon lifetime. The quadratic part of the tunnelon–phonon interaction results in dephasing processes in the TLS.

Phonons manifest themselves in the optical spectrum via the electron–phonon interaction. Tunnelons can manifest themselves in the optical spectrum of the guest molecule via the electron–tunnelon interaction. How does this interaction emerge?

The mechanism for the electron–tunnelon interaction is similar to that for the electron–phonon interaction. In fact, it was shown in Sect. 4.2 that the electron–phonon interaction of Franck–Condon type emerges via a change in the adiabatic potential, when the electronic state of a guest molecule is changed. The difference between the adiabatic potentials in the ground and excited electronic states is the FC electron–phonon interaction. The adiabatic potential in (6.54) is a 2×2 matrix. The Pauli matrices relate to an additional degree of freedom whose quantum number can be proportional to -1 or $+1$. This degree of freedom is ascribed to a tunnelon. The situation is similar to what happens with the electron spin.

The adiabatic potential of matrix type depends on the index f of the electronic state. Therefore the Hamiltonian of the system depends on the same electronic index f, i.e.,

$$H^f = H^f_{\mathrm{ph}} + H^f_{\mathrm{TLS}} + H^f_{\mathrm{TLS\text{-}ph}} \,. \tag{6.62}$$

If the ground and excited electronic states are denoted by indices g and e, the difference

$$H^{\mathrm{e}}_{\mathrm{ph}} - H^{\mathrm{g}}_{\mathrm{ph}} = \boldsymbol{a}\frac{U^{\mathrm{e}}}{2}\boldsymbol{a} + \boldsymbol{a}U\boldsymbol{R} + \boldsymbol{R}\frac{W}{2}\boldsymbol{R} \,, \tag{6.63}$$

where

$$\boldsymbol{a} = \boldsymbol{a}^{\mathrm{e}} - \boldsymbol{a}^{\mathrm{g}} \,, \qquad W = U^{\mathrm{e}} - U^{\mathrm{g}} \,, \tag{6.64}$$

describes an electron–phonon interaction of FC type, and the difference

$$H^{\mathrm{e}}_{\mathrm{TLS}} - H^{\mathrm{g}}_{\mathrm{TLS}} = (\varepsilon^{\mathrm{e}} - \varepsilon^{\mathrm{g}})c^{+}c = \Delta c^{+}c \tag{6.65}$$

describes an electron–tunnelon interaction of FC type. Because the tunnelon–phonon interaction yielding the tunneling transition is very weak, we shall neglect the small difference $H^{\mathrm{e}}_{\mathrm{TLS\text{-}ph}} - H^{\mathrm{g}}_{\mathrm{TLS\text{-}ph}}$, which is also strictly speaking a part of the electron–tunnelon interaction.

Part III

Spectroscopy of a Single Impurity Center

7. Density Matrix for an Impurity Center

In experiments with a single molecule we deal with a guest molecule doping a polymer or a glassy solid matrix cooled down to 4.2 K and even less. Despite such cooling of the solid and guest molecule, we cannot remove the interaction of the molecule with phonons and tunnelons, because creation of phonons and tunnelons during optical excitation is possible even at $T = 0$. Such a single molecule interacting with low frequency excitations in a solid is called an impurity center.

The theory of two-photon correlators discussed in Part I does not take into account electron–phonon and electron–tunnelon interactions. In Part III, we generalize the theory of two-photon correlators so as to account for these interactions. This enables us to derive an infinite set of equations for the density matrix of the whole system consisting of two- or three-level chromophores, photons, phonons and tunnelons. We discuss the approximations that have to be made in order to reduce the infinite set of equations for the whole density matrix to the four optical Bloch equations. These are equations for a reduced density matrix that is able to describe the dynamics of a two-level chromophore interacting with phonons and tunnelons. In the process of finding the infinite set of equations for the whole density matrix and the optical Bloch equations for the reduced density matrix, we also find an expression for the full two-photon correlator of a single molecule interacting with phonons and tunnelons. This expression for the full two-photon correlator will be used later in the analysis of experimental data.

7.1 Transition Amplitudes in the Electron–Phonon–Tunnelon System

When measuring photons emitted spontaneously by a single molecule, we are dealing with a system that has an infinite number of degrees of freedom. Such a system is described by an infinite set of equations for the density matrix. It was shown in Part I that, despite this fact, the theoretical expression for the two-photon correlator can be found using just four equations which involve a single relaxation time T_1. These equations are similar to the optical Bloch equations in which the rate constant $1/T_2$ is replaced by $1/2T_1$. This reduction of the infinite set of equations with respect to the indices of emitted photons is made possible by the four approximations discussed in Sects. 3.2 and 3.3. In this and the next section it will be shown that these approximations are valid in the case when interactions with phonons and tunnelons are allowed for. Consider the ground electronic state and one excited electronic state of a molecule. Then (4.4) takes the form

$$\begin{pmatrix} H^{\mathrm{e}}(R)+E & U(R) \\ U(R) & H^{\mathrm{g}}(R) \end{pmatrix} \begin{pmatrix} \Phi^{\mathrm{e}}(R) \\ \Phi^{\mathrm{g}}(R) \end{pmatrix} = E_n \begin{pmatrix} \Phi^{\mathrm{e}}(R) \\ \Phi^{\mathrm{g}}(R) \end{pmatrix} . \tag{7.1}$$

Here the energy of the ground electronic state is zero. This set of equations is similar to (6.52) for the electron–tunnelon system. These two equations are the Schrödinger equation for a two-level chromophore interacting with phonons and tunnelons. Using the Pauli matrix and neglecting the operator U for the non-adiabatic interaction, we can transform the Hamiltonian in (7.1) to the form

$$H_{\mathrm{ch}} = \frac{H^{\mathrm{e}}(R)+H^{\mathrm{g}}(R)+E}{2} + \frac{E+V(R)}{2}\sigma_z . \tag{7.2}$$

Here $V(R) = H^{\mathrm{e}}(R) - H^{\mathrm{g}}(R)$ is the FC interaction with phonons and tunnelons and H_{ch} is the Hamiltonian of a chromophore interacting with phonons and tunnelons in the adiabatic approximation. The coordinate $R = (x, q)$ describes both vibration and tunneling degrees of freedom. Let us introduce the Fermi-type operators related to the Pauli matrices as follows:

$$\begin{aligned} &B^+ = \frac{\sigma_x + \mathrm{i}\sigma_y}{2} , \quad B = \frac{\sigma_x - \mathrm{i}\sigma_y}{2} , \quad \sigma_z = 2B^+B - 1 , \\ &B^+B + BB^+ = 1 . \end{aligned} \tag{7.3}$$

Operators B^+ and B are creation and annihilation operators for the electronic excitation in a chromophore, i.e.,

$$B^+|0\rangle = |1\rangle , \quad B|1\rangle = |0\rangle , \quad B^+|1\rangle = B|0\rangle = 0 . \tag{7.4}$$

Using (7.3) one can easily find the following expression for the Hamiltonian of a chromophore:

$$H_{\mathrm{ch}} = H^{\mathrm{g}}(R) + [E + V(R)]\, B^+B . \tag{7.5}$$

It should be borne in mind that the coordinates R are related to both coordinates q of tunneling systems and coordinates x of vibrations. Therefore the adiabatic Hamiltonians can be rewritten in the form

$$H^{\mathrm{g,e}}(R) = H^{\mathrm{g,e}}_{\mathrm{ph}}(x) + H^{\mathrm{g,e}}_{\mathrm{TLS}}(q) + H^{\mathrm{g,e}}_{\mathrm{TLS-ph}}(x,q) . \tag{7.6}$$

The Hamiltonian of the system consisting of optical electrons, phonons, tunnelons and photons can be written as

$$H = H_{\mathrm{ch}} + H_{\perp} + \Lambda \,, \tag{7.7}$$

where $H_{\perp}$ is the Hamiltonian for the transverse electromagnetic field and Λ is the operator representing the electron–photon interaction. We assume that solutions of the Schrödinger equations,

$$H^{\mathrm{g}}(R)|a\rangle = \hbar\Omega_a|a\rangle \,, \quad H^{\mathrm{e}}(R)|b\rangle = \hbar\Omega_b|b\rangle \,, \quad H_{\perp}|\boldsymbol{n}\rangle = \hbar\omega_{\boldsymbol{n}}|\boldsymbol{n}\rangle \,, \tag{7.8}$$

are known. Then solutions of the Schrödinger equation with Hamiltonian H_{ch} are given by

$$H_{\mathrm{ch}}\,|0\rangle\,|a\rangle = \hbar\Omega_a\,|0\rangle\,|a\rangle \,, \quad H_{\mathrm{ch}}\,|1\rangle\,|b\rangle = \hbar\left(\Omega + \Omega_b\right)|1\rangle\,|b\rangle \,, \tag{7.9}$$

where $\Omega = E/\hbar$ is the resonant frequency of the chromophore. The eigenfunctions of the system consisting of non-interacting chromophore and electromagnetic field are described by the products

$$|a) = |0\rangle|\boldsymbol{n}\rangle|a\rangle \,, \quad |b) = |1\rangle|\boldsymbol{n}'\rangle|b\rangle \,, \tag{7.10}$$

i.e., they are given by the following infinite sets of functions:

$$\begin{aligned}
&|0\rangle|n\rangle|a\rangle = \left|\begin{matrix} a \\ 0 \end{matrix}\right\rangle , && |1\rangle|n-1\rangle|a\rangle = \left|\begin{matrix} b \\ 0 \end{matrix}\right\rangle , \\
&|0\rangle|n-1\rangle|1_{\boldsymbol{k}}\rangle|a\rangle = \left|\begin{matrix} a \\ \boldsymbol{k} \end{matrix}\right\rangle , && |1\rangle|n-2\rangle|1_{\boldsymbol{k}}\rangle|b\rangle = \left|\begin{matrix} b \\ \boldsymbol{k} \end{matrix}\right\rangle , \\
&|0\rangle|n-2\rangle|1_{\boldsymbol{k}}, 1_{\boldsymbol{k}'}\rangle|a\rangle = \left|\begin{matrix} a \\ \boldsymbol{k}\boldsymbol{k}' \end{matrix}\right\rangle , && |1\rangle|n-3\rangle|1_{\boldsymbol{k}}, 1_{\boldsymbol{k}'}\rangle|b\rangle = \left|\begin{matrix} b \\ \boldsymbol{k}\boldsymbol{k}' \end{matrix}\right\rangle , \\
&\vdots && \vdots
\end{aligned} \tag{7.11}$$

where n is the number of photons in the laser mode. These functions describe the stationary states of the system with $\Lambda = 0$.

$$\left|\begin{matrix} a \\ 0 \end{matrix}\right\rangle \overset{\Lambda}{\longleftrightarrow} \left|\begin{matrix} b \\ 0 \end{matrix}\right\rangle \overset{\lambda_{\boldsymbol{k}}}{\longleftrightarrow} \left|\begin{matrix} a \\ \boldsymbol{k} \end{matrix}\right\rangle \overset{\Lambda'}{\longleftrightarrow} \left|\begin{matrix} b \\ \boldsymbol{k} \end{matrix}\right\rangle \overset{\lambda_{\boldsymbol{k}'}}{\longleftrightarrow} \left|\begin{matrix} a \\ \boldsymbol{k}\boldsymbol{k}' \end{matrix}\right\rangle \overset{\Lambda''}{\longleftrightarrow} \left|\begin{matrix} b \\ \boldsymbol{k}\boldsymbol{k}' \end{matrix}\right\rangle \longleftrightarrow \cdots$$

Fig. 7.1. Quantum states mixed by the operator Λ in the resonant approximation

When the interaction Λ is 'switched on', these functions no longer describe stationary states of the system and transitions take place between them. The transitions are indicated by arrows in Fig. 7.1. The situation is similar to that for a single atom, as was discussed in connection with Fig. 2.4. We arrive at small and large matrix elements, as in the atomic case:

$$\begin{aligned}(b|\Lambda/\hbar|a) &= \Lambda_{ba} = \Lambda\langle b|a\rangle\,, & (a|\Lambda/\hbar|b) &= \Lambda_{ab} = \Lambda^*\langle a|b\rangle = \Lambda^*_{ba}\,,\\ (b|\Lambda/\hbar|a) &= \lambda^{ba}_{\boldsymbol{k}} = \lambda_{\boldsymbol{k}}\langle b|a\rangle\,, & (a|\Lambda/\hbar|b) &= \lambda^{ab}_{\boldsymbol{k}} = \lambda^*_{\boldsymbol{k}}\langle a|b\rangle = \lambda^{*ba}_{\boldsymbol{k}}\,,\end{aligned} \tag{7.12}$$

where

$$\Lambda = -\mathrm{i}\Omega\sqrt{\frac{4\pi}{\hbar\omega_0}}\sqrt{\frac{n}{V}}\boldsymbol{d}\cdot\boldsymbol{e}_0\,, \qquad \lambda_{\boldsymbol{k}} = -\mathrm{i}\Omega\sqrt{\frac{4\pi}{\hbar\omega_{\boldsymbol{k}}V}}\boldsymbol{d}\cdot\boldsymbol{e}_{\boldsymbol{k}}\,. \tag{7.13}$$

Here upper case Λ denotes a large matrix element because it includes the large number n of laser photons. Here and elsewhere, indices a and b have to be related to the phonon–tunnelon subsystem in the ground and excited electronic states of the chromophore, respectively. Therefore the integrals of overlapping $\langle a \mid b\rangle = \langle b \mid a\rangle \neq \delta_{ab}$ determine amplitudes of the electron–phonon–tunnelon transitions.

Let us now find a set of equations for the transition amplitudes in the electron–phonon–tunnelon–photon system. Using only quantum states shown in Fig. 7.1 in the general (1.59), we arrive at the following set of equations:

$$\begin{aligned}
\dot G^a_0 &= -\mathrm{i}\omega^a_0 G^a_0 - \mathrm{i}\sum_b \Lambda_{ab}G^b_0\,,\\
\dot G^b_0 &= -\mathrm{i}\omega^b_0 G^b_0 - \mathrm{i}\sum_a \Big[\Lambda_{ba}G^a_0 + \sum_{\boldsymbol{k}}\lambda^{ba}_{\boldsymbol{k}}G^a_{\boldsymbol{k}}\Big]\,,\\
\dot G^a_{\boldsymbol{k}} &= -\mathrm{i}\omega^a_{\boldsymbol{k}} G^a_{\boldsymbol{k}} - \mathrm{i}\sum_b \big[\lambda^{ab}_{\boldsymbol{k}}G^b_0 + \Lambda'_{ab}G^b_{\boldsymbol{k}}\big]\,,\\
\dot G^b_{\boldsymbol{k}} &= -\mathrm{i}\omega^b_{\boldsymbol{k}} G^b_{\boldsymbol{k}} - \mathrm{i}\sum_a \Big[\Lambda'_{ba}G^a_{\boldsymbol{k}} + \sum_{\boldsymbol{k}'}\lambda^{ba}_{\boldsymbol{k}'}G^a_{\boldsymbol{k}\boldsymbol{k}'} + \underline{\lambda^{ba}_{\boldsymbol{k}}\sqrt{2}G^a_{2\boldsymbol{k}}}\Big]\,,\\
\dot G^a_{2\boldsymbol{k}} &= -\mathrm{i}\omega^a_{2\boldsymbol{k}} G^a_{2\boldsymbol{k}} - \mathrm{i}\sum_b \Big[\lambda^{ab}_{\boldsymbol{k}}\sqrt{2}G^b_{\boldsymbol{k}} + \Lambda''_{ab}G^b_{2\boldsymbol{k}}\Big]\,,\\
\hline
\dot G^a_{\boldsymbol{k}\boldsymbol{k}'} &= -\mathrm{i}\omega^a_{\boldsymbol{k}\boldsymbol{k}'} G^a_{\boldsymbol{k}\boldsymbol{k}'} - \mathrm{i}\sum_b \big[\lambda^{ab}_{\boldsymbol{k}}G^b_{\boldsymbol{k}'} + \lambda^{ab}_{\boldsymbol{k}'}G^b_{\boldsymbol{k}} + \Lambda''_{ab}G^b_{\boldsymbol{k}\boldsymbol{k}'}\big]\,,\\
\dot G^b_{\boldsymbol{k}\boldsymbol{k}'} &= -\mathrm{i}\omega^b_{\boldsymbol{k}\boldsymbol{k}'} G^b_{\boldsymbol{k}\boldsymbol{k}'} - \mathrm{i}\sum_a \Big[\Lambda''_{ba}G^a_{\boldsymbol{k}\boldsymbol{k}'} + \sum_{\boldsymbol{k}''}\lambda^{ba}_{\boldsymbol{k}''}G^a_{\boldsymbol{k}\boldsymbol{k}'\boldsymbol{k}''}\\
&\qquad + \underline{\lambda^{ba}_{\boldsymbol{k}}\sqrt{2}G^a_{2\boldsymbol{k}\boldsymbol{k}'} + \lambda^{ba}_{\boldsymbol{k}'}\sqrt{2}G^a_{\boldsymbol{k}2\boldsymbol{k}'}}\Big]\,,\\
&\;\;\vdots
\end{aligned} \tag{7.14}$$

The frequencies are given by

$$\begin{aligned}
\omega^a_0 &= \Omega_a + \omega_{\boldsymbol{n}}\,, & \omega^b_0 &= \Omega_b + \omega_{\boldsymbol{n}} + \Delta\,,\\
\omega^a_{\boldsymbol{k}} &= \omega^a_0 + \Delta_{\boldsymbol{k}}\,, & \omega^b_{\boldsymbol{k}} &= \omega^b_0 + \Delta_{\boldsymbol{k}}\,,\\
\omega^a_{\boldsymbol{k}\boldsymbol{k}'} &= \omega^a_0 + \Delta_{\boldsymbol{k}} + \Delta_{\boldsymbol{k}'}\,, & \omega^b_{\boldsymbol{k}\boldsymbol{k}'} &= \omega^b_0 + \Delta_{\boldsymbol{k}} + \Delta_{\boldsymbol{k}'}\,, \quad \text{etc.}\,,
\end{aligned} \tag{7.15}$$

where $\Delta = \Omega - \omega_0$ and $\Delta_{\boldsymbol{k}} = \omega_{\boldsymbol{k}} - \omega_0$.

The set (7.14) generalizes the set (3.4) studied in Sect. 3.2. Three approximations were formulated in that section. These approximations enable one to uncouple the infinite set of equations and reduce it to pairs of equations. These approximations can also be made in the case under consideration here. According to the first approximation, we may neglect the underlined amplitudes and equations for these amplitudes. According to the second approximation, we may omit primes in the matrix element Λ. The third approximation is given by the following equations:

$$\begin{aligned} \sum_{\boldsymbol{k}} \lambda_{\boldsymbol{k}}^{ba} G_{\boldsymbol{k}}^{a'} &= -\frac{\mathrm{i}}{2T_1} \langle b|a\rangle \sum_{b'} \langle a'|b'\rangle G_0^{b'} , \\ \sum_{\boldsymbol{k}} \lambda_{\boldsymbol{k}}^{ba} G_{\boldsymbol{k}\boldsymbol{k}'}^{a'} &= -\frac{\mathrm{i}}{2T_1} \langle b|a\rangle \sum_{b'} \langle a'|b'\rangle G_{\boldsymbol{k}'}^{b'} , \\ \sum_{\boldsymbol{k}} \lambda_{\boldsymbol{k}}^{ba} G_{\boldsymbol{k}\boldsymbol{k}'\boldsymbol{k}''}^{a'} &= -\frac{\mathrm{i}}{2T_1} \langle b|a\rangle \sum_{b'} \langle a'|b'\rangle G_{\boldsymbol{k}'\boldsymbol{k}''}^{b'} , \quad \text{etc.} \end{aligned} \tag{7.16}$$

These equations are proven in Appendix F. Once these three approximations have been implemented, the infinite set (7.14) is reduced to the following pairs of equations:

$$\begin{aligned} \dot{G}_0^a &= -\mathrm{i}\omega_0^a G_0^a - \mathrm{i} \sum_b \Lambda_{ab} G_0^b , \\ \dot{G}_0^b &= -\mathrm{i}\left(\omega_0^b - \frac{\mathrm{i}}{2T_1}\right) G_0^b - \mathrm{i} \sum_a \Lambda_{ba} G_0^a , \\ \dot{G}_{\boldsymbol{k}}^a &= -\mathrm{i}\omega_{\boldsymbol{k}}^a G_{\boldsymbol{k}}^a - \mathrm{i} \sum_b \left[\lambda_{\boldsymbol{k}}^{ab} G_0^b + \Lambda_{ab} G_{\boldsymbol{k}}^b\right] , \\ \dot{G}_{\boldsymbol{k}}^b &= -\mathrm{i}\left(\omega_{\boldsymbol{k}}^b - \frac{\mathrm{i}}{2T_1}\right) G_{\boldsymbol{k}}^b - \mathrm{i} \sum_a \Lambda_{ba} G_{\boldsymbol{k}}^a , \\ \dot{G}_{\boldsymbol{k}\boldsymbol{k}'}^a &= -\mathrm{i}\omega_{\boldsymbol{k}\boldsymbol{k}'}^a G_{\boldsymbol{k}\boldsymbol{k}'}^a - \mathrm{i} \sum_b \left[\lambda_{\boldsymbol{k}}^{ab} G_{\boldsymbol{k}'}^b + \lambda_{\boldsymbol{k}'}^{ab} G_{\boldsymbol{k}}^b + \Lambda_{ab} G_{\boldsymbol{k}\boldsymbol{k}'}^b\right] , \\ \dot{G}_{\boldsymbol{k}\boldsymbol{k}'}^b &= -\mathrm{i}\left(\omega_{\boldsymbol{k}\boldsymbol{k}'}^b - \frac{\mathrm{i}}{2T_1}\right) G_{\boldsymbol{k}\boldsymbol{k}'}^b - \mathrm{i} \sum_a \Lambda_{ba} G_{\boldsymbol{k}\boldsymbol{k}'}^a , \quad \text{etc.} \end{aligned} \tag{7.17}$$

Every pair of equations involves a definite number of emitted photons: no emitted photons in the first pair, one emitted photon $\boldsymbol{k}$ in the second pair, two photons $\boldsymbol{k}$ and $\boldsymbol{k}'$ in the third pair, and so on. These pairs of equations will be used for deriving equations for the density matrix in the next section.

7.2 Equations for the Density Matrix

The density matrix for the whole system enables one to calculate any physical characteristics of the system. The equations for the density matrix of the system consisting of a two-level chromophore, phonons, tunnelons and photons can be written with the help of (1.76), if we determine the meaning of the quantum indices l, s and p in the system considered here. However, following this approach, we arrive at a rather complicated infinite set of linked equations. It would be difficult to make approximations in this infinite set of linked equations for the density matrix. Therefore we have adopted another approach: the equations for the density matrix are found with the help of (7.17) for the transition amplitudes, where approximations have already been made.

There is an additional feature that facilitates the calculation of the optical band shape or the echo signal intensity. In order to find these optical characteristics, we do not need an expression for the full density matrix. It is enough to know expressions for the elements of the density matrix that is reduced with respect to indices of the emitted photons. It was shown in Part I, where we neglected phonons and tunnelons, that after summing the equations for the density matrix with respect to the indices of emitted photons, we arrive at the four equations of (3.12). These are similar to the optical Bloch equations. They differ from those equations in that the rate constant in the equations for the off-diagonal elements is not the same: the equations found involve a dephasing rate $1/2T_1$ rather than the dephasing rate of $1/T_2$ that occurs in the true optical Bloch equations. It will be shown later that the true dephasing time T_2 emerges in the optical Bloch equations for a single molecule due to quadratic FC electron–phonon and electron–tunnelon interactions. It will also be shown that it is impossible to generalize the four optical Bloch equations in such a way as to take into account the whole FC electron–phonon and electron–tunnelon interaction. The task can only be solved if we deal with the infinite set of equations for the full density matrix of the system. These equations will be derived in this section.

The equations for the density matrix that take into account both electron–phonon and electron–tunnelon interactions can be found using (7.17), in the same way as they were found using (3.6) in Part I when we neglected these interactions. The appearance of additional quantum numbers a and b does not significantly influence the derivation.

Our task is to find equations for the elements of the density matrix which are determined by the following infinite series:

$$
\begin{aligned}
\rho_{ba} &= G_0^b G^{*a}_0 + \sum_{\boldsymbol{k}} G_{\boldsymbol{k}}^b G^{*a}_{\boldsymbol{k}} + \frac{1}{2!}\sum_{\boldsymbol{k}\boldsymbol{k}'} G_{\boldsymbol{k}\boldsymbol{k}'}^b G_{\boldsymbol{k}\boldsymbol{k}'}^a + \cdots\,, \\
\rho_{ab} &= \rho_{ba}^*\,, \\
\rho_{bb'} &= G_0^b G^{*b'}_0 + \sum_{\boldsymbol{k}} G_{\boldsymbol{k}}^b G^{*b'}_{\boldsymbol{k}} + \frac{1}{2!}\sum_{\boldsymbol{k}\boldsymbol{k}'} G_{\boldsymbol{k}\boldsymbol{k}'}^b G_{\boldsymbol{k}\boldsymbol{k}'}^{b'} + \cdots\,, \\
\rho_{aa'} &= G_0^a G^{*a'}_0 + \sum_{\boldsymbol{k}} G_{\boldsymbol{k}}^a G^{*a'}_{\boldsymbol{k}} + \frac{1}{2!}\sum_{\boldsymbol{k}\boldsymbol{k}'} G_{\boldsymbol{k}\boldsymbol{k}'}^a G_{\boldsymbol{k}\boldsymbol{k}'}^{a'} + \cdots\,.
\end{aligned}
\tag{7.18}
$$

Using the equation

$$\dot{\rho}_{ls} = \dot{G}_l G_s^* + G_l \dot{G}_s^* \,, \tag{7.19}$$

together with the first pair of (7.17) and their conjugates, we find the following four equations for the first term in each infinite series:

$$\begin{aligned}
\frac{\mathrm{d}}{\mathrm{d}t} G_0^b G^{*a}_0 &= -\mathrm{i}\left[\omega_{ba} - \frac{\mathrm{i}}{2T_1}\right] G_0^b G^{*a}_0 - \mathrm{i}\sum_{a'} \Lambda_{ba'} G_0^{a'} G^{*a}_0 \\
&\quad + \mathrm{i}\sum_{b'} G_0^b G^{*b'}_0 \Lambda_{b'a} \,, \\
\frac{\mathrm{d}}{\mathrm{d}t} G_0^a G^{*b}_0 &= \frac{\mathrm{d}}{\mathrm{d}t}\left(G_0^b G^{*a}_0\right)^* \,, \\
\frac{\mathrm{d}}{\mathrm{d}t} G_0^a G^{*a'}_0 &= -\mathrm{i}\omega_{aa'} G_0^a G^{*a'}_0 - \mathrm{i}\sum_b \left[\Lambda_{ab} G_0^b G^{*a'}_0 - G_0^a G^{*b}_0 \Lambda_{ba'}\right] \,, \\
\frac{\mathrm{d}}{\mathrm{d}t} G_0^b G^{*b'}_0 &= -\mathrm{i}\left[\omega_{bb'} - \frac{\mathrm{i}}{T_1}\right] G_0^b G^{*b'}_0 - \mathrm{i}\sum_a \left[\Lambda_{ba} G_0^a G^{*b'}_0 - G_0^b G^{*a}_0 \Lambda_{ab'}\right] \,,
\end{aligned} \tag{7.20}$$

where

$$\omega_{ba} = \omega_0^b - \omega_0^a \,, \quad \omega_{bb'} = \omega_0^b - \omega_0^{b'} \,, \quad \omega_{aa'} = \omega_0^a - \omega_0^{a'} \,. \tag{7.21}$$

Let us now inspect the second pair of equations in (7.17). This pair with one emitted photon and the pair of equations conjugate to them are given by

$$\begin{aligned}
\dot{G}_{\boldsymbol{k}}^a &= -\mathrm{i}\omega_{\boldsymbol{k}}^a G_{\boldsymbol{k}}^a - \mathrm{i}\sum_b \left[\lambda_{\boldsymbol{k}}^{ab} G_0^b + \Lambda_{ab} G_{\boldsymbol{k}}^b\right] \,, \\
\dot{G}_{\boldsymbol{k}}^b &= -\mathrm{i}\left(\omega_{\boldsymbol{k}}^b - \frac{\mathrm{i}}{2T_1}\right) G_{\boldsymbol{k}}^b - \mathrm{i}\sum_a \Lambda_{ba} G_{\boldsymbol{k}}^a \,, \\
\dot{G}^{*a}_{\boldsymbol{k}} &= \mathrm{i}\omega_k^a G^{*a}_{\boldsymbol{k}} + \mathrm{i}\sum_b \left[\lambda_{\boldsymbol{k}}^{ba} G^{*b}_0 + \Lambda_{ba} G^{*b}_{\boldsymbol{k}}\right] \,, \\
\dot{G}^{*b}_{\boldsymbol{k}} &= \mathrm{i}\left(\omega_{\boldsymbol{k}}^b + \frac{\mathrm{i}}{2T_1}\right) G^{*b}_{\boldsymbol{k}} + \mathrm{i}\sum_a \Lambda_{ab} G^{*a}_{\boldsymbol{k}} \,.
\end{aligned} \tag{7.22}$$

From these we can find four equations for the terms which are described by a single sum over $\boldsymbol{k}$ in (7.18). However, we now face a problem that did not arise when we derived (7.20): additional terms emerge that are absent in the infinite series described by (7.18). For instance, the equation

$$\frac{\mathrm{d}}{\mathrm{d}t}\sum_{\boldsymbol{k}} G^a_{\boldsymbol{k}} G^{*a'}_{\boldsymbol{k}} = -\mathrm{i}\omega_{aa'}\sum_{\boldsymbol{k}} G^a_{\boldsymbol{k}} G^{*a'}_{\boldsymbol{k}}$$

$$-\mathrm{i}\sum_b \underline{\left(\sum_{\boldsymbol{k}} \lambda^{ab}_{\boldsymbol{k}} G^b_0 G^{*a'}_{\boldsymbol{k}} - \sum_{\boldsymbol{k}} G^a_{\boldsymbol{k}} G^{*b}_0 \lambda^{ba'}_{\boldsymbol{k}}\right)}$$

$$-\mathrm{i}\sum_b\left(\Lambda_{ab}\sum_{\boldsymbol{k}} G^b_{\boldsymbol{k}} G^{*a'}_{\boldsymbol{k}} - \sum_{\boldsymbol{k}} G^a_{\boldsymbol{k}} G^{*b}_{\boldsymbol{k}}\Lambda_{ba'}\right) \quad (7.23)$$

can be obtained from the first and third equations of the set described by (7.22). The underlined term includes elements of the density matrix which are absent in the series of (7.18). Fortunately, with the help of (7.16), they can be transformed as follows:

$$\sum_b\left(\sum_{\boldsymbol{k}} \lambda^{ab}_{\boldsymbol{k}} G^b_0 G^{*a'}_{\boldsymbol{k}} - \sum_{\boldsymbol{k}} G^a_{\boldsymbol{k}} G^{*b}_0 \lambda^{ba'}_{\boldsymbol{k}}\right) = \frac{1}{T_1}\sum_{bb'}\langle a|b\rangle G^b_0 G^{*b'}_0 \langle b'|a'\rangle\,. \quad (7.24)$$

The term on the right hand side already exists in the above-mentioned infinite series. By inserting (7.24) into (7.23), we find the equation

$$\frac{\mathrm{d}}{\mathrm{d}t}\sum_{\boldsymbol{k}} G^a_{\boldsymbol{k}} G^{*a'}_{\boldsymbol{k}} = -\mathrm{i}\omega_{aa'}\sum_{\boldsymbol{k}} G^a_{\boldsymbol{k}} G^{*a'}_{\boldsymbol{k}} + \frac{1}{T_1}\sum_{bb'}\langle a|b\rangle G^b_0 G^{*b'}_0 \langle b'|a'\rangle$$

$$-\mathrm{i}\sum_b\left(\Lambda_{ab}\sum_{\boldsymbol{k}} G^b_{\boldsymbol{k}} G^{a'}_{\boldsymbol{k}} - \sum_{k} G^a_{\boldsymbol{k}} G^b_{\boldsymbol{k}}\Lambda_{ba'}\right), \quad (7.25)$$

which only involves terms existing in the infinite series described by (7.18).

Another undesirable feature emerges in the equation for the off-diagonal elements:

$$\frac{\mathrm{d}}{\mathrm{d}t}\sum_{\boldsymbol{k}} G^b_{\boldsymbol{k}} G^{*a}_{\boldsymbol{k}} = -\mathrm{i}\left(\omega_{ba} - \frac{\mathrm{i}}{2T_1}\right)\sum_{\boldsymbol{k}} G^b_{\boldsymbol{k}} G^{*a}_{\boldsymbol{k}} - \mathrm{i}\sum_{a'}\Lambda_{ba'}\sum_{\boldsymbol{k}} G^{a'}_{\boldsymbol{k}} G^{*a}_{\boldsymbol{k}}$$

$$+\mathrm{i}\sum_{b'}\sum_{\boldsymbol{k}} G^b_{\boldsymbol{k}} G^{*b'}_{\boldsymbol{k}}\Lambda_{b'a} + \mathrm{i}\sum_{b'}\sum_{\boldsymbol{k}} \underline{G^b_{\boldsymbol{k}} G^{*b'}_0 \lambda^{b'a}_{\boldsymbol{k}}}\,. \quad (7.26)$$

The underlined term includes such an element. Fortunately, it is possible to neglect terms of this type because of their small value. Indeed we have

$$\sum_{\boldsymbol{k}} G^b_{\boldsymbol{k}} \lambda^{b'a}_{\boldsymbol{k}} \propto \sum_{\boldsymbol{k}} \lambda_{\boldsymbol{k}} G^1_{\boldsymbol{k}}\,, \quad (7.27)$$

The right hand side of this relation is small and so can be omitted on the basis of the fourth approximation discussed in Sect. 3.3. Hence, the undesirable terms can either be transformed to a desirable form or omitted because of their small value. Therefore the equations for the matrix elements described by single sums in the infinite series are given by

$$\begin{aligned}
\frac{\mathrm{d}}{\mathrm{d}t}\sum_{\boldsymbol{k}} G^b_{\boldsymbol{k}} G^{*a}_{\boldsymbol{k}} &= -\mathrm{i}\left(\omega_{ba} - \frac{\mathrm{i}}{2T_1}\right)\sum_{\boldsymbol{k}} G^b_{\boldsymbol{k}} G^{*a}_{\boldsymbol{k}} - \mathrm{i}\sum_{a'} \Lambda_{ba'} \sum_{\boldsymbol{k}} G^{a'}_{\boldsymbol{k}} G^{*a}_{\boldsymbol{k}} \\
&\quad + \mathrm{i}\sum_{b'}\sum_{\boldsymbol{k}} G^b_{\boldsymbol{k}} G^{*b'}_{\boldsymbol{k}} \Lambda_{b'a} , \\
\frac{\mathrm{d}}{\mathrm{d}t}\sum_{\boldsymbol{k}} G^a_{\boldsymbol{k}} G^{*b}_{\boldsymbol{k}} &= \frac{\mathrm{d}}{\mathrm{d}t}\left(\sum_{\boldsymbol{k}} G^b_{\boldsymbol{k}} G^{*a}_{\boldsymbol{k}}\right)^* , \\
\frac{\mathrm{d}}{\mathrm{d}t}\sum_{\boldsymbol{k}} G^b_{\boldsymbol{k}} G^{*b'}_{\boldsymbol{k}} &= -\mathrm{i}\left(\omega_{bb'} - \frac{\mathrm{i}}{T_1}\right)\sum_{\boldsymbol{k}} G^b_{\boldsymbol{k}} G^{*b'}_{\boldsymbol{k}} \\
&\quad - \mathrm{i}\sum_{a}\left(\Lambda_{ba}\sum_{\boldsymbol{k}} G^a_{\boldsymbol{k}} G^{*b'}_{\boldsymbol{k}} - \sum_{\boldsymbol{k}} G^b_{\boldsymbol{k}} G^{*a}_{\boldsymbol{k}} \Lambda_{ab'}\right) , \\
\frac{\mathrm{d}}{\mathrm{d}t}\sum_{\boldsymbol{k}} G^a_{\boldsymbol{k}} G^{*a'}_{\boldsymbol{k}} &= -\mathrm{i}\omega_{aa'}\sum_{\boldsymbol{k}} G^a_{\boldsymbol{k}} G^{*a'}_{\boldsymbol{k}} + \frac{1}{T_1}\sum_{bb'}\langle a|b\rangle G^b_0 G^{*b'}_0 \langle b'|a'\rangle \\
&\quad - \mathrm{i}\sum_{b}\left(\Lambda_{ab}\sum_{\boldsymbol{k}} G^b_{\boldsymbol{k}} G^{*a'}_{\boldsymbol{k}} - \sum_{\boldsymbol{k}} G^a_{\boldsymbol{k}} G^{*b}_{\boldsymbol{k}} \Lambda_{ba'}\right) .
\end{aligned} \tag{7.28}$$

Similarly the equations for the terms described by the double sums in the infinite series can be derived and so on. New equations will not include new types of undesirable terms.

By summing the first, second, third and fourth equations from the sets described by (7.20), (7.28), and so on, we arrive at the following set of equations for the elements of the density matrix when it is reduced with respect to the indices of emitted photons:

$$\begin{aligned}
\dot{\rho}_{ba} &= -\mathrm{i}\left(\omega_{ba} - \frac{\mathrm{i}}{2T_1}\right)\rho_{ba} - \mathrm{i}\sum_{a'}\Lambda_{ba'}\rho_{a'a} + \mathrm{i}\sum_{b'}\rho_{bb'}\Lambda_{b'a} , \\
\dot{\rho}_{ab} &= \dot{\rho}^*_{ba} , \\
\dot{\rho}_{bb'} &= -\mathrm{i}\left(\omega_{bb'} - \frac{\mathrm{i}}{T_1}\right)\rho_{bb'} - \mathrm{i}\sum_{a}\left(\Lambda_{ba}\rho_{ab'} - \rho_{ba}\Lambda_{ab'}\right) , \\
\dot{\rho}_{aa'} &= -\mathrm{i}\omega_{aa'}\rho_{aa'} + \frac{1}{T_1}\sum_{bb'}\langle a|b\rangle\rho_{bb'}\langle b'|a'\rangle - \mathrm{i}\sum_{b}\left(\Lambda_{ab}\rho_{ba'} - \rho_{ab}\Lambda_{ba'}\right) .
\end{aligned} \tag{7.29}$$

This is the final result. This set of equations describes the temporal evolution of the density matrix of the system consisting of a two-level chromophore, phonons, tunnelons and the laser driving field. Each element of the density matrix is described by an infinite series in accordance with (7.18). The indices a and b characterize phonons and tunnelons in the ground and excited electronic state, respectively. Equation (7.29) comprises an infinite set of equations. Nevertheless, (7.29) allows us to carry out calculations of various spectroscopic characteristics in an analytical form, as will be demonstrated in the following chapters.

7.3 Reducing Equations for the Full Density Matrix to the Optical Bloch Equations

The commonly-used optical Bloch equations can be derived from the infinite set of equations (7.29) if two approximations are made. The first approximation can be written mathematically as

$$\rho_{aa'} = \rho_{bb'} = 0 \quad \text{if} \quad a \neq a' \, , \quad b \neq b' \, . \tag{7.30}$$

This means that we neglect dephasing in the phonon–tunnelon subsystem. The approximation is incorrect if the laser pumping included in the matrix element Λ is large, i.e., if $\Lambda T_2 > 1$. Indeed, there is experimental evidence that the optical Bloch equations work badly if $\Lambda T_2 > 1$ [53]. Therefore we shall consider the case $\Lambda T_2 < 1$.

In this approximation, the equations in (7.29) simplify to

$$\begin{aligned}
\dot{\rho}_{ba} &= -\mathrm{i}\left(\omega_{ba} - \frac{\mathrm{i}}{2T_1}\right)\rho_{ba} - \mathrm{i}\Lambda_{ba}\left(\rho_{aa} - \rho_{bb}\right) \, , \\
\dot{\rho}_{ab} &= \dot{\rho}^{*}_{ba} \, , \\
\dot{\rho}_{bb} &= -\frac{\rho_{bb}}{T_1} - \mathrm{i}\sum_{a}\left(\Lambda_{ba}\rho_{ab} - \rho_{ba}\Lambda_{ab}\right) \, , \\
\dot{\rho}_{aa} &= \frac{1}{T_1}\sum_{b}\langle a|b\rangle\rho_{bb}\langle b|a\rangle - \mathrm{i}\sum_{b}\left(\Lambda_{ab}\rho_{ba} - \rho_{ab}\Lambda_{ba}\right) \, .
\end{aligned} \tag{7.31}$$

Now elements of the reduced density matrix can be introduced as follows:

$$\rho_0 = \sum_{a}\rho_{aa} \, , \quad \rho_1 = \sum_{b}\rho_{bb} \, . \tag{7.32}$$

They determine the probabilities of finding a chromophore in the ground and excited electronic state, respectively. It is obvious that

$$p_a = \frac{\rho_{aa}}{\rho_0} \, , \quad p_b = \frac{\rho_{bb}}{\rho_1} \tag{7.33}$$

are the probabilities of finding the electron–tunnelon subsystem in the a and b quantum states. If the electron–tunnelon subsystem is in thermal equilibrium the probabilities are described by the Boltzmann function. However, the tunnelon subsystem does not reach thermal equilibrium as fast as the phonon subsystem does. Therefore, for the time of an experiment, the probabilities

p_a and p_b will approach their equilibrium values. In this case the probabilities are functions of both temperature and time. Nevertheless, the equations

$$\sum_a p_a = \sum_b p_b = 1 \tag{7.34}$$

have to be fulfilled at arbitrary time and temperature. By summing over the indices a and b in the third and fourth equations of (7.31) and taking into account the fact that $\Lambda_{ba} = \Lambda\langle b \mid a\rangle = -\mathrm{i}\chi\langle b \mid a\rangle$, the following set of equations can be derived:

$$\begin{aligned} \dot{\rho}_{ba} &= -\mathrm{i}\left(\omega_{ba} - \frac{\mathrm{i}}{2T_1}\right)\rho_{ba} - \chi\langle b|a\rangle\left(p_a\rho_0 - p_b\rho_1\right) , \\ \dot{\rho}_{ab} &= \dot{\rho}^*_{ba} , \\ \dot{\rho}_1 &= -\frac{\rho_1}{T_1} - \chi\sum_{ab}\left[\langle b|a\rangle\rho_{ab} + \rho_{ba}\langle a|b\rangle\right] , \\ \dot{\rho}_0 &= -\dot{\rho}_1 . \end{aligned} \tag{7.35}$$

In order to discuss the second approximation, we need an expression for the probability of light absorption. Let us turn back to the full two-photon correlator, which is a function of the laser detuning Δ and time, and the following expression from Part I:

$$p(t) = \frac{\rho_1(t)}{T_1} . \tag{7.36}$$

This equation was derived by neglecting electron–phonon and electron–tunnelon interactions. If these interactions are taken into account, the physical meaning of the function $\rho_1(\Delta, t)$ is not changed. However, this function differs from that of an atom that is not interacting with phonons and tunnelons. Now this function should be a solution of (7.35). In Sect. 3.5 it was shown that the two-photon correlator at infinite time, i.e., the function $\rho_1(\Delta, \infty)$, describes the absorption line shape. Therefore this function found from (7.35) describes the absorption line shape of a chromophore interacting with phonons and tunnelons.

The function $\rho_1(\Delta, \infty)$ relates to a stationary regime when all derivatives in (7.35) are zero. Then the following expression can be found for the off-diagonal element of the density matrix:

$$\rho_{ba} = \frac{\mathrm{i}\chi}{\omega_{ba} - \mathrm{i}/2T_1}\langle b|a\rangle\left(p_a\rho_0 - p_b\rho_1\right) = \rho^*_{ab} . \tag{7.37}$$

Inserting this into the third equation of (7.35), we arrive at the following result for the diagonal elements of the density matrix:

$$0 = -\left(\frac{1}{T_1} + k^{\mathrm{e}}(\Delta)\right)\rho_1(\infty) + k^{\mathrm{g}}(\Delta)\,\rho_0(\infty) , \tag{7.38}$$

where the functions

$$
\begin{aligned}
k^{\mathrm{g}}(\Delta) &= \chi^2 \sum_{a,b} p_a \langle a|b\rangle\langle b|a\rangle \frac{1/T_1}{(\Delta + \Omega_b - \Omega_a)^2 + (1/2T_1)^2}\,, \\
k^{\mathrm{e}}(\Delta) &= \chi^2 \sum_{a,b} p_b \langle a|b\rangle\langle b|a\rangle \frac{1/T_1}{(\Delta + \Omega_b - \Omega_a)^2 + (1/2T_1)^2}
\end{aligned}
\tag{7.39}
$$

have a simple physical meaning. The function $k^{\mathrm{g}}(\Delta)$ determines the probability of absorption of one photon per second if the photon frequency ω_0 differs by an amount $\Delta = \Omega - \omega_0$ from the molecular resonant frequency Ω. The function $k^{\mathrm{e}}(\Delta)$ means the same only for the emission of one photon. Taking into account the fact that $\rho_0 = 1 - \rho_1$, we find the following expression for two-photon correlator:

$$
p\,(\Delta, \infty) = \frac{\rho_1(\infty)}{T_1} = \frac{k^{\mathrm{g}}(\Delta)}{1 + T_1 k^{\mathrm{g}}(\Delta) + T_1 k^{\mathrm{e}}(\Delta)}\,. \tag{7.40}
$$

In linear spectroscopy, the intensity of the laser light is weak and the rate of absorption is therefore much less than the rate $1/T_1$ of spontaneous transitions. Hence, to the first nonvanishing approximation with respect to the Rabi frequency χ, the following simple formula is correct: $p(\Delta, \infty) \approx k^{\mathrm{g}}(\Delta)$. The two-photon correlator coincides with the rate of light absorption by a chromophore interacting with phonons and tunnelons. The function $k^{\mathrm{g}}(\Delta)$ determines the shape of the absorption band if the laser light intensity is rather weak.

The interaction of FC type with phonons and tunnelons is determined by the integrals $\langle a \mid b\rangle = \langle b \mid a\rangle \neq \delta_{ab}$. If we neglect this interaction, then $\langle a \mid b\rangle = \langle b \mid a\rangle = \delta_{ab}$ and (7.39) takes the form

$$
k^{\mathrm{g}}(\Delta) = k^{\mathrm{e}}(\Delta) = \chi^2 \frac{1/T_1}{\Delta^2 + (1/2T_1)^2}\,. \tag{7.41}
$$

The probability of absorption and emission of one photon is described by a Lorentzian. Its full width at half maximum (FWHM) equals $1/T_1$. It will be shown in Sects. 12.3, 12.5 and 18.3 that interaction with phonons and tunnelons yields an additional broadening of the molecular optical line and therefore, instead of (7.41), we should use

$$
k^{\mathrm{g}}(\Delta) = 2\chi^2 \frac{1/T_2}{\Delta^2 + (1/T_2)^2}\,, \tag{7.42}
$$

where $T_2 < 2T_1$. The substitution of a true electron–phonon–tunnelon band by the optical line described in (7.42) is in fact the second approximation in the derivation of the optical Bloch equations. In such an approximation the influence of phonons is only allowed for line broadening. The existence of electron–phonon and electron–tunnelon optical transitions is not taken into account. The new relaxation constant T_2 takes into account the influence of electron–phonon and electron–tunnelon interactions on the equations for the density matrix in the simplest form.

The set (7.35) includes equations for the off-diagonal elements ρ_{ab} and ρ_{ba} of the density matrix. Let us consider the first equation for ρ_{ba}. After Laplace transformation of this equation, we find that

$$-\mathrm{i}\left(\omega - \omega_{ba} + \mathrm{i}/2T_1\right)\rho_{ba}(\omega) = -\chi\langle b|a\rangle\left[p_a\rho_0(\omega) - p_b\rho_1(\omega)\right] . \tag{7.43}$$

Here we have used the initial condition $\rho_{ba}(0) = 0$. Using this equation, we find a new function,

$$\begin{aligned}\rho_{10}(\omega) &= \sum_{a,b}\langle a|b\rangle\rho_{ba} \\ &= -\mathrm{i}\chi\sum_{a,b}\left[p_a\frac{\langle a|b\rangle\langle b|a\rangle}{\omega - \Delta - \Omega_b + \Omega_a + \mathrm{i}/2T_1}\rho_0\right. \\ &\qquad\left. - p_b\frac{\langle a|b\rangle\langle b|a\rangle}{\omega - \Delta - \Omega_b + \Omega_a + \mathrm{i}/2T_1}\rho_1\right] .\end{aligned} \tag{7.44}$$

The substitution of (7.39) by a Lorentzian with enlarged half-width $2/T_2$ is equivalent to replacing

$$\sum_{a,b}\frac{p_a\langle a|b\rangle\langle b|a\rangle}{\omega - \Delta - \Omega_b + \Omega_a + \mathrm{i}/2T_1} = \sum_{a,b}\frac{p_b\langle a|b\rangle\langle b|a\rangle}{\omega - \Delta - \Omega b + \Omega a + \mathrm{i}/2T_1} \tag{7.45}$$

by

$$\frac{1}{\omega - \Delta + \mathrm{i}/T_2}$$

in (7.44). After such a substitution, (7.44) takes the form

$$-\mathrm{i}\left(\omega - \Delta + \mathrm{i}/T_2\right)\rho_{10}(\omega) = -\chi\left[\rho_0(\omega) - \rho_1(\omega)\right] . \tag{7.46}$$

Carrying out the inverse Laplace transformation, we find

$$\dot\rho_{10} = -\mathrm{i}\left(\Delta - \mathrm{i}/T_2\right)\rho_{10} - \chi\left(\rho_0 - \rho_1\right) . \tag{7.47}$$

The equation for ρ_{01} can be obtained from (7.47) by Hermitian conjugation. Thus, instead of infinite sets of equations for ρ_{ba} and ρ_{ab}, we find two equations for off-diagonal elements. Therefore, starting from (7.29) and making the two approximations, we arrive at the optical Bloch equations:

$$\begin{aligned}\dot\rho_{10} &= -\mathrm{i}\left(\Delta - \frac{\mathrm{i}}{T_2}\right)\rho_{10} - \chi\left(\rho_0 - \rho_1\right) , \\ \dot\rho_{01} &= \dot\rho_{10}^* , \\ \dot\rho_1 &= -\frac{\rho_1}{T_1} - \chi\left(\rho_{10} + \rho_{01}\right) , \\ \dot\rho_0 &= \frac{\rho_1}{T_1} + \chi\left(\rho_{10} + \rho_{01}\right) .\end{aligned} \tag{7.48}$$

These four equations are in fact approximate equations for the density matrix when it is reduced with respect to the indices of emitted photons and indices of the phonon–tunnelon subsystem. Similar equations were discovered fifty years ago by F. Bloch for describing the temporal evolution of magnetic moments [41].

The optical Bloch equations include two relaxation constants, $1/T_1$ and $1/T_2$. The constant $1/T_1$ describes the rate of electronic energy decay due to spontaneous decay of the excited electronic state. Therefore T_1 is called the energy relaxation time. The constant $1/T_2$ describes the rate of decay of off-diagonal elements of the density matrix. Since these elements contain information about the phase $\mathrm{i}\Delta t$, the constant T_2 is called the phase relaxation time or optical dephasing time. The optical dephasing time of a single guest molecule is determined by the interaction of its optical electron with phonons and tunnelons in the solvent. Therefore the dephasing time T_2 depends on temperature. Sometimes the optical Bloch equations are unable to describe experimental data [53]. This happens because of their approximate nature.

8. One- and Two-Photon Counting Methods in the Spectroscopy of a Single Impurity Center

8.1 Two-Photon Correlator for a Two-Level Impurity Center

It was shown in the last chapter that relaxation of off-diagonal elements of the density matrix is determined by a rate constant $1/T_2$. This constant results from the interaction with phonons and tunnelons. In Part VI, when we study coherent effects like optical free induction decay or photon echo, we shall see that these off-diagonal elements contain information concerning the phase of the electronic excitation. Therefore T_2 is called the phase relaxation time or dephasing time. The constant T_2 depends on temperature and this dephasing time increases when the temperature decreases. However, the dephasing time T_2 is less than the fluorescence lifetime T_1 by a significant factor, even at $T = 4.2$ K.

It was shown in Sect. 3.5 that the temporal behavior of the two-photon correlator depends strongly on whether the Rabi frequency χ is larger or smaller than the dephasing rate $1/2T_1$. After the 'switching on' of the electron–phonon and electron–tunnelon interactions, the Rabi frequency should be compared to the dephasing rate $1/T_2$, which is larger than $1/2T_1$.

Temporal Behavior of the Two-Photon Correlator $p(t) = \rho_1(t)/T_1$. The probability $\rho_1(t)$ is a solution of the optical Bloch equations which differ from (3.12) solely by the rate constant in the equations for the off-diagonal elements of the density matrix. Therefore we may use the mathematical results obtained in Sect. 3.5, where a solution of (3.12) was found. According to (3.27) and (3.28), the Laplace transform of the two-photon correlator is given by

$$p(\omega) = \frac{2\mathrm{i}\chi^2}{T_1} \frac{\omega + \frac{\mathrm{i}}{T_2}}{\omega \left\{ 4\chi^2 \left(\omega + \frac{\mathrm{i}}{T_2}\right) - \left(\omega + \frac{\mathrm{i}}{T_1}\right) \left[\left(\omega + \frac{\mathrm{i}}{T_2}\right)^2 - \Delta^2 \right] \right\}} . \tag{8.1}$$

Here the substitution $\Gamma \rightarrow 1/T_2$ is made. In the case of strong resonance when $\Delta = 0$, (8.1) simplifies to

$$p(\omega) = \frac{2i\chi^2}{T_1} \frac{1}{\omega \left[4\chi^2 - \left(\omega + \frac{i}{T_1}\right)\left(\omega + \frac{i}{T_2}\right)\right]}$$

$$= \frac{2i\chi^2}{T_1} \frac{1}{\omega(\omega - \omega_1)(\omega - \omega_2)}, \tag{8.2}$$

where

$$\omega_{1,2} = -i\frac{1}{2}\left(\frac{1}{T_1} + \frac{1}{T_2}\right) \pm \sqrt{4\chi^2 - \frac{1}{4}\left(\frac{1}{T_1} - \frac{1}{T_2}\right)^2} = -i\gamma_0 \pm R. \tag{8.3}$$

By transforming a product of the pole terms into a sum of the pole terms, as in Sect. 3.5, we find

$$p(\omega) = \frac{-2i\chi^2}{T_1} \frac{1}{\omega_1\omega_2} \left[\frac{1}{\omega + i0} + \frac{\omega_2}{\omega_1 - \omega_2} \frac{1}{\omega - \omega_1} + \frac{\omega_1}{\omega_2 - \omega_1} \frac{1}{\omega - \omega_2}\right]. \tag{8.4}$$

When the inverse Laplace transformation is carried out with the help of (1.63), every pole term is replaced by an exponential function with an exponent that includes the appropriate pole. Using (8.3) for the poles, we find the following expression for the two-photon correlator:

$$p(t) = \frac{2\chi^2}{T_1} \frac{1}{\gamma_0^2 + R^2} \left[1 - e^{-\gamma_0 t}\left(\cos Rt + \frac{\gamma_0}{R} \sin Rt\right)\right]. \tag{8.5}$$

It can be seen from (8.3) and (8.5) that the two-photon correlator is an oscillating function at large values of the Rabi frequency and a monotonic damping function at small values of the Rabi frequency. The boundary value for the Rabi frequency can be found from the equation $R = 0$. Since

$$1/2T_1 \ll 1/T_2, \tag{8.6}$$

the equation for the bounding intensity of the laser pumping can be rewritten in the form

$$4\chi = 1/T_2. \tag{8.7}$$

Hence, electron–phonon and electron–tunnelon interactions result in the fact that the Rabi frequency should be compared with the dephasing rate $1/T_2$, which is one order of magnitude larger than the dephasing rate $1/2T_1$ in the case when these interactions are neglected. Figure 8.1 shows the temporal dependence of the two-photon correlator at various intensities of the laser pumping. Figures 8.1b, c and d enable us to clarify a very important matter: when may we use the rate equations rather than the optical Bloch equations? Using the optical Bloch equations, we can determine the type of inaccuracies that arise when we use the rate equations

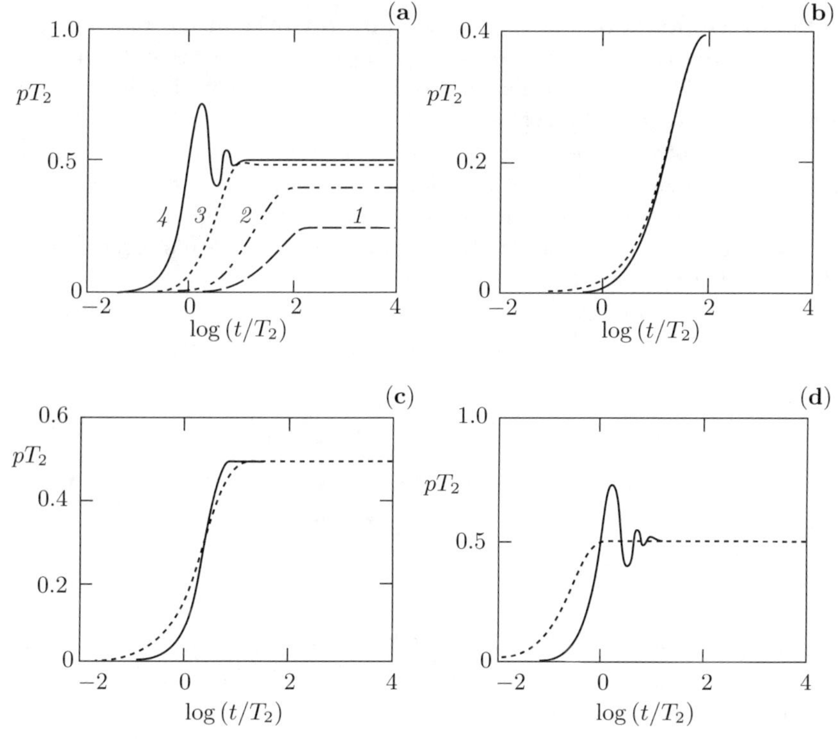

Fig. 8.1. (**a**) Temporal dependence of the full two-photon correlator at various intensities of exciting light: $\chi T_2 = 0.05$ (curve 1), 0.1 (curve 2), 0.3 (curve 3) and 1 (curve 4). (**b**)–(**d**) Comparison of solutions of the Bloch equations (*solid lines*) with solutions of the rate equations at various intensities of exciting light: $\chi T_2 = 0.1$ (**b**), 0.3 (**c**) and 1 (**d**). $T_1/T_2 = 100$

$$\begin{aligned} \dot{\rho}_1 &= -\left[\frac{1}{T_1} + k(\varDelta)\right]\rho_1 + k(\varDelta)\rho_0 \,, \\ \dot{\rho}_0 &= \left[\frac{1}{T_1} + k(\varDelta)\right]\rho_1 - k(\varDelta)\rho_0 \,, \end{aligned} \tag{8.8}$$

where

$$k(\varDelta) = 2\chi^2 \frac{1/T_2}{\varDelta^2 + (1/T_2)^2} \,. \tag{8.9}$$

A solution of these rate equations is given by

$$\frac{\rho_1(t)}{T_1} = \frac{k(\varDelta)}{1 + 2T_1 k(\varDelta)}\left[1 - \mathrm{e}^{-[1/T_1 + 2k(\varDelta)]t}\right] \,. \tag{8.10}$$

The temporal dependence of this approximate expression for the two-photon correlator is depicted by the dashed lines in Figs. 8.1b, c and d. It is clear that the smaller the Rabi frequency, the closer the solid and dashed lines, i.e., the closer the approximate and true formulas for the two-photon correlator.

Frequency Dependence of the Two-Photon Correlator. The function $p(\Delta, t)$ describes the line shape of a chromophore interacting with phonons and tunnelons. The line shape depends on time. This aspect of the line shape problem is absent in the spectroscopy of molecular ensembles and emerges only in single-molecule spectroscopy. The measurement of the line shape in a molecular ensemble yields the function $p(\Delta, \infty)$, which is only a limiting case of the more general expression $p(\Delta, t)$ measured in single-molecule spectroscopy. This shows the impressive potential of single-molecule spectroscopy.

In accordance with (8.10), the correlator is described by $p(\Delta, t)$, or $\rho_1(t)/T_1$, in the short time limit, whereas the function $p(\Delta, \infty)$ which describes the line shape in the long time limit is given by

$$p(\Delta, \infty) = \frac{\rho_1(\infty)}{T_1} = \frac{2\chi^2(1/T_1)}{\Delta^2 + (1/T_2)^2 + 4\chi^2(T_1/T_2)} \,. \tag{8.11}$$

It is clear that the line shape is described by a Lorentzian whose FWHM changes from the value $2/T_2$ in the short time limit to the value described by

$$\Delta_{1/2} = \frac{2}{T_2}\sqrt{1 + 4\chi^2 T_1 T_2} \tag{8.12}$$

in the long time limit. The value of the FWHM in the long time limit increases with growing intensity of the laser driving field. This phenomenon is called light-induced line-broadening.

8.2 Influence of a Triplet Level on the Two-Photon Correlator. Photon Bunching and Antibunching

Every complex organic molecule has a set of triplet electronic levels in addition to a set of singlet electronic levels. The ground electronic state of almost all organic molecules is of singlet type, i.e., its spin equals zero. The energy of the lowest triplet level is as a rule less than the energy of the first excited singlet level. This situation is shown in Figure 8.2. Although direct optical

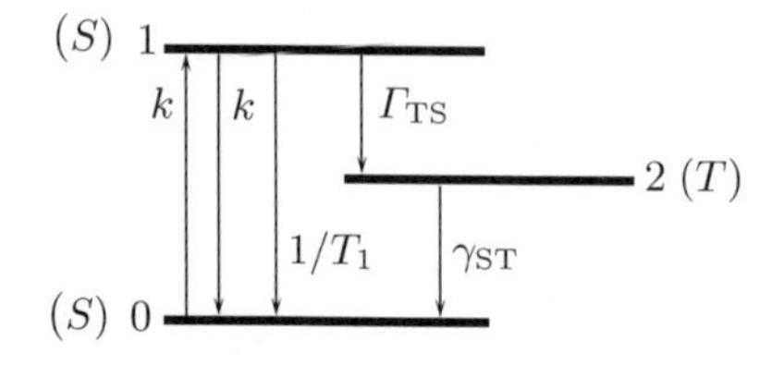

Fig. 8.2. Typical diagram for the electronic levels of a chromophore and relevant rate constants

transitions between levels of singlet and triplet type are forbidden, the triplet level will strongly influence the optical dynamics of singlet levels because of the large value of the nonradiative transition between excited singlet and triplet levels. This nonradiative transition from level 1 to level 2 is called intersystem crossing. For instance, the quantum yield $\Gamma_{\mathrm{TS}}/(1/T_1 + \Gamma_{\mathrm{TS}})$ of intersystem crossing for complex molecules of the porphyrin family is of the order of 0.8–0.9. This means that the rate constant Γ_{TS} describing this crossing is larger than the rate $1/T_1$ of spontaneous emission. In these conditions the majority of molecules that reach the singlet level 1 go into the triplet level. The ground electronic singlet level 0 will therefore be depopulated because the transition $0 \leftarrow 2$ is hampered. Typical values of the rate constants are

$$1/T_1 \approx 10^9\text{–}10^8\,\mathrm{s}^{-1}\,, \quad \Gamma_{\mathrm{TS}} \approx 10^{11}\text{–}10^{10}\,\mathrm{s}^{-1}\,, \quad \gamma_{\mathrm{ST}} \approx 10^6\text{–}10^0\,\mathrm{s}^{-1}\,. \tag{8.13}$$

Upper and lower case characters denote large and small rate constants, respectively. The influence of the triplet level on the optical dynamics of singlet electronic levels is the subject under consideration in this section.

Since the optical transition between the ground singlet level and the triplet level is forbidden, the optical Bloch equations for such a three-level chromophore can be written in the form

$$\begin{aligned}
\dot{\rho}_{10} &= -\mathrm{i}\left(\Delta - \frac{\mathrm{i}}{T_2}\right)\rho_{10} - \chi\left(\rho_0 - \rho_1\right)\,, \\
\dot{\rho}_{01} &= \dot{\rho}_{10}^{*}\,, \\
\dot{\rho}_1 &= -\left(\frac{1}{T_1} + \Gamma_{\mathrm{TS}}\right)\rho_1 - \chi\left(\rho_{10} + \rho_{01}\right)\,, \\
\dot{\rho}_0 &= \frac{\rho_1}{T_1} + \chi\left(\rho_{10} + \rho_{01}\right) + \gamma_{\mathrm{ST}}\rho_2\,, \\
\dot{\rho}_2 &= \Gamma_{\mathrm{TS}}\rho_1 - \gamma_{\mathrm{ST}}\rho_2\,.
\end{aligned} \tag{8.14}$$

The fifth equation takes into account the kinetics of the triplet level, whilst the third and fourth equations allow for possible transitions to and from the triplet level.

The determinant of this system has five roots. Therefore the system can only be solved numerically. However, it was shown in the last section that it is possible to study the electronic dynamics using the rate equations if the intensity of the driving field is weak. Making this assumption, we can set $\mathrm{d}\rho_{10}/\mathrm{d}t = \mathrm{d}\rho_{01}/\mathrm{d}t = 0$. With this condition, we can deduce the following rate equations from the optical Bloch equations described by (8.14):

$$\begin{aligned}
\dot{\rho}_1 &= -\left(\Gamma + k\right)\rho_1 + k\rho_0\,, \\
\dot{\rho}_0 &= \left(\frac{1}{T_1} + k\right)\rho_1 - k\rho_0 + \gamma_{\mathrm{ST}}\rho_2\,, \\
\dot{\rho}_2 &= \Gamma_{\mathrm{TS}}\rho_1 - \gamma_{\mathrm{ST}}\rho_2\,,
\end{aligned} \tag{8.15}$$

where

$$k = 2\chi^2 \frac{1/T_2}{\Delta^2 + (1/T_2)^2}, \quad \Gamma = \frac{1}{T_1} + \Gamma_{\mathrm{TS}} . \tag{8.16}$$

Equation (8.15) can be solved exactly. Carrying out the Laplace transformation of both sides of (8.15) and remembering that

$$\left[\frac{\mathrm{d}\rho(t)}{\mathrm{d}t}\right]_\omega = -\rho(t=0) - \mathrm{i}(\omega + \mathrm{i}0)\rho(\omega) ,$$

we find

$$\begin{aligned}
&(\mathrm{i}\omega - \Gamma - k)\,\rho_1 + k\rho_0 = 0 , \\
&\left(\frac{1}{T_1} + k\right)\rho_1 + (\mathrm{i}\omega - k)\,\rho_0 + \gamma_{\mathrm{ST}}\rho_2 = -1 , \\
&\Gamma_{\mathrm{TS}}\rho_1 + (\mathrm{i}\omega - \gamma_{\mathrm{ST}})\,\rho_2 = 0 .
\end{aligned} \tag{8.17}$$

The determinant of this set of equations is given by

$$\begin{aligned}
\mathrm{Det} &= (\mathrm{i}\omega)\left[(\mathrm{i}\omega)^2 - (\Gamma + 2k + \gamma_{\mathrm{ST}})(\mathrm{i}\omega) + \Gamma_{\mathrm{TS}}k + (\Gamma + 2k)\,\gamma_{\mathrm{ST}}\right] \\
&= -\mathrm{i}\,(\omega + \mathrm{i}0)\,(\omega - \omega_1)\,(\omega - \omega_2) ,
\end{aligned} \tag{8.18}$$

where

$$\begin{aligned}
&\omega_{1,2} = -\mathrm{i}\,(\gamma_0 \mp R) , \quad \gamma_0 = \frac{\Gamma + 2k + \gamma_{\mathrm{ST}}}{2} , \\
&R = \sqrt{\left(\frac{\Gamma + 2k - \gamma_{\mathrm{ST}}}{2}\right)^2 - \Gamma_{\mathrm{TS}}k} .
\end{aligned} \tag{8.19}$$

Solving (8.17), we arrive at the following expression for the Laplace transform of the two-photon correlator:

$$\begin{aligned}
p(\omega) &= \frac{\rho_1(\omega)}{T_1} = \frac{\mathrm{i}k}{T_1} \frac{\mathrm{i}\omega - \gamma_{\mathrm{ST}}}{(\omega + \mathrm{i}0)\,(\omega - \omega_1)\,(\omega - \omega_2)} \\
&= \frac{\mathrm{i}k}{T_1}\left[\frac{-\gamma_{\mathrm{ST}}}{\omega_1\omega_2}\frac{1}{\omega + \mathrm{i}0} + \frac{\mathrm{i}\omega_1 - \gamma_{\mathrm{ST}}}{\omega_1\,(\omega_1 - \omega_2)}\frac{1}{\omega - \omega_1} - \frac{\mathrm{i}\omega_2 - \gamma_{\mathrm{ST}}}{\omega_2\,(\omega_1 - \omega_2)}\frac{1}{\omega - \omega_2}\right] .
\end{aligned} \tag{8.20}$$

After inverse Laplace transformation of this expression, each pole term is replaced by an exponential function with the exponent that includes the appropriate pole. Allowing for $\omega_{1,2} = -\mathrm{i}(\gamma_0 \mp R)$ and using (8.21), the expression for the two-photon correlator becomes

$$\begin{aligned}
p\,(t) = \frac{k}{T_1}\Bigg[&\frac{\gamma_{\mathrm{ST}}}{\gamma_0^2 - R^2} + \left(1 - \frac{\gamma_{\mathrm{ST}}}{\gamma_0 - R}\right)\frac{\mathrm{e}^{-(\gamma_0 - R)t}}{2R} - \\
&- \left(1 - \frac{\gamma_{\mathrm{ST}}}{\gamma_0 + R}\right)\frac{\mathrm{e}^{-(\gamma_0 + R)t}}{2R}\Bigg] .
\end{aligned} \tag{8.21}$$

The driving laser field does not populate the triplet level if $\Gamma_{TS} = 0$. It is easy to verify that the expression in the first brackets of (8.21) equals zero so that this formula transforms to (8.10), as derived in the last section. The correlator reaches a constant value, as shown in Figs. 8.1a and b. The rate of growth of the correlator is determined by the intensity of the laser light.

Temporal Behavior of the Two-Photon Correlator. If $\Gamma_{TS} \neq 0$, the transition to the triplet level is possible and we have to use (8.21). The temporal behavior of the two-photon correlator changes considerably. Let us examine the situation when the following hierarchy of rate constants exists:

$$\Gamma \gg k \gg \gamma_{ST} \,. \tag{8.22}$$

The correlator equals zero at $t = 0$. Since $\gamma_0 + R \sim \Gamma$ and $\gamma_0 - R \sim k$, the temporal behavior of the correlator is easily predicted. It begins to increase in value with time and then decreases, approaching a finite value. Figure 8.3 exhibits this behavior of the two-photon correlator. The growth of the correlator is determined by the fast decrease of the function $\exp[-(\gamma_0 + R)t]$. This growth is almost independent of the laser pumping k. The decrease of the correlator is determined by the slow decay of the function $\exp[-(\gamma_0 - R)t]$. The decrease becomes slower when k decreases. This can be seen in Figure 8.3. With the logarithmic scale used on the abscissa, any exponential decay looks like a smooth step extended over one order of magnitude on the time scale. Figure 8.3 shows that the rate of correlator decay increases when the laser pumping k increases. Because the rate of depopulation of the excited singlet level 1 is proportional to k, the depopulation of level 1 becomes faster when k increases. The decreased correlator for long time delays t shows the decreased photon pair count rate for large delays. The curves in Fig. 8.3 are calculated for the resonant case when $\Delta = 0$. In the off-resonant case the situation is of the same type.

The time of emission of the first photon is taken as 0. The probability of a single molecule emitting a second photon at time t approaches zero when $t \to 0$. Therefore the correlator approaches zero in the short time limit. This is a general feature of all the two-photon correlators describing single-molecule dynamics. This phenomenon is called photon antibunching. Antibunching describes a strong correlation existing between successively emitted photons. Figure 8.1 shows that the photon pair count rate reaches a maximum at large delays between photons and that the count rate becomes independent of the value of the time delay. This means that the average distribution of photons in a photon train is homogeneous.

The triplet level dramatically changes the number of photon pairs emitted by a single molecule with large delay between the photons. The changes are obvious if we compare the long time behavior of the correlators shown in Figs. 8.1 and 8.3. Indeed, Fig. 8.3 shows that the probability of finding a photon pair with a large time delay is smaller then that with a short delay. This means that the distribution of time intervals between photons in a photon

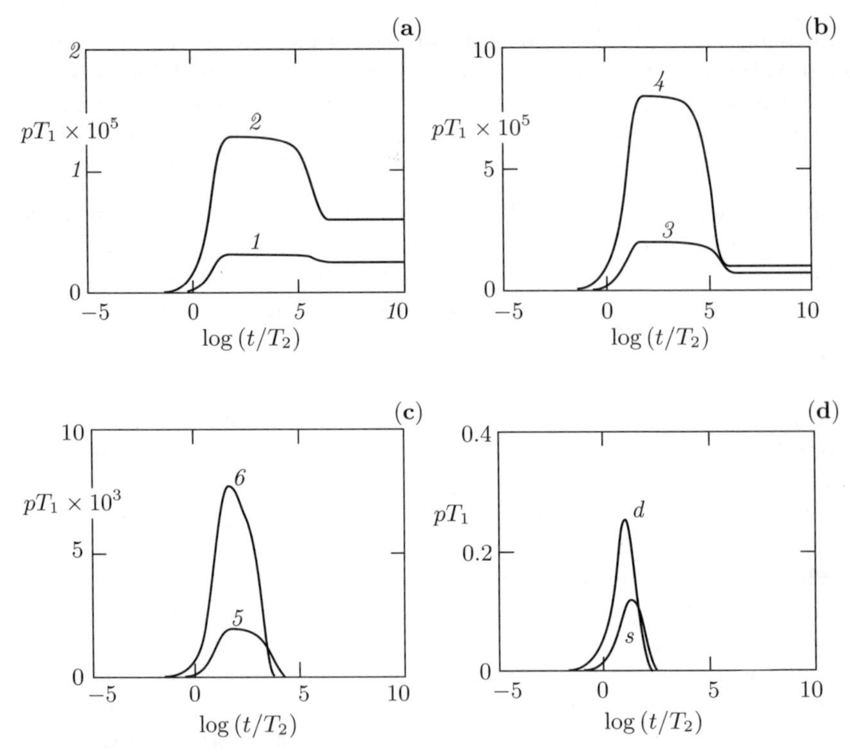

Fig. 8.3. (a–d) Temporal dependence of the full two-photon correlator for the chromophore with a triplet level at various intensities of the exciting light, calculated using (8.21) with $T_2/T_1 = 10^{-2}$, $\gamma_{\mathrm{ST}}T_2 = 10^{-6}$, $\Gamma_{\mathrm{TS}}T_1 = 9$ and $\chi T_2 = 10^{-4}$ (curve 1), $\chi T_2 = 2 \times 10^{-4}$ (curve 2), $\chi T_2 = 10^{-3}$ (curve 3), $\chi T_2 = 2 \times 10^{-3}$ (curve 4), $\chi T_2 = 10^{-2}$ (curve 5), $\chi T_2 = 2 \times 10^{-2}$ (curve 6), $\chi T_2 = 10^{-1}$ (curve 7) and $\chi T_2 = 2 \times 10^{-1}$ (curve 8)

train emitted by a single molecule with a triplet level is inhomogeneous. In this case the whole train of photons consists of bunches, as shown in Fig. 8.4. This phenomenon is called photon bunching. Indeed, if the average extent of the bunch on the time scale equals t_0 and the average time interval between bunches equals $t_1 \gg t_0$, the count rate of photon pairs with time delay t_1 is less than that with $t < t_0$.

The appearance of photon bunching can also be explained as follows. A single molecule irradiated by CW laser light jumps many times at random instants of time between the ground and excited singlet electronic states.

$h\nu$

Fig. 8.4. Photon bunching in light emission due to a long-lived triplet level in a single molecule irradiated by CW laser light

Hence, for a time interval of order of t_0, the molecule emits photons. Suddenly the molecule goes into the triplet state with a long lifetime. Being in the triplet state, the molecule cannot emit photons. Therefore, for a long time t_1 comparable with the lifetime of the triplet level, the molecule does not radiate light. After going back into the singlet ground state, the molecule starts to jump with photon emission again. This photon bunching has been observed experimentally in single-molecule spectroscopy on many occasions.

Frequency Dependence of the Two-Photon Correlator. Equation (8.21) also enables us to examine the two-photon correlator as a function of the detuning Δ at a fixed time delay t. In the short time limit we may expand both exponential functions in (8.21) in a series. We then arrive at the expression

$$p(\Delta, t) \approx \frac{t}{T_1} k = \frac{t}{T_1} \frac{2\chi^2 (1/T_2)}{\Delta^2 + (1/T_2)^2} , \tag{8.23}$$

i.e., the two-photon correlator is proportional to the probability of light absorption. The probability is described by the Lorentzian with FWHM

$$\Delta_{1/2} = \frac{2}{T_2} . \tag{8.24}$$

Let us now inspect the frequency dependence of the correlator in the long time limit when the correlator no longer depends on time. The correlator in the long time limit is given by

$$p(\Delta, \infty) = \frac{k}{T_1} \frac{\gamma_{\mathrm{ST}}}{\gamma_0^2 - R^2} . \tag{8.25}$$

Inserting k, γ_0 and R given by (8.16) and (8.19), we find that

$$p(\Delta, \infty) = \frac{1}{T_1 \Gamma} \frac{2\chi^2 (1/T_2)}{\Delta^2 + (1/T_2)^2 \left(1 + 2\chi^2 \eta T_2 \tau_{\mathrm{T}}\right)} , \tag{8.26}$$

where

$$\eta = \frac{\gamma_{\mathrm{ST}} + \Gamma_{\mathrm{TS}}}{\Gamma} \approx \frac{\Gamma_{\mathrm{TS}}}{\Gamma} , \quad \tau_{\mathrm{T}} = \frac{1}{\gamma_{\mathrm{ST}}} \tag{8.27}$$

are the intersystem crossing quantum yield and the lifetime of the triplet state, respectively. In accordance with (8.26), the two-photon correlator is described by the Lorentzian with FWHM given by

$$\Delta_{1/2} = \frac{2}{T_2} \sqrt{1 + 2\chi^2 \eta T_2 \tau_{\mathrm{T}}} . \tag{8.28}$$

This half-width depends on the intensity of the laser light. Comparison of (8.28) with (8.12) derived for a two-level chromophore enables us to conclude that the existence of a triplet level in a molecule yields the changes described by the following substitutions:

$$T_1 \to \tau_{\mathrm{T}} , \qquad \chi^2 \to \chi^2 \frac{\eta}{2} . \tag{8.29}$$

The substitution of T_1 by a triplet lifetime τ_{T} that is a few orders of magnitude longer facilitates light-induced line-broadening. The FWHM begins to depend on the laser intensity at weaker laser light levels. By summing, we can write the following inequalities involving the Rabi frequency:

$$4\chi T_2 \geqslant 1 , \qquad 2\chi\sqrt{T_2 T_1} \geqslant 1 , \qquad \chi\sqrt{2\eta T_2 \tau_{\mathrm{T}}} \geqslant 1 . \tag{8.30}$$

If the first inequality is fulfilled, we are faced with a case of strong laser pumping when substitution of the optical Bloch equations by the rate equations is incorrect. Then the second and third inequalities are fulfilled as well. The second and third inequalities determine values of the Rabi frequency at which light-induced line-broadening can occur in two- and three-level chromophores. However, our considerations in this book are based on the opposite inequality

$$4\chi T_2 < 1 . \tag{8.31}$$

Since $T_2 \ll T_1$, there is a range for the variable χ in which the inequality (8.31) and the second inequality in (8.30) can both be fulfilled. This means that light-induced line-broadening can be studied with the help of the rate equations even for the case of a two-level chromophore. Moreover, this statement is correct for a molecule with a triplet level because in this case T_2 exceeds τ_{T} by a few orders of magnitude. There is no problem in simultaneously satisfying the inequality (8.31) and the third inequality in (8.30).

Equation (8.21) does not only enable us to examine the FWHM in the short and long time limits. It also provides a way of studying the temporal behavior of the optical line shape on a time scale extending over a few orders

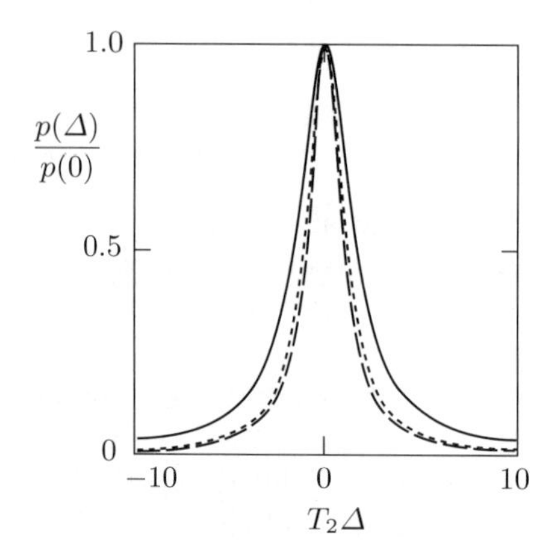

Fig. 8.5. Dependence of full two-photon correlator on frequency detuning at $\chi T_2 = 10^{-4}$ and at $t/T_2 = 10^{-5}$ (*dashed line*), 10^5 (*dotted line*) and 10^7 (*solid line*)

of magnitude. The result of numerical calculations with (8.21) for the ratio $\tau_T/T_2 = 10^5$ is shown in Fig. 8.5. This figure shows that, at $t/T_2 < 10^5$, the FWHM is determined by $2/T_2$ and there is no light-induced line-broadening. The latter occurs at times when the two-photon correlator decreases, approaching its long time limit $p(\Delta, \infty)$.

8.3 One-Photon Counting Method. Quantum Trajectories

A single molecule irradiated by CW laser light emits photons. The method whereby they are registered can be of one-photon type, when we count all emitted photons, or of two-photon type, when we count pairs of photons with a definite delay between the photons in the pair. Each method has advantages and disadvantages relative to the other. In order to compare the possibilities of the one- and two-photon methods, let us first discuss what can be done with the two-photon correlator.

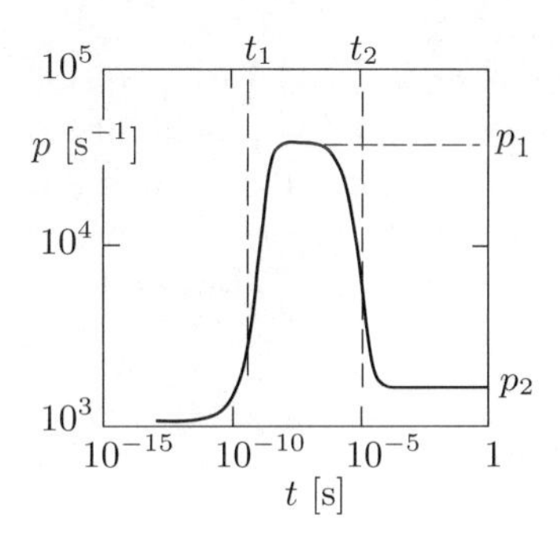

Fig. 8.6. Parameters measured with the help of the two-photon correlator

By counting photon pairs with various fixed delays over a long time, we obtain a function $p(t)$ like the one described by curve 4 in Fig. 8.3. This curve is shown in Fig. 8.6. Various details of the curve have a particular physical meaning. The approach of the correlator to zero in the short time limit is called antibunching. This means that emission of photons by single atoms is a quantum process. Two values of the coordinate p_1 and p_2 characterize the count rate of photon pairs in a bunch and the count rate of bunches, respectively. Two values of the abscissa t_1 and t_2 enable us to determine a rate $R_1 = 1/t_1$ of population of the first singlet level due to excitation by the laser field and a rate $R_2 = 1/t_2$ of transition from the excited singlet level to the triplet level of the molecule.

However, some physical characteristics cannot be found with the help of the two-photon correlator. In accordance with Fig. 8.4, time intervals when a single molecule emits light (on-intervals) are separated by dark intervals (off-intervals) when the molecule is in the triplet state. A two-photon correlator fails to find the distribution of on- and off-intervals and average values τ_{on} and τ_{off} of these intervals. However, these physical characteristics can be found if the one-photon counting method is used.

Indeed, let us count all photons emitted by a single molecule for a time interval Δt. If this time interval and the times shown in Fig. 8.6 satisfy the inequality $t_1 \ll \Delta t \ll t_2$, the fluorescence signal will appear as shown in Fig. 8.7. It displays some kind of quantum trajectory of a random type. In the case under consideration, the quantum trajectory describes fluctuations in the fluorescence intensity. However, the quantum trajectory can describe other physical characteristics of a single molecule. For instance, in the case of a single impurity center of a polymer, the resonant molecular frequency will fluctuate. The temporal dependence of the resonant frequency will display another type of quantum trajectory – a spectral trajectory. The spectral trajectory cannot be observed in molecular ensembles due to the average over spectral trajectories of several molecules. Therefore, a spectral trajectory measured in an experiment is a telling feature of single-molecule spectroscopy. A common property of all quantum trajectories is their random character. Quantum trajectories reflect the quantum nature of a molecule and so are not reproducible from one experiment to another.

Using the methods of quantum mechanics, we cannot calculate a quantum trajectory. Quantum mechanics describes how to calculate the probability of measuring some physical value in a certain quantum state. It is impossible to measure these probabilities immediately in one-photon experiments where random characteristics like quantum trajectories are measured. However, we

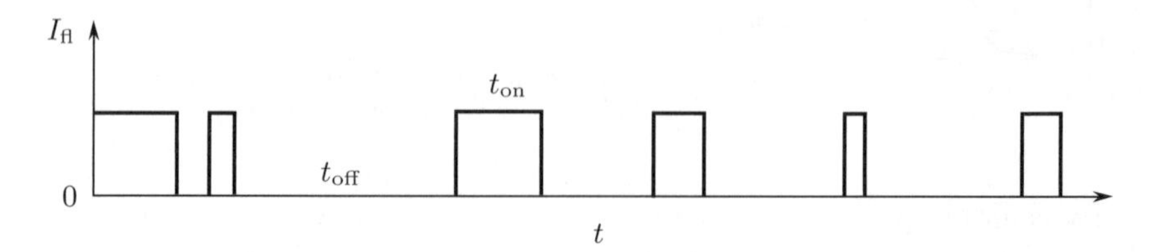

Fig. 8.7. Quantum trajectory for the fluorescence intensity

can find these probabilities by carrying out a statistical treatment of quantum trajectories measured in one-photon experiments.

It is obvious that by measuring a quantum trajectory over a long time, we can find the distribution of light and dark intervals. Then, with the help of the simple formula

$$\tau = \frac{1}{N} \sum_{i}^{N} t_i \,, \tag{8.32}$$

we can find average values for the on- and off-intervals. The question arises as to how we can calculate these average values of on- and off-intervals using quantum mechanics.

It is clear that, when a molecule emits, it is in a singlet state. During off-intervals, the molecule is in the triplet state. The dynamics of singlet states irradiated by laser light is described by the following equations:

$$\begin{aligned}\dot{\rho}_1 &= -\left(\frac{1}{T_1} + \Gamma_{\mathrm{TS}} + k\right)\rho_1 + k\rho_0 \,, \\ \dot{\rho}_0 &= \left(\frac{1}{T_1} + k\right)\rho_1 - k\rho_0 \,,\end{aligned} \tag{8.33}$$

where the physical meaning of all the constants is obvious from Fig. 8.2. The function

$$\rho_{\mathrm{on}} = \rho_1 + \rho_0 \tag{8.34}$$

is the probability of finding the molecule in a singlet state, i.e., in the on-state. Summation of both equations yields the result

$$\dot{\rho}_{\mathrm{on}} = -\Gamma_{\mathrm{TS}}\rho_1 \,. \tag{8.35}$$

In the long time limit when $t > 1/\Gamma_{\mathrm{TS}}$, we may set $\mathrm{d}\rho_1/\mathrm{d}t = 0$ in (8.33). In this quasi-stationary case the following relation can be found from (8.33):

$$\rho_1 = \frac{kT_1}{1 + kT_1 + \Gamma_{\mathrm{TS}}T_1}\rho_0 \,. \tag{8.36}$$

Inserting this relation into (8.34), we express the probability ρ_1 via the probability ρ_{on} of finding the molecule in the on-state:

$$\rho_1 = \frac{kT_1}{1 + (2k + \Gamma_{\mathrm{TS}})\,T_1}\rho_{\mathrm{on}} \approx k\frac{T_1}{1 + \Gamma_{\mathrm{TS}}T_1}\rho_{\mathrm{on}} \,. \tag{8.37}$$

Using this equation and (8.35), we can easily find the following equation, which describes the dynamics of the on-state:

$$\dot{\rho}_{\mathrm{on}} = -\frac{\rho_{\mathrm{on}}}{\tau_{\mathrm{on}}} \,, \tag{8.38}$$

where

$$\frac{1}{\tau_{\mathrm{on}}} = k\frac{\Gamma_{\mathrm{TS}}T_1}{1 + \Gamma_{\mathrm{TS}}T_1} \,. \tag{8.39}$$

Here the ratio determines the quantum yield of intersystem crossing. A solution of (8.38) is given by

$$w_{\mathrm{on}} = \frac{1}{\tau_{\mathrm{on}}}\exp\left(-\frac{t}{\tau_{\mathrm{on}}}\right) \,. \tag{8.40}$$

This function describes the density of the probability of measuring a definite value of $t_{\mathrm{on}} = t$. This is a function that describes a distribution of random on-intervals. It is obvious that τ_{on} determined by (8.40) is an average value of the random value t_{on}.

When the molecule goes into the triplet state, it stops emitting photons. This means that the molecule is in an off-state. Hence, $\rho_2 = \rho_{\mathrm{off}}$. The dynamics of the off-state is described by the equation

$$\dot{\rho}_2 = -\gamma_{\mathrm{ST}}\rho_2 , \tag{8.41}$$

from which we deduce that

$$\frac{1}{\tau_{\mathrm{off}}} = \gamma_{\mathrm{ST}} . \tag{8.42}$$

A solution of (8.41) is given by

$$w_{\mathrm{off}} = \gamma_{\mathrm{ST}} \exp\left(-\gamma_{\mathrm{ST}} t\right) . \tag{8.43}$$

This function describes the distribution of off-intervals. It is easy to verify that the small root of the two-photon correlator that describes the transition from the excited singlet state to the triplet state is the sum of the reciprocal average on- and off-intervals, i.e.,

$$\gamma_0 - R \approx \frac{1}{\tau_{\mathrm{on}}} + \frac{1}{\tau_{\mathrm{off}}} . \tag{8.44}$$

Part IV

Optical Band Shape Theory for Impurity Centers

9. Stochastic Theories of Line Broadening

Optical bands of impurity centers can consist of well-resolved optical lines or they can look like broad bands without any structure. All details of the optical bands are due to the interaction of the electronic degrees of freedom of a guest molecule with phonons and tunnelons. It was shown in the last two chapters that a system consisting of a two-level chromophore, phonons and tunnelons has an infinite number of degrees of freedom and therefore that the elements of the density matrix of the system can be found from an infinite set of equations. In the special case when all influence from phonons and tunnelons can be reduced to their influence only on line broadening, we can reduce the infinite set of equations for the density matrix to four equations called the optical Bloch equations.

Also in Part III, we derived (7.39) for the functions $k^{\mathrm{g}}(\Delta)$ and $k^{\mathrm{e}}(\Delta)$. These functions describe probabilities of absorbing or emitting one photon and they take into account the influence of phonons and tunnelons on these probabilities. This influence is included in the integrals $\langle a \mid b \rangle$, which depend on the overlap of phonon–tunnelon functions in different electronic states. It was also shown that the substitution of the functions $k^{\mathrm{g}}(\Delta)$ and $k^{\mathrm{e}}(\Delta)$ by a Lorentzian with half-width $2/T_2$ enables us to deduce the optical Bloch equations.

However, the functions $k^{\mathrm{g}}(\Delta)$ and $k^{\mathrm{e}}(\Delta)$ are rather complicated in realistic systems and cannot therefore be described by a single Lorentzian. In the next four chapters, the calculation of these functions will be carried out using stochastic and dynamic approaches.

9.1 Dynamic and Stochastic Approaches to the Line-Broadening Problem

Optical line broadening due to the interaction with phonons can be considered within the framework of either a dynamic or a stochastic approach.

Dynamic Approach. This approach is based on the idea that the temporal evolution of any physical system is completely determined by its initial state and its Hamiltonian. All physical characteristics of the system, including the optical band shape can be calculated using the methods of quantum statistics.

These methods enable one to calculate the probability of any event. It has already been shown that the probabilities of emission and absorption of one photon manifest themselves in the two-photon correlator. The latter can be measured experimentally as the count rate of photon pairs.

Stochastic Approach. This approach supposes that the probabilities of quantum jumps and quantum states are not calculated, but rather that they are introduced a priori. The success of the approach depends strongly on the extent to which the probabilities introduced take into account peculiarities of the interactions existing in the system [54]. The probabilities introduced 'by hand' are simpler than the probabilities calculated quantum mechanically using the Hamiltonian of the system. Therefore it is impossible within the framework of the stochastic approach to describe the physical system in as much detail as the dynamic theory does. For instance, a realistic optical band consists of the zero-phonon line and phonon side band. This important physical fact has not yet been described within the framework of the stochastic theory. In contrast, the dynamical theory is able to successfully explain this fact. However, calculations within the framework of a stochastic approach are simpler and this makes it a very attractive option. Various modifications of the stochastic approach are used in practice. The optical line-broadening problem is one of the problems that can be solved by both the stochastic and dynamic methods. We therefore begin by considering the stochastic approach.

9.2 Stochastic Theory Due to Anderson and Weiss

Let the temporal behavior of an electronic dipolar moment be described by the following equation:

$$\ddot{d} + \Omega^2 d = 0 . \tag{9.1}$$

A solution of the equation is given by

$$d\,(t) = \mathrm{e}^{-\mathrm{i}\Omega t} d(0) . \tag{9.2}$$

The oscillating dipolar moment is a source of electromagnetic radiation. The electromagnetic field oscillates at the same frequency Ω, i.e., the spectral density of the radiation is given by the simple formula

$$I\,(\omega) \propto \delta\,(\omega - \Omega) = \mathrm{Re} \int_0^\infty \mathrm{e}^{\mathrm{i}(\omega-\Omega)t} \frac{dt}{\pi} . \tag{9.3}$$

Hence, harmonic oscillation of the dipolar moment results in a delta-like emission line. The line is a Fourier transform of the function that describes oscillations of the dipolar moment. Up to now we have neglected the influence of phonons on the optical line.

If phonons are taken into account they influence the frequency Ω of the oscillations of the dipolar moment. The electronic frequency will be a random function of time,

$$\Omega(t) = \Omega + \Delta(t) \, . \tag{9.4}$$

For example fluctuations in the frequency can look like those shown in Fig. 9.1. If this fact is allowed for in (9.1), the solution of this equation is given by

$$d(t) = d(0) \exp\left[-\mathrm{i}\Omega t - \mathrm{i} \int_0^t \Delta(x) \mathrm{d}x \right] \, , \tag{9.5}$$

where $\Delta(t)$ is a random function of time. The dipolar moment can oscillate with various frequencies. A spectral trajectory like the one shown in Fig. 9.1,

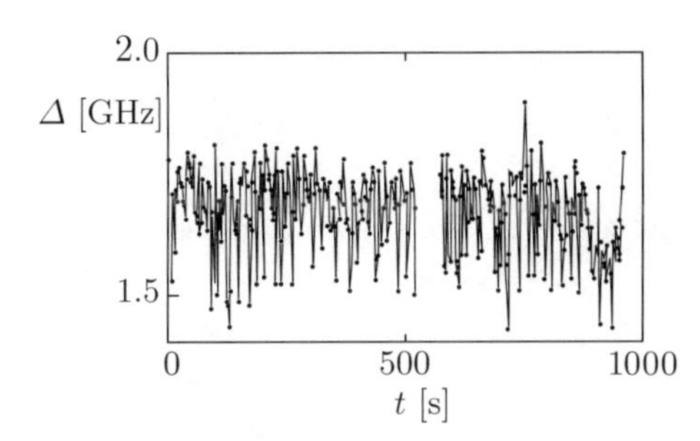

Fig. 9.1. Fluctuation of photon frequency absorbed by a terylene molecule. $\lambda = 574.60$ nm, $T = 1.5$ K [28]

being a random function of time, includes information about probabilities of finding oscillations with a definite value of the frequency. In stochastic theories such a probability is postulated. For instance we may postulate the probability of finding a definite value for the phase of the oscillating dipolar moment. Then the spectral density of the radiation is given by

$$I(\omega) = \mathrm{Re} \int_0^\infty \frac{\mathrm{d}t}{\pi} \, \mathrm{e}^{-\mathrm{i}(\omega - \Omega)t} \left\langle \exp\left[-\mathrm{i} \int_0^t \Delta(x) \, \mathrm{d}x \right] \right\rangle \, , \tag{9.6}$$

where

$$\left\langle \mathrm{e}^{-\mathrm{i}\varphi} \right\rangle = \int_{-\infty}^{\infty} P(\varphi) \, \mathrm{e}^{-\mathrm{i}\varphi} \mathrm{d}\varphi \, . \tag{9.7}$$

Here $P(\varphi)$ is the probability of finding a definite value of the phase φ. By choosing a function for the probability, we choose a model to describe the system under consideration.

Anderson and Weiss [55] chose a probability of Gaussian type, viz.,

$$P(\varphi) = \frac{1}{\sqrt{2\pi \langle \varphi^2 \rangle}} \exp - \frac{\varphi^2}{2 \langle \varphi^2 \rangle} \,. \tag{9.8}$$

By inserting this formula into (9.7) and carrying out the integration, they found the following equation for the average value of the dipolar moment:

$$\left\langle \mathrm{e}^{-\mathrm{i}\varphi} \right\rangle = \mathrm{e}^{\langle \varphi^2 \rangle / 2} = \exp \left[- \int_0^t \mathrm{d}x \int_0^t \mathrm{d}y \, \langle \Delta(x) \Delta(y) \rangle \right] . \tag{9.9}$$

Taking into account the fact that the correlation function of the frequency displacements is a function of the time difference, i.e.,

$$\langle \Delta(x) \, \Delta(y) \rangle = \Delta^2 C(x - y) \,, \tag{9.10}$$

where $C(0) = 1$, we can transform (9.9) to the form

$$\langle \mathrm{e}^{-\mathrm{i}\varphi} \rangle = \exp \left[-\Delta^2 \int_0^t \mathrm{d}z (t - z) C(z) \right] . \tag{9.11}$$

The spectral line shape is expressed in terms of the correlation function of the frequency displacements. Anderson and Weiss chose this function in the Gaussian form,

$$C(t) = \mathrm{e}^{-(t/\tau)^2} \,, \tag{9.12}$$

where $1/\tau$ determines the rate of decay of the correlation function. Two limiting cases are possible.

- Fast Frequency Jumping. In this case $\tau\Delta \ll 1$ and therefore we may set $C(z) = C(0) = 1$ in (9.11). The average value of the dipolar moment is given by

$$\left\langle \mathrm{e}^{-\mathrm{i}\varphi} \right\rangle = \mathrm{e}^{-\Delta^2 t^2 / 2} \,. \tag{9.13}$$

  By inserting this equation into (9.6), we arrive at the spectral line of Gaussian type.
- Slow Frequency Jumping. In this case $\tau\Delta \gg 1$ and therefore the average dipolar moment is given by

$$\left\langle \mathrm{e}^{-\mathrm{i}\varphi} \right\rangle \approx \exp \left(-\Delta^2 t \int_0^\infty \mathrm{d}z \exp \left[- \left(\frac{z}{\tau} \right)^2 \right] \right) = \exp \left(-\frac{\sqrt{\pi}}{2} \tau \Delta^2 t \right) . \tag{9.14}$$

  By inserting this equation into (9.6), we arrive at the spectral line of Lorentzian type with half-width $\sqrt{\pi}\tau\Delta$.

Thus, Anderson and Weiss found that the line shape depends on the rate of decay of the correlation function for frequency displacements. Fast frequency jumping results in a Gaussian line shape, whilst slow frequency jumping yields a Lorentzian line shape. These formulas for line shape were derived for magnetic resonance.

9.3 Anderson Theory for Optical Lines

The theory discussed in the last section explains broadening of a single optical line. However, in single-molecule spectroscopy, the optical band may consist of several optical lines as shown in Fig. 9.2. The theory must explain not only line broadening, but also the appearance of several lines. The Anderson–Weiss theory cannot explain the spectral picture shown in Fig. 9.2. Anderson therefore developed a more comprehensive approach to the optical band shape problem [57]. This approach serves as a basis for many theories for the optical band shape developed later. It therefore deserves a detailed discussion.

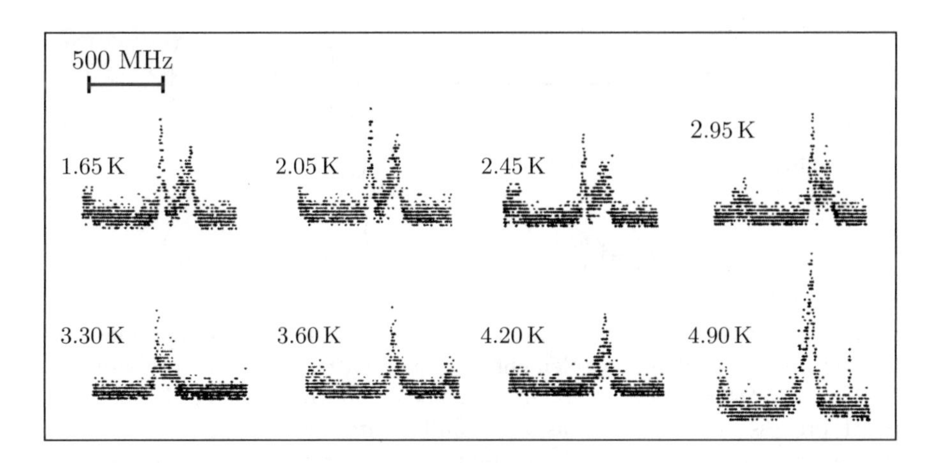

Fig. 9.2. Two lines of the excitation fluorescence band of a pentacene molecule doping a para-terphenyl crystal [56]

The Anderson theory investigates the temporal evolution of the frequency of an electronic excitation. The theory is based on the assumption that fluctuations $\Delta(t)$ in the resonant frequency are of Markovian type. The temporal evolution is of Markovian type if the probability of each infinitesimal step includes no information about preceding steps. This assumption is expressed by the fact that the probability of N steps is a product of the probabilities of the infinitesimal steps, i.e.,

$$P(\Delta_N, \ldots, \Delta_3, \Delta_2, \Delta_1, t) = \tag{9.15}$$
$$P\left(\Delta_N, \Delta_{N-1}, \frac{t}{N}\right) \ldots P\left(\Delta_3, \Delta_2, \frac{t}{N}\right) P\left(\Delta_2, \Delta_1, \frac{t}{N}\right) P(\Delta_1) .$$

Time increases from right to left in (9.15). Here Δ_j is the change in the resonant frequency Ω in the jth time interval t/N, and $P(\Delta_1)$ is the probability of finding an initial state. The function $P(\Delta_{j+1}, \Delta_j, t/N)$ determines the probability of the frequency changing by the value $\Delta_{j+1} - \Delta_j$ in the time interval t/N. This method is certainly closer to the dynamical approach than the stochastic approach discussed in the last section. In any dynamical approach, the probability of temporal evolution is calculated using the Hamiltonian of the system, whereas the probability $P(\Delta_{j+1}, \Delta_j, t/N)$ is chosen in accordance with a plausible physical model.

The Fourier transform of the average in (9.6) describes the optical band shape. The average is given by

$$I(t) = \left\langle \exp\left[-\mathrm{i}\int_0^t \Delta(x)\mathrm{d}x\right]\right\rangle \tag{9.16}$$
$$= \lim_{N\to\infty} \sum_{\Delta_1,\Delta_2,\ldots\Delta_N} P(\Delta_1, \Delta_2, \ldots, \Delta_N, t) \exp\left(-\mathrm{i}\sum_{j=1}^{N} \Delta_j \frac{t}{N}\right)$$
$$= \lim_{N\to\infty} \sum_{\Delta_1,\Delta_2,\ldots,\Delta_N} \ldots \exp\left(-\mathrm{i}\frac{t}{N}\Delta_3\right) P\left(\Delta_3, \Delta_2, \frac{t}{N}\right)$$
$$\times \exp\left(-\mathrm{i}\frac{t}{N}\Delta_2\right) P\left(\Delta_2, \Delta_1, \frac{t}{N}\right) \exp\left(-\mathrm{i}\frac{t}{N}\Delta_1\right) P(\Delta_1) .$$

The sum describes a product of matrices. The exponential functions can be considered as diagonal elements of one matrix and the probabilities $P(\Delta_{j+1}, \Delta_j, t/N)$ as off-diagonal elements of another matrix. If the temporal process is considered as an infinite number of infinitesimal steps, the matrices discussed include an infinite number of elements. By setting $\exp(-\mathrm{i}\Delta_N t/N) = 1$, we can transform (9.16) to the form

$$I(t) = \lim_{N\to\infty} \mathbf{1}\left[\exp\left(-\mathrm{i}\frac{t}{N}\hat{\Delta}\right) \hat{P}\left(\frac{t}{N}\right)\right]^{N-1} \boldsymbol{P} , \tag{9.17}$$

where $\boldsymbol{P}$ is the column vector whose components are the probabilities $P(\Delta_j) = \rho_j$ of finding the frequency $\Omega_j = \Omega + \Delta_j$ in the initial state, and $\mathbf{1}$ is the row vector whose components are 1.

Since $t/N \to 0$ the matrix $P(t/N)$ can be written in the form

$$\hat{P}\left(\frac{t}{N}\right) = 1 + \hat{p}\frac{t}{N} . \tag{9.18}$$

If the matrix p equals zero, the frequency of the electronic excitation cannot change with time. Hence the change in the resonant frequency is due to the matrix p. The elements of this matrix are the probabilities of electronic frequency change per second. Due to the infinitesimal character of the value t/N, we may transform (9.17) to

$$I(t) = \lim_{N\to\infty} \mathbf{1} \left[1 + (\hat{p} - \mathrm{i}\hat{\Delta})\frac{t}{N}\right]^{N-1} \boldsymbol{P} = \mathbf{1}\mathrm{e}^{(\hat{p}-\mathrm{i}\hat{\Delta})t}\boldsymbol{P} = \mathbf{1}\boldsymbol{P}(t) \;, \qquad (9.19)$$

where the jth component of the vector given by

$$\boldsymbol{P}(t) = \mathrm{e}^{(\hat{p}-\mathrm{i}\hat{\Delta})t}\boldsymbol{P} \qquad (9.20)$$

determines the probability of finding the system in the jth state at time t. Inserting (9.19) into (9.6), we arrive at the following expression for the optical band shape

$$\begin{aligned} I(\omega) &= \frac{1}{\pi}\,\mathrm{Re}\int_0^\infty \mathrm{d}t\,\mathrm{e}^{\mathrm{i}(\omega-\Omega)t}\mathbf{1}\mathrm{e}^{(\hat{p}-\mathrm{i}\hat{\Delta})t}\boldsymbol{P} \\ &= \frac{1}{\pi}\,\mathrm{Re}\int_0^\infty \mathrm{d}t\,\mathrm{e}^{\mathrm{i}(\omega-\Omega)t}I_{\mathrm{A}}(t) \\ &= -\frac{1}{\pi}\,\mathrm{Im}\,\mathbf{1}\frac{1}{\omega-\Omega-\hat{\Delta}-\mathrm{i}\hat{p}}\boldsymbol{P} \;. \end{aligned} \qquad (9.21)$$

9.4 Exchange Model for Line Broadening

Let us apply the Anderson formula to the simplest case, when the resonant frequency can jump between two values $\omega_1 = \Omega$ and $\omega_2 = \Omega + \Delta$. Then the matrix in (9.21) is the matrix with two rows and columns:

$$\omega - \hat{A} = \omega - \Omega - \hat{\Delta} - \mathrm{i}\hat{p} = \begin{pmatrix} a & b \\ c & d \end{pmatrix} \;. \qquad (9.22)$$

By calculating the inverse matrix, we arrive at the expression

$$\begin{aligned} \mathbf{1}\frac{1}{\omega-\hat{A}}\boldsymbol{P} &= (1\;1)\,\frac{1}{ad-bc}\begin{pmatrix} d & -b \\ -c & a \end{pmatrix}\begin{pmatrix} \rho_2 \\ \rho_1 \end{pmatrix} \\ &= \frac{\rho_1\,(a-b) + \rho_2\,(d-c)}{ad-bc} = J(\omega-\Omega) \;. \end{aligned} \qquad (9.23)$$

The elements of the matrix $\hat{A}$ must satisfy the natural condition that the sum of the probabilities of finding the system in the initial and final states equals unity. In order to satisfy this condition, the elements of the transition matrix $P(\Delta_j, \Delta_{j'}, t/N) = P_{jj'}$ must satisfy

$$P_{11} + P_{21} = 1\,, \qquad P_{12} + P_{22} = 1\,. \tag{9.24}$$

By taking into account (9.18) and (9.24), we arrive at the following expression for the transition matrix:

$$\hat{P} = 1 + \hat{p}\frac{t}{N} = 1 + \begin{pmatrix} -P & p \\ P & -p \end{pmatrix} \frac{t}{N}\,. \tag{9.25}$$

Values P and p have a simple physical meaning. They are the probabilities of the transitions $2 \leftarrow 1$ and $1 \leftarrow 2$ per second. The diagonal matrix describes the frequency jump. The matrix is given by

$$\hat{\Delta} = \begin{pmatrix} \Delta & 0 \\ 0 & 0 \end{pmatrix}\,. \tag{9.26}$$

Let us consider the temporal evolution of the vector $\boldsymbol{P}$ whose components ρ_2 and ρ_1 determine the probabilities of finding the system in the states 2 and 1 with frequencies ω_2 and ω_1, respectively. By differentiating (9.20) with respect to time, we find the following set of coupled equations for the probabilities:

$$\begin{aligned} \dot{\rho}_2(t) &= -\left(P + \mathrm{i}\Delta\right)\rho_2\left(t\right) + p\rho_1(t)\,, \\ \dot{\rho}_1\left(t\right) &= P\rho_2(t) - p\rho_1(t)\,. \end{aligned} \tag{9.27}$$

Let the states 1 and 2 correspond to the two states of the TLS, i.e., the frequencies $\omega_1 = \Omega$ and $\omega_2 = \Omega + \Delta$ relate to the two electronic transitions shown in Fig. 9.3.

This figure allows us to find a relation between the Anderson theory and the dynamical theory for line broadening to be developed shortly. Indeed the parameter Δ determines the change of the splitting in the TLS or, in other words, a change of the tunnelon energy. Hence, Δ is a parameter of the quadratic FC interaction, to use the terminology of the dynamical theory. The Anderson theory does not take into account the transition probability in the excited electronic state.

With the help of the equations for the matrices p and Δ, we can find the following expression for the matrix $\hat{A}$:

$$\hat{A} = \begin{pmatrix} \Omega + \Delta - \mathrm{i}P & \mathrm{i}p \\ \mathrm{i}P & \Omega - \mathrm{i}p \end{pmatrix}\,. \tag{9.28}$$

By inserting this expression into (9.23) and allowing for the obvious equation $\rho_1 + \rho_2 = 1$, we arrive at the following expression for the function that describes the optical band shape:

$$I(\nu) = -\frac{1}{\pi}\,\mathrm{Im}\,J(\nu) = -\frac{1}{\pi}\,\mathrm{Im}\,\frac{\nu - \Delta + \mathrm{i}R + \Delta\rho_2}{\nu(\nu - \Delta + \mathrm{i}R) - \mathrm{i}p\Delta}\,, \tag{9.29}$$

where $\nu = \omega - \Omega$ and $R = p + P$.

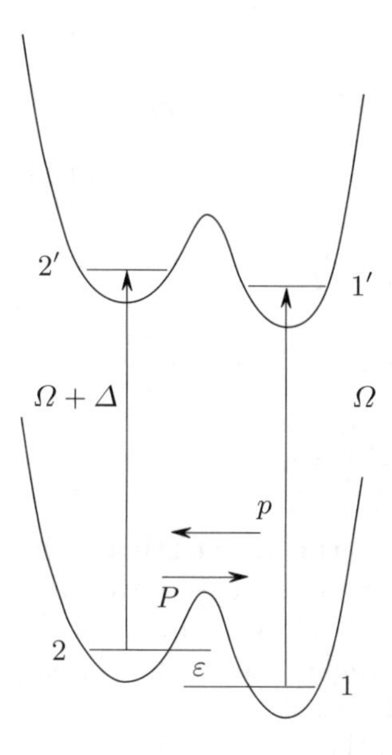

Fig. 9.3. Optical transitions in a TLS

The quadratic polynomial in the denominator can be written in the form $(\nu - \nu_1)(\nu - \nu_2)$, where

$$\nu_{1,2} = \frac{\Delta - \mathrm{i}R}{2}\left(1 \pm \sqrt{1 + \mathrm{i}\frac{4p\Delta}{(\Delta - \mathrm{i}R)^2}}\right) . \tag{9.30}$$

The ratio under the square root sign is less than unity. Indeed, its modulus satisfies the inequality

$$\frac{4p\Delta}{\Delta^2 + R^2} < \frac{2R\Delta}{\Delta^2 + R^2} \leqslant 1 \tag{9.31}$$

for any values of the parameters Δ and R. At low temperatures this value is much less than unity because $p \ll R/2$. Therefore, the square root can be expanded in a power series in this small ratio. Then the roots of the polynomial in the denominator take the form

$$\omega_1 \approx \Omega + \delta - \mathrm{i}\frac{\gamma}{2} , \qquad \omega_2 \approx \Omega + \Delta - \mathrm{i}R , \tag{9.32}$$

where

$$\delta = \frac{p}{R}\frac{R^2\Delta}{\Delta^2 + R^2} , \qquad \frac{\gamma}{2} = \frac{p}{R}\frac{R\Delta^2}{\Delta^2 + R^2} . \tag{9.33}$$

The right hand side of (9.29) can be expressed in terms of the roots of the polynomial in the denominator. Then taking into account the fact that

$$\delta \ll \Delta\,, \qquad \frac{\gamma}{2} \ll R\,, \tag{9.34}$$

we find the following final expression for the optical band shape function:

$$\begin{aligned} I(\omega) &= -\frac{1}{\pi}\,\mathrm{Im}\,J(\omega - \Omega) \\ &= -\frac{1}{\pi}\,\mathrm{Im}\left[\left(1 - \frac{\Delta}{\Delta - \mathrm{i}R}\rho_2\right)\frac{1}{\omega - \Omega - \delta + \mathrm{i}\gamma/2}\right. \\ &\qquad\qquad \left. + \frac{\Delta}{\Delta - \mathrm{i}R}\rho_2 \frac{1}{\omega - \Omega - \Delta + \mathrm{i}R}\right]\,. \end{aligned} \tag{9.35}$$

The optical band consists of two Lorentzians corresponding to the transitions 1–1′ and 2–2′ shown in Fig. 9.3. This situation is shown in Fig. 9.4 for the case $\Delta/R \gg 1$.

Dependence of the Optical Band Shape on the Strength of the Electron–Tunnelon Coupling. The figure shows that the optical band shape depends strongly on the value of the ratio Δ/R. This parameter determines the strength of the electron–tunnelon coupling. Indeed, line broadening disappears and the optical band consists of a single δ-like optical line at $\Delta = 0$. The parameter Δ describes a change in the adiabatic double-well potential upon electronic excitation of the chromophore. This is the parameter of the quadratic term in the FC-type tunnelon interaction operator. The dimensionless parameter Δ/R characterizes the strength of the quadratic electron–tunnelon interaction.

The optical bands relating to the cases of weak, intermediate and strong coupling are shown in Fig. 9.4. The figure allows us to check that the approximations made in the course of the transformation of the exact (9.29) to the approximate (9.35) are correct, regardless of the strength of the electron–tunnelon coupling. This conclusion can be reached by comparing the left hand side of Fig. 9.4 with the right hand side of the same figure.

Temperature Dependence of the Optical Band Shape. The dependence on temperature results from the probabilities p and P, which correspond to the $2 \leftarrow 1$ and $1 \leftarrow 2$ transitions shown in Fig. 9.3. Therefore, in thermal equilibrium these probabilities must satisfy the relation

$$\frac{p}{P} = \frac{\rho_2}{\rho_1} = \mathrm{e}^{-\varepsilon/kT}\,, \tag{9.36}$$

where ε is the tunnelon energy in the ground electronic state. Hence, we find the expressions

$$\rho_2 = f(T) = \left[\exp\left(\frac{\varepsilon}{kT}\right) + 1\right]^{-1}\,, \qquad \rho_1 = 1 - f(T)\,. \tag{9.37}$$

With the help of these equations, (9.33) can be transformed to

$$\delta(T) = f(T)\frac{R^2\Delta}{\Delta^2 + R^2}\,, \qquad \frac{\gamma(T)}{2} = f(T)\frac{R\Delta^2}{\Delta^2 + R^2}\,. \tag{9.38}$$

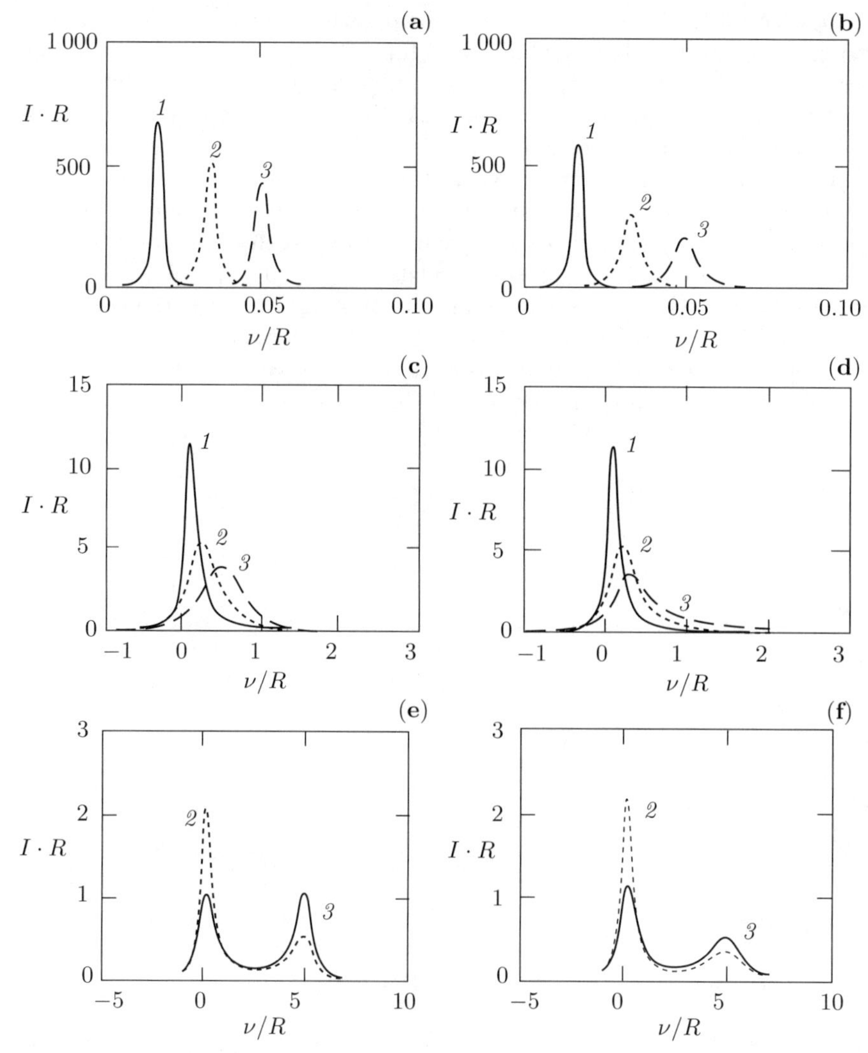

Fig. 9.4. Band shape function calculated using the exact (9.29) (*left*) and approximate (9.35) (*right*) for various values of the electron–phonon coupling and $kT/\varepsilon = 0.625$ (curve 1), 1.45 (curve 2) and 100 (curve 3). (**a**) and (**b**) $\Delta/R = 0.1$, (**c**) and (**d**) $\Delta/R = 1$, (**e**) and (**f**) $\Delta/R = 5$

These functions of temperature determine a thermal frequency shift and line broadening of the optical line relating to the 1–1′ transition. The larger the parameter Δ, the larger the spectral distance between optical lines relating to the 1–1′ and 2–2′ transitions. The FWHM of the line 2–2′ equals $2R$. It is almost independent of temperature. Curves 1, 2 and 3 in Fig. 9.4 correspond to various values of temperature. The temperature line shift dominates at weak coupling. In contrast to that, temperature-broadening dominates at strong electron–tunnelon coupling. Both line broadening and line shift disappear if $p = 0$. Because this probability results in a jump between two values of the resonant frequency Ω and $\Omega + \Delta$, the mechanism discussed here is called the exchange mechanism.

Comparison of the Anderson–Weiss theory with the Anderson theory discussed in this section demonstrates that the theories give different predictions concerning the shape of the 1–1′ optical line. The first theory predicts a Gaussian line shape for fast decay of the correlation function. It is obvious that fast decay relates to strong electron–tunnelon coupling. However, the stochastic Anderson theory predicts a Lorentzian line shape for any value of the coupling. It will be shown later that this result of the stochastic Anderson theory is similar to the result of the dynamical theory for line broadening and shift.

10. Dynamical Theory of Electron–Phonon Bands

Any stochastic theory for the optical band fails to describe a number of details found in realistic optical bands of impurity centers. For instance, stochastic theories are unable to explain why the optical band of a guest molecule has a rather complicated shape and, as a rule, consists of a zero-phonon line and a phonon side band. Nor can they explain why conjugate absorption and fluorescence bands whose zero-phonon lines are in resonance with each other can differ in shape. Real electron–phonon bands consist of an infinite set of electron–phonon transitions. From the point of view of the Anderson theory discussed above, the frequency of the optical transition in such a band jumps between an infinite number of spectral positions. The stochastic approach loses simplicity when applied to such a system. Therefore, electron–phonon optical bands and vibronic spectra of complex organic molecules with well-resolved phonon and vibrational structure can only be examined with the help of a dynamical theory.

In the beginning of the 1950s, the main principles of the dynamical approach to the calculation of electron–phonon optical bands were formulated. Two very powerful mathematical methods were proposed for calculation of the optical bands of impurity centers. The first mathematical method, proposed by Lax [58], was based on calculation of the quantum statistical average of the time-ordered perturbation operators. The method was borrowed from quantum field theory, where it worked very well. The second method, proposed by Kubo and Toyozawa [59], was based on the direct calculation of a quantum statistical average with the help of a multidimensional density matrix of the electron-vibrational system. The method by Kubo and Toyozawa enables one to calculate the phonon structure of the optical band shape using the linear and quadratic FC interaction in a non-perturbative way. Unfortunately, this method fails to calculate the broadening of the zero-phonon line.

The dynamical method of calculation of the optical band shape proposed by Lax was generalized by the present author and the method is used throughout this book. It enables one to take into account both FC and HT interactions without any restrictions on the values of these interactions. The method also enables one to include an interaction with tunneling systems of polymers and glasses and to develop a dynamical theory for spectral diffusion. However, in this chapter we shall still ignore the existence of tunneling systems and consider only electron–phonon spectra.

10.1 Absorption Cross-Section and the Probability of Light Emission

In Part III we have already derived (7.39) for the probabilities of light-induced transitions with absorption or emission of one photon. Let us insert the explicit expression $\chi^2 = (4\pi\omega_k/\hbar)(n_k/V)d^2\cos^2\alpha_k$ for the square of the Rabi frequency into these formulas. Here α_k is the angle between the molecular dipolar moment and the light polarization vector. The photon density n_k/V in the laser beam can be expressed in terms of the number of photons I arriving per second and per unit area (cm^2) by $n_k/V = I/c$. Then the Rabi frequency is given by

$$\chi^2 = \frac{4\pi\omega_{\boldsymbol{k}}}{\hbar}\frac{I}{c}\,d^2\cos^2\alpha_{\boldsymbol{k}}\ . \tag{10.1}$$

Inserting this equation into (7.39), we arrive at the following expressions:

$$k^{\mathrm{g}}(\Delta_{\boldsymbol{k}}) = \sigma^{\mathrm{g}}(\Delta_{\boldsymbol{k}})\,I\ , \qquad k^{\mathrm{e}}(\Delta_{\boldsymbol{k}}) = \sigma^{\mathrm{e}}(\Delta_{\boldsymbol{k}})\,I\ , \tag{10.2}$$

where

$$\sigma^{\mathrm{g}}(\Delta_{\boldsymbol{k}}) = \frac{8\pi\omega_{\boldsymbol{k}}}{\hbar c}d^2\cos^2\alpha_{\boldsymbol{k}}\sum_{a,b}\rho_a\left|\langle b|a\rangle\right|^2\frac{(1/2T_1)}{(\Delta_{\boldsymbol{k}}+\Omega_b-\Omega_a)^2+(1/2T_1)^2} \tag{10.3}$$

is the cross-section for photon absorption. The dimensions of the absorption cross-section are cm^2.

The expression for σ^{e} describes emission of a photon with frequency ω_k. The expression can be derived from (10.3) by substituting ρ_b for ρ_a. However, the dependence of the fluorescence intensity on the frequency of the emitted photon is measured as a rule with the help of a spontaneous emission.

The rate of spontaneous emission of one photon is given by

$$\frac{1}{T_1} = \left[\sum_{\boldsymbol{k}}k^{\mathrm{e}}(\Delta_{\boldsymbol{k}})\right]_{n_k=1} = \frac{4\pi}{\hbar}d^2\sum_{a,b}\rho_b\left|\langle b|a\rangle\right|^2 S_{ab}\ , \tag{10.4}$$

where

$$S_{ab} = \frac{1}{V}\sum_{\boldsymbol{k}}\omega_{\boldsymbol{k}}\cos^2\alpha_{\boldsymbol{k}}\frac{(1/T_1)}{(\Delta_{\boldsymbol{k}}+\Omega_b-\Omega_a)^2+(1/2T_1)^2} \tag{10.5}$$

$$= \int_0^\infty \omega\frac{(1/T_1)}{(\Delta+\Omega_b-\Omega_a)^2+(1/2T_1)^2}\frac{1}{V}\sum_{\boldsymbol{k}}\cos^2\alpha_{\boldsymbol{k}}\delta(\omega-\omega_{\boldsymbol{k}})\mathrm{d}\omega\ .$$

The sum has been calculated in Appendix A and is given by

$$\frac{1}{V}\sum_{\boldsymbol{k}}\cos^2\alpha_{\boldsymbol{k}}\delta\left(\omega-\omega_{\boldsymbol{k}}\right)=N(\omega)=\frac{\omega^2}{6\pi^2c^3}\,. \tag{10.6}$$

Therefore the equation for the rate of spontaneous emission can be transformed to

$$\begin{aligned}\frac{1}{T_1}&=\frac{8\pi}{\hbar}d^2\sum_{a,b}\rho_b\left|\langle b|a\rangle\right|^2\int_0^\infty\frac{(1/2T_1)}{\left(\Delta+\Omega_b-\Omega_a\right)^2+\left(1/2T_1\right)^2}N\left(\omega\right)\omega\mathrm{d}\omega\\&=\int_0^\infty I^{\mathrm{e}}\left(\omega\right)\mathrm{d}\omega,\end{aligned} \tag{10.7}$$

where the dimensionless function of the frequency

$$\begin{aligned}I^{\mathrm{e}}(\omega)&=cN\left(\omega\right)\frac{8\pi\omega}{\hbar c}d^2\sum_{a,b}\rho_b\left|\langle b|a\rangle\right|^2\frac{(1/2T_1)}{\left(\Delta+\Omega_b-\Omega_a\right)^2+\left(1/2T_1\right)^2}\\&=cN\left(\omega\right)\sigma^{\mathrm{e}}(\omega)\end{aligned} \tag{10.8}$$

describes the line shape function of the spontaneous fluorescence. This function differs by the factor $cN(\omega)$ from the cross-section $\sigma^{\mathrm{e}}(\omega)$ of induced fluorescence.

The absorption cross-section $\sigma^{\mathrm{g}}(\omega)$ and the fluorescence line shape function $I^{\mathrm{e}}(\omega)$ measured experimentally can be written in the form

$$\sigma^{\mathrm{g}}\left(\omega\right)=\frac{8\pi^2}{\hbar}\frac{\omega}{c}\left(d\cos\alpha\right)^2S^{\mathrm{g}}\left(\omega\right)\,,\quad I^{\mathrm{e}}\left(\omega\right)=\frac{4}{3\hbar}\left(\frac{\omega}{c}\right)^3d^2S^{\mathrm{e}}\left(\omega\right)\,, \tag{10.9}$$

where

$$\begin{aligned}S^{\mathrm{g}}\left(\omega\right)&=\frac{1}{\pi}\sum_{a,b}\rho_a\left|\langle b|a\rangle\right|^2\frac{(1/2T_1)}{\left(\Delta+\Omega_b-\Omega_a\right)^2+\left(1/2T_1\right)^2}\,,\\S^{\mathrm{e}}\left(\omega\right)&=\frac{1}{\pi}\sum_{a,b}\rho_b\left|\langle b|a\rangle\right|^2\frac{(1/2T_1)}{\left(\Delta+\Omega_b-\Omega_a\right)^2+\left(1/2T_1\right)^2}\end{aligned} \tag{10.10}$$

are the line shape functions of absorption and emission, which can be calculated theoretically. These functions of the detuning $\Delta=\Omega-\omega$ describe the influence of FC electron–phonon interactions on the absorption and fluorescence bands. The calculation of these functions is carried out in the next section.

10.2 Electron–Phonon Optical Transitions in a Condon Approximation at $T=0$

The realistic optical band of an impurity center consists of many electron–phonon transitions. The linear FC interaction is the main reason for the appearance of electron–phonon transitions. The linear FC interaction is characterized by displacements of the equilibrium positions of the multidimensional Franck–Condon adiabatic potentials. We shall neglect modulation of the dipolar moment by molecular vibrations. This means that we neglect the HT interaction. This approximation is called the Condon approximation.

The natural coordinate systems in the ground and excited electronic states coincide with each other because the linear FC interaction does not mix up natural coordinates. Therefore, the multidimensional Franck–Condon integrals $\langle a \mid b\rangle$ in (10.10) are products of the one-dimensional FC integrals, and we may study the influence of the linear FC interaction with the help of a chromophore interacting with the one-phonon mode. In this case, if the temperature equals zero, one can write $\rho_a = \rho_b = \delta_{n0}$. One summation in (10.10) disappears and these equations take the form

$$\begin{aligned} S^{\mathrm{g}}(\Delta) &= \sum_{n=0}^{\infty} \langle n|0\rangle^2 \frac{1/2\pi T_1}{(\Delta - n\omega_0)^2 + (1/2T_1)^2}\,, \\ S^{\mathrm{e}}(\Delta) &= \sum_{n=0}^{\infty} \langle n|0\rangle^2 \frac{1/2\pi T_1}{(\Delta + n\omega_0)^2 + (1/2T_1)^2}\,. \end{aligned} \tag{10.11}$$

The FC integrals are given by

$$\begin{aligned} \langle n|0\rangle &= \int_{-\infty}^{\infty} \Phi_n(R+R_0)\Phi_0(R)\mathrm{d}R \\ &= \exp\left(-\frac{a^2}{4}\right)\frac{1}{\sqrt{n!}}\left(\frac{a}{\sqrt{2}}\right)^n\,, \end{aligned} \tag{10.12}$$

where $a = R_0/\sqrt{\hbar/\mu\omega_0}$ is a dimensionless shift in the equilibrium position upon electronic excitation of the chromophore, and $\Phi_n(R)$ is a normalized harmonic oscillator function. By inserting (10.12) into the expression for the absorption band, we find the expression

$$S^{\mathrm{g}}(\Delta) = \mathrm{e}^{-(a/\sqrt{2})^2}\frac{1/2\pi T_1}{\Delta^2 + (1/2T_1)^2} + \Psi^{\mathrm{g}}(\Delta)\,, \tag{10.13}$$

where

$$\Psi^{\mathrm{g}}(\Delta) = \mathrm{e}^{-(a/\sqrt{2})^2}\sum_{n=1}^{\infty}\frac{1}{n!}\left(\frac{a}{\sqrt{2}}\right)^{2n}\frac{1/2\pi T_1}{(\Delta - n\omega_0)^2 + (1/2T_1)^2}\,. \tag{10.14}$$

Here $\Delta = \omega_{\mathrm{L}} - \Omega$ is the difference between the laser and resonant frequencies of electronic excitation. This detuning differs in sign from the detuning used in the equations for the density matrix. The first term in (10.13) describes a photo-transition without creation of phonons. This optical line is called the zero-phonon line (ZPL). The second term in (10.14) determines a contribution to the optical band from photo-transitions with creation of one, two and more phonons.

The fluorescence spectrum is given by the second formula of (10.11). After calculating the Franck–Condon integrals, the formula takes the form

$$S^{\mathrm{e}}(\Delta) = \mathrm{e}^{-(a/\sqrt{2})^2} \frac{1/2\pi T_1}{\Delta^2 + (1/2T_1)^2} + \Psi^{\mathrm{e}}(\Delta) \ , \tag{10.15}$$

where

$$\Psi^{\mathrm{e}}(\Delta) = \Psi^{\mathrm{g}}(-\Delta) \ . \tag{10.16}$$

Hence the ZPLs of the conjugate absorption and fluorescence spectra are in resonance with each other, whereas the electron–phonon lines of the absorption and fluorescence spectra extend to the blue and red frequency domains from the ZPL, as shown in Fig. 10.1.

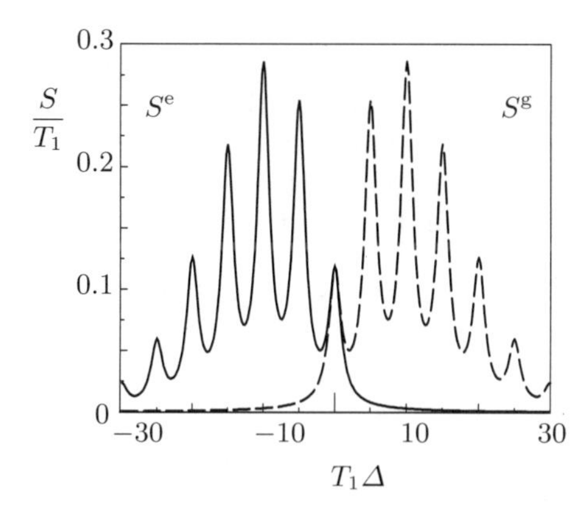

Fig. 10.1. Electron–phonon absorption band (*dashed line*) and fluorescence band (*solid line*). Zero-phonon lines are in exact resonance

A dimensionless shift a in the equilibrium positions determines the strength of the linear FC interaction. Absorption spectra at weak and strong electron–phonon coupling are shown in Fig. 10.2. It is obvious that the strength of the linear electron–phonon interaction can be found using

$$\frac{a^2}{2} = \frac{J_1}{J_0} \ , \tag{10.17}$$

where J_0 and J_1 are the integrated intensities of the zero-phonon and one-phonon transitions. If the electron–phonon coupling is increased, the intensity of the ZPL decreases and the intensity of the electron–phonon lines increases. If the electron–phonon coupling parameter satisfies the simple equation

$$\frac{a^2}{2} = n \ , \tag{10.18}$$

where n is an integer, the electron–phonon transition with creation of n vibrations has the highest intensity, i.e., the maximum of the optical band is shifted by $n\hbar\nu$ from the ZPL.

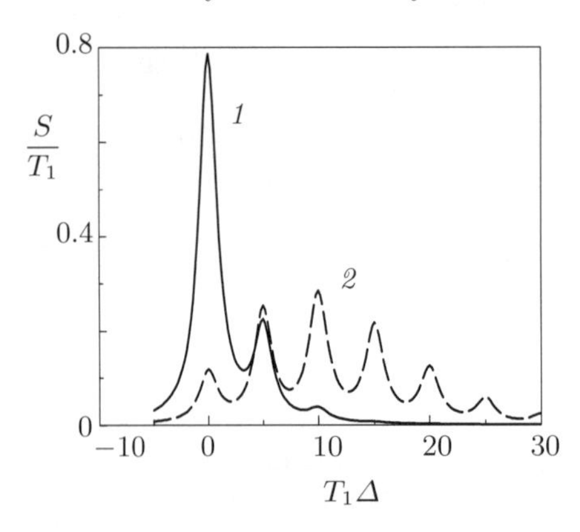

Fig. 10.2. Absorption band at weak (*solid line*) and strong (*dashed line*) electron–phonon coupling, calculated using (10.15) with $a = 0.6$ (curve 1) and 3 (curve 2)

10.3 Zero-Phonon Line and Phonon Side Band

The simple relations (10.17) and (10.18), which enable one to estimate the strength of the linear FC interaction, can be generalized to the more realistic case when an optical electron of the chromophore interacts with a huge number of phonon modes. In such a case the calculation with (10.10) is a more complicated task. Nevertheless, the problem can be solved using the ordered operator method proposed by Lax [58].

Let us consider first the function $S^{\mathrm{g}}(\omega)$. By taking into account the integral representation of a Lorentzian,

$$\frac{\gamma}{\omega^2+\gamma^2} = \frac{1}{2}\int_{-\infty}^{\infty} \mathrm{e}^{-\mathrm{i}\omega t-\gamma|t|}\mathrm{d}t \; , \tag{10.19}$$

we can transform (10.10) to

$$S^{\mathrm{g}}(\omega) = \frac{1}{2\pi}\int_{-\infty}^{\infty} \mathrm{e}^{\mathrm{i}(\omega-\Omega)t-|t|/2T_1} S^{\mathrm{g}}(t)\,\mathrm{d}t \; , \tag{10.20}$$

where

$$S^{\mathrm{g}}(t) = \sum_a \rho_a(T)\sum_b \langle a|\mathrm{e}^{-\mathrm{i}\Omega_b t}|b\rangle\langle b|\mathrm{e}^{\mathrm{i}\Omega_a t}|a\rangle \; . \tag{10.21}$$

In accordance with (7.8), the frequencies Ω_a and Ω_b are eigenvalues of the adiabatic Hamiltonians H^{g} and H^{e}. The elements of the density matrix are given by

$$\begin{aligned}\rho_a(T) &= \langle a|\hat{\rho}^{\mathrm{g}}(T)|a\rangle \\ &= \langle a|\exp\left(\frac{F^{\mathrm{g}}-\hbar\Omega_a}{kT}\right)|a\rangle = \langle a|\exp\left(\frac{F^{\mathrm{g}}-H^{\mathrm{g}}}{kT}\right)|a\rangle \; ,\end{aligned} \tag{10.22}$$

where the free energy F^{g} is determined by

$$\exp\left(-\frac{F^{\mathrm{g}}}{kT}\right) = \sum_a \langle a| \exp\left(-\frac{H^{\mathrm{g}}}{kT}\right)|a\rangle \, . \tag{10.23}$$

After inserting (10.22) into (10.21) and substituting the Hamiltonians for the frequencies, we arrive at

$$S^{\mathrm{g}}(t) = \mathrm{Tr}\left[\hat{\rho}^{\mathrm{g}}(T) \exp\left(-\mathrm{i}t\frac{H^{\mathrm{e}}}{\hbar}\right) \exp\left(\mathrm{i}t\frac{H^{\mathrm{g}}}{\hbar}\right)\right] = \left\langle \hat{S}\left(t\right)\right\rangle_{\mathrm{g}} \, . \tag{10.24}$$

Similar transformations can be carried out on the formula for the fluorescence band in (10.10). The result is given by

$$S^{\mathrm{e}}\left(\omega\right) = \frac{1}{2\pi}\int\limits_{-\infty}^{\infty} \mathrm{e}^{\mathrm{i}(\omega-\Omega)t-t/(2T_1)} S^{\mathrm{e}}\left(t\right)\mathrm{d}t \, , \tag{10.25}$$

where

$$\begin{aligned} S^{\mathrm{e}}\left(t\right) &= \sum_b \rho_b\left(T\right)\sum_a \langle b|\mathrm{e}^{-\mathrm{i}\Omega_b t}|a\rangle\langle a|\mathrm{e}^{\mathrm{i}\Omega_a t}|b\rangle \\ &= \mathrm{Tr}\left\{\hat{\rho}^{\mathrm{e}}\left(T\right)\exp\left(-\mathrm{i}\frac{H^{\mathrm{e}}}{\hbar}t\right)\exp\left(\mathrm{i}\frac{H^{\mathrm{g}}}{\hbar}t\right)\right\} = \langle\hat{S}\left(t\right)\rangle_{\mathrm{e}} \, . \end{aligned} \tag{10.26}$$

The subscripts g and e on the brackets show what type of adiabatic Hamiltonian is used in the density matrix.

The functions $S^{\mathrm{g}}(t)$ and $S^{\mathrm{e}}(t)$ describe the temporal evolution of the dipolar correlator. These functions can be calculated even in the case when the multidimensional Franck–Condon integrals $\langle a \mid b\rangle$ in (10.10) are not factorized. We face a situation of this kind if a quadratic FC interaction is taken into account. The influence of the quadratic FC interaction will be discussed in detail in Chap. 12.

Here we consider the influence of the linear FC interaction. This interaction is determined by displacements a of the equilibrium positions of the natural vibrations of the system. The natural coordinate systems in the ground and excited electronic states are the same. Therefore the vibrational Hamiltonian can be written as a sum over indices q of the normal modes: $H^{\mathrm{g,e}} = \sum_{q=1}^{N} H_q^{\mathrm{g,e}}$. The natural modes are independent of one another. This means that the operators of the various modes commute. Therefore the function $S^{\mathrm{g}}(t)$ factorizes:

$$S^{\mathrm{g}}\left(t\right) = \prod_{q=1}^{N}\left\langle \exp\left(-\mathrm{i}\frac{H_q^{\mathrm{e}}}{\hbar}t\right)\exp\left(\mathrm{i}\frac{H_q^{\mathrm{g}}}{\hbar}t\right)\right\rangle_{\mathrm{g}} = \prod_{q=1}^{N} S_q^{\mathrm{g}}\left(t\right) . \tag{10.27}$$

It suffices to consider a single mode by omitting the mode index.

The adiabatic Hamiltonian for the excited electronic state can be expressed in terms of the adiabatic Hamiltonian for the ground state,

$$H^{\mathrm{e}}(Q) = H^{\mathrm{g}}(Q - a) , \tag{10.28}$$

where

$$Q = \frac{b + b^{+}}{\sqrt{2}} \tag{10.29}$$

is an operator for the dimensionless coordinate of the harmonic oscillator. Here b and b^{+} are phonon annihilation and creation operators.

Let us introduce an operator for the coordinate shift, viz.,

$$\exp \hat{L} = \exp\left[\frac{a}{\sqrt{2}}\left(b - b^{+}\right)\right] . \tag{10.30}$$

By differentiating the expression $b(a) = \exp(-L) b \exp L$ with respect to a, we easily find

$$\frac{\mathrm{d}}{\mathrm{d}a} b(a) = -\frac{1}{\sqrt{2}} . \tag{10.31}$$

Then we find that

$$b(a) = b - \frac{a}{\sqrt{2}} , \qquad b^{+}(a) = b^{+} - \frac{a}{\sqrt{2}} . \tag{10.32}$$

Hence for the operator for the coordinate, we find the expression

$$Q(a) = \exp(-\hat{L}) Q \exp \hat{L} = Q - a , \tag{10.33}$$

which proves that the operator $\exp L$ is indeed the operator for the coordinate shift. With the help of this operator, the relation between the adiabatic Hamiltonians is

$$\begin{aligned} \exp\left(-\mathrm{i}\frac{H^{\mathrm{e}}}{\hbar}t\right) &= \exp\left[-\mathrm{i}\frac{H(Q-a)}{\hbar}t\right] \\ &= \exp(-\hat{L}) \exp\left[-\mathrm{i}\frac{H(Q)}{\hbar}t\right] \exp \hat{L} . \end{aligned} \tag{10.34}$$

Let us insert this equation into (10.24). The equation takes the form

$$S^{\mathrm{g}}(t) = \left\langle \mathrm{e}^{-\hat{L}(t)} \mathrm{e}^{\hat{L}} \right\rangle , \tag{10.35}$$

where

$$\mathrm{e}^{-\hat{L}(t)} = \mathrm{e}^{\mathrm{i}Ht/\hbar} \mathrm{e}^{-\hat{L}} \mathrm{e}^{-\mathrm{i}Ht/\hbar} . \tag{10.36}$$

Here $H = H^{\mathrm{g}} \equiv H(Q)$ is the Hamiltonian of the harmonic oscillator. The last equation is easily calculated if we take into account the following relations:

$$\begin{aligned} b\,(t) &= \mathrm{e}^{-\mathrm{i}Ht/\hbar} b \mathrm{e}^{\mathrm{i}Ht/\hbar} = \mathrm{e}^{-\mathrm{i}\nu t} b\,, \\ b^{+}\,(t) &= \mathrm{e}^{-\mathrm{i}Ht/\hbar} b^{+} \mathrm{e}^{\mathrm{i}Ht/\hbar} = \mathrm{e}^{\mathrm{i}\nu t} b^{+}\,, \end{aligned} \tag{10.37}$$

which can be derived by differentiating the operators with respect to time and using the relation $H = \hbar\nu(b^{+}b + 1/2)$.

In order to calculate (10.35), we use the Weyl formula

$$\mathrm{e}^{\hat{F}} \mathrm{e}^{\hat{G}} = \mathrm{e}^{\hat{F}+\hat{G}} \mathrm{e}^{\frac{1}{2}[\hat{F},\hat{G}]}\,, \tag{10.38}$$

which is valid for any operators $\hat{F}$ and $\hat{G}$ whose commutator is not an operator. Using the Weyl formula, (10.35) can be transformed to

$$S^{\mathrm{g}}\,(t) = \left\langle \mathrm{e}^{-\hat{L}(t)+\hat{L}} \right\rangle \mathrm{e}^{-\frac{1}{2}[\hat{L}(t),\hat{L}]}\,. \tag{10.39}$$

The average of the product of Bose operators can be calculated with the help of the Wick–Bloch–Dominicis theorem proven in Appendix G. Using this theorem, we find that

$$\left\langle \left[\hat{L} - \hat{L}(t)\right]^{2p} \right\rangle = (2p-1)!! \left\langle \left[\hat{L} - \hat{L}(t)\right]^{2} \right\rangle\,, \quad \left\langle \left[\hat{L} - \hat{L}(t)\right]^{2p-1} \right\rangle = 0\,. \tag{10.40}$$

Then (10.39) takes the form

$$\begin{aligned} S^{\mathrm{g}}\,(t) &= \exp\left\{ \frac{1}{2}\langle(\hat{L} - \hat{L}(t))^2\rangle + \frac{1}{2}[\hat{L}, \hat{L}(t)] \right\} \\ &= \exp\left\{ \frac{1}{2}\left(\langle\hat{L}^2\rangle + \langle\hat{L}^2\,(t)\rangle - \langle\hat{L}\hat{L}(t)\rangle - \langle\hat{L}(t)\hat{L}\rangle + [\hat{L}, \hat{L}(t)] \right) \right\}\,. \end{aligned} \tag{10.41}$$

Calculation of the averages and the commutator gives the following result:

$$\begin{aligned} \langle\hat{L}\hat{L}(t)\rangle &= -\frac{a^2}{2}\left[n\mathrm{e}^{-\mathrm{i}\nu t} + (n+1)\mathrm{e}^{\mathrm{i}\nu t}\right]\,, \\ \langle\hat{L}^2\rangle = \langle\hat{L}^2(t)\rangle &= -\frac{a^2}{2}\,(2n+1)\,, \\ \langle\hat{L}(t)\hat{L}\rangle &= -\frac{a^2}{2}\left[n\mathrm{e}^{\mathrm{i}\nu t} + (n+1)\,\mathrm{e}^{-\mathrm{i}\nu t}\right]\,, \\ [\hat{L}, \hat{L}(t)] &= \frac{a^2}{2}\left(\mathrm{e}^{-\mathrm{i}\nu t} - \mathrm{e}^{\mathrm{i}\nu t}\right)\,, \end{aligned} \tag{10.42}$$

where

$$n = \langle b^{+}b\rangle = \frac{1}{\exp\left(\hbar\nu/kT\right) - 1} \tag{10.43}$$

is the average number of phonons at temperature T. By inserting these equations into (10.41), we arrive at the following expression for the function $S_q^{\mathrm{g}}(t)$:

$$S_q^{\mathrm{g}}(t) = \exp g_q(t) = \exp\left[\varphi_q\left(t,T\right) - \varphi_q\left(0,T\right)\right] , \tag{10.44}$$

where

$$\varphi_q\left(t,T\right) = \frac{a_q^2}{2}\left[n_q \exp\left(\mathrm{i}\nu_q t\right) + \left(n_q + 1\right)\exp\left(-\mathrm{i}\nu_q t\right)\right] . \tag{10.45}$$

Here we reintroduce the mode subscript. Substituting (10.44) for S_q^{g} in (10.27), we arrive at the final expression for the function that describes the temporal evolution of the dipolar correlator:

$$S^{\mathrm{g}}\left(t\right) = \exp g\left(t\right) = \exp\left[\varphi\left(t,T\right) - \varphi\left(0,T\right)\right] , \tag{10.46}$$

where

$$\varphi\left(t,T\right) = \sum_{q=1}^{N} \frac{a_q^2}{2}\left[n_q \exp\left(\mathrm{i}\nu_q t\right) + \left(n_q + 1\right)\exp\left(-\mathrm{i}\nu_q t\right)\right] . \tag{10.47}$$

Let us turn back to the fluorescence case. We need to calculate an average in (10.26). It can be transformed to

$$\langle\hat{S}\left(t\right)\rangle_{\mathrm{e}} = \left\langle\exp\left(-\mathrm{i}H^{\mathrm{e}}t\right)\exp\left(\mathrm{i}H^{\mathrm{g}}t\right)\right\rangle_{\mathrm{e}} = \left\langle\exp\left(-\mathrm{i}H^{\mathrm{g}}t\right)\exp\left(\mathrm{i}H^{\mathrm{e}}t\right)\right\rangle_{\mathrm{e}}^{*} . \tag{10.48}$$

The formula for light emission differs from the formula for light absorption by the permutation of the electronic indices g and e in the adiabatic Hamiltonians. Taking into account the fact that $H^{\mathrm{g}} = H(R)$, $H^{\mathrm{e}} = H(R-a)$ and the fact that the integration over variable R is carried out from $-\infty$ to ∞, we may substitute the variable $R' + a$ for the variable R. Then we find that $H^{\mathrm{g}} = H(R' + a)$ and $H^{\mathrm{e}} = H(R')$. Hence the average $\langle\exp(-\mathrm{i}H^{\mathrm{g}}t)\exp(\mathrm{i}H^{\mathrm{e}}t)\rangle_{\mathrm{e}}$ differs from the average $\langle\exp(-\mathrm{i}H^{\mathrm{e}}t)\exp(\mathrm{i}H^{\mathrm{g}}t)\rangle_{\mathrm{g}}$ solely by the sign of the shift a. However, the function $\varphi(t,T)$ does not feel the sign of a. Therefore, the following simple relation can be derived from (10.48):

$$S^{\mathrm{e}}\left(t\right) = \left[S^{\mathrm{g}}\left(t\right)\right]^{*} = \exp g^{*}\left(t,T\right) = \exp g\left(-t,T\right) . \tag{10.49}$$

The fluorescence band is described by (10.20), where the index g is replaced by the index e. Using (10.46) and (10.49), we can find the following expression for the absorption and fluorescence bands:

$$S^{\mathrm{g,e}}(\omega) = \frac{1}{2\pi}\int_{-\infty}^{\infty} \mathrm{e}^{\mathrm{i}(\omega-\Omega)t-|t|/(2T_1)}\mathrm{e}^{\varphi(\pm t,T)-\varphi(0,T)}\mathrm{d}t , \tag{10.50}$$

where we should choose the + sign and the − sign for the absorption and fluorescence bands, respectively.

In order to calculate the band shape, we expand the function $\exp(t,T)$ in (10.50) as a power series in the function $\varphi(t,T)$ and carry out the integration of all terms in the series. The first integral, whose integrand equals unity, can be calculated with the help of (10.19). The integrals whose integrands are various powers of the function $\varphi(t,T)$ can be calculated with the help of the same (10.19), once each function $\varphi(t,T)$ has been expressed in the form $\varphi(t,T) = \int_{-\infty}^{\infty} \mathrm{d}\nu\varphi(v,T)\exp(-\mathrm{i}\nu t)$. The result is given by

$$S^{\mathrm{g,e}}(\omega) = \mathrm{e}^{-\varphi(0,T)} \frac{1/2\pi T_1}{\Delta^2 + (1/2T_1)^2} + \Psi^{\mathrm{g,e}}(\Delta) \,, \tag{10.51}$$

where the first term describes the zero-phonon line (ZPL) and the second term describes all possible electron–phonon transitions. It is given by

$$\begin{aligned} \Psi^{\mathrm{g,e}}(\Delta) &= \sum_{m=1}^{\infty} \Psi_m^{\mathrm{g,e}}(\Delta) \\ &= \mathrm{e}^{-\varphi(0,T)} \sum_{m=1}^{\infty} \frac{1}{m!} \int_{-\infty}^{\infty} \mathrm{d}\nu_1 \varphi(\nu_1,T) \ldots \int_{-\infty}^{\infty} \mathrm{d}\nu_m \varphi(\nu_m,T) \\ &\qquad \times \frac{1/2\pi T_1}{(\Delta \mp \nu_1 \mp \ldots \mp \nu_m)^2 + (1/2T_1)^2} \,, \end{aligned} \tag{10.52}$$

where

$$\begin{aligned} \varphi(\nu,T) &= \int_{-\infty}^{\infty} \varphi(t,T)\,\mathrm{e}^{\mathrm{i}\nu t}\mathrm{d}t = \varphi(\nu)\left[n(\nu)+1\right] + \varphi(-\nu)\,n(-\nu) \,, \\ \varphi(\nu) &= \sum_{q=1}^{N} \frac{a_q^2}{2}\delta(\nu-\nu_q) \,, \qquad n(\nu) = \frac{1}{\exp(\hbar\nu/kT)-1} \,. \end{aligned} \tag{10.53}$$

Equation (10.51) generalizes a similar equation derived in the last section for the one-mode case and for zero temperature. The function $\Psi_m^{\mathrm{g,e}}$ describes photo-transitions with creation or annihilation of m phonons. The sum of all m-phonon transitions described by the function $\Psi^{\mathrm{g,e}}$ is called the phonon side band (PSB). The simple relation

$$S^{\mathrm{e}}(\Delta) = S^{\mathrm{g}}(-\Delta) \tag{10.54}$$

can be derived for the functions which describe the shape of the absorption and emission bands. This means that the ZPL of the absorption and fluorescence spectra coincide, and the mirror symmetry with respect to the ZPL for the absorption and fluorescence PSB exists in the case when only the linear FC interaction is taken into account. The shape of the PSB is determined by a single function $\varphi(\nu)$, containing all information concerning the phonon spectrum and linear FC interaction.

Since phonon frequencies fill the frequency interval $(0, \nu_D)$, this function is nonvanishing at each point of this interval. It is similar to the phonon density of states function. Indeed, if $a_q^2/2$ is replaced by $1/N$, where N is the number of phonon modes, the function $\varphi(\nu)$ is transformed to the phonon density of states function. Therefore, the function $\varphi(\nu)$ is called the weighted density of phonon states. It is weighted by the values $a_q^2/2$, which play the role of coupling constants in the linear FC interaction.

After integrating over frequency ν_1, the first term with $m = 1$ in (10.53) takes the form

$$\Psi_1(\Delta) = \mathrm{e}^{-\varphi(0,T)} \varphi(\Delta, T) \ . \tag{10.55}$$

It describes the probability of electron–phonon transitions with creation and annihilation of one phonon. This function extends over the one-phonon interval $(-\nu_D, \nu_D)$. The part of this function in the interval $(-\nu_D, 0)$ which lies to the red side of the ZPL is due to electron–phonon transitions with annihilation of one phonon. Therefore this part disappears when the temperature approaches zero. The part of the function in the interval $(0, \nu_D)$ which lies to the blue side of the ZPL is due to electron–phonon transitions with creation of one phonon. This part of the function does not disappear at $T = 0$. Electron–phonon transitions from the red and the blue side of the ZPL are called anti-Stokes and Stokes transitions, respectively. The probability of an m-phonon transition is completely determined by the probability of the one-phonon transition. This is a telling feature of the linear FC interaction. It is obvious that m-phonon transitions fill the frequency interval $(-m\nu_D, m\nu_D)$.

In the last section, we derived (10.17) and (10.18) relating to the coupling constant for the linear FC interaction. Here we can generalize these equations to the case when an optical electron of the molecule interacts with an infinite number of phonon modes:

$$\frac{S_1}{S_0} = \sum_{q=1}^{N} a_q^2 \left(n_q + \frac{1}{2} \right) = \varphi(0, T) \ , \qquad n = \sum_{q=1}^{N} a_q^2 \left(n_q + \frac{1}{2} \right) \ , \tag{10.56}$$

where

$$n_q = \frac{1}{\exp(\hbar\nu_q / kT) - 1} \ . \tag{10.57}$$

The strength of the linear coupling is increased when the temperature rises. The function $\varphi(0, T)$ characterizes the strength of the linear electron–phonon coupling. It is called the Huang–Rhys or Pekar–Huang factor [60, 61].

10.4 Influence of Temperature on the ZPL Intensity

The integrated intensity of ZPL + PSB does not depend on temperature in the Condon approximation. Indeed, by carrying out the integration over frequency in (10.51), we arrive at the following result:

$$S^{\mathrm{g,e}} = \mathrm{e}^{-\varphi(0,T)} \left[1 + \sum_{m=1}^{\infty} \frac{\varphi^m(0,T)}{m!} \right] = 1 \,. \tag{10.58}$$

The first factor describes the integrated intensity of the ZPL and the sum describes the integrated intensity of the PSB. Since the Huang–Rhys factor $\varphi(0,T)$ increases when the temperature increases, the integrated intensity $\exp[-\varphi(0,T)]$ of the ZPL decreases when the sample is heated. The integrated intensity of the PSB, in contrast to that of the ZPL, increases if the temperature is raised. Such a redistribution of intensity is indeed observed in experiments, as can be seen from Fig. 10.3. The decrease in the ZPL intensity with increase in temperature is a telling feature of the linear FC interaction. The ratio of the integrated intensities of the ZPL and the whole electron–phonon band is called the Debye–Waller factor. It is given by

$$\begin{aligned} \alpha(T) &= \frac{J_0}{J_0 + J_{\mathrm{PSB}}} \qquad (10.59) \\ &= \exp\left(-\varphi(0,T)\right) = \exp\left[-\int_0^{\infty} [2n(\nu)+1]\,\varphi(\nu)\,\mathrm{d}\nu \right] . \end{aligned}$$

The Debye–Waller factor is expressed in terms of the same weighted density function of phonon states $\varphi(\nu)$ as was established earlier for the phonon side band. Hence, if this function is found from analysis of the phonon side band, the temperature behavior of the Debye–Waller factor can be calculated with the help of (10.59) for any temperature. Its temperature behavior can thus be compared with experimental results. The temperature behavior of the optical band shape shown in Fig. 10.3 was treated theoretically in [62] by following this very prescription. Good agreement between theory and experimental

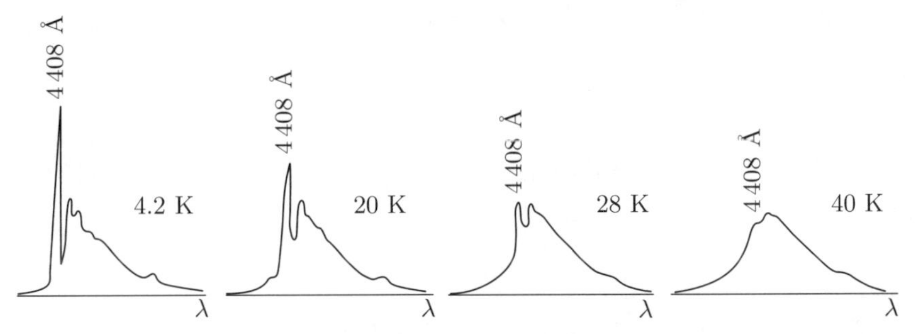

Fig. 10.3. Temperature decrease of the zero-phonon line (ZPL) intensity in the fluorescence band of a perylene molecule doping heptane [62]

data proved that the optical band shown in Fig. 10.3 results from the linear FC interaction. In order to estimate the strength of the electron–phonon coupling, we can use the very simple relation

$$-\ln\alpha(T) = \varphi(0,T)\,, \tag{10.60}$$

which shows how the strength of the electron–phonon coupling can be found with the help of the measured Debye–Waller factor.

10.5 Optical Bands for Strong Electron–Phonon Coupling

It is clear from (10.56) that, in the case when the Huang–Rhys factor is much larger than unity, processes involving the creation and annihilation of many phonons will provide the main contribution to the PSB. This PSB is of multiphonon type and in this case we speak of strong electron–phonon coupling. What is the band shape in this case?

The function for the optical band shape described by (10.50) can be expressed in the form

$$S^{\mathrm{g,e}}(\Delta) = \frac{1}{\pi}\,\mathrm{Re}\int\limits_0^\infty \mathrm{e}^{\mathrm{i}[\Delta+\mathrm{i}/(2T_1)]t}\,\mathrm{e}^{\varphi(\pm t,T)-\varphi(0,T)}\mathrm{d}t\,. \tag{10.61}$$

The function $\varphi(\nu)$ equals zero beyond the frequency interval $(0,\nu_{\mathrm{D}})$, where ν_{D} is the Debye frequency. In accordance with the general properties of the Fourier transformation, the function $\varphi(t,T)$ approaches zero at $t > \nu_{\mathrm{D}}^{-1} \approx 10^{-13}$ s. The interval $(0,\infty)$ in the integral in (10.61) can be divided as follows: $(0,\nu_{\mathrm{D}}^{-1})$ and $(\nu_{\mathrm{D}}^{-1},\infty)$. Then the integral in (10.61) is a sum of two integrals. In the first integral, the time is near zero and we can carry out the expansion

$$\varphi(\pm t,T) - \varphi(0,T) = \mp\mathrm{i}At - B\frac{t^2}{2} + \dots\,, \tag{10.62}$$

where

$$A = \sum_{q=1}^{N}\frac{a_q^2}{2}\nu_q = \int\limits_0^\infty \nu\varphi(\nu)\mathrm{d}\nu\,, \tag{10.63}$$

$$B = \sum_{q=1}^{N}\frac{a_q^2}{2}\nu_q^2\,(2n_q+1) = \int\limits_0^\infty \nu^2\,[2n(\nu)+1]\,\varphi(\nu)\,\mathrm{d}\nu\,. \tag{10.64}$$

In the second integral where the time is large, we may set $\varphi(t,T)=0$. Then (10.61) can be transformed to

$$S^{\mathrm{g,e}}(\Delta)=\frac{1}{\pi}\,\mathrm{Re}\int\limits_0^{1/\nu_{\mathrm{D}}}\mathrm{e}^{\mathrm{i}[\Delta\mp A]t-Bt^2/2}\mathrm{d}t+\mathrm{e}^{-\varphi(0,T)}\frac{1}{\pi}\,\mathrm{Re}\int\limits_{1/\nu_{\mathrm{D}}}^{\infty}\mathrm{e}^{\mathrm{i}[\Delta+\mathrm{i}/(2T_1)]t}\mathrm{d}t\,. \tag{10.65}$$

At strong electron–phonon coupling the following strong inequalities are fulfilled:

$$\frac{1}{\sqrt{B}}\ll\frac{1}{\nu_{\mathrm{D}}}\ll T_1\,. \tag{10.66}$$

Using the first inequality we can substitute ∞ for the upper limit in the first integral. Then,

$$\frac{1}{\pi}\int\limits_0^{\infty}\mathrm{e}^{-Bt^2/2}\cos\left(\Delta\mp A\right)t\mathrm{d}t=\frac{1}{\sqrt{2\pi B}}\exp\left[-\frac{(\Delta\mp A)^2}{2B}\right]. \tag{10.67}$$

Using the second inequality we can substitute zero for the lower limit in the second integral. After such a substitution, (10.65) takes the form

$$S^{\mathrm{g,e}}(\Delta)=\mathrm{e}^{-\varphi(0,T)}\frac{(1/2T_1)\pi}{\Delta^2+(1/2T_1)^2}+\frac{1}{\sqrt{2\pi B}}\exp\left[-\frac{(\Delta\mp A)^2}{2B}\right]. \tag{10.68}$$

In accordance with this equation, the optical band consists of a narrow ZPL with Lorentzian shape and a very broad electron–phonon band with Gaussian shape. Although the integrated intensity of the ZPL is weak at strong coupling, the peak intensity of the narrow ZPL is large and can be measured experimentally if the spectral resolution of the setup is high enough.

It is found from (10.68) that the constant A determines the distance between the maximum of the PSB and the resonant frequency of the ZPL. PSBs in the absorption and fluorescence spectra lie lower and higher than the ZPL on the frequency scale. The distance $2A$ between the maximum of the PSBs in the absorption band and the maximum of the PSB in the fluorescence band is called the Stokes shift. The position of the PSB maximum does not depend on temperature. The half-width of the PSB is given by

$$\Delta\omega_{1/2}=\sqrt{B2\ln 2}\,. \tag{10.69}$$

The half-width is almost independent of temperature for $kT<\hbar\nu_{\mathrm{D}}$. The half-width of the PSB increases as the square-root of the temperature in the high temperature domain, when $kT>\hbar\nu_{\mathrm{D}}$:

$$\Delta\omega_{1/2}\left(T\right)=\Delta\omega_{1/2}\left(0\right)\sqrt{1+\frac{A}{B\left(0\right)}\frac{kT}{\hbar}}\,. \tag{10.70}$$

11. Vibronic Spectra of Complex Molecules

Vibrations in a solid matrix doped with guest molecules can be of two types. The first is the intermolecular type, when each molecule vibrates with respect to the others. The quanta of these intermolecular vibrations are called phonons. Phonon frequencies occupy a frequency interval from 0 to 100 cm^{-1}. Phonons are an important characteristic of a solvent.

The other type of vibration is when the atoms of a molecule vibrate. These intramolecular vibrations of atoms are characterized by frequencies of the order of a few hundred cm^{-1}. Intramolecular vibrations of a guest molecule play the most important role, because these vibrations interact with the electronic degrees of freedom of a guest molecule excited by laser light. It is obvious that we can neglect the intramolecular vibrations of the host molecules because they do not interact with the optical electron of the guest molecule and hence will not manifest themselves in the optical spectrum of the guest molecule. The quantum of such an intramolecular vibration will be called a vibron and electron–vibron spectra will be called vibronic spectra. The theory developed in the last chapter can be applied not only to phonons, but also to the analysis of vibronic spectra.

11.1 Vibronic Spectra in the Condon Approximation. Similarity Rule in Vibronic Spectra

Intramolecular vibrations of a guest molecule are vibrations of a localized type. The localized mode is characterized by a narrow peak in the vibrational density of states $N_{\mathrm{ph}}(\nu)$. For simplicity, we consider the case when a single guest molecule with a single mode l and frequency Ω_l dopes the solvent. Then the vibrational density of states is given by

$$N_{\mathrm{ph}}(\nu) = \frac{1}{N} \sum_{q=1}^{N-1} \delta(\nu - \nu_q) + \frac{1}{N} \delta(\nu - \Omega_l) , \tag{11.1}$$

where N is the total number of vibrational degrees of freedom. It is obvious that the contribution of such a mode l to the vibrational density of states is proportional to the concentration of guest molecules, i.e., it is negligible. Therefore the vibration of the guest molecule will not manifest itself in non-optical bulk experiments like measurement of the heat capacity or IR absorption.

We face another situation when we examine the weighted density function of the vibrational states. It is given by

$$\varphi(\nu) = \sum_{q=1}^{N-1} \frac{a_q^2}{2} \delta(\nu - \nu_q) + \frac{a_l^2}{2} \delta(\nu - \Omega_l) , \tag{11.2}$$

where the dimensionless shift a_l of the equilibrium position does not depend on the concentration of the guest molecule and the shift can even exceed unity. Therefore a vibronic line is shifted by the frequency Ω_l from the origin of the optical spectrum of the guest molecule. It will be clearly seen in the spectrum despite the very low concentration of guest molecules. The difference between (11.1) and (11.2) demonstrates that the vibrational density of states found in non-optical experiments should be used in simulation of optical spectra by taking this fact into account.

Let us consider the optical spectrum of the guest molecule whose weighted density of vibrational states is described by (11.2). Then the function $\varphi(t, T)$ is a sum of two terms:

$$\begin{aligned} \varphi(t, T) &= \sum_{q=1}^{N} \frac{a_q^2}{2} \left[n_q \exp(\mathrm{i}\nu_q t) + (n_q + 1) \exp(-\mathrm{i}\nu_q t) \right] + \frac{a_l^2}{2} \exp(-\mathrm{i}t\Omega_l) \\ &= \varphi_{\mathrm{ph}}(t, T) + \varphi_\nu(t) . \end{aligned} \tag{11.3}$$

The first term describes a contribution from phonons and the second term relates to a vibron. As a rule the vibronic spectra are studied at low temperature when the energy of the vibronic quanta exceeds kT. This is allowed for in (11.3).

By inserting (11.3) into (10.50), we arrive at the expression

$$S^{\mathrm{g,e}}(\omega) = \frac{1}{2\pi} \int_{-\infty}^{\infty} \mathrm{e}^{\mathrm{i}(\omega - \Omega)t - |t|/(2T_1)} \mathrm{e}^{\varphi_{\mathrm{ph}}(\pm t, T) - \varphi_{\mathrm{ph}}(0, T) + \varphi_\nu(\pm t) - \varphi_\nu(0)} \mathrm{d}t . \tag{11.4}$$

If $a_l = 0$, the function $\varphi_\nu(t) = 0$. In this limiting case the vibronic mode does not manifest itself and therefore (11.4) can be transformed to (10.51). We denote the left hand side of this equation by $S^{\mathrm{g,e}}_{\mathrm{el-ph}}(\Delta)$. If $a_l \neq 0$, we can expand $\exp \varphi_\nu(t)$ as a power series in the function $\varphi_\nu(t)$. By carrying out the integration of all terms in the series, we arrive at the expression

$$S^{\mathrm{g,e}}(\Delta) = \mathrm{e}^{-\varphi_\nu(0)} S^{\mathrm{g,e}}_{\mathrm{el-ph}}(\Delta) + \Psi^{\mathrm{g,e}}(\Delta) , \tag{11.5}$$

where

$$\Psi^{\mathrm{g,e}}(\Delta) = \sum_{m=1}^{\infty} \Psi_m^{\mathrm{g,e}}(\Delta) \tag{11.6}$$

$$= \mathrm{e}^{-\varphi_\nu(0)} \sum_{m=1}^{\infty} \frac{1}{m!} \int_{-\infty}^{\infty} \mathrm{d}\nu_1 \varphi_\nu(\nu_1) \ldots \int_{-\infty}^{\infty} \mathrm{d}\nu_m \varphi_\nu(\nu_m) \times S_{\mathrm{el-ph}}^{\mathrm{g,e}}(\Delta \mp \nu_1 \ldots \mp \nu_m) \, ,$$

These equations are not only correct in the case of a single vibronic mode in a guest molecule. However, when

$$\varphi_\nu(\nu) = \frac{a_l^2}{2} \delta(\nu - \Omega_l) \, , \tag{11.7}$$

the integrals in (11.6) can be calculated and (11.5) takes the form

$$S^{\mathrm{g,e}}(\Delta) = \mathrm{e}^{-a_l^2/2} \left[S_{\mathrm{el-ph}}^{\mathrm{g,e}}(\Delta) + \sum_{m=1}^{\infty} \frac{1}{m!} \left(\frac{a_l^2}{2} \right)^m S_{\mathrm{el-ph}}^{\mathrm{g,e}}(\Delta \mp m\Omega_l) \right] . \tag{11.8}$$

If the function $S_{\mathrm{el-ph}}^{\mathrm{g,e}}(\Delta)$ is replaced by a Lorentzian, (11.8) coincides with (10.13) and (10.14) shown in Fig. 10.1. The function $S_{\mathrm{el-ph}}^{\mathrm{g,e}}(\Delta)$ differs from

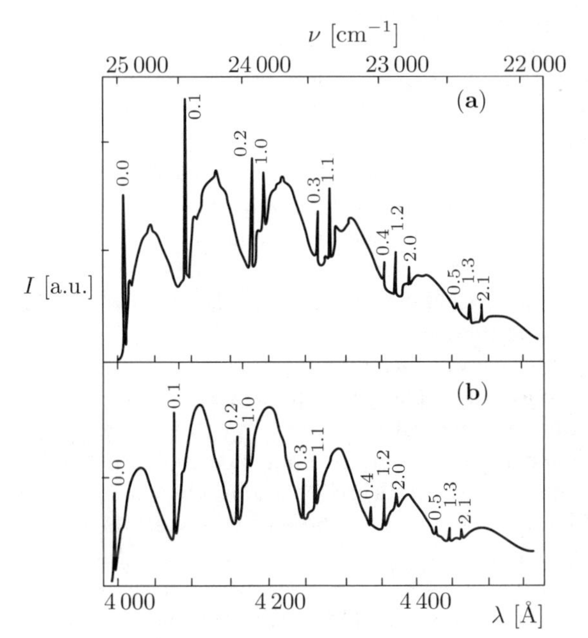

Fig. 11.1. Similarity rule in the fluorescence spectra of a PO_2 ion doping KCl (**a**) and KBr (**b**) at $T = 4.2$ K [63]

the Lorentzian by the existence of PSBs. Therefore, calculation with (11.8) yields Fig. 10.1 again, but every Lorentzian should be replaced by the optical band consisting of the ZPL and PSB. Hence, the vibronic spectrum consists of the pure electronic ZPL and the so-called vibronic ZPL. These vibronic ZPL relate to the vibronic transitions with creation of one, two, and so on, quanta of intramolecular vibrations. Every vibronic ZPL is accompanied by a PSB. This means that every vibronic spectrum consists of a number of copies of the electron–phonon band. We say that a similarity rule is satisfied in the vibronic spectrum. The rule is also satisfied in the case when there are many vibronic modes. The similarity rule is clearly seen in the measured vibronic spectra shown in Fig. 11.1 [63].

11.2 Influence of the HT Interaction on Optical Bands

In the last section, the electron–phonon and vibronic spectra were studied in the Condon approximation. This means that the influence of the FC interaction on the optical band shape was taken into account and the influence of the HT interaction was ignored. The HT interaction can play an important role in some cases although it is, as a rule, weaker than the linear FC interaction.

The HT interaction plays a decisive role in the case of forbidden pure electronic transitions. For example, the benzene molecule has D_{6h} symmetry. The pure electronic transition in the benzene molecule is forbidden due to the high symmetry of the electronic states. In the optical spectrum of a benzene molecule, there is very weak pure electronic optical line. It appears due to deformation of the benzene molecule in a crystal lattice. The role of the band origin in the absorption vibronic spectrum of the benzene molecule is played by the very intensive vibronic line shifted by the frequency 520 cm^{-1} from weak 0–0 transition. This frequency relates to the one-quantum vibration of the E_{2g} symmetry. This one-quantum vibronic peak emerges due to the HT interaction. It should be noted that all nonsymmetric vibrations of a complex molecule manifest themselves in its electronic spectrum solely due to the HT interaction. Let us discuss this statement in more detail.

Complex organic molecules generally have low symmetry, i.e., they only have axes of symmetry of second order. Therefore the representations of such a symmetry group are of one-dimensional type and cannot be degenerate. The shape of such a molecule cannot be changed upon electronic excitation. This means that displacements of the equilibrium positions which accompany electronic excitation must have a symmetry of the molecule, i.e., they must be of a fully symmetric type. Therefore $a_q = a_s \neq 0$, where s is the index of a fully symmetric vibration, whereas $a_q = a_n = 0$, where n is the index of a nonsymmetric vibration of the molecule. This means that nonsymmetric vibrations cannot manifest themselves in the electronic optical spectrum due to the FC interaction.

The HT interaction is due to modulation of the electronic dipolar moment $\boldsymbol{d}$ by atomic vibrations. This means that $\boldsymbol{d}(Q)$ is a function of the vibration coordinates Q. In this case the dipolar moment cannot be factored out of the integrals for the overlap of the oscillator functions. Therefore, instead of the functions $d^2 S^{\mathrm{g,e}}(\omega)$ for the band shape, we shall have in (10.9) for the measured values the functions $J^{\mathrm{g,e}}(\omega)$ given by

$$I^{\mathrm{g}}(\omega) = \frac{1}{\pi}\sum_{a,b} \rho_a \left|\langle b|d|a\rangle\right|^2 \frac{(1/2T_1)}{(\Delta + \Omega_b - \Omega_a)^2 + (1/2T_1)^2}\,,$$
$$I^{\mathrm{e}}(\omega) = \frac{1}{\pi}\sum_{a,b} \rho_b \left|\langle b|d|a\rangle\right|^2 \frac{(1/2T_1)}{(\Delta + \Omega_b - \Omega_a)^2 + (1/2T_1)^2} \tag{11.9}$$

where $\Delta = \Omega - \omega$. These formulas can be transformed in the same way that the formula (10.21) was transformed to (10.24). It should be taken into account that $H^{\mathrm{g}} = H^{\mathrm{e}} = H(Q)$, if we neglect the FC interaction. The result is given by

$$J^{\mathrm{g,e}}(\omega) = \frac{1}{2\pi}\int_{-\infty}^{\infty} \mathrm{e}^{\mathrm{i}(\omega-\Omega)t - t/(2T_1)} J^{\mathrm{g,e}}(t)\,\mathrm{d}t\,, \tag{11.10}$$

where

$$J^{\mathrm{g}}(t) = \left\langle d\big(Q(t)\big) d(Q)\right\rangle\,, \quad J^{\mathrm{e}}(t) = \left\langle d(Q)\, d\big(Q(t)\big)\right\rangle = \left[J^{\mathrm{g}}(t)\right]^*\,, \tag{11.11}$$

$$Q(t) = \exp\left[\mathrm{i}H(Q)\,t\right] Q \exp\left[-\mathrm{i}H(Q)\,t\right]\,. \tag{11.12}$$

The dimensionless coordinate Q_q is a ratio of the vibration coordinate R_q and the amplitude $\sqrt{\hbar/\mu\nu_q}$ of the zero-point vibrations. The amplitude is less than the distance between atoms in a molecule. On such a scale, the dipolar moment as a function of the space coordinates can be approximated by a linear function of the coordinates:

$$d(Q) = d(0) + \sum_q d_q Q_q = d(0)\left(1 + \sum_q \alpha_q Q_q\right)\,, \tag{11.13}$$

where

$$\alpha_q = d_q / d(0) \tag{11.14}$$

is the parameter of the linear HT interaction. By inserting this linear function into (11.11), we arrive at the following expression for the dipolar correlator:

$$J^{\mathrm{g}}(t) = d^2(0)\left(1 + \left\langle \sum_q \alpha_q Q_q(t) \sum_q \alpha_q Q_q \right\rangle\right) = d^2(0)\left[1 + h(t,T)\right]\,, \tag{11.15}$$

where the one-phonon function of time is given by

$$h(t,T)=\sum_q \frac{\alpha_q^2}{2}\left[(n_q+1)\,\mathrm{e}^{-\mathrm{i}\nu_q t}+n_q\mathrm{e}^{\mathrm{i}\nu_q t}\right] . \qquad (11.16)$$

This function is similar to the one-phonon function $\varphi(t,T)$ determined by (10.47), which emerges due to the linear FC interaction. If we substitute the HT parameter α_q for the FC parameter a_q in the function $\varphi(t,T)$, we arrive at the function $h(t,T)$. Replacing the function $J^{\mathrm{g}}(t)$ in (11.10) by the right hand side of (11.15) and integrating over time, we find that

$$J^g(\Delta)=d^2(0)\frac{(1/2T_1)\pi}{\Delta^2+(1/2T_1)^2} \qquad (11.17)$$

$$+\,d^2(0)\sum_q\frac{\alpha_q^2}{2}\left[(n_q+1)\frac{(1/2T_1)\pi}{(\Delta+\nu_q)^2+(1/2T_1)^2}+n_q\frac{(1/2T_1)\pi}{(\Delta-\nu_q)^2+(1/2T_1)^2}\right] ,$$

i.e., the optical band consists of the ZPL and one-phonon lines in the Stokes and anti-Stokes frequency domain. Here, the index q labels both phonons and vibrons. The expression for the fluorescence band can be obtained by substituting $-\Delta$ for Δ, i.e., $J^{\mathrm{e}}(\Delta)=J^{\mathrm{g}}(-\Delta)$.

In contrast to the FC interaction, the HT interaction does not yield the temperature dependence of the ZPL, but rather the temperature dependence of the whole electron–phonon band:

$$J^{\mathrm{g,e}}=d^2(0)\left[1+h(0,T)\right] . \qquad (11.18)$$

The Debye–Waller factors for the cases of HT and FC interactions are given by

$$\alpha_{\mathrm{HT}}(T)=\frac{1}{1+h(0,T)} ,\quad \alpha_{\mathrm{FC}}(T)=\exp\left[-\varphi(0,T)\right] . \qquad (11.19)$$

When the temperature rises, both Debye–Waller factors approach zero. However, the Debye–Waller factor α_{FC} approaches zero more rapidly.

11.3 Interference of HT and FC Amplitudes. Breakdown of the Mirror Symmetry of Conjugate Absorption and Fluorescence Bands

Each molecule has some elements of the symmetry which generate a symmetry group of the molecule. Each intramolecular natural vibration can be classified by its behavior with respect to the symmetry operation from the symmetry group. The displacements of the atoms in the natural vibration can keep a symmetry of the molecule or can infringe this symmetry. In accordance with this fact the natural vibration can be of symmetric or asymmetric type. For example a natural vibration of a linear three-atom molecule with atomic displacements along the molecular axis is a vibration of symmetric type, whereas a natural vibration with atomic displacements of the transverse type will break the linear symmetry of this molecule, so that this vibration is of asymmetric type. More information about classifying natural vibrations with the help of the symmetry group can be found in [38].

It has been already shown in the last section that symmetric vibrations manifest themselves in vibronic spectra due to the FC interaction, whereas asymmetric vibrations cannot manifest themselves due to the FC interaction. Instead they manifest themselves in vibronic spectra through the HT interaction. Indeed symmetric modes manifest themselves in realistic vibronic spectra via a sequence of vibronic lines with creation of one, two, three, and even more quanta of the symmetric vibration. In contrast to this, asymmetric modes manifest themselves, as a rule, via one-quantum vibronic lines. This fact proves the suitability of the linear approximation taken in (11.13) for the HT interaction. When a vibration mode manifests itself either via the FC interaction or via the HT interaction, the model predicts mirror symmetry of the conjugate absorption and fluorescence spectra. However, in realistic vibronic spectra, breakdown of the m irror symmetry is often observed. One possible reason for this fact can be the joint influence of FC and HT interactions on the vibronic mode.

Indeed, in contrast to the equilibrium displacements a_q, which can be of purely symmetric type, the HT parameter α_q can be nonvanishing both for symmetric and asymmetric modes. If $a_q \neq 0$ and $\alpha_q \neq 0$ for some symmetric mode, this means that both the FC and HT interactions contribute to the amplitude of the vibronic line, i.e., the full amplitude consists of FC and HT amplitudes. If both amplitudes have one sign in the absorption spectrum, these amplitudes must have opposite signs in the emission spectrum.

Let us consider this statement in more detail. One-quantum vibronic amplitudes in conjugate absorption and emission spectra are given by

$$\begin{aligned}\langle 1|d(R)|0\rangle &= \int_{-\infty}^{\infty} \Phi_1\,(R-a)\, d\,(R)\, \Phi_0\,(R)\,\mathrm{d}R\,, \\ \langle 0|d(R)|1\rangle &= \int_{-\infty}^{\infty} \Phi_0\,(R-a)\, d\,(R)\, \Phi_1\,(R)\,\mathrm{d}R\,.\end{aligned} \tag{11.20}$$

Here Φ is the harmonic oscillator function. The conjugate transitions described by these amplitudes are shown in Fig. 11.2. Carrying out the substitution $R = \rho - a/2$ for the integration variable and taking into account the fact that $\Phi_1(R) = 2^{1/2} R\Phi_0(R)$, we can transform the integrals to the form

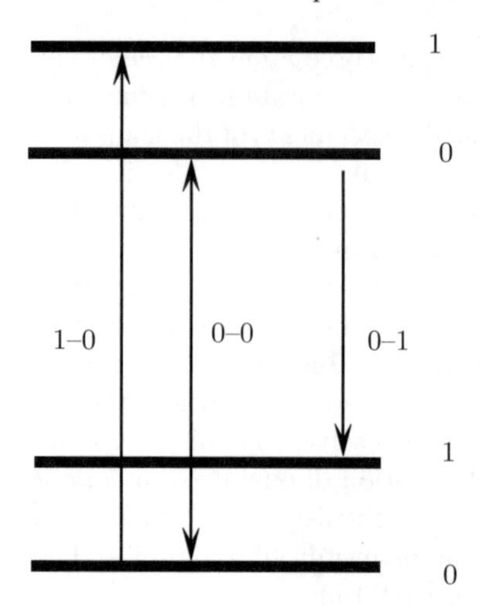

Fig. 11.2. One-phonon conjugate transitions. The 0–0 transition is the one without phonons

$$\begin{aligned}\langle 1|d|0\rangle &= \sqrt{2}\int_{-\infty}^{\infty}\left(\rho-\frac{a}{2}\right)\Phi_0\left(\rho-\frac{a}{2}\right)d\left(\rho+\frac{a}{2}\right)\Phi_0\left(\rho+\frac{a}{2}\right)\mathrm{d}\rho\,,\\ \langle 0|d|1\rangle &= \sqrt{2}\int_{-\infty}^{\infty}\left(\rho+\frac{a}{2}\right)\Phi_0\left(\rho-\frac{a}{2}\right)d\left(\rho+\frac{a}{2}\right)\Phi_0\left(\rho+\frac{a}{2}\right)\mathrm{d}\rho\,.\end{aligned} \tag{11.21}$$

Making the linear approximation $d(\rho + a/2) = d_0 + d_1\rho$ for the dipolar moment, we can transform the expressions for the conjugate amplitudes to give

$$\begin{aligned}\langle 1|d|0\rangle &= d_0\frac{-a}{\sqrt{2}}\int_{-\infty}^{\infty}\Phi_0\left(\rho-\frac{a}{2}\right)\Phi_0\left(\rho+\frac{a}{2}\right)\mathrm{d}\rho\\ &\qquad +\sqrt{2}d_1\int_{-\infty}^{\infty}\Phi_0\left(\rho-\frac{a}{2}\right)\rho^2\Phi_0\left(\rho+\frac{a}{2}\right)\mathrm{d}\rho\,,\\ \langle 0|d|1\rangle &= d_0\frac{a}{\sqrt{2}}\int_{-\infty}^{\infty}\Phi_0\left(\rho-\frac{a}{2}\right)\Phi_0\left(\rho+\frac{a}{2}\right)\mathrm{d}\rho\\ &\qquad +\sqrt{2}d_1\int_{-\infty}^{\infty}\Phi_0\left(\rho-\frac{a}{2}\right)\rho^2\Phi_0\left(\rho+\frac{a}{2}\right)\mathrm{d}\rho\,.\end{aligned} \tag{11.22}$$

Here the first terms describe FC amplitudes and the second terms describe HT amplitudes. It is clear that the intensities of the conjugate amplitudes are quite different. The mirror symmetry of one-quantum vibronic lines of the conjugate spectra will be broken.

This example demonstrates that calculation even of one-quantum amplitudes is no simple task if both interactions contribute to the amplitude.

Nevertheless, we can successfully apply the technique developed in Sect. 10.3 for the FC interaction and generalize it so that it can take into account both the FC and HT interactions. The generalization is carried out in Appendix H. The result of the calculation is

$$J^{g,e}(\Delta) = d^2 \left[J_0(\Delta) + J_1(\Delta) + J_2(\Delta) + \cdots\right] , \tag{11.23}$$

where the functions of the detuning are given by

$$J_0(\Delta) = e^{-\varphi(0,T)} \frac{(1/2T_1)\pi}{\Delta^2 + (1/2T_1)^2} + \Psi^{g,e}(\Delta) , \tag{11.24}$$

$$J_1(\Delta) = \frac{1}{2} \sum_q (\mp a_q + \alpha_q)^2 J_0(\Delta \mp \nu_q) ,$$

$$J_2(\Delta) = \frac{1}{2!} \frac{1}{2^2} \sum_{q,q'} (a_q a_{q'} \mp a_q \alpha_{q'} \mp a_{q'} \alpha_q + \alpha_{qq'})^2 J_0(\Delta \mp \nu_q \mp \nu_{q'}) ,$$

$$J_3(\Delta) = \frac{1}{3!} \frac{1}{2^3} \sum_{q,q',q''} (\mp a_q a_{q'} a_{q''} + a_q a_{q'} \alpha_{q''} + a_q a_{q''} \alpha_{q'} + a_{q'} a_{q''} a_q \mp$$
$$\mp a_q \alpha_{q'q''} \mp a_{q'} \alpha_{qq''} \mp a_{q''} \alpha_{qq'} + \alpha_{qq'q''})^2 J_0(\Delta \mp \nu_q \mp \nu_{q'} \mp \nu_{q''}) ,$$

and so on. They determine the pure electron–phonon band, and the vibronic bands with creation of one, two, three, and so on, intramolecular vibrations. The upper sign is for the absorption band and the lower sign is for the fluorescence band. Here,

$$\alpha_q = \left(\frac{\partial}{\partial Q_q} D(Q) \Big/ d(Q) \right)_{a/2} ,$$
$$\alpha_{qq'} = \left(\frac{\partial^2}{\partial Q_q \partial Q_{q'}} D(Q) \Big/ d(Q) \right)_{a/2} , \tag{11.25}$$
$$\alpha_{qq'q''} = \left(\frac{\partial^3}{\partial Q_q \partial Q_{q'} Q_{q''}} D(Q) \Big/ d(Q) \right)_{a/2}$$

are the parameters of the linear, quadratic and cubic HT interaction. The most important role is played by the parameters α_q of the linear interaction. For the vibronic lines which combine a symmetric and an asymmetric mode, the parameter α_{sa} can play an important role.

11.4 Theoretical Treatment of Electron–Phonon and Vibronic Spectra

The theory developed in the last section can be used to obtain information concerning phonons and vibrons, and electron–phonon and vibronic interactions from the electron–phonon and vibronic spectra measured in experiments. Such an analysis was carried out in [64–69]. Let us now discuss two examples.

Analysis of the Electron–Phonon Band. Figure 11.3 shows the conjugate electron–phonon bands of 3,4,5,6-dibenzopyrene in normal octane [65]. They consist of a ZPL and a PSB. This PSB is rather narrow because it is formed by a quasi-localized mode. The mode is created by a guest molecule. These spectra are inhomogeneously broadened. The inhomogeneous broadening is of the order of 1 cm^{-1}. These spectra were measured and analyzed in [65].

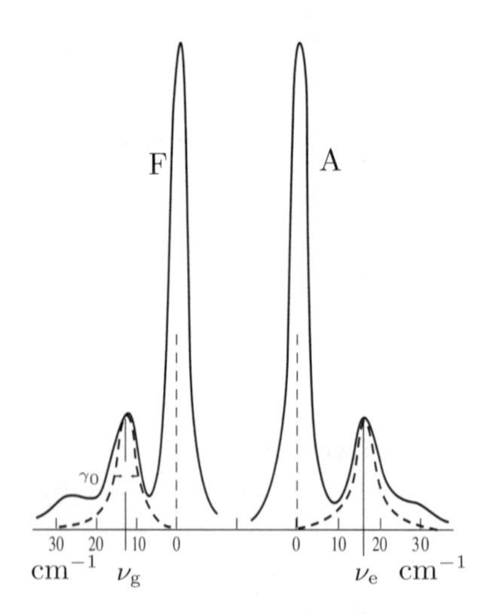

Fig. 11.3. Fluorescence (F) and excitation fluorescence (A) electron–phonon bands of a 3,4,5,6-dibenzopyrene molecule doping octane at 4.2 K. Observed bands (*solid line*) and bands after removal of inhomogeneous broadening (*dashed line*) [65]

First of all, the mirror symmetry of these bands is clearly visible. Therefore the Debye–Waller factors of both spectra are characterized by one value. The measurement yields $\alpha = 0.72$. We conclude that there is no interference between FC and HT amplitudes. Therefore the PSBs emerge due either to the pure FC or the pure HT interaction. The sample was heated and the decrease in the Debye–Waller factor was measured. The relevant experimental data are plotted in Fig. 11.4. The temperature behavior of the Debye–Waller factor given by (11.19) was calculated with the help of the functions

$$
\begin{aligned}
h(0,T) &= \int_0^\infty h(\nu)[2n(\nu)+1]\mathrm{d}\nu \,, \\
\varphi(0,T) &= \int_0^\infty \varphi(\nu)[2n(\nu)+1]\mathrm{d}\nu \,.
\end{aligned}
\tag{11.26}
$$

These formulas correspond to the pure HT and FC interactions. The one-phonon functions $h(\nu)$ and $\varphi(\nu)$ can be found with the help of the PSB

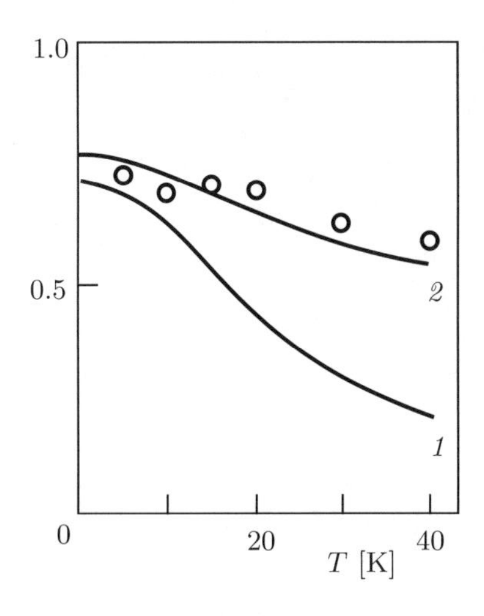

Fig. 11.4. Temperature dependence of the Debye–Waller factor. *Circles* indicate experimental data. Curve 1 is calculated with the Franck–Condon (FC) interaction, curve 2 with the Herzberg–Teller (HT) interaction [65]

because the one-phonon part of this PSB at $T = 0$ is the function $h(\nu)$ or $\varphi(\nu)$. Two possibilities were considered: the PSB results from the FC or the HT interaction. With the functions $h(\nu)$ or $\varphi(\nu)$ found from the PSB, the temperature dependence of the Debye–Waller factor was determined and the result of the calculation is shown in Fig. 11.4. Theoretical curves were calculated without any adjustable parameters. The curve relating to the HT interaction fits the experimental data much better. It proves that the PSB results from the pure HT interaction.

In the case considered, the PSB consists of a localized mode and there is no problem in extracting one-phonon transitions from the PSB because two-phonon transitions are shifted on the frequency scale. If the PSB consists of acoustic modes, as happens for the PSB shown in Fig. 10.3, the problem of finding the one-phonon function from the PSB becomes somewhat complicated. However, even in this complicated case, the problem of extracting the one-phonon function from the PSB can be solved. This was carried out in [62].

Analysis of Vibronic Spectra. It is convenient to study vibronic spectra of complex organic molecules when these molecules dope Shpol'ski matrices at low temperature. Inhomogeneous broadening and electron–phonon coupling in these matrices are weak, so that the vibronic spectra consist of a narrow vibronic ZPL without PSB. This fact is clearly visible in Fig. 11.5, which shows the fluorescence spectrum of a naphthalene molecule doping normal pentane [66]. Analysis of such a spectrum can be carried out using (11.24) without the terms describing the PSB.

Further, we denote the symmetric and asymmetric modes by the indices s and n, respectively. Recall that symmetric and asymmetric modes manifest themselves in vibronic spectra due to the FC and HT interactions, respectively. This means that

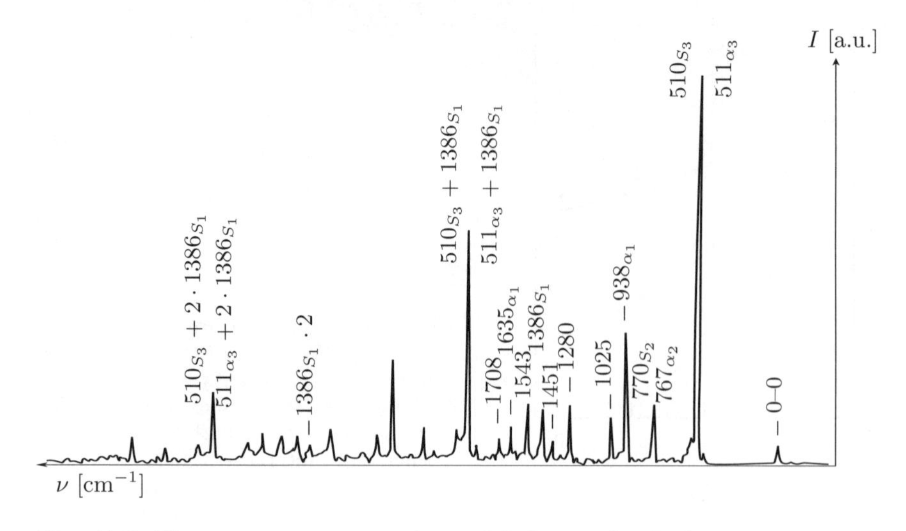

Fig. 11.5. Fluorescence spectrum of a naphthalene molecule doping pentane. $T = 4.2$ K, $C = 10^4$ m/l, $\lambda_{\text{exc}} = 288$ nm [66]

$$a_q = \begin{cases} a_s \,, & q = s \,, \\ 0 \,, & q = n \,, \end{cases} \qquad a_q = \begin{cases} 0 \,, & q = s \,, \\ a_n \,, & q = n \,. \end{cases} \tag{11.27}$$

According to this model, (11.23) takes the form

$$J^{\text{g,e}}(\Delta) = d^2(0) J_{\text{FC}}(\Delta) + \sum_n \left[\frac{\partial}{\partial Q_n} d\,(Q) \right]^2_{Q=0} J_{\text{FC}}\,(\Delta \mp \Omega_n) \,, \tag{11.28}$$

where

$$J_{\text{FC}}(\Delta) = \mathrm{e}^{-\varphi(0,T)} \left[\frac{(1/2T_1)\pi}{\Delta^2 + (1/2T_1)^2} + \sum_s \frac{a_s^2}{2} \frac{(1/2T_1)\pi}{(\Delta \mp \Omega_s)^2 + (1/2T_1)^2} \right.$$
$$\left. + \frac{1}{2!} \sum_{s,s'} \frac{a_s^2}{2} \frac{a_{s'}^2}{2} \frac{(1/2T_1)\pi}{(\Delta \mp \Omega_s \mp \Omega_{s'})^2 + (1/2T_1)^2} + \cdots \right] \tag{11.29}$$

is a part of the vibronic spectrum consisting solely of vibronic lines with symmetric modes. Here Ω_s and Ω_n are the frequencies of the symmetric and asymmetric modes. In accordance with (11.28), the vibronic spectrum is one of FC type.

The pure electronic ZPL and the ZPL relating to one-quantum asymmetric modes play the role of the origins. If the pure electronic transition is forbidden, i.e., $d(0) = 0$, only one-quantum asymmetric modes play the role of the origin. This situation is realized in the benzene molecule. Since the pure electronic transition is forbidden, there is a gap between the conjugate absorption and fluorescence spectra.

Breakdown of the mirror symmetry in the sequences of symmetric modes results from the joint influence of FC and HT interactions. The simplest model which is able to describe the conjugate vibronic spectra takes the mathematical form

$$a_q = \begin{cases} a_s \,, & q = s \,, \\ 0, & q = n \,, \end{cases} \qquad \alpha_q = \begin{cases} \alpha_s \,, & q = s \,, \\ \alpha_n \,, & q = n \,. \end{cases} \tag{11.30}$$

In this case the general formula (11.23) takes the form

$$J^{\mathrm{g,e}}(\Delta) = d^2(0)\, J'(\Delta) + \sum_n \left[\frac{\partial}{\partial Q_n} d(Q) \right]^2_{Q=0} J_{\mathrm{FC}}(\Delta \mp \Omega_n) \,. \tag{11.31}$$

This equation is similar to (11.28). However, the expression for the function $J'(\Delta)$ which describes the sequence of vibronic lines with the pure electronic ZPL as the origin is given by

$$J'(\Delta) = \mathrm{e}^{-\varphi(0,T)} \left[\frac{(1/2T_1)\pi}{\Delta^2 + (1/2T_1)^2} + \sum_s (\mp a_s + \alpha_s)^2 \frac{(1/4T_1)\pi}{(\Delta \mp \Omega_s)^2 + (1/2T_1)^2} \right.$$
$$\left. + \frac{1}{2!} \sum_{s,s'} (a_s a_{s'} \mp a_s \alpha_{s'} \mp a_s \alpha_s)^2 \frac{(1/4T_1)\pi}{(\Delta \mp \Omega_s \mp \Omega_{s'})^2 + (1/2T_1)^2} + \cdots \right] . \tag{11.32}$$

This function only includes vibronic lines of symmetric modes. However, vibronic lines from these sequences in the conjugate absorption and emission spectra have different intensities, whereas the intensities of the vibronic lines in the function $J_{\mathrm{FC}}(\Delta \mp \Omega_n)$ describe equal parts of vibronic absorption and fluorescence spectra in both spectra.

As an example, let us consider the analysis of the vibronic spectrum of a naphthalene molecule carried out in [66] and [67]. The analysis of the absorption spectrum shows that the distance between the first and second singlet electronic transitions is about 3 000 cm^{-1}. Since the final quantum states of the transition with the absorption of light quanta lie in the region where two electronic excited levels are involved, the absorption spectrum involves non-adiabatic effects and cannot be described by our theory. The final quantum states of the transition with light emission belong to the ground electronic state. Therefore we can neglect non-adiabatic effects when considering the fluorescence vibronic spectrum.

The intensities of the vibronic lines are given by

$$J_n = \frac{\alpha_n^2}{2} \,, \quad J_s = \left[\mp \frac{a_s}{\sqrt{2}} + \frac{\alpha_s}{\sqrt{2}} \right]^2 , \qquad J_{2s} = \frac{1}{2} \left[\frac{a_s^2}{2} \mp a_s \alpha_s \right]^2 ,$$
$$J_{n+s} = \frac{a_s^2}{2} \frac{\alpha_n^2}{2} \,, \quad J_{n+2s} = \frac{1}{2} \left[\frac{a_s^2}{2} \frac{\alpha_n^2}{2} \right]^2 . \tag{11.33}$$

The analysis of the fluorescence vibronic spectrumshown in Fig. 11.5 was carried out in [66] as follows. Values of the parameters a_s, α_s, and α_n were chosen with the help of the measured intensities of the most intense lines, where inaccuracy in the measurement was minimal. The intensities of other lines were calculated with the help of the parameters found previously. Since the total number of the measured vibronic lines was three times greater than the number of lines used to find the unknown parameters, the good agreement between calculated and measured intensities proves the suitability of the model used for the vibronic interaction.

To what extent is the set of parameters found unique? The following example demonstrates that the analysis does indeed yield the unique set of parameters and that it enables one to explain very fine features of the vibronic spectrum.

The vibronic line with frequency 767 cm^{-1} in the fluorescence spectrum is usually identified as a transition with creation of one quantum of an asymmetric mode. Analysis with the model described by (11.33) enables one to conclude that this vibronic line consists of two unresolved peaks of asymmetric and symmetric modes. It was found that the value of the FC parameter $a^2/2$ for this symmetric mode should be 0.17 in order to get a good fit between the calculated intensity of the vibronic peak at frequency 767 cm^{-1} and the measured intensity.

When the authors of [66] came to this conclusion, they were not aware that a few years earlier another group had studied the shape of the same vibronic line in the excitonic absorption spectra of the pure naphthalene crystal [70]. This group came to the conclusion that the shape of this excitonic line fits the measured shape well if one assumes that the FC interaction can contribute to this line for the asymmetric vibration, and the value of the parameter $a^2/2$ for the FC interaction was found to be 0.2. Thus, two groups independently studying different optical characteristics of a naphthalene molecule, arrived at similar conclusions about the physical nature of the peak with frequency 700 cm^{-1}. Moreover, the values of the parameter $a^2/2$ found independently from the shape of the excitonic band and from the intensity of the vibronic line of the guest molecule are very close to one another. The fact that the conclusion and values of the parameter found in independent experiments should coincide is very impressive and demonstrates the great potential of the theory of vibronic interactions presented here.

12. Dynamical Theory of Line Broadening

The theory of line broadening studied in Chap. 9 is based on a stochastic approach in which the probabilities of jumps between various spectral positions are modeled. This approach enables one to find an expression for the line width. However, it is unable to explain the appearance of the PSB or the vibronic spectrum. It has been shown in the last few chapters that all peculiarities of the PSB and vibronic spectrum can be successfully explained within the framework of a dynamic approach based on the Hamiltonian of the impurity center. However, within the framework of the dynamic approach, we have not yet discussed why the FWHM of the ZPL is one or two orders of magnitude larger than the fluorescence FWHM $1/T_1$.

The problem of the phonon and vibron structure of the optical spectrum was studied with the help of the HT interaction and the linear FC interaction. Both interactions determine values of electron–phonon and vibronic transitions, but they do not yield line broadening. The quadratic FC interaction only weakly influences the intensities of optical transitions and we thus omitted this interaction in Chaps. 10 and 11. However the quadratic FC interaction must be taken into account if we are interested in line broadening because, within the framework of the adiabatic approximation, only the quadratic electron–phonon or electron–tunnelon FC interaction results in line broadening. The influence of the quadratic electron–phonon interaction on the ZPL is considered in this chapter.

The dynamical theory for the temperature broadening of the ZPL was first developed using perturbation theory [71, 72]. This approach has serious shortcomings because, in contrast to the linear FC interaction, the quadratic FC interaction yields a renormalization of the phonon spectrum of the system, i.e., it changes phonon frequencies. Therefore, phonon frequencies of the ground electronic state differ from the frequencies of the excited electronic state. It is well known that the frequencies of the final electronic state manifest themselves in the optical spectrum. For instance, the frequencies of the excited electronic state must manifest themselves in the absorption spectrum. However, if we consider the FC interaction using perturbation theory, the formulas involve the phonon frequencies of the ground electronic state. This serious shortcoming can be overcome only by summing the infinite series in the coupling constant. After this summation, a renormalization of the phonon frequencies occurs and the final expression for the absorption band includes the phonon frequencies of the excited electronic state. Such a theory that goes beyond the framework of perturbation theory to account for the quadratic FC interaction was developed by the author [47, 73–75] and is presented in this chapter.

The theory enables one to find a number of exact relations between the absorption and fluorescence spectra. For instance, the ZPL of the absorption and fluorescence spectra are described by different mathematical expressions because they depend on the phonon frequencies of the excited and ground electronic states, respectively. However, the non-perturbative theory allows us to prove the identity of the mathematical expressions for both ZPLs.

The results of the theory [73–75] contradict the results of other theories and therefore Hsu and Skinner [76, 77] undertook a special investigation to find out which theory is correct. They found that the formulas of [73–75] are correct. The formulas for the temperature broadening and shift of the ZPL are more general than the formulas obtained using the stochastic theory. Moreover, no stochastic theory takes into account the fact that the correlation functions of the absorption and fluorescence spectra are different.

The dynamical approach developed in [73–75] for the electron–phonon bands enables one to generalize it in such a way as to be able to develop subsequently a dynamical theory for electron–tunnelon bands [78, 79] and a dynamical theory for spectral diffusion, presented in Part VII of this book.

12.1 Specific Features of the Quadratic FC Interaction

The difference between two adiabatic potentials which are quadratic functions of the phonon coordinates determines a quadratic FC interaction. We shall neglect tunneling degrees of freedom in this section and consider only the electron–phonon FC interaction, given by

$$\hat{W} = R\frac{U^{\mathrm{e}} - U^{\mathrm{g}}}{2}R = R\frac{W}{2}R\,. \tag{12.1}$$

In the one-mode approximation, W is the change in the force constant upon electronic excitation, i.e., the quadratic interaction describes the change in the slope of the parabolic curves upon electronic excitation.

In the realistic multimode case, U is a force matrix and W is the change in this matrix upon electronic excitation. This change is given by

$$\hat{W} = \sum_{\boldsymbol{n}i,\boldsymbol{m}j} R_{\boldsymbol{n}i}\frac{W_{\boldsymbol{n}i,\boldsymbol{m}j}}{2}R_{\boldsymbol{m}j}\,, \tag{12.2}$$

where $W_{\boldsymbol{n}i,\boldsymbol{m}j}$ is the change in the force constant connecting molecules in the $\boldsymbol{n}$th and $\boldsymbol{m}$th crystal nodes. It is obvious that this change decreases rapidly when the distance between the guest molecules and these nodes increases. If the guest molecule occupies node 0, one of the indices of the matrix $W_{\boldsymbol{n}i,\boldsymbol{m}j}$ equals 0. Let us consider the cases of the optical and acoustic modes separately.

Optical Modes. The displacements R can be of orientational type. The simplest model is described by the matrix $W_{\boldsymbol{n}i,\boldsymbol{m}j} = \delta_{\boldsymbol{n}0}\delta_{\boldsymbol{m}0}W$, where W describes the change in the elastic constant. Using (5.16), which describes a relation between the coordinates in the node representation with the natural coordinates $Q_q = (b_q + b_q^+)/\sqrt{2}$, we arrive at the expression

$$\hat{W} = \frac{\hbar W}{2} R_0^2 = \frac{\hbar W}{2} \left(b + b^+\right)^2 \tag{12.3}$$

for the quadratic FC interaction, where

$$b = \sum_q l_q^{\mathrm{opt}} b_q \,, \qquad l_q^{\mathrm{opt}} = \sum_j \sqrt{\frac{1}{2\nu_q}} u(0j,q) \,, \tag{12.4}$$

and the dimensionless coefficients $u(0j,q)$ are solutions of (5.11). Hereafter, the parameter W has the dimensions of a square of the frequency.

Acoustic Modes. It was shown at the beginning of Sect. 5.1 that the elements of the force matrix satisfy the equations

$$\sum_{\boldsymbol{n}i} U_{\boldsymbol{n}i,\boldsymbol{m}j} = \sum_{\boldsymbol{m}j} U_{\boldsymbol{n}i,\boldsymbol{m}j} = 0 \,, \tag{12.5}$$

which express the fact that the potential energy cannot be changed by a translation of all the molecules. It seems that the same condition will be fulfilled for the matrix W. For the simplest case, we find that $W_{00} = -W_{10} = -W_{01} = W_{11}$. The quadratic FC interaction is given by

$$\hat{W} = \frac{\hbar W}{2} \left(R_0 - R_1\right)^2 = \frac{\hbar W}{2} \left(b + b^+\right)^2 \,, \tag{12.6}$$

where

$$b = \sum_q l_q^{\mathrm{ac}} b_q \,, \qquad l_q^{\mathrm{ac}} = \sum_j \sqrt{\frac{1}{2\nu_q}} \left[u\left(0j,q\right) - u\left(1j,q\right)\right] . \tag{12.7}$$

Comparison of the FC interaction for the cases of acoustic and optical modes shows that only the coupling function l_q differs in the two cases. Therefore we can take the quadratic FC interaction as

$$\hat{W} = \frac{\hbar W}{2} R^2 = \frac{\hbar W}{2} \sum_{q,q'} l_q l_{q'} \left(b_q + b_q^+\right)\left(b_{q'} + b_{q'}^+\right) . \tag{12.8}$$

This interaction includes terms which are off-diagonal in the indices of the natural coordinates. This means that the natural phonon coordinates in the ground and excited electronic state are different. Therefore the multidimensional FC integrals cannot be expressed as a product of one-dimensional integrals, as was possible in the case of the linear FC interaction. This fact considerably complicates the calculation of the optical band shape when we allow for the quadratic FC interaction.

12.2 Cumulant Expansion of the Dipolar Correlator

The mathematical method used to account for the linear FC interaction was based on the fact that multidimensional FC integrals can be calculated with the help of a one-dimensional integral. In the case of the quadratic FC interaction, this is impossible. A new method for calculating FC integrals was therefore developed by the author. This method will be used hereafter.

The optical band shape is described by the expression

$$J^{\mathrm{g,e}}(\omega) = \frac{1}{2\pi} \int_{-\infty}^{\infty} \mathrm{e}^{\mathrm{i}(\omega-\Omega)t-|t|/(2T_1)} J^{\mathrm{g,e}}(t)\,\mathrm{d}t\,, \tag{12.9}$$

where the dipolar correlators in the Condon approximation can be expressed as

$$J^{\mathrm{g}}(t) = d^2\,\mathrm{Tr}\left[\hat{\rho}^{\mathrm{g}}(T)\exp\left(-\mathrm{i}t\frac{H^{\mathrm{e}}}{\hbar}\right)\exp\left(\mathrm{i}t\frac{H^{\mathrm{g}}}{\hbar}\right)\right] = d^2\langle\hat{S}(t)\rangle_{\mathrm{g}}\,, \tag{12.10}$$

$$J^{\mathrm{e}}(t) = d^2\,\mathrm{Tr}\left[\hat{\rho}^{\mathrm{e}}(T)\exp\left(-\mathrm{i}t\frac{H^{\mathrm{e}}}{\hbar}\right)\exp\left(\mathrm{i}t\frac{H^{\mathrm{g}}}{\hbar}\right)\right] = d^2\langle\hat{S}(t)\rangle_{\mathrm{e}}\,. \tag{12.11}$$

It is obvious that the dipolar correlator $J^{\mathrm{e}}(t)$ can be derived from the dipolar correlator $J^{\mathrm{g}}(t)$ by complex conjugating and swapping indices e and g. Therefore one need only study in detail one of the dipolar correlators, for instance $J^{\mathrm{g}}(t)$.

We can carry out circular permutations under signTr in (12.10) and (12.11) and introduce an operator $\hat{S}(t) = \exp(\mathrm{i}tH^{\mathrm{g}}/\hbar)\exp(-\mathrm{i}tH^{\mathrm{e}}/\hbar)$. In order to express the operator $\hat{S}(t)$ in a form which is more convenient for calculating quantum statistical averages, we differentiate it with respect to time. This yields

$$\frac{\mathrm{d}}{\mathrm{d}t}\hat{S}(t) = -\frac{\mathrm{i}}{\hbar}\hat{W}(t)\hat{S}(t)\,, \tag{12.12}$$

where

$$\hat{W}(t) = \exp\left(\frac{\mathrm{i}tH^{\mathrm{g}}}{\hbar}\right)(H^{\mathrm{e}} - H^{\mathrm{g}})\exp\left(-\frac{\mathrm{i}tH^{\mathrm{g}}}{\hbar}\right)\,. \tag{12.13}$$

Integrating this differential equation yields the integral equation

$$\hat{S}(t) = 1 - \frac{\mathrm{i}}{\hbar} \int_0^t \mathrm{d}t_1 \hat{W}(t_1) \hat{S}(t_1) . \tag{12.14}$$

The equation can be solved by an iteration procedure:

$$\hat{S}(t) = 1 + \sum_{m=1}^{\infty} \left(-\frac{\mathrm{i}}{\hbar}\right)^m \int_0^t \mathrm{d}t_1 \int_0^{t_1} \mathrm{d}t_2 \ldots \int_0^{t_{m-1}} \mathrm{d}t_m \hat{W}(t_1) \hat{W}(t_2) \ldots \hat{W}(t_m) . \tag{12.15}$$

The quantum statistical average of both sides of this equation can be calculated using the Wick–Bloch–Dominicis (WBD) theorem proven in Appendix G. However, the integration of the the expressions obtained after applying the WBD theorem is a rather complicated task. In order to facilitate this integration, the right hand side of (12.15) is symmetrized with the help of the so-called time-ordering operation.

If the operators $W(t_j)$ were commutative at different times, any permutation of the operators in the integrand would not change the expression on the right hand side of (12.15). Therefore we could write the right hand side in the symmetric form

$$\begin{aligned} &\int_0^t \mathrm{d}t_1 \int_0^{t_1} \mathrm{d}t_2 \ldots \int_0^{t_{m-1}} \mathrm{d}t_m \hat{W}(t_1)\hat{W}(t_2)\ldots\hat{W}(t_m) \\ &\qquad = \frac{1}{m!} \int_0^t \mathrm{d}t_1 \int_0^t \mathrm{d}t_2 \ldots \int_0^t \mathrm{d}t_m \hat{W}(t_1)\hat{W}(t_2)\ldots\hat{W}(t_m) . \end{aligned} \tag{12.16}$$

It is obvious that the right hand side is the mth term of a series representing an exponential. However, the operators $W(t_j)$ are not commutative. Nevertheless, some kind of symmetrization can be carried out in this case as well. Indeed the left hand side of the last equation can be written in the form

$$\begin{aligned} &\int_0^t \mathrm{d}t_1 \int_0^{t_1} \mathrm{d}t_2 \ldots \int_0^{t_{m-1}} \mathrm{d}t_m \hat{W}(t_1)\hat{W}(t_2)\ldots\hat{W}(t_m) \\ &\quad = \int_0^t \mathrm{d}t_1 \int_0^t \mathrm{d}t_2 \ldots \int_0^t \mathrm{d}t_m \theta(t_1, t_2, \ldots, t_m) \hat{W}(t_1)\hat{W}(t_2)\ldots\hat{W}(t_m) , \end{aligned} \tag{12.17}$$

where the function $\theta(t_1, t_2, \dots, t_m)$ equals unity only if $t_1 > t_2 > \dots > t_m$. For other time orders, the function equals zero. Any permutation of times on the right hand side of the last equation does not change the integral. If we carry out all $m!$ possible permutations of times on the left hand side, then sum all terms and divide the sum by $m!$, we find

$$\int_0^t \mathrm{d}t_1 \int_0^{t_1} \mathrm{d}t_2 \dots \int_0^{t_{m-1}} \mathrm{d}t_m \hat{W}(t_1)\hat{W}(t_2)\dots\hat{W}(t_m)$$
$$= \frac{1}{m!}\int_0^t \mathrm{d}t_1 \int_0^t \mathrm{d}t_2 \dots \int_0^t \mathrm{d}t_m \hat{T}\hat{W}(t_1)\hat{W}(t_2)\dots\hat{W}(t_m)\,, \quad (12.18)$$

where the T product is determined by

$$\hat{T}\hat{W}(t_1)\hat{W}(t_2)\dots\hat{W}(t_m)$$
$$= \sum_P \hat{P}\theta(t_1, t_2, \dots, t_m)\hat{W}(t_1)\hat{W}(t_2)\dots\hat{W}(t_m)\,, \quad (12.19)$$

and $\hat{P}$ is a permutation operator. Times in each term decrease from left to right. Therefore the T product of the operators is called the time-ordered product of the operators. Inserting (12.18) into (12.15), we can express the operator $S(t)$ via the time-ordered exponential:

$$\hat{S}(t) = 1 + \sum_{m=1}^{\infty} \frac{1}{m!}\left(-\frac{\mathrm{i}}{\hbar}\right)^m \int_0^t \mathrm{d}t_1 \int_0^t \mathrm{d}t_2 \dots \int_0^t \mathrm{d}t_m \hat{T}\hat{W}(t_1)\,\hat{W}(t_2)\dots\hat{W}(t_m)$$
$$= \hat{T}\exp\left[-\frac{\mathrm{i}}{\hbar}\int_0^t \hat{W}(t_1)\,\mathrm{d}t_1\right]. \quad (12.20)$$

Here, in contrast to (12.15), the integration with respect to any time is carried out in a similar way. The prescription for calculating the quantum statistical average from the time-ordered product is, fortunately, very similar to that established by the WBD theorem for the simple product of Bose operators. It is given by

$$\langle \hat{T}a_1a_2a_3a_4\dots a_{2N}\rangle = \langle \hat{T}a_1a_2\rangle\langle \hat{T}a_3a_4\dots a_{2N}\rangle$$
$$+ \langle \hat{T}a_1a_3\rangle\langle \hat{T}a_2a_4\dots a_{2N}\rangle + \dots + \langle \hat{T}a_1a_{2N}\rangle\langle \hat{T}a_2a_3a_4\dots a_{2N-1}\rangle\,, \quad (12.21)$$

i.e., averages of simple products are replaced by averages of time-ordered products. Using (12.21), one can calculate the so-called T-average, i.e., the average of any time-ordered product of the Bose operators.

It is shown in Appendix I that the average of the operator $S(t)$ can be transformed to

$$\langle \hat{S}(t) \rangle = \exp g(t) , \tag{12.22}$$

where the so-called cumulant function

$$\begin{aligned} g(t) &= \langle \hat{S}(t) - 1 \rangle_{\mathrm{c}} \\ &= \sum_{m=1}^{\infty} \frac{1}{m!} \left(-\frac{\mathrm{i}}{\hbar} \right)^m \int_0^t \mathrm{d}t_1 \int_0^t \mathrm{d}t_2 \ldots \int_0^t \mathrm{d}t_m \left\langle \hat{T} \hat{W}(t_1) \hat{W}(t_2) \ldots \hat{W}(t_m) \right\rangle_{\mathrm{c}} \end{aligned} \tag{12.23}$$

is expressed solely in terms of 'coupled' averages of the operators. The subscript c denotes this fact. Examples of the coupled and uncoupled averages are given in Appendix I. Equations (12.22) and (12.23) express the dipolar correlator via the cumulant function. The prescription for calculating the terms of the cumulant function is given by (12.21).

12.3 Broadening and Shift of the ZPL at Weak Coupling with Acoustic and Localized Phonons

The cumulant function described by (12.23) is a power series

$$g(t) = \sum_{m=1}^{\infty} g_m(t) \tag{12.24}$$

in the quadratic FC interaction. If the interaction is weak, one can take into account the first and second terms of this series. In this case we use perturbation theory.

First Order Approximation. In this case only the first term of the series is kept,

$$g_1(t) = -\frac{\mathrm{i}}{\hbar} \int_0^t \mathrm{d}x \, \langle \hat{W}(x) \rangle = -\mathrm{i}\frac{W}{2} \int_0^t \mathrm{d}x \, \langle R^2(x) \rangle = -\mathrm{i}\delta(T)t , \tag{12.25}$$

where

$$\delta(T) = W \sum_q l_q^2 \left(n_q + \frac{1}{2} \right) . \tag{12.26}$$

The dipolar correlator is described by the exponential function with this exponent. The Laplace transform of the dipolar correlator is the line shape function. It is obvious that (12.26) describes a frequency shift of the ZPL due to the quadratic FC interaction. The linear FC interaction does not yield a line shift.

Second Order Approximation. In this case the second term of the series is taken into account,

$$
\begin{aligned}
g_2(t) &= \frac{1}{2}\left(-\frac{\mathrm{i}}{\hbar}\right)^2 \int_0^t \mathrm{d}x \int_0^t \mathrm{d}y\, \langle \hat{T}\hat{W}(x)\hat{W}(y)\rangle_{\mathrm{c}} \\
&= -\frac{1}{2}\left(\frac{W}{2}\right)^2 \int_0^t \mathrm{d}x \int_0^t \mathrm{d}y\, \langle \hat{T}R^2(x)R^2(y)\rangle_{\mathrm{c}} \\
&= -\left(\frac{W}{2}\right)^2 \int_0^t \mathrm{d}x \int_0^t \mathrm{d}y\, \langle \hat{T}R(x)R(y)\rangle^2 \, .
\end{aligned}
\tag{12.27}
$$

Since the average is given by

$$
\langle \hat{T}R(x)R(y)\rangle = \theta(x-y)\langle R(x)R(y)\rangle + \theta(y-x)\langle R(y)R(x)\rangle \, , \tag{12.28}
$$

where

$$
\theta(t) = \begin{cases} 1\, , & t > 0\, , \\ 0\, , & t < 0\, , \end{cases} \tag{12.29}
$$

is a step function, (12.27) can be transformed to

$$
g_2(t) = -\frac{W^2}{2} \int_0^t \mathrm{d}x \int_0^x \mathrm{d}y\, \langle R(x)\, R(y)\rangle^2 \, . \tag{12.30}
$$

Inserting the equation

$$
R(x) = \sum_q l_q \left(b_q \mathrm{e}^{-\mathrm{i}\nu_q x} + b_q^+ \mathrm{e}^{\mathrm{i}\nu_q x}\right) \tag{12.31}
$$

into (12.28), we find the following expression for the average:

$$
\langle R(x)\, R(y)\rangle = \sum_q l_q^2 \left[(n_q + 1)\, \mathrm{e}^{-\mathrm{i}\nu_q(x-y)} + n_q \mathrm{e}^{\mathrm{i}\nu_q(x-y)}\right] . \tag{12.32}
$$

Taking into account the relation

$$
\int_0^t \mathrm{d}x \int_0^x \mathrm{d}y\, \mathrm{e}^{-\mathrm{i}\nu(x-y)} = -\left(\frac{\mathrm{i}t}{\nu} + \frac{\mathrm{e}^{-\mathrm{i}\nu t} - 1}{\nu^2}\right) , \tag{12.33}
$$

the following expression can be derived for the cumulant function in the second order approximation:

$$
g_2(t) = -\mathrm{i}\delta_2(T)\, t + \Psi'(t, T) + \Psi''(t, T) \, , \tag{12.34}
$$

where

$$\delta_2(T) = -\frac{W^2}{2}\sum_{q,q'} l_q^2 l_{q'}^2 \left[\frac{n_q + n_{q'} + 1}{\nu_q + \nu_{q'}} + \frac{n_{q'} - n_q}{\nu_q - \nu_{q'}}\right],$$

$$\Psi'(t,T) = W^2 \sum_{q,q'} l_q^2 l_{q'}^2 n_q (n_{q'} + 1) \frac{e^{i(\nu_q - \nu_{q'})t} - 1}{(\nu_q - \nu_{q'})^2},$$

$$\Psi''(t,T) = \frac{W^2}{2}\sum_{q,q'} l_q^2 l_{q'}^2 \left[(n_q+1)(n_{q'}+1)\frac{e^{-i(\nu_q+\nu_{q'})t} - 1}{(\nu_q + \nu_{q'})^2} + n_q n_{q'} \frac{e^{i(\nu_q+\nu_{q'})t} - 1}{(\nu_q + \nu_{q'})^2}\right]. \tag{12.35}$$

Here δ_2 describes the second order contribution to the shift of the ZPL. The function Ψ'' describes photo-transitions with creation and annihilation of two phonons. The function is bounded and it contributes to the PSB and the Debye–Waller factor. The function Ψ' plays the most important role. It approaches infinity as time increases. Indeed, at large times the following substitution is possible:

$$\frac{1 - e^{iat}}{a^2} \to \pi\delta(a)\,|t| . \tag{12.36}$$

Therefore in the long time limit we may replace the function Ψ' by the function

$$\Psi'(t) = -\frac{\gamma_{\mathrm{ph}}(T)}{2}|t| , \tag{12.37}$$

where

$$\gamma_{\mathrm{ph}}(T) = 2\pi W^2 \sum_{q,q'} l_q^2 l_{q'}^2 n_q (n_{q'} + 1)\,\delta(\nu_q - \nu_{q'}) \tag{12.38}$$

describes a contribution to the ZPL half-width from the quadratic FC interaction. In contrast to the linear FC interaction, the quadratic interaction yields additional broadening of the ZPL. This contribution to the half-width depends on temperature, because it is due to creation and annihilation of phonons. The FWHM of the ZPL is determined by the expression

$$\frac{2}{T_2} = \frac{1}{T_1} + \gamma_{\mathrm{ph}}(T) . \tag{12.39}$$

Let us examine (12.38) more closely.

Temperature Broadening of the ZPL. Let us replace the sum in (12.38) by an integration. For this purpose we introduce the following function of the frequency:

$$\Gamma(\nu) = \pi \sum_q l_q^2 \delta(\nu - \nu_q) = \pi \frac{g(\nu)}{\nu}, \tag{12.40}$$

where the function $g(\nu)$ looks like the one described by (5.52). With the help of this formula, the equation for the half-width can be transformed to

$$\gamma_{\mathrm{ph}}(T) = 2W^2 \int_0^\infty \frac{\mathrm{d}\nu}{\pi} \Gamma^2(\nu) n(\nu)[n(\nu)+1]. \tag{12.41}$$

The spectral function $\Gamma(\nu)$ is a nonvanishing function within the one-phonon frequency interval. In accordance with (5.52) and (5.53), the spectral function must satisfy the condition

$$\int_0^\infty \Gamma(\nu)\nu \frac{\mathrm{d}\nu}{\pi} = 1. \tag{12.42}$$

We consider the interaction with acoustic and localized phonon modes separately.

Interaction with Acoustic Modes. If the guest molecule takes part in the acoustic phonon modes of a solvent, the spectral function $\Gamma(\nu)$ is given by

$$\Gamma(\nu) = 5\pi \frac{\nu^3}{\nu_{\mathrm{D}}^5}, \qquad 0 < \nu < \nu_{\mathrm{D}}, \tag{12.43}$$

in the Debye model. After inserting this formula into (12.41), we arrive at the expression

$$\gamma_{\mathrm{ph}}(T) = \nu_{\mathrm{D}} \left(\frac{W}{\nu_{\mathrm{D}}^2}\right)^2 \frac{25\pi}{2} \left(\frac{kT}{\hbar\nu_{\mathrm{D}}}\right)^7 \int_0^{\hbar\nu_{\mathrm{D}}/kT} \frac{x^6 \mathrm{d}x}{\sinh^2(x/2)}. \tag{12.44}$$

If the inequality $kT \ll \hbar\nu_{\mathrm{D}}$ is satisfied, the upper limit in the integral approaches infinity and then the integral equals 3 000. If the opposite inequality $kT \gg \hbar\nu_{\mathrm{D}}$ is satisfied, the upper limit is very small and then the integral equals $(4/5)(\hbar\nu_{\mathrm{D}}/kT)^5$. Therefore the phonon half-width is given by

$$\gamma_{\mathrm{ph}}(T) = \begin{cases} \nu_{\mathrm{D}} \left(\dfrac{W}{\nu_{\mathrm{D}}^2}\right)^2 37\,500\pi \left(\dfrac{kT}{\hbar\nu_{\mathrm{D}}}\right)^7, & kT \ll \hbar\nu_{\mathrm{D}}, \\ \nu_{\mathrm{D}} \left(\dfrac{W}{\nu_{\mathrm{D}}^2}\right)^2 10\pi \left(\dfrac{kT}{\hbar\nu_{\mathrm{D}}}\right)^2, & kT \gg \hbar\nu_{\mathrm{D}}. \end{cases} \tag{12.45}$$

It should be noted that the parameter W has dimensions of the square of the frequency.

Interaction with a Localized Mode. If the guest molecule is coupled weakly with host molecules, the amplitude of vibration of this guest molecule is large at a definite frequency ν_0. This vibration is of localized type. In such a case the spectral function $\Gamma(\nu)$ is large in the vicinity of the frequency ν_0. The spectral function can be approximated by a Lorentzian:

$$\Gamma(\nu) = \frac{1}{\nu_0} \frac{\gamma_0/2}{(\nu - \nu_0)^2 + (\gamma_0/2)^2} . \tag{12.46}$$

After inserting this formula into the integral of (12.41), we find

$$\int_0^\infty \frac{\mathrm{d}\nu}{\pi} \Gamma^2(\nu) \approx \int_{-\infty}^\infty \frac{\mathrm{d}\nu}{\pi} \Gamma^2(\nu) = \frac{1}{\gamma_0 \nu_0^2} . \tag{12.47}$$

Then (12.41) for the phonon half-width of the ZPL takes the form

$$\gamma_{\mathrm{ph}}(t) = \gamma_0 \left(\frac{W}{\gamma_0 \nu_0} \right)^2 \frac{1}{2 \sinh^2 (\hbar \nu_0 / 2kT)} . \tag{12.48}$$

Let us analyze (12.45) and (12.48). In the high temperature domain, both formulas predict a quadratic T^2 temperature law for line broadening. This temperature behavior of the line width emerges at $kT \gg \hbar\nu_0$ if the broadening results from the interaction with a quasi-localized mode, and at $kT \gg \hbar\nu_{\mathrm{D}}$ if the broadening results from interaction with acoustic phonons. In the low temperature domain, (12.48) yields line broadening in accordance with the Arrhenius law, i.e., $\exp(-\hbar\nu_0/kT)$. Equation (12.45) yields a T^7 temperature law. The boundary temperature between the low and high temperature domains depends on the type of phonons involved.

12.4 Quantum Phonon Green Functions

In the next section we consider line broadening and shift at arbitrary strength of the quadratic FC interaction. In the course of the derivation we shall use averages of the time-ordered product involving an arbitrary number of operators W. Since such T averages can be converted into products of T averages of two-phonon operators, we should examine in detail the time-ordered average of two-phonon operators. Let us introduce the following average, which is called the phonon Green function:

$$D(t) = -\mathrm{i} \langle \hat{T} R(t) R(0) \rangle = D_+(t) + D_-(t) , \tag{12.49}$$

where

$$D_+(t) = -\mathrm{i}\theta(t) \langle R(t) R(0) \rangle , \quad D_-(t) = -\mathrm{i}\theta(-t) \langle R(0) R(t) \rangle . \tag{12.50}$$

The functions $D_+(t)$ and $D_-(t)$ differ from zero at positive and negative times, respectively. They are called the retarded Green function and the advanced Green function. Their sum is called the causal Green function. These types of quantum Green functions were used first in quantum field theory. By carrying out the calculations in (12.50), we arrive at the result

$$\begin{aligned}\langle R(t)\, R(0)\rangle &= \sum_q l_q^2 \left[(n_q+1)\,\mathrm{e}^{-\mathrm{i}\nu_q t} + n_q \mathrm{e}^{\mathrm{i}\nu_q t}\right] , \\ \langle R(0)\, R(t)\rangle &= \langle R(-t)\, R(0)\rangle \ .\end{aligned} \tag{12.51}$$

The transition from a sum to an integral can be carried out with the help of the functions

$$\Gamma(\nu) = \pi \sum_q l_q^2 \delta(\nu - \nu_q) = \pi \frac{g(\nu)}{\nu} , \tag{12.52}$$

$$\Gamma(\nu, T) = [n(\nu) + 1]\, \Gamma(\nu, T) + n(-\nu)\, \Gamma(-\nu) \ . \tag{12.53}$$

The first of these functions has already been considered in Sect. 5.2. The second spectral function resembles the function $\varphi(\nu, T)$, which is the weighted density of phonon states. However, it does not depend on the shifts in the equilibrium positions of the oscillators. With the help of the functions introduced, one can express the retarded, advanced, and causal Green functions in the form

$$D_\pm(t) = \theta(\pm t) \int_{-\infty}^{\infty} \frac{\mathrm{d}\nu}{\mathrm{i}\pi} \Gamma(\nu, T) \mathrm{e}^{\mp \mathrm{i}\nu t} ,$$

$$\begin{aligned}D(t) &= \int_{-\infty}^{\infty} \frac{\mathrm{d}\nu}{\mathrm{i}\pi} \Gamma(\nu, T) \left[\theta(t) \mathrm{e}^{-\mathrm{i}\nu t} + \theta(-t) \mathrm{e}^{\mathrm{i}\nu t}\right] \\ &= D_+(t) + D_-(t) \ .\end{aligned} \tag{12.54}$$

The causal Green function is an even function of time. The Laplace transforms of these functions can be calculated if we take into account the relation

$$\int_{-\infty}^{\infty} \theta(\pm t)\, \mathrm{e}^{\mathrm{i}(\omega-\nu)t} \mathrm{d}t \equiv \int_{-\infty}^{\infty} \theta(\pm t)\, \mathrm{e}^{\mathrm{i}(\omega-\nu\pm \mathrm{i}0)t} \mathrm{d}t = \frac{\pm \mathrm{i}}{\omega - \nu \pm \mathrm{i}0} \ . \tag{12.55}$$

Using the last equation, one can easily find the Laplace transforms of the Green functions:

$$D_+(\omega) = \int_0^{\infty} D(t)\,\mathrm{e}^{\mathrm{i}\omega t}\mathrm{d}t = \int_{-\infty}^{\infty} \frac{\mathrm{d}\nu}{\pi} \frac{\Gamma(\nu,T)}{\omega-\nu+\mathrm{i}0}\,,$$

$$D_-(\omega) = \int_0^{-\infty} D(t)\,\mathrm{e}^{\mathrm{i}\omega t}\mathrm{d}t = \int_{-\infty}^{\infty} \frac{\mathrm{d}\nu}{\pi} \frac{\Gamma(\nu,T)}{\omega+\nu-\mathrm{i}0}\,,$$

$$D(\omega) = \int_{-\infty}^{\infty} D(t)\,\mathrm{e}^{\mathrm{i}\omega t}\mathrm{d}t = \int_{-\infty}^{\infty} \frac{\mathrm{d}\nu}{\pi}\Gamma(\nu,T)\left(\frac{1}{\omega-\nu+\mathrm{i}0} - \frac{1}{\omega+\nu-\mathrm{i}0}\right)$$
$$= D_+(\omega) - D_-(\omega)\,. \tag{12.56}$$

The function $D(\omega)$ is an even function of the frequency. At $T = 0$ this function coincides with the function described by (5.50), which was used to discuss the way the guest molecule takes part in vibrations with various frequencies. Figure 5.1 shows the function $\Gamma(\nu)$.

Analytic Properties of Green Functions. Later we shall calculate integrals whose integrands include the Laplace transforms of the Green functions. The integration can be carried out much more easily if we know the analytic properties of the functions $D(\omega)$. The retarded Green function and the advanced Green function have no poles in the upper and the lower half plane of the complex frequency ω, respectively. They are analytic functions in the upper and lower half plane of the complex frequency. It is obvious that the causal Green function is not an analytic function either in the upper or the lower half plane of the complex variable ω.

12.5 Temperature Broadening and Shift of the ZPL at Arbitrary Strength of the Quadratic FC Interaction

The mth term of the series for the cumulant function is given by

$$g_m(t) = \frac{1}{m!}\left(-\frac{\mathrm{i}}{\hbar}\right)^m \int_0^t \mathrm{d}t_1 \int_0^t \mathrm{d}t_2 \ldots \int_0^t \mathrm{d}t_m \left\langle \hat{T}\hat{W}(t_1)\,\hat{W}(t_2)\ldots\hat{W}(t_m)\right\rangle_{\mathrm{c}}$$
$$= \frac{1}{m!}\left(\frac{W}{2}\right)^m (-\mathrm{i})^m \int_0^t \mathrm{d}t_1 \int_0^t \mathrm{d}l_2 \ldots \int_0^t \mathrm{d}t_m \left\langle \hat{T}R_1R_1R_2R_2\ldots R_mR_m\right\rangle_{\mathrm{c}}\,,$$
$$\tag{12.57}$$

where $R_j = R(t_j)$. The average in the integrand can be expressed as a sum of terms, each of which relates to a definite type of pairing of the phonon operators. For instance,

$$
\begin{aligned}
&(-\mathrm{i})^m \langle \hat{T} R_1 R_1 R_2 R_2 \ldots R_m R_m \rangle_{\mathrm{c}} \qquad (12.58)\\
&\quad = (-\mathrm{i})^m \left\langle \hat{T} R_1 \underbrace{\overbrace{R_1 R_2}\overbrace{R_2 R_3}\overbrace{R_3 R_4}\overbrace{R_4 \quad \ldots \quad R_m} R_m} \right\rangle + \ldots\\
&\quad = D(t_1 - t_2) D(t_2 - t_3) \ldots D(t_{m-1} - t_m) D(t_m - t_1) + \ldots .
\end{aligned}
$$

It is obvious that $D(t_1 - t_m) = D(t_m - t_1)$. The product

$$
D\left(t_1 - t_3\right) D\left(t_3 - t_2\right) \ldots D\left(t_{m-1} - t_m\right) D\left(t_m - t_1\right) \qquad (12.59)
$$

relates to another type of pairing. Such products are integrated with respect to all times in a similar way, in accordance with (12.57). After permutation of times t_2 and t_3, (12.59) coincides with (12.58). This means that both products yield one result after integration over times. It is easy to see that all possible types of pairing yield the same result when the time integration has been carried out. Therefore, bearing this fact in mind, we may replace (12.58) by the expression

$$
\begin{aligned}
&(-\mathrm{i})^m \left\langle \hat{T} R_1 R_1 R_2 R_2 \ldots R_m R_m \right\rangle_{\mathrm{c}} \qquad (12.60)\\
&\quad = N(m) D\left(t_1 - t_2\right) D\left(t_2 - t_3\right) \ldots D\left(t_{m-1} - t_m\right) D\left(t_m - t_1\right) ,
\end{aligned}
$$

where $N(m)$ is the number of different types of phonon operator pairing. Let us find this number.

Times of all operators from the product can be denoted by points on a time axis. Pairing of two operators which depend on different times can be shown by a solid line connecting these points on the time axis. The solid line enters and leaves each time point because two phonon operators depend on one time variable. Therefore each type of pairing of the operator from the product looks like some kind of diagram where a solid line enters and leaves each time point. If all times are connected by one continuous line, such a diagram is called the coupled diagram. The subscript c in (12.60) shows that we must only take into account coupled diagrams.

Starting from time point 1, we can reach the other $m - 1$ time points in $m-1$ ways. Then we can reach the other $m-2$ time points in $m-2$ ways, and so on. The total number of the various types of pairing is $(m - 1)!$. However, we should take into account the fact that two-phonon operators depend on one time. Therefore there are two possibilities for connecting two time points, and the number of all possible types of coupled pairing should be multiplied by a factor 2^{m-1}, giving $N(m) = 2^{m-1}(m - 1)!$. Inserting this number into (12.60) and integrating over the times, we find the following expression for the mth term of the cumulant function:

$$
g_m(t) = \frac{W^m}{2m} \int_0^t \mathrm{d}t_1 \int_0^t \mathrm{d}t_2 \ldots \int_0^t \mathrm{d}t_m D\left(t_1 - t_2\right) D\left(t_2 - t_3\right) \ldots D(t_m - t_1) . \qquad (12.61)
$$

Summing all the terms, we arrive at the following expression for the cumulant function:

$$g(t) = \sum_{m=1}^{\infty} \frac{W^m}{2m} \int_0^t \mathrm{d}t_1 \int_0^t \mathrm{d}t_2 \ldots \int_0^t \mathrm{d}t_m D(t_1 - t_2) D(t_2 - t_3) \ldots D(t_m - t_1) . \tag{12.62}$$

This function determines the temporal behavior of the dipolar correlator at arbitrary strength of the quadratic FC interaction.

It has already been shown in the last section that the cumulant function is not a bounded function of time if the quadratic FC interaction is taken into account. Equation (12.62) enables one to find the line broadening and shift for strong quadratic FC interaction. Pursuing this purpose, we need to find the function $g_\infty(t)$ that describes the asymptotic behavior of the cumulant function in the long time limit. Using the expression

$$D(t) = \int_{-\infty}^{\infty} \frac{\mathrm{d}\omega}{2\pi} D(\omega, T) \mathrm{e}^{-\mathrm{i}\omega t} \tag{12.63}$$

for the phonon Green function, we are able to carry out the integration of the exponential functions in (12.61). The result is given by

$$\begin{aligned} g_m(t) = \frac{W^m}{2m} \int_{-\infty}^{\infty} \frac{\mathrm{d}\omega_1}{2\pi} D(\omega_1) \int_{-\infty}^{\infty} \frac{\mathrm{d}\omega_2}{2\pi} D(\omega_2) \ldots \int_{-\infty}^{\infty} \frac{\mathrm{d}\omega_m}{2\pi} D(\omega_m) \\ \times \Delta_t(1-2) \Delta_t(2-3) \ldots \Delta_t(m-1) , \end{aligned} \tag{12.64}$$

where

$$\Delta_t(k-j) = \frac{\sin \dfrac{\omega_k - \omega_j}{2} t}{\dfrac{\omega_k - \omega_j}{2}} . \tag{12.65}$$

The functions Δ_t are integrated over frequencies. In the long time limit the following substitution is possible:

$$\begin{aligned} &\Delta_t(k-j) \to 2\pi\delta(\omega_k - \omega_j) , \\ &\Delta_t(k-j)\Delta_t(j-k) \to |t| 2\pi\delta(\omega_k - \omega_j) . \end{aligned} \tag{12.66}$$

After such a substitution, the integration of the delta functions can be carried out and the result is given by

$$g_m(t) = |t| \int_{-\infty}^{\infty} \frac{\mathrm{d}\omega_1}{4\pi} \frac{1}{m} \left[W D(\omega_1, T) \right]^m . \tag{12.67}$$

Summing over all m and allowing for the relation $D(\omega, T) = D(-\omega, T)$, we find the final expression for the cumulant function in the long time limit:

$$g_{\infty}(t) = -t \int_0^{\infty} \frac{\mathrm{d}\omega}{2\pi} \ln\left[1 - WD^{\mathrm{g}}(\omega, T)\right] = -t\left[\mathrm{i}\delta^{\mathrm{g}}_{\mathrm{ph}}(T) + \frac{\gamma^{\mathrm{g}}_{\mathrm{ph}}(T)}{2}\right], \tag{12.68}$$

where the real functions of temperature $\delta^{\mathrm{g}}_{\mathrm{ph}}(T)$ and $\gamma^{\mathrm{g}}_{\mathrm{ph}}(T)$ describe the shift and half-width of the ZPL in the absorption spectrum.

The relation $g(-t) = g^*(t)$ can be derived from (12.10). Therefore the dipolar correlator in the long time limit can be written in the form

$$\left\langle \hat{S}(t) \right\rangle_{\mathrm{g},\infty} = \exp\left[-\mathrm{i}\delta^{\mathrm{g}}_{\mathrm{ph}}(T)\, t - \frac{\gamma^{\mathrm{g}}_{\mathrm{ph}}(T)}{2}|t|\right]. \tag{12.69}$$

From this and (12.9), the following expression is found for the ZPL:

$$J^{\mathrm{g}}(t) = d^2 \frac{\gamma^{\mathrm{g}}(T)/2}{[\omega - \Omega - \delta^{\mathrm{g}}_{\mathrm{ph}}(T)]^2 + [\gamma^{\mathrm{g}}(T)/2]^2}, \tag{12.70}$$

where the FWHM of the ZPL is

$$\gamma^{\mathrm{g}}(T) = \frac{2}{T_2(T)} = \frac{1}{T_1} + \gamma^{\mathrm{g}}_{\mathrm{ph}}(T), \tag{12.71}$$

i.e., the FWHM is the sum of the natural half-width $1/T_1$ and the half-width $\gamma^{\mathrm{g}}_{\mathrm{ph}}(T)$ resulting from the quadratic FC interaction with phonons.

Zero Temperature Limit. The Green function can be written in the form

$$D^{\mathrm{g}}(\omega, 0) = \Delta^{\mathrm{g}}(\omega) - \mathrm{i}\left[\Gamma^{\mathrm{g}}(\omega) + \Gamma^{\mathrm{g}}(-\omega)\right], \tag{12.72}$$

where

$$\Delta^{\mathrm{g}}(\omega) = \int_0^{\infty} \frac{\mathrm{d}\nu}{\pi} \Gamma^{\mathrm{g}}(\nu) \frac{2\nu}{\omega^2 - \nu^2}, \tag{12.73}$$

Making use of (12.68) and the formula

$$\ln(x + \mathrm{i}y) = (1/2)\ln(x^2 + y^2) + \mathrm{i}\arctan(y/x),$$

we find

$$\delta^{\mathrm{g}}_{\mathrm{ph}}(0) = \int_0^{\infty} \frac{\mathrm{d}\omega}{2\pi} \arctan \frac{W\Gamma^{\mathrm{g}}(\omega)}{1 - W\Delta^{\mathrm{g}}(\omega)}, \tag{12.74}$$

$$\gamma_{\mathrm{ph}}^{\mathrm{g}}(0) = \int_0^\infty \frac{\mathrm{d}\omega}{2\pi} \ln\left|1 - WD^{\mathrm{g}}(\omega,0)\right|^2 . \tag{12.75}$$

Equation (12.41), found using perturbation theory, predicts that the phonon half-width approaches zero at $T = 0$, i.e., $\gamma_{\mathrm{ph}}^{g}(0) = 0$. Equation (12.75) allows us to prove this result for an arbitrary strength of the quadratic coupling. This and some other results are based on the equation

$$D^{\mathrm{e}}(\omega,0) = \frac{D^{\mathrm{g}}(\omega,0)}{1 - WD^{\mathrm{g}}(\omega,0)} , \tag{12.76}$$

proven in Appendix J. The last equation is similar to (5.42), derived with the simplified interaction in connection with the problem of the influence of a guest molecule on the vibrational system of a pure crystal. We said in that context that (5.42) remains correct in the more general case if we carry out the relevant substitution of the D functions. Indeed the influence of the change ΔU in the force matrix due to the appearance of a guest molecule in the pure crystal can be treated in the same way as the influence of the change W in the force matrix upon electronic excitation has been considered in Appendix J.

Let us consider (12.75). We can write the following chain of transformations:

$$\begin{aligned}\gamma_{\mathrm{ph}}^{\mathrm{g}}(0) &= \int_0^\infty \frac{\mathrm{d}\omega}{4\pi} \ln\left|1 - WD^{\mathrm{g}}(\omega,0)\right|^2 = \mathrm{Re}\int_0^\infty \frac{\mathrm{d}\omega}{2\pi} \ln\left[1 - WD^{\mathrm{g}}(\omega,0)\right] \\ &= -\,\mathrm{Re}\int_0^\infty \frac{\mathrm{d}\omega}{2\pi}\int_0^W \mathrm{d}x \frac{D^{\mathrm{g}}(\omega,0)}{1 - xD^{\mathrm{g}}(\omega,0)} = -\,\mathrm{Re}\int_0^W \mathrm{d}x \int_0^\infty \frac{\mathrm{d}\omega}{2\pi} D^{\mathrm{e}}(\omega,0,x) \\ &= \int_0^W \mathrm{d}x\, \mathrm{Re}\left[\mathrm{i}\int_0^\infty \frac{\mathrm{d}\omega}{2\pi}\, \mathrm{Im}\, D^{\mathrm{e}}(\omega,0,x)\right] = 0 . \end{aligned} \tag{12.77}$$

Here we took into account (12.56) and the fact that $D(\omega) = D(-\omega)$. The contribution to the FWHM from the quadratic FC interaction results from the process of creation and annihilation of phonons. The annihilation of phonons is forbidden at $T = 0$. It is shown in Appendix K that the shift is proportional to the difference of the energies of zero point vibrations in the excited and ground electronic states. Therefore it does not equal zero at $T = 0$.

The ZPL of the absorption and fluorescence spectra are in resonance with each other and their half-widths coincide. This statement can be proven with the help of the relation

$$\left[1 + WD^{\mathrm{e}}(\omega,T)\right]^* = \frac{1 - WD^{\mathrm{g}}(\omega,T)}{\left|1 - WD^{\mathrm{g}}(\omega,0)\right|^2} , \tag{12.78}$$

which can be derived from (12.76) and

$$D^{\mathrm{g,e}}(\omega,T)=\int_0^\infty \frac{\mathrm{d}\nu}{\pi}\Gamma^{\mathrm{g,e}}(\nu,T)\left(\frac{1}{\omega-\nu+\mathrm{i}0}-\frac{1}{\omega+\nu-\mathrm{i}0}\right). \tag{12.79}$$

Indeed, if we take into account the fact that the dipolar correlator $J^{\mathrm{e}}(t)$ can by obtained from the dipolar correlator $J^{\mathrm{g}}(t)$ by complex conjugation and by permutation of the superscripts e and g, we arrive at the following expression for the shift and half-width of the fluorescence line:

$$\mathrm{i}\delta^{\mathrm{e}}_{\mathrm{ph}}(T)+\frac{\gamma^{\mathrm{e}}_{\mathrm{ph}}(T)}{2}=\int_0^\infty \frac{\mathrm{d}\omega}{2\pi}\ln\left[1+WD^{\mathrm{e}}(\omega,T)\right]^*. \tag{12.80}$$

Here we used the fact that the sign of the parameter W differs in the fluorescence and absorption lines. Inserting (12.78) into the right hand side of (12.80) results in the equation

$$\mathrm{i}\delta^{\mathrm{e}}_{\mathrm{ph}}(T)+\frac{\gamma^{\mathrm{e}}_{\mathrm{ph}}(T)}{2}=\mathrm{i}\delta^{\mathrm{g}}_{\mathrm{ph}}(T)+\frac{\gamma^{\mathrm{g}}_{\mathrm{ph}}(T)}{2}. \tag{12.81}$$

This proves the statement made above. The statement is only correct for the 0–0′ transition. If the transitions 1–1′, 2–2′, and so on, are included in the ZPL, the resonance of the absorption and emission lines is broken. Sometimes temperature broadening of the ZPL is explained by the increased probabilities of transitions 1–1′, 2–2′, and so on. This explanation of ZPL broadening in not correct because such a model for line broadening fails to explain temperature broadening of the pure 0–0′ ZPL.

Using the equation

$$D^{\mathrm{g}}(\omega,T)=D^{\mathrm{g}}(\omega,0)-2\mathrm{i}\,n(\omega)\,\Gamma^{\mathrm{g}}(\omega), \tag{12.82}$$

which is correct at $\omega>0$, we can transform (12.68) to

$$\mathrm{i}\delta_{\mathrm{ph}}(T)+\frac{\gamma_{\mathrm{ph}}(T)}{2}=\mathrm{i}\delta_{\mathrm{ph}}(0)+\int_0^\infty \frac{\mathrm{d}\omega}{2\pi}\ln\left[1+2\mathrm{i}\,n(\omega)\frac{W\Gamma^{\mathrm{g}}(\omega)}{1-WD^{\mathrm{g}}(\omega,0)}\right]. \tag{12.83}$$

Finding the modulus and the argument of the integrand, we arrive at the following final result for the shift and half-width of the ZPL:

$$\delta_{\mathrm{ph}}(T)=\delta_{\mathrm{ph}}(0)+\int_0^\infty \frac{\mathrm{d}\omega}{2\pi}\arctan\frac{2n(\omega)W\Gamma^{\mathrm{e}}(\omega)[1-W\Delta^{\mathrm{g}}(\omega)]}{1+2n(\omega)W^2\Gamma^{\mathrm{e}}(\omega)\Gamma^{\mathrm{g}}(\omega)}, \tag{12.84}$$

$$\gamma_{\mathrm{ph}}(T) = \int_0^\infty \frac{\mathrm{d}\omega}{2\pi} \ln\left[1 + 4n(\omega)[n(\omega)+1] W^2 \Gamma^{\mathrm{e}}(\omega) \Gamma^{\mathrm{g}}(\omega)\right] , \tag{12.85}$$

where

$$\Gamma^{\mathrm{e}}(\omega) = \frac{\Gamma^{\mathrm{g}}(\omega)}{[1 - W\Delta^{\mathrm{g}}(\omega)]^2 + [W\Gamma^{\mathrm{g}}(\omega)]^2} \tag{12.86}$$

is a spectral function for phonons in the excited electronic state. It is expressed in terms of the spectral function $\Gamma^{\mathrm{g}}(\omega)$ for phonons in the ground electronic state. Comparison of the two functions demonstrates the changes in the phonon system caused by the quadratic FC interaction.

It is found from the last equation that $W^2\Gamma^{\mathrm{g}}(\omega)\Gamma^{\mathrm{e}}(\omega) < 1$ at arbitrary values of the coupling constant W. Hence at low temperature, $\ln(1+x)$ in (12.85) can be replaced by x. In the low temperature approximation, (12.85) takes the form

$$\gamma_{\mathrm{ph}}(T) = \int_0^\infty \frac{\mathrm{d}\omega}{2\pi} \frac{W^2\Gamma^{\mathrm{e}}(\omega)\Gamma^{\mathrm{g}}(\omega)}{\sinh^2(\hbar\omega/2kT)} . \tag{12.87}$$

Here the equation $4n(\omega)[n(\omega)+1] = \sinh^{-2}(\hbar\omega/2kT)$ has been used. If perturbation theory is used, the spectral function $\Gamma^{\mathrm{e}}(\omega)$ should be replaced by the spectral function $\Gamma^{\mathrm{g}}(\omega)$ for the absorption spectrum. In this case (12.87) coincides with (12.41).

In the high temperature limit, the expansion of the logarithm is not correct. Hence the high temperature law T^2 resulting from (12.87) should be replaced by a more moderate temperature dependence resulting from (12.85). In the high temperature domain, it is difficult to separate the broad ZPL from the PSB accompanying it. Therefore the low temperature domain is more attractive for investigation of the ZPL.

Up to now the perturbation matrix W has been taken to be first rank, i.e., a scalar. However, all the calculations up to (12.68) can be carried out if the matrix W has arbitrary rank. In this case, instead of (12.68), the following expression is found [64, 77]:

$$\begin{aligned}\mathrm{i}\delta_{\mathrm{ph}}(T) + \frac{\gamma_{\mathrm{ph}}(T)}{2} &= \mathrm{Tr}\int_0^\infty \frac{\mathrm{d}\omega}{2\pi} \ln\left[1 - \hat{W}\hat{D}^{\mathrm{g}}(\omega, T)\right] \\ &= \int_0^\infty \frac{\mathrm{d}\omega}{2\pi} \ln \mathrm{Det}\left[1 - \hat{W}\hat{D}^{\mathrm{g}}(\omega, T)\right] ,\end{aligned} \tag{12.88}$$

where the matrix W consists of the elements $W_{ni,mj}$ and the matrix D^{g} is determined by (12.79), with the spectral function Γ replaced by the matrix

$$\Gamma_{nimj}(\nu) = \frac{\pi}{2\nu} \sum_q \left[u(0i,q) - u(\boldsymbol{n}i,q)\right]\left[u(0i,q) - u(\boldsymbol{m}j,q)\right]\delta(\nu - \nu_q) \ . \tag{12.89}$$

It is expedient to use the more complex (12.88) if the simpler expression does not fit the experimental data.

12.6 General Expression for the Cumulant Function of an Electron–Phonon System

The infinite series for the cumulant function described by (12.62) enables one to find its long time behavior and calculate the shift and half-width of the ZPL. However, if we know the temporal behavior of the cumulant function over the whole time scale, we can calculate the whole electron–phonon band including the PSB. For this purpose the infinite series must be summed. Let us express (12.62) in the form

$$g(t) = \frac{W}{2} \int_0^t \mathrm{d}\tau \left[D(+0) + \sum_{m=2}^{\infty} \frac{W^{m-1}}{m} \int_0^t \mathrm{d}t_2 \int_0^t \mathrm{d}t_3 \ldots \int_0^t \mathrm{d}t_m \right. \\ \left. \times D(\tau - t_2) D(t_2 - t_3) \ldots D(t_m - \tau) \right] . \tag{12.90}$$

Differentiating this expression with respect to time, we find

$$\dot{g}(t) = \frac{W}{2} \left[D(+0) + \sum_{m=2}^{\infty} W^{m-1} \int_0^t \mathrm{d}t_2 \int_0^t \mathrm{d}t_3 \ldots \int_0^t \mathrm{d}t_m \right. \\ \left. \times D(t - t_2) D(t_2 - t_3) \ldots D(t_m - t) \right] . \tag{12.91}$$

The infinite series on the right hand side of this expression resembles a solution of an integral equation found by means of an iteration procedure. If this equation is found, we can try to find its solution instead of summing the infinite series.

Let us examine the following integral equation:

$$S(x,y,t) = D(x-y) + W \int_0^t \mathrm{d}z D(x-z) S(z,y,t) \ . \tag{12.92}$$

The solution of this equation by means of the iteration procedure is given by the following series:

$$S(x,y,t) = D(x-y) + \sum_{m=2}^{\infty} W^{m-1} \int_0^t \mathrm{d}t_2 \int_0^t \mathrm{d}t_3 \ldots \int_0^t \mathrm{d}t_m$$
$$\times D(x-t_2) D(t_2-t_3) \ldots D(t_m-y) \ . \quad (12.93)$$

Comparing (12.91) and (12.93), we can write

$$\dot{g}(t) = \frac{W}{2} S(t,t,t) \ . \quad (12.94)$$

Making use of a variable substitution and the fact that $D(z) = D(-z)$, it is possible to prove that

$$S(t-x, t-y, t) \equiv S(y,x,t) \ . \quad (12.95)$$

The two functions do indeed satisfy one integral equation and as a result, $S(t,t,t) = S(0,0,t)$ is satisfied. Using (12.94), the following expression for the cumulant function is found:

$$g(t) = \frac{W}{2} \int_0^t S(0,0,\tau)\, \mathrm{d}\tau \ . \quad (12.96)$$

In order to calculate the cumulant function, we need to know the function $S(0,0,t)$.

In the long time limit, this expression takes the form

$$g_\infty(t) = \frac{W}{2} S(0,0,\infty)\, t \ . \quad (12.97)$$

Hence the shift and half-width of the ZPL are related to the value $S(0,0,\infty)$ by

$$\frac{W}{2} S(0,0,\infty) = -\mathrm{i}\delta_{\mathrm{ph}}(T) - \frac{\gamma_{\mathrm{ph}}(T)}{2} \ . \quad (12.98)$$

The calculation of the cumulant function for arbitrary time and for zero temperature was carried out in [75]. The cumulant function has been found in [78] and [79] for arbitrary temperature and for an FC interaction with a localized phonon mode.

12.7 Theoretical Treatment of Experimental Data for Temperature Broadening of the ZPL

Temperature broadening of the ZPL has been studied experimentally in many works. However, in the majority of these studies, the measured broadening was fitted by theoretical formulas in which unknown parameters were found from the fit. It was found that the measured line broadening is described by an Arrhenius-type law $\exp(-E_0/kT)$ at low temperatures and by a T^2 law at high temperatures. These experimental data can be explained by the theoretical model which takes into account the interaction with the localized mode with energy E_0. Perhaps this mode emerges when a guest molecule is substituted for a host molecule in a matrix. The logarithm of the Arrhenius-type dependence is a linear function of $1/T$. The slope of the straight line allows us to find the energy E_0 of a localized mode.

Once the responsibility of the localized mode for the line broadening has been established, the question arises: is the coupling weak or strong? The answer to this question allows us to choose the formula for weak or strong coupling. The theoretical analysis should start with the formula for weak coupling.

Experimental data for line broadening allows us to find the constant $C = \gamma_0(W/\gamma_0\nu_0)$. In the case of weak coupling, the inequality $W/\gamma_0\nu_0 \ll 1$ must be fulfilled. In order to check whether this inequality is satisfied, we need to find the lifetime $1/\gamma_0$ of the localized mode from independent measurements. This lifetime can be measured from the half-width of the localized mode peak. However, this peak manifests itself in the PSB only if the linear FC or HT interaction takes place for the localized mode involved in the line-broadening mechanism. There are examples when the linear interaction is absent, whereas the quadratic FC interaction manifests itself. An example of this type will be considered later.

Fortunately the localized mode manifests itself in the PSB as one quantum peak. It enables us to measure the lifetime $1/\gamma_0$ directly. However, we should keep in mind that the shape of the one-quantum Stokes peak is described by the expression

$$\Psi_1(\Delta) = \mathrm{e}^{-\varphi(0,T)}[n(\nu_0)+1]\int_{-\infty}^{\infty} \mathrm{d}\nu\varphi(\nu)\frac{\gamma(T)/2\pi}{(\Delta-\nu)^2+[\gamma(T)/2]^2}\,, \qquad (12.99)$$

i.e., the peak is a convolution of the ZPL and the one-quantum spectral function $\varphi(\nu)$. If this spectral function is approximated by a Lorentzian with half-width γ_0, the half-width of the peak measured experimentally is given by

$$\Delta_{1/2} = \gamma_0 + \gamma(T)\ . \qquad (12.100)$$

The width of the measured one-quantum peak can depend on temperature. However, this fact does not mean that we are witnessing anharmonicity as claimed in some works. Figure 12.1 shows an example of this type. The temperature broadening of the full half-width is well-described by (12.100) if the lifetime is independent of the temperature.

If the quadratic FC interaction is taken into account in a non-perturbative way, the infinite series of the phonon Green functions $D^{\mathrm{g}}(\omega)$ which depend

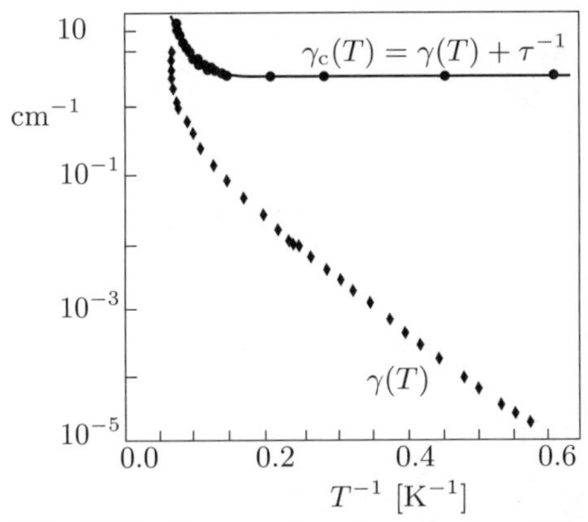

Fig. 12.1. Temperature broadening of the ZPL of a pentacene molecule doping benzoic acid (*filled circles*) and broadening of the pseudo-localized peak with frequency 16.7 cm^{-1} (*open circles*). The calculation was carried out using (12.100) with $\gamma_0 = 2.2$ cm^{-1} and $\gamma(T)$ taken from this figure [80]

on the phonon frequencies of the initial electronic state is summed. As a result of this summation, new phonon Green functions $D^{\mathrm{e}}(\omega)$ emerge. Equation (12.76) exhibits this fact. The new phonon Green functions depend on the phonon frequencies of the final electronic state. Therefore, spectral positions of the calculated one-quantum phonon peaks correspond to phonon frequencies of the final electronic state in accordance with the experimental observations.

Let us consider the theoretical treatment of experimental data by Hesselink and Wiersma [81], using the non-perturbative theory. Their experimental data relating to the conjugate absorption and fluorescence electron–phonon bands of a pentacene molecule doping a naphthalene crystal are shown in Fig. 12.2. The ZPL of the two spectra are in resonance. The conjugate spectra extend on the frequency scale in different directions from the resonant ZPL. However, comparison of the bands is more convenient if the PSB of both spectra are plotted as shown in Fig. 12.2. A single peak is seen in each PSB, as marked by arrows. Hesselink and Wiersma believed that these peaks were the electron–phonon transitions with creation of two quanta of a localized mode. When they assumed that these peaks were one-quantum transitions, Hesselink and Wiersma could not explain their experimental data on temperature broadening of the ZPL. The difference between the frequencies of the localized mode in the ground and excited electronic states is a measure of the quadratic FC interaction. Therefore, when we deal with a localized mode, we can find the coupling constant as follows

$$W = \nu_{\mathrm{e}}^2 - \nu_{\mathrm{g}}^2 \,. \tag{12.101}$$

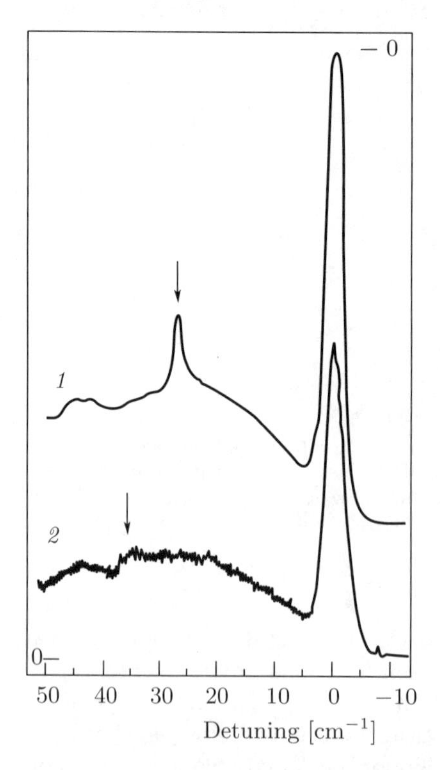

Fig. 12.2. Absorption (curve 1) and fluorescence (curve 2) bands of a pentacene molecule doping naphthalene. $T = 1.5$ K [81]

From Fig. 12.2, we find $\nu_{\mathrm{e}} = 27.6\ \mathrm{cm}^{-1}/2 = 13.8\ \mathrm{cm}^{-1}$ and $\nu_{\mathrm{g}} = 36\ \mathrm{cm}^{-1}/2 = 18\ \mathrm{cm}^{-1}$. The spectral function for the one-quantum peak is given by

$$\Gamma^{\mathrm{g}}(\omega) = \frac{1}{\nu_{\mathrm{g}}}\left[\frac{(\gamma_0/2)}{(\omega-\nu_{\mathrm{g}})^2+(\gamma_0/2)^2} - \frac{(\gamma_0/2)}{(\omega+\nu_{\mathrm{g}})^2+(\gamma_0/2)^2}\right] . \tag{12.102}$$

Why did we take the difference of the Lorentzians? As a matter of fact the exact expression for the spectral function $\Gamma^{\mathrm{g}}(\omega)$ must approach zero at $\omega = 0$. One Lorentzian does not satisfy this condition. The calculation of $\Delta^{\mathrm{g}}(\omega)$ yields

$$\begin{aligned}\Delta^{\mathrm{g}}(\omega) &= \int_0^{\infty} \frac{\mathrm{d}\nu}{\pi}\Gamma^{\mathrm{g}}(\nu)\left(\frac{1}{\omega-\nu} - \frac{1}{\omega+\nu}\right) = \int_{-\infty}^{\infty}\frac{\mathrm{d}\nu}{\pi}\Gamma^{\mathrm{g}}(\nu)\frac{1}{\omega-\nu} \\ &= \frac{1}{2\nu_{\mathrm{g}}}\left[\frac{\omega-\nu_{\mathrm{g}}}{(\omega-\nu_{\mathrm{g}})^2+(\gamma_0/2)^2} - \frac{\omega+\nu_{\mathrm{g}}}{(\omega+\nu_{\mathrm{g}})^2+(\gamma_0/2)^2}\right] .\end{aligned} \tag{12.103}$$

Using (12.102), (12.86) and (12.103), we find that

$$W^2 \Gamma^{\mathrm{g}}(\omega) \Gamma^{\mathrm{e}}(\omega) = \frac{\left[\omega\left(\nu_{\mathrm{e}}^2 - \nu_{\mathrm{g}}^2\right)\gamma_0\right]^2}{\left(\omega^2 - \nu_1^2\right)^2 \left(\omega^2 - \nu_2^2\right)^2 + \left[\omega\left(\nu_{\mathrm{e}}^2 - \nu_{\mathrm{g}}^2\right)\gamma_0\right]^2}, \tag{12.104}$$

where

$$\nu_{1,2}^2 = \frac{\nu_{\mathrm{g}}^2 + \nu_{\mathrm{e}}^2}{2} - \left(\frac{\gamma_0}{2}\right)^2 \pm \sqrt{\left(\frac{\nu_{\mathrm{g}}^2 - \nu_{\mathrm{e}}^2}{2}\right)^2 - \gamma_0 \frac{\nu_{\mathrm{g}}^2 + \nu_{\mathrm{e}}^2}{2}}. \tag{12.105}$$

These two frequencies give the resonances of the function $W^2\Gamma^{\mathrm{g}}(\omega)\Gamma^{\mathrm{e}}(\omega) = F(\omega)$. This function is shown in Fig. 12.3 for the frequencies $\nu_{\mathrm{e}} = 13.8\ \mathrm{cm}^{-1}$ and $\nu_{\mathrm{g}} = 18\ \mathrm{cm}^{-1}$ found from the spectra, and for two values of the parameter γ_0. The key point is the determination of the half-width γ_0 of the localized mode peak. The peaks in Fig. 12.2 are inhomogeneously broadened. Therefore their half-width is less than $2.4\ \mathrm{cm}^{-1}$. Even if we take this value for the half-width, we arrive at the following value of the parameter which characterizes the strength of the coupling: $W/\gamma_0\nu_{\mathrm{g}} = (\nu_{\mathrm{e}}^2 - \nu_{\mathrm{g}}^2)/\gamma_0\nu_{\mathrm{g}} \approx -3$, i.e., we are dealing with a case of strong quadratic coupling.

In the low temperature case, the general (12.85) can be transformed to (12.87). Curve 2 in Fig. 12.3 can be approximated by a sum of two Lorentzians with $\gamma_0 = 0.5\ \mathrm{cm}^{-1}$. In this strong coupling case, we can carry out the integration in (12.87) and obtain the simple expression

$$\gamma_{\mathrm{ph}}(T) = \frac{\gamma_0}{4}\left[\sinh^{-2}(h\nu_{\mathrm{g}}/kT) + \sinh^{-2}(\hbar\nu_{\mathrm{e}}/kT)\right]. \tag{12.106}$$

It is clear that the value of the half-width of the ZPL does not depend on the coupling constant W at strong coupling. It is possible to deduce the following conclusion from (12.106): the contribution to the half-width of the ZPL from the localized mode approaches zero as the lifetime $1/\gamma_0$ of the localized mode increases. Hence, interaction with a few long-lived localized modes is unable to broaden the ZPL.

Figure 12.4 shows experimental data for the reciprocal half-width of the ZPL and the theoretical curve calculated using (12.106). Unfortunately,

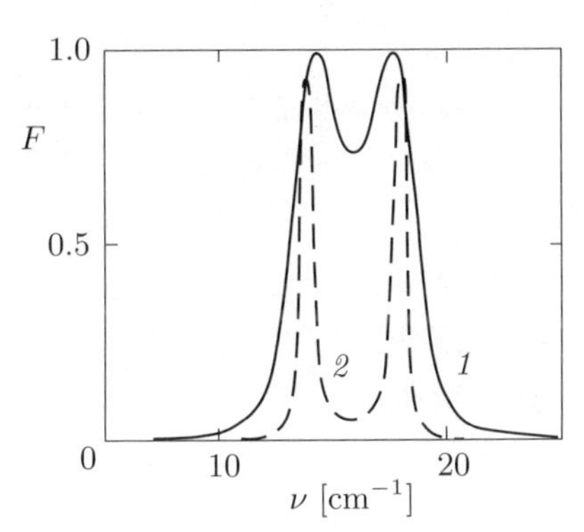

Fig. 12.3. Function F at $\gamma_0 = 2.4\ \mathrm{cm}^{-1}$ (curve 1) and $0.5\ \mathrm{cm}^{-1}$ (curve 2)

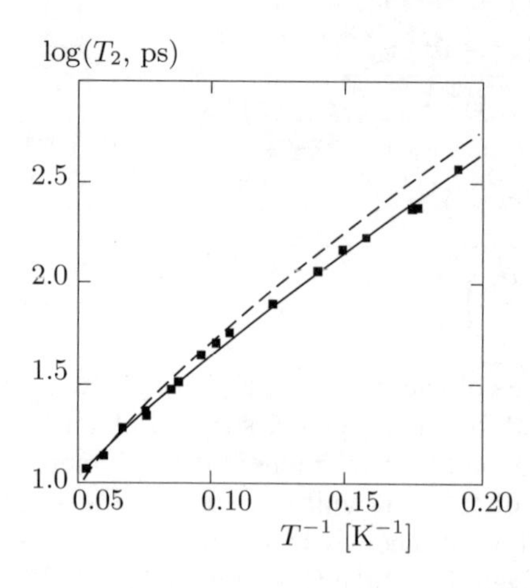

Fig. 12.4. Temperature dependence of the dephasing time of a pentacene molecule in a naphthalene crystal [81] and results of our calculation with (12.85) at $\nu_e = 13.8$ cm^{-1}, $\nu_g = 18$ cm^{-1} and $\gamma_0 = 0.5$ cm^{-1} (*solid line*) and with (12.106) at $\nu_e = 13.8$ cm^{-1}, $\nu_g = 18$ cm^{-1} and $\gamma_0 = 0.35$ cm^{-1} (*dashed line*)

the parameter γ_0 was not measured in the experiments by Hesselink and Wiersma. Instead, it was fitted. The theoretical curve fits experimental data well at $\gamma_0 = 0.5$ cm^{-1}.

In accordance with the theory, there is also a temperature shift of the ZPL. The expressions for the half-width and shift of the ZPL depend on the same parameters. Therefore, if parameters are found from the fitted experimental data for the half-width, the shift of the ZPL can be calculated without using any adjustable parameters. Such a comparison between the calculated and measured shift of the ZPL has been carried out in a few studies. Agreement between experimental and theoretical data for the shift has not been found. It is not yet clear what the reason is for this disagreement.

Part V

Methods of Selective Spectroscopy

13. Fluorescence Line Narrowing

Although single-molecule spectroscopy is a very promising method, it is not accessible to all experimental groups. Most laboratories working in the field of solid spectroscopy deal with guest molecule ensembles in crystalline or amorphous solids. Amorphous solids like polymers and glasses are convenient solvents for many guest molecules. However, optical bands in amorphous solids have large inhomogeneous broadening. Laser excitation of the molecular ensemble enables one to remove inhomogeneous broadening and to make the first step towards single-molecule spectroscopy. Progress in the field of selective spectroscopy is discussed in this and the next chapter.

Spectra of molecules doping amorphous solids like a polymer or glass undergo significant inhomogeneous broadening which hides all details of the molecular spectrum. Cooling such systems does not increase spectral resolution because cooling cannot remove inhomogeneous broadening. Laser excitation of such solid solutions enables one to remove inhomogeneous broadening and to increase spectral resolution considerably [9, 10, 82]. Inhomogeneous broadening arises because the frequency of the electronic transition in the guest molecule depends strongly on the configuration of its local environment. The various local environments of guest molecules in an amorphous solid are a source of strong inhomogeneous broadening in the optical spectra of guest molecule ensembles. The high resolution optical spectrum which emerges after inhomogeneous broadening has been removed is our topic in this chapter.

13.1 Frequency Selection of Molecules by Laser Excitation

The fluorescence of various guest molecule ensembles in amorphous solids obtained upon ordinary excitation by light from a mercury lamp is shown by dashed lines in Fig. 13.1. This figure shows that sample cooling does not yield high resolution spectra. The theory discussed in Part IV predicts the existence of narrow ZPL and PSB in the homogeneous optical band of a molecule. The spectra shown by dashed lines consist of the coalesced homogeneous optical bands with various resonant electronic frequencies. This situation is shown

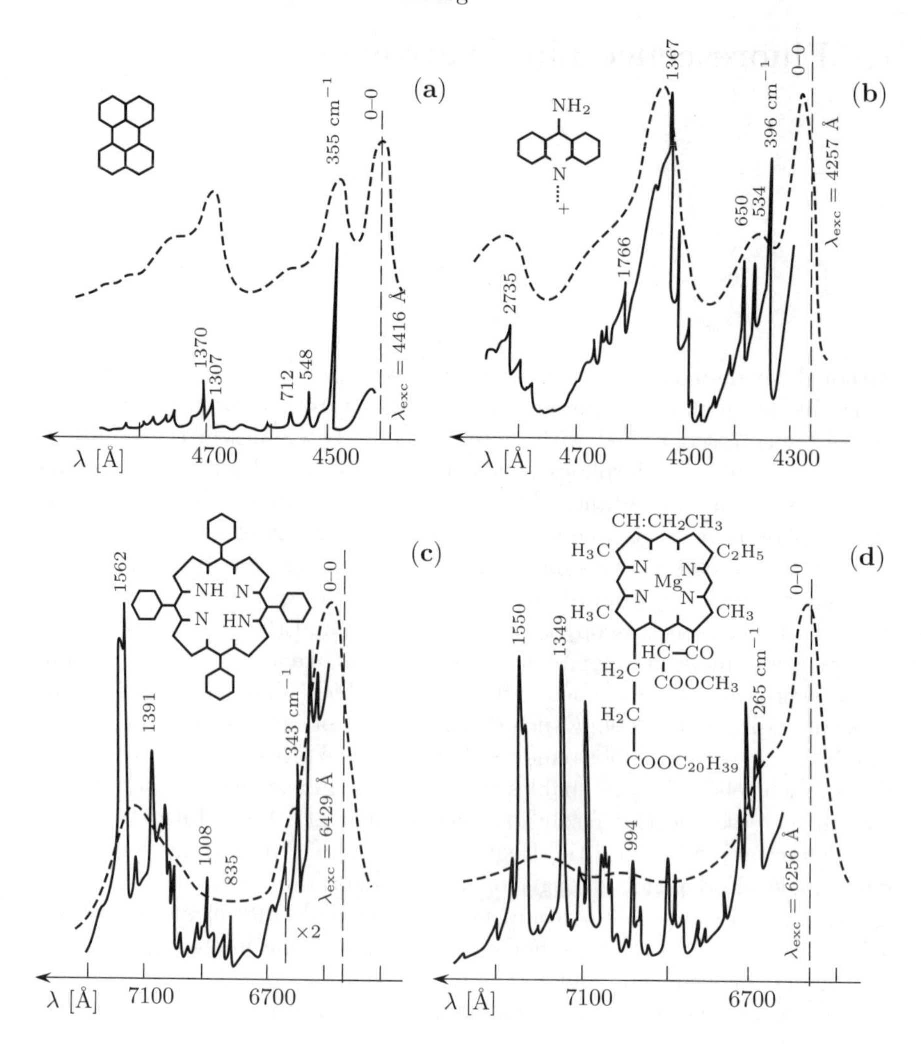

Fig. 13.1. Fluorescence bands of organic solutions doped with various molecules under ordinary ultraviolet excitation (*dashed lines*) and under laser excitation (*solid lines*). (**a**) Perylene in ethanol, (**b**) protonated 9-aminoacridine in ethanol with HCl, (**c**) tetraphenylporphyrin in polystyrol, (**d**) protochlorophyl in ether. Temperature $T = 4.2$ K [82]

schematically in Fig. 13.2. The question to be addressed here is: how should we excite samples with the broadened optical bands in order to obtain high resolution guest molecule spectra? These spectra are shown by solid lines in Fig. 13.1.

The inhomogeneous optical band consists of homogeneous bands of the guest molecules with various resonant frequencies. Those molecules whose

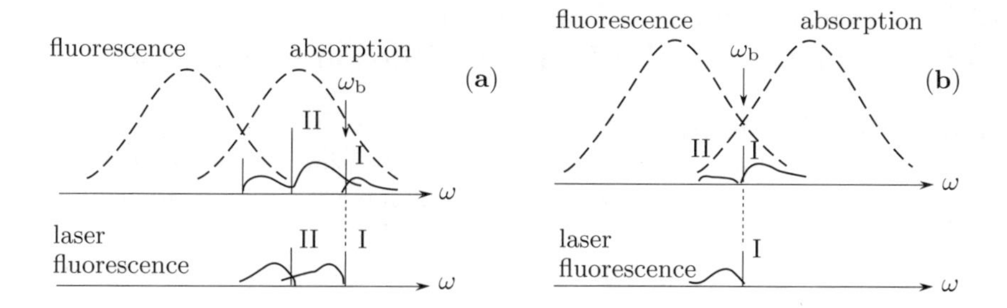

Fig. 13.2. **(a,b)** Fluorescence line narrowing (FLN) due to laser excitation in the resonant frequency domain. Bands with inhomogeneous broadening are shown by *dashed lines*

ZPL coincides with the laser frequency will be called type I molecules. Molecules excited via their PSB will be called type II molecules. It is obvious that type I molecules are more efficiently excited than type II molecules. The difference in excitation efficiency is proportional to the difference in the peak intensity of the ZPL and PSB. Hence, in the ensemble of the excited molecule, type I molecules will predominate. However, this frequency selection will only be effective under an additional condition. What is this condition?

The spectral region where absorption and fluorescence bands overlap is called the resonant region. If we excite on the blue side of the resonant region, as happens with the mercury lamp, type I molecules are more efficiently excited than type II molecules. However, the number of type II molecules is larger than the number of type I molecules. This situation is shown in Fig. 13.2a. Therefore, both types of molecule give a comparable contribution to the intensity of fluorescence light. It is obvious that inhomogeneous broadening in the fluorescence spectrum is not removed under such excitation conditions. The fluorescence spectra shown by dashed lines in Fig. 13.1 relate to this very method of excitation.

However, we face a different situation if the excitation is carried out via the resonant region. This situation is shown in Fig. 13.2b. In this case, type I molecules are more efficiently excited and they are more numerous than type II molecules excited via the PSB. Therefore the light emitted by type I molecules will predominate in the fluorescence, i.e., inhomogeneous broadening will be removed from the fluorescence spectrum. The fluorescence spectra shown by solid lines in Fig. 13.1 relate to excitation via the resonant spectral region. This effect whereby the fluorescence spectrum narrows under excitation in the resonant spectral region is called fluorescence line narrowing.

13.2 Fluorescence Line Narrowing and Its Relation to the Full Two-Photon Correlator

In the last section, we presented a qualitative picture of well resolved fluorescence spectra. We now derive a mathematical expression for the fluorescence spectrum obtained under laser excitation, explaining its relation to the full two-photon correlator considered in Chaps. 2 and 3.

The fluorescence band shape function of an ensemble of the excited molecules is given by

$$J_{\mathrm{F}}(\omega_{\mathrm{r}}) = \int_{0}^{\infty} J^{\mathrm{e}}(\omega_{\mathrm{r}} - \omega_0) n_1(\omega_0, \omega_{\mathrm{b}}) \, \mathrm{d}\omega_0 \, , \tag{13.1}$$

where $J^{\mathrm{e}}(\omega_{\mathrm{r}} - \omega_0)$ is the fluorescence band shape function of an individual molecule with resonant frequency ω_0 and $n_1(\omega_0, \omega_{\mathrm{b}})$ is the number of molecules with resonant frequency ω_0 excited by laser light with frequency ω_{b}. There is a simple relation between this formula and the full two-photon correlator. The population of the excited electronic level n_1 is expressed via the probability ρ_1 as follows: $n_1 = \rho_1 n$, where n is the number of molecules in the ensemble. The full two-photon correlator is described by the expression $p = \rho_1 / T_1$. The correlator depends on the detuning and the time delay between photons emitted upon CW excitation. If this time delay exceeds all relaxation times of the system, the full two-photon correlator is a function of the detuning $\omega_{\mathrm{b}} - \omega_0$ alone and does not depend on time. In this case one can write

$$J_{\mathrm{F}}(\omega_{\mathrm{r}}, \omega_{\mathrm{b}}) = T_1 \int_{0}^{\infty} J^{\mathrm{e}}(\omega_{\mathrm{r}} - \omega_0) p(\omega_{\mathrm{b}} - \omega_0) n(\omega_0) \, \mathrm{d}\omega_0 \, . \tag{13.2}$$

The fluorescence band shape function of the molecular ensemble is a convolution of the fluorescence band shape function of individual molecules and the full two-photon correlator taken for large time delay.

Let us examine the last expression for the case of excitation by an ordinary ultraviolet light source like a mercury lamp or nitrogen laser. Under such conditions all molecules are excited via the upper electron–phonon–vibron states and therefore each molecule has equal probability of occurring in the first electronic state independently of the resonant transition ω_0 of the first electronic transition. In this event, the function $n(\omega_{\mathrm{b}}, \omega_0)$ does not depend on the frequency ω_{b} and this function is proportional to the function $n(\omega_0)$ which describes the distribution of the molecules over the resonant frequencies. If we take this into account and neglect the PSB, i.e., we set

$$J^{\mathrm{e}}(\omega) = (\gamma/2\pi) \left[\omega^2 + (\gamma/2)^2 \right]^{-1} ,$$

we find, after integrating (13.2),

$$J_{\mathrm{F}}(\omega_{\mathrm{r}}) \approx n(\omega_{\mathrm{r}}) \, . \tag{13.3}$$

The fluorescence band shape function of the ensemble molecule is similar to a very broad distribution function if excitation is carried out much higher than the first singlet transition.

Let us examine (13.2) for the case when the detuning $\Delta = \omega_{\mathrm{b}} - \omega_0$ is small. In this case and under CW excitation, the ratio of the populations of the excited and ground electronic states is given by

$$\frac{n_1}{n_0} = \frac{k^{\mathrm{g}}(\Delta)}{k^{\mathrm{e}}(\Delta) + 1/T_1} \, , \tag{13.4}$$

where the numerator is the count rate of absorbed photons and the denominator is the count rate of spontaneously and induced emitted photons. The probabilities $k^{\mathrm{g,e}}(\Delta)$ of the induced transitions are determined by (10.2) and (10.3). The number of excited molecules is expressed in terms of the total number of molecules $n = n_0 + n_1$ as follows:

$$n_1 = \frac{k^{\mathrm{g}}(\Delta)}{k^{\mathrm{g}}(\Delta) + k^{\mathrm{e}}(\Delta) + 1/T_1} n \, . \tag{13.5}$$

At weak laser light the probabilities of induced transitions are smaller than the probability $1/T_1$ of spontaneous transitions. In this case one finds, instead of (13.5), the expression

$$n_1(\omega_{\mathrm{b}}, \omega_0) \approx T_1 k^{\mathrm{g}}(\omega_{\mathrm{b}} - \omega_0) \, n(\omega_0) \propto J^{\mathrm{g}}(\omega_{\mathrm{b}} - \omega_0) n(\omega_0) \, . \tag{13.6}$$

Inserting the last expression into (13.2), we derive the following equation for the fluorescence band shape function of the molecular ensemble:

$$J_{\mathrm{FLN}}(\omega_{\mathrm{r}}, \omega_{\mathrm{b}}) \propto \int_0^\infty J^{\mathrm{e}}(\omega_{\mathrm{r}} - \omega_0) \, J^{\mathrm{g}}(\omega_{\mathrm{b}} - \omega_0) \, n(\omega_0) \, \mathrm{d}\omega_0 \, . \tag{13.7}$$

This is the final formula describing the so-called FLN spectrum, i.e., fluorescence line narrowing due to laser excitation of the inhomogeneous optical band via the resonant spectral region.

Let us apply the last equation to the case when the homogeneous optical band consists of only the ZPL, i.e., $J^{\mathrm{e}}(\omega) = (\gamma/2\pi)[\omega^2 + (\gamma/2)^2]^{-1}$. Upon ultraviolet excitation, we find that the fluorescence band is proportional to the distribution function. Upon laser excitation via the resonant spectral region, (13.7) takes the form

$$J_{\mathrm{FLN}}(\omega_{\mathrm{r}}, \omega_{\mathrm{b}}) \approx n(\omega_{\mathrm{b}}) \frac{\gamma}{(\omega_{\mathrm{r}} - \omega_{\mathrm{b}})^2 + \gamma^2} \, , \tag{13.8}$$

i.e., the FLN spectrum consists of a Lorentzian with half-width twice the Lorentzian halfwidth of the homogeneous spectrum. The intensity of fluorescence is proportional to the number of molecules whose ZPL coincides with the laser frequency. However, it is difficult to measure this fluorescence line because it coincides with the laser excitation light. It becomes possible to record this fluorescence after the excitation has been switched off. The fluorescence will decay exponentially with the lifetime of the excited electronic level. For instance the R-line of a ruby crystal radiates the electronic level with lifetime equal to a few milliseconds. There is no problem in detecting such slow fluorescence decay. However, as a rule, the electronic transition to the first electronic excited state of organic molecules is not forbidden, in contrast to what happens in the ruby crystal. In such molecules, the fluorescence lifetime is of nanosecond order. The measurement of the fluorescence line shape on a nanosecond time scale is a rather complicated task. Therefore, fluorescence line narrowing is used to study the vibronic spectra of complex organic molecules.

The shape of realistic optical absorption and emission bands for a guest molecule are described by the expressions

$$\begin{aligned} J^{\mathrm{g}}\left(\omega_{\mathrm{b}}-\omega_0\right) &= \alpha \frac{(\gamma/2\pi)}{\left(\omega_{\mathrm{b}}-\omega_0\right)^2+(\gamma/2)^2} + \Psi^{\mathrm{g}}\left(\omega_{\mathrm{b}}-\omega_0\right) , \\ J^{\mathrm{e}}\left(\omega_{\mathrm{r}}-\omega_0\right) &= \alpha \frac{(\gamma/2\pi)}{\left(\omega_{\mathrm{r}}-\omega_0\right)^2+(\gamma/2)^2} + \Psi^{\mathrm{e}}\left(\omega_0-\omega_{\mathrm{r}}\right) , \end{aligned} \tag{13.9}$$

where α is the Debye–Waller factor which determines the integrated intensity of the ZPL. The functions $\Psi^{\mathrm{g,e}}$ describe the electron–vibration parts of the optical bands. If we neglect the HT interaction, $\Psi^{\mathrm{g,e}} = (1-\alpha)\Phi^{\mathrm{g,e}}$. Let us ignore the first intramolecular vibrations of the molecules. Then the functions $\Phi^{\mathrm{g,e}}$ describe the PSB of the absorption and fluorescence bands. By inserting (13.9) into (13.7), we arrive at the expression

$$\begin{aligned} J_{\mathrm{FLN}}\left(\omega_{\mathrm{r}},\omega_{\mathrm{b}}\right) = {} & \alpha^2 \frac{\gamma}{\left(\omega_{\mathrm{r}}-\omega_{\mathrm{b}}\right)^2+\gamma^2} n\left(\omega_{\mathrm{b}}\right) \\ & +\alpha\left(1-\alpha\right)\Phi(\omega_{\mathrm{b}}-\omega_{\mathrm{r}})[n(\omega_{\mathrm{b}})+n(\omega_{\mathrm{r}})] \\ & +\left(1-\alpha\right)^2 \int_0^\infty \Phi\left(\omega_{\mathrm{b}}-\omega_0\right)\Phi\left(\omega_0-\omega_{\mathrm{r}}\right) n(\omega_0)\mathrm{d}\omega_0 . \end{aligned} \tag{13.10}$$

Here the function $\Phi(\omega)$ describes the shape of the PSB of the absorption band. The first term in (13.10) describes the ZPL with the double half-width. The second term describes the PSB which extends to the red side from the ZPL as in the ordinary fluorescence spectrum. However, this PSB consists of two terms which are proportional to $n(\omega_{\mathrm{b}})$ and $n(\omega_{\mathrm{r}})$. The third term is a convolution of two PSBs. It is obvious that the structure of the fluorescence band is determined by the first and second terms in (13.10). The shape of the PSB in the FLN spectrum is described by the second term. The shape is similar to that in the homogeneous spectrum. However, the ratio of the integrated intensity of the ZPL and the integrated intensity of the whole

FLN band, including the structureless background described by the third term in (13.10), is given by

$$\frac{(J_{\mathrm{ZPL}})_{\mathrm{FLN}}}{J_{\mathrm{FLN}}} = \alpha^2 \,. \tag{13.11}$$

The Debye–Waller factor of the FLN spectrum equals the square of the Debye–Waller factor of the homogeneous band. This means that the ZPL in the FLN spectrum can emerge only at weak electron–phonon coupling, when $\alpha \leq 1$.

Let us include one vibronic mode with frequencies Ω^{e} and Ω^{g}. In this case the electron–vibration functions are given by

$$\begin{aligned}
\Psi^{\mathrm{g}}(\omega_{\mathrm{b}} - \omega_0) &= \alpha F \frac{\gamma_1/2\pi}{(\omega_{\mathrm{b}} - \omega_0 - \Omega^{\mathrm{e}})^2 + (\gamma_1/2)^2} + (1-\alpha)\,\Phi\,(\omega_{\mathrm{b}} - \omega_0) + \dots\,, \\
\Psi^{\mathrm{e}}(\omega_0 - \omega_{\mathrm{r}}) &= \alpha F \frac{\gamma_1/2\pi}{(\omega_0 - \omega_{\mathrm{r}} - \Omega^{\mathrm{g}})^2 + (\gamma_1/2)^2} + (1-\alpha)\Phi(\omega_0 - \omega_{\mathrm{r}}) + \dots\,.
\end{aligned} \tag{13.12}$$

Here $F = a^2/2$ is the parameter of the FC interaction and γ_1 is the half-width of the vibronic line. The lines relating to two or more vibrons are not written explicitly here. Inserting the last expressions into (13.7) yields

$$\begin{aligned}
J_{\mathrm{FLN}} \propto &\left[\frac{\gamma}{(\omega_{\mathrm{b}} - \omega_{\mathrm{r}})^2 + \gamma^2} + F \frac{(\gamma + \gamma_1)/2}{(\omega_{\mathrm{b}} - \omega_{\mathrm{r}} - \Omega^{\mathrm{g}})^2 + [(\gamma + \gamma_1)/2]^2}\right] n\,(\omega_{\mathrm{b}}) \\
+F &\left[\frac{(\gamma + \gamma_1)/2}{(\omega_{\mathrm{b}} - \omega_{\mathrm{r}} - \Omega^{\mathrm{e}})^2 + [(\gamma + \gamma_1)/2]^2} + F \frac{\gamma_1}{(\omega_{\mathrm{b}} - \omega_{\mathrm{r}} - \Omega^{\mathrm{g}} - \Omega^{\mathrm{e}})^2 + \gamma_1^2}\right] \\
&\times n\,(\omega_{\mathrm{b}} - \Omega^{\mathrm{e}}) + \dots\,.
\end{aligned} \tag{13.13}$$

The expression in the first square brackets describes the main fluorescence spectrum, emerging due to excitation via those molecules whose ZPL coincides with the laser line. The expression in the second square brackets describes an additional fluorescence spectrum which emerges due to excitation via molecules whose vibronic lines coincide with the laser line. The additional spectrum is shifted to the red side from the main spectrum. This spectral shift of the additional fluorescence spectrum equals the vibronic frequency in the excited electronic state. This means that vibronic frequencies of the excited electronic state can be measured in the FLN spectrum. It should be borne in mind that the frequencies of the ground electronic state manifest themselves in the homogeneous fluorescence spectrum.

Obviously, when many vibronic modes are taken into account, the number of additional vibronic spectra is large. These additional vibronic spectra will hamper vibration analysis in the main fluorescence spectrum. Therefore it would be better to get rid of such additional fluorescence spectra. In accordance with (13.13), the intensity of the additional spectrum is proportional to $n(\omega_b - \Omega^e)$. If the laser frequency ω_b coincides with the maximum of the distribution function $n(\omega)$, the value $n(\omega_b - \Omega^e)$ of this function is small. Therefore the additional fluorescence spectrum is weak compared to the main fluorescence spectrum. In this case the main fluorescence spectrum dominates the FLN spectrum. Indeed, Fig. 13.1 shows that an additional spectrum II is suppressed if the excitation is realized via the resonant spectral region where the function $n(\omega)$ has a maximum. If we change the laser frequency, the FLN spectrum follows the laser frequency. This property is typical for Raman spectra but not for ordinary fluorescence.

13.3 Laser Fluorescence Analysis

The FLN spectrum can serve as a good way of identifying the individual substances in an organic mixture. As a matter of fact the vibronic spectrum is connected with the chemical structure of the organic molecule. Therefore, by measuring a vibronic spectrum, we are able to determine which organic substance emits this light. This substance can thus be identified in an organic mixture by its FLN spectrum. Such an identification is shown in Fig. 13.3.

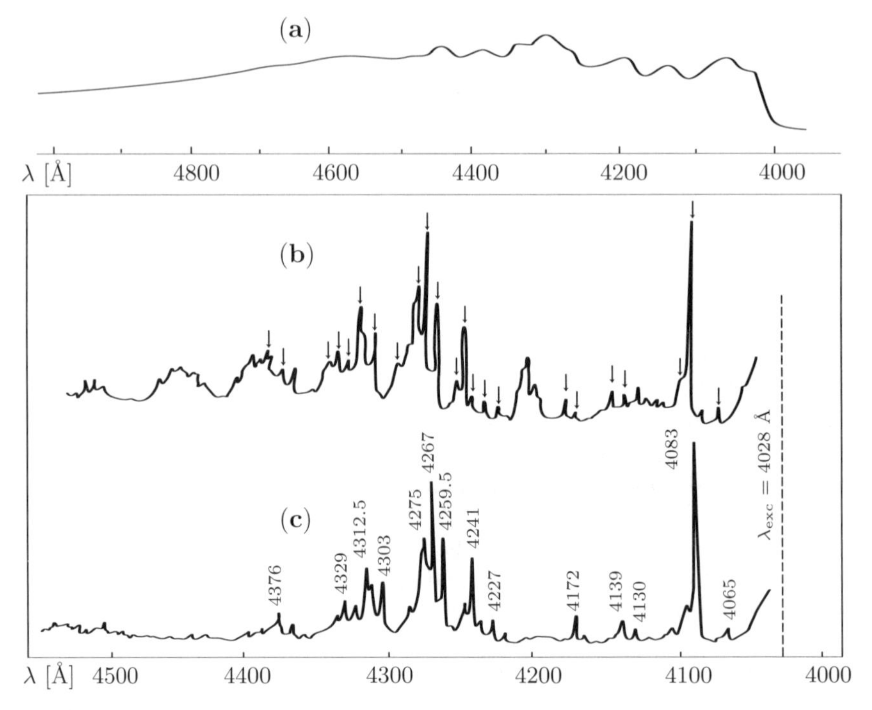

Fig. 13.3. Fluorescence of ordinary gasoline under (**a**) ultraviolet excitation and (**b**) laser excitation. (**c**) Spectrum of 3,4-benzopyrene in the same gasoline. $T = 4.2$ K [82]

The fluorescence spectrum of the ordinary gasoline A-92 upon ultraviolet excitation is shown in Fig. 13.3a. The spectrum is structureless, although the temperature is rather low. By means of this spectrum, it is impossible to determine which of the chemical substances radiate this light. However, if this gasoline is irradiated by a laser line with wavelength 402.8 nm, the fluorescence spectrum changes dramatically, as shown in Fig. 13.3b. Figure 13.3c shows the homogeneous fluorescence spectrum of a 3,4-benzopyrene molecule, measured in a specially prepared low temperature solution of these molecules. Comparing the FLN spectrum in Fig. 13.3b with the spectrum in Fig. 13.3c, we can conclude that the FLN spectrum in Fig. 13.3b belongs to 3,4-benzopyrene molecules dissolved in the gasoline. 3,4-benzopyrene is a highly cancerogenic substance. Therefore, the possibility of identifying it in an organic mixture is very important for ecological purposes.

It is possible to obtain various FLN spectra by changing the excitation frequency, as shown in Fig. 13.4. These spectra belong to various organic substances dissolved in gasoline. By using special reference books with the measured fluorescence spectra of various organic molecules, we can carry out an identification of the molecules in the mixture by comparing the measured FLN spectrum of the mixture with the spectrum from the book.

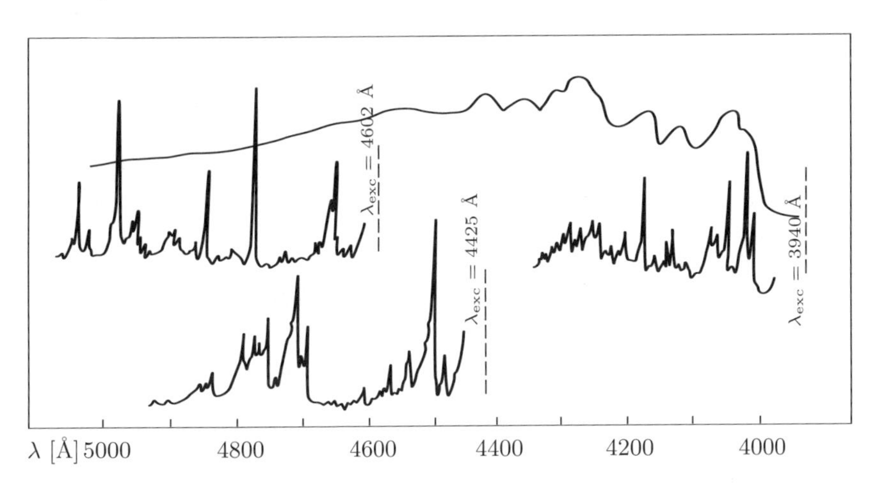

Fig. 13.4. Dependence of gasoline fluorescence on λ_{exc}. Various observed spectra belong to different kinds of impurity molecule present in the gasoline. $T = 4.2$ K [83]

14. Spectral Hole Burning in Inhomogeneous Optical Bands

It was shown in the last chapter that laser excitation of inhomogeneous optical bands enables one to obtain well resolved FLN–vibronic spectra. However, it is not convenient to study the ZPL in fluorescence spectra because of the short lifetime of the allowed electronic transition. Fluorescence line narrowing is not the only method available for removing inhomogeneous broadening. There is another method of selective spectroscopy called spectral hole burning (SHB). The shape of the ZPL is studied by means of the SHB method [11,12,84].

14.1 Transient Spectral Hole Burning. Relation to the Full Two-Photon Correlator

Let us consider the absorption band of the ensemble of guest molecules in an amorphous solid. The distribution of frequencies ω_0 of the ZPL is described by the function $n(\omega_0)$. Each guest molecule can absorb laser light with probability $J^{\mathrm{g}}(\omega_{\mathrm{r}} - \omega_0)$, where ω_{r} is the laser light frequency. If the laser light intensity is extremely weak, so that we can neglect the number of excited molecules, then the absorption band shape function of this ensemble is given by

$$J_{\chi=0}(\omega_{\mathrm{r}}) = \int_0^{\infty} J^{\mathrm{g}}(\omega_{\mathrm{r}} - \omega_0) n(\omega_0)\, \mathrm{d}\omega_0 \ , \tag{14.1}$$

where the subscript $\chi = 0$ shows that the intensity of the probing laser light should be extremely weak. Let us neglect the PSB. Then the absorption band is described by the ZPL, i.e., $J^{\mathrm{g}}(\omega) = (\gamma/2\pi)[\omega^2 + (\gamma/2)^2]^{-1}$. Inserting this equation into (14.1), we find the following expression for the absorption band of the molecular ensemble:

$$J_{\chi=0}(\omega_{\mathrm{r}}) \approx n(\omega_{\mathrm{r}}) \ . \tag{14.2}$$

i.e., even if the molecular spectrum consists of the ZPL alone, the ensemble band is described by the broad band function $n(\omega_{\mathrm{r}})$.

In the example considered, we neglected the fact that laser light transmits some molecules from the ground to the excited electronic state. If we take into account the change in the absorption coefficient due to some depopulation of the ground state, this depopulation is described by $\Delta n_0 = n_0(0) - n_0(t)$. In this case the experiment has two stages. In the first stage, called hole burning, we use intensive light to excite a great deal of molecules $\Delta n_0(\omega_b, \omega_0, t)$ in the ground electronic state. In the second stage, called hole probing, we irradiate the optical band with a hole by the probing laser light whose frequency ω_r is scanned. The intensity of the probing laser light must be as weak as possible in order to be able to neglect any additional change in the population of the ground electronic state due to this additional irradiation of the sample. The PMT signal from the probing light is given by

$$h(\omega_r, \omega_b, t) = \int_0^\infty J^g(\omega_r - \omega_0)\Delta n_0(\omega_b, \omega_0, t)\,d\omega_0 \ . \tag{14.3}$$

This formula describes the shape of the hole detected by a photoreceiver. It resembles (13.1) for the FLN spectrum. However, there are two differences.

The first difference is that the fluorescence function J^e is replaced by the absorption function J^g. The second difference concerns the physical meaning of the function $\Delta n_0(\omega_b, \omega_0, t)$. In the case of the FLN spectrum, the function n_1 describes the number of molecules transferred by light from the ground to the excited electronic state. This function determines the full two-photon correlator $p(t) = n_1(t)/T_1$ as well. In the case of hole burning spectroscopy, we deal with the function $\Delta n_0(\omega_b, \omega_0, t)$, which describes the deficit of molecules in the ground electronic state.

In the case of a two level molecule, there is no difference between the functions n_1 and $\Delta n_0(\omega_b, \omega_0, t)$. However, as a rule, each organic molecule has a triplet electronic level between the ground and first excited singlet electronic levels. This situation is shown in Fig. 8.2. In such a system the singlet excitation gets to triplet level due to intersystem crossing. The probability of intersystem crossing in the excited electronic state is of the order of inverse nanoseconds, whereas the lifetime of the triplet level is of millisecond order. Therefore, after excitation, the molecule quickly reaches the triplet level and then sits there for a long time. Hence, although the molecule has left the ground state, it is absent from the excited singlet state because it occupies the triplet level. Therefore $n_1 = 0$, but $\Delta n_0(\omega_b, \omega_0, t) \neq 0$. The lifetime of this depopulation of the ground electronic state equals the triplet lifetime and can reach several seconds. The long lifetime of the depopulation $\Delta n_0(\omega_b, \omega_0, t)$ in the ground electronic state and the absence of the population n_1 in the excited singlet state constitutes the main difference between FLN and spectral hole burning (SHB) and it allows us to measure the ZPL shape with the help of SHB.

In Sect. 13.2, we found the simple relation (13.2) between the function for the FLN spectrum and the full two-photon correlator. However, (14.3) for the spectral hole cannot be expressed in terms of the full two-photon correlator.

Let us find an expression for the spectral hole shape function in the case of the molecule with triplet level shown in Fig. 8.2. The results obtained in Sect. 8.2 can be used in this section. The dynamics of the molecule with one triplet and two singlet levels can by described by the three rate equations in (8.15). Our task is to find the probability $\rho_0(t)$. In accordance with the equation $n_0(\omega_b, \omega_0, t) = \rho_0(t)n(\omega_0)$, this probability determines the ground state population at arbitrary times. We shall calculate this probability in the same way as we found the probability ρ_1 and the two-photon correlator p in Sect. 8.2. Using (8.17), the Laplace transform of this probability is

$$\begin{aligned}\rho_0(\omega) &= \frac{-(\mathrm{i}\omega - \Gamma - k)(\mathrm{i}\omega - \gamma_{\mathrm{ST}})}{\mathrm{Det}} = \mathrm{i}\frac{-(\mathrm{i}\omega - \Gamma - k)(\mathrm{i}\omega - \gamma_{\mathrm{ST}})}{(\omega + \mathrm{i}0)(\omega - \omega_1)(\omega - \omega_2)} \\ &= \left[\frac{(\Gamma + k)\gamma_{\mathrm{ST}}}{\mathrm{i}\omega_1\omega_2}\frac{1}{\omega + \mathrm{i}0} + \frac{(\mathrm{i}\omega_1 - \Gamma - k)(\mathrm{i}\omega_1 - \gamma_{\mathrm{ST}})}{\mathrm{i}\omega_1(\omega_1 - \omega_2)}\frac{1}{\omega - \omega_1}\right. \\ &\qquad \left. - \frac{(\mathrm{i}\omega_2 - \Gamma - k)(\mathrm{i}\omega_2 - \gamma_{\mathrm{ST}})}{\mathrm{i}\omega_2(\omega_1 - \omega_2)}\frac{1}{\omega - \omega_2}\right]. \end{aligned} \tag{14.4}$$

The roots of the determinant are determined by (8.19). The physical meaning of the rate constants is determined in Fig. 8.2. Inserting the roots and taking the inverse Laplace transform yields

$$\begin{aligned}\Delta\rho_0(t) &= \frac{(\gamma_0 + R - \Gamma - k)(\gamma_0 + R - \gamma_{\mathrm{ST}})}{2R(\gamma_0 + R)}\left[1 - \mathrm{e}^{-(\gamma_0 + R)t}\right] \\ &\qquad - \frac{(\gamma_0 - R - \Gamma - k)(\gamma_0 - R - \gamma_{ST})}{2R(\gamma_0 - R)}\left[1 - \mathrm{e}^{-(\gamma_0 - R)t}\right]. \end{aligned} \tag{14.5}$$

This function of the detuning and time is expressed in terms of the same rate constants as the full two-photon correlator. Here time is measured from the beginning of the hole-burning process. Equation (14.5) enables one to calculate the temporal evolution of the spectral hole. The parameters take the following values: $1/T_1 \approx 10^9$–$10^8\ \mathrm{s}^{-1}$, $\Gamma_{\mathrm{TS}} \approx 10^{11}$–$10^{10}\ \mathrm{s}^{-1}$, $\gamma_{\mathrm{ST}} \approx 10^6$–$10^0\ \mathrm{s}^{-1}$. As usual we shall consider the case when the inequalities

$$\Gamma \gg k \gg \gamma_{\mathrm{ST}} \tag{14.6}$$

are satisfied. This allows us to compare the case of hole burning with the full two-photon correlator calculated under the same assumptions as in Sect. 8.2.

Figure 14.1 shows the temporal dependence of the hole depth in the ground state population at various intensities of the burning laser. The dependence of the rate of burning on the intensity of the laser light is clearly visible. Since $\gamma_0 + R \sim \Gamma$ and $\gamma_0 - R \sim k$, the first exponential function falls off faster than the second. The coefficient of the first exponential factor is very small and this term can therefore be omitted. The increase in the hole depth is thus determined by the slow exponential factor, i.e., it depends on the laser pumping k.

Let us compare the temporal behavior of the hole depth and the full two-photon correlator at the same parameter values. Both functions are shown

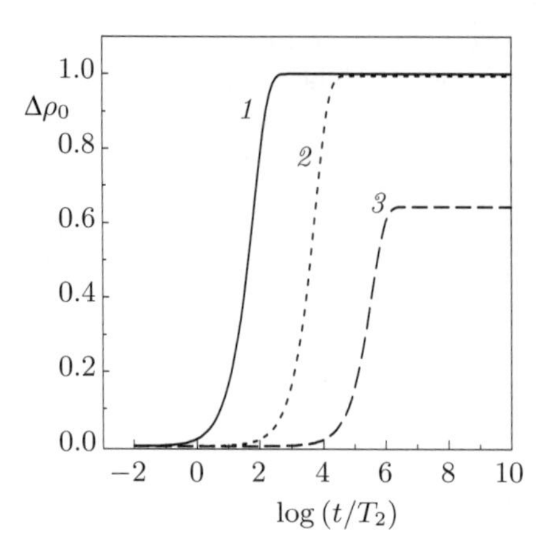

Fig. 14.1. Dependence of a hole burnt in the ground state population on the burning time at $T_2/T_1 = 10^{-2}$, $T_2\gamma_{\mathrm{ST}} = 10^{-6}$, $T_1\Gamma_{ST} = 9$, $T_2\Delta = 0$ and $\chi T_2 = 10^{-1}$ (curve 1), 10^{-2} (curve 2) and 10^{-3} (curve 3)

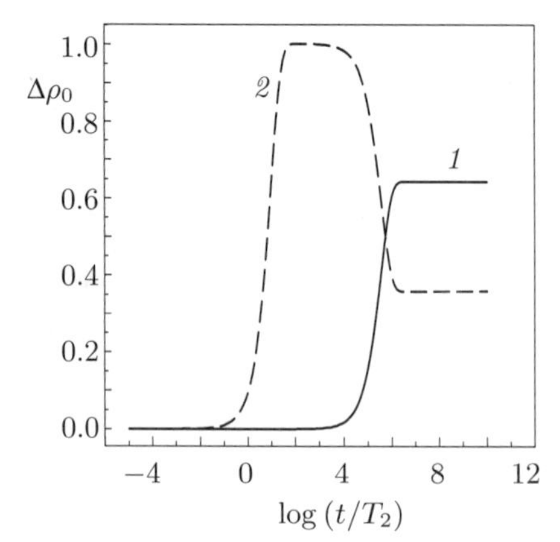

Fig. 14.2. Temporal dependence of a hole burnt in the ground state population (*solid line*) and of the full two-photon correlator (*dashed line*) at the parameter values taken in Fig. 14.1. (Curve 1) $\Delta\rho_0$ and (curve 2) $p(t/T_2)/p(10^3)$

in Fig. 14.2. Both slow and fast exponential behavior manifest themselves in the dependence of the two-photon correlator on the time delay t. It is clearly seen that the hole reaches saturation at large times when the two-photon correlator becomes small. Hence, single-molecule measurements and hole-burning experiments allow us to obtain information on various time scales.

Let us consider the temporal evolution of holes. Equation (14.5) shows that the hole in the ground state population increases with time and its shape changes. This equation takes a very simple form in the short time limit:

$$\Delta\rho_0\,(t) \approx k\,(\Delta)\,t\,, \tag{14.7}$$

i.e., the hole in the ground state population displays the homogeneous absorption line shape. In the case when we can neglect the PSB, the absorption band is described by the function (8.16), i.e., it is a Lorentzian with half-width $2/T_2$.

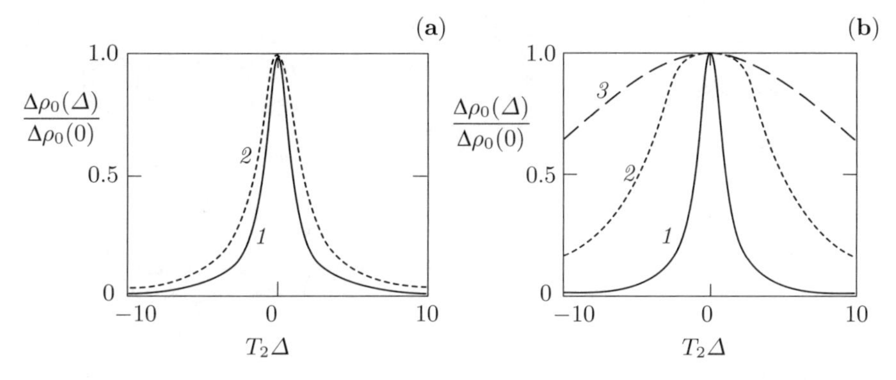

Fig. 14.3. Temporal dependence of hole shape in (**a**) the short-lived ($T_2/\gamma_{\mathrm{ST}} = 10^{-6}$) and (**b**) the long-lived ($T_2/\gamma_{\mathrm{ST}} = 10^{-8}$) triplet levels. $T_2/T_1 = 10^{-2}$, $T_1\Gamma_{\mathrm{TS}} = 9$, $\chi T_2 = 10^{-3}$ and $t/T_2 = 10^5$ (curve 1), 10^7 (curve 2), 10^9 (curve 3)

Calculation using (14.5) yields the result shown in Fig. 14.3. The half-width starts growing from the moment the triplet level begins to populate and stops once equilibrium is established between molecules coming into and leaving the triplet state. If the rate of triplet state population exceeds the rate of triplet state depopulation, the hole width can greatly exceed the homogeneous line width. This is shown in Fig. 14.3b. Spectral holes whose lifetime coincides with the lifetime of the excited electronic state are called transient spectral holes.

14.2 Persistent Spectral Holes

The lifetime of transient spectral holes cannot extend to hours or days because there are no excited electronic states with such a long lifetime. Impressive progress in hole-burning spectroscopy is connected with the discovery that spectral holes with lifetimes of several days can be burnt in the inhomogeneous optical bands of many guest–host systems. Such long lifetimes can only exist if the molecule reaches the ground electronic state. Spectral holes with long lifetime are called persistent spectral holes. Persistent holes have the following physical properties:

- the lifetime of spectral holes can exceed days at $T = 4.2$ K,
- a spectral hole can be erased by heating, or by irradiating the sample with white light,
- after erasure has been carried out, the hole can be burnt again at the same spectral point of the inhomogeneous optical band of the guest molecule.

Persistent spectral holes with these properties have been burnt in the inhomogeneous optical bands of molecules of various types including proteins,

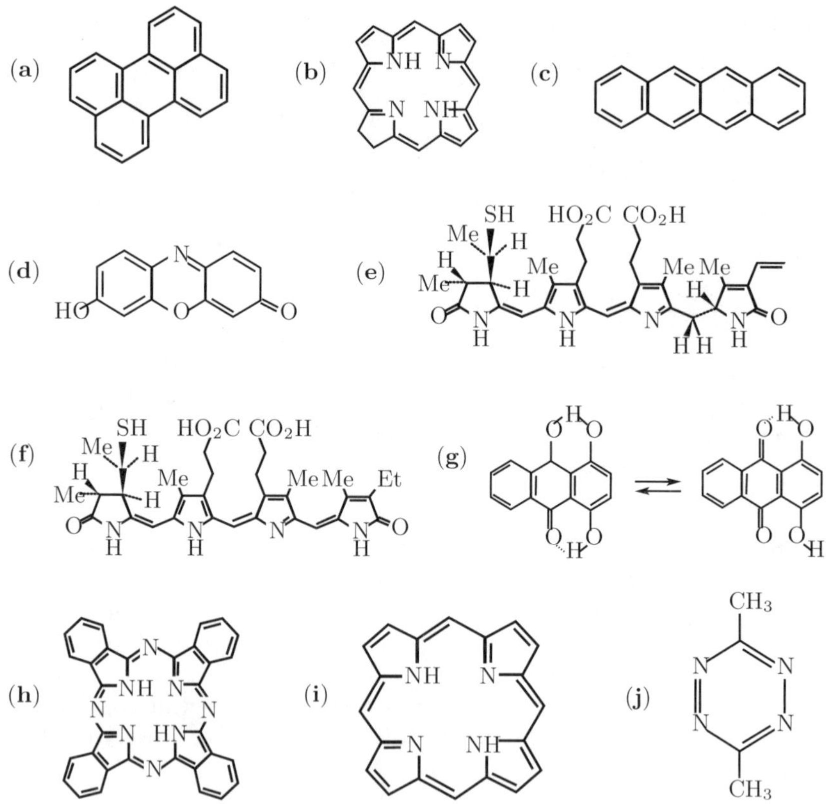

Fig. 14.4. Molecules undergoing persistent spectral hole-burning. (**a**) perylene, (**b**) chlorine, (**c**) tetracene, (**d**) resorufin, (**e**) phycoerythrobilin, (**f**) phycocyanobilin, (**g**) quinizarin, (**h**) phthalocyanine, (**i**) porphyrin, (**j**) dimethyl-*s*-tetrazine [85]

chlorophyll, and so on. These molecules are of great significance for molecular biology. This explains why so much attention has been paid to hole burning and why such great progress has been observed in persistent spectral hole-burning spectroscopy over the last decade. The chemical structure of molecules for which persistent spectral holes can be burnt in the inhomogeneous singlet optical band is shown in Fig. 14.4.

Persistent spectral hole burning can be explained on a qualitative level as shown in Fig. 14.5. In accordance with this figure, a molecule persisting in the ground electronic state because of a high potential barrier can become unstable in one of the excited electronic states reached after absorption. After S_0–S_1 singlet excitation, the molecule can change its configuration, e.g., going into the triplet state, as shown in Fig. 14.5, moving through a potential barrier which is low enough in the triplet electronic state. Finally, the molecule goes into a new electronic ground state. In fact, a new molecular form is created after light absorption. This form can be called the photoproduct. However,

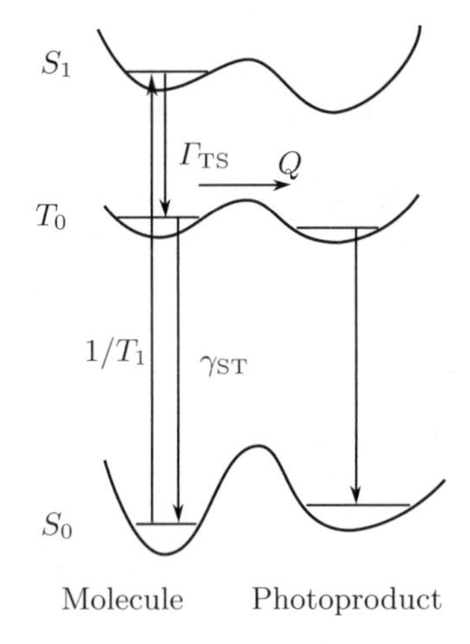

Fig. 14.5. Energy diagram of a molecule and a photoproduct providing persistent spectral hole burning

such a photoproduct is unable to absorb photons with the former frequency ω_b. Therefore, after irradiation of the sample by laser light with frequency ω_b, some molecules lose the ability to absorb photons with this frequency. The optical density of the sample becomes smaller at the frequency ω_b. This means that a spectral hole with bottom at ω_b emerges in the inhomogeneous optical band of the guest molecules.

The lifetime of this spectral hole is determined by the lifetime of the photoproduct in its ground electronic state. The lifetime can be very long at low temperatures. The ground state of the photoproduct is separated by a high potential barrier from the ground state of the molecule. Heating of the sample or irradiation by white light facilitates the overcoming of this potential barrier in the ground electronic state. This leads to erasure of the hole. It is obvious that the hole-burning procedure can be repeated after erasure. The reversibility of hole burning is a valuable property in applications of this effect.

Let us find a formula for the deficit $\Delta n_0(\omega_b, \omega_0, t)$ in the ground state population. The kinetics of the population of molecular levels shown in Fig. 14.5 can be found from the following equations:

$$\begin{aligned}
\dot{n}_0 &= -k^{g} n_0 + (k^{e} + 1/T_1)\, n_1 + \gamma_{ST} n_2 \,, \\
\dot{n}_1 &= k^{g} n_0 - (k^{e} + \Gamma)\, n_1 \,, \\
\dot{n}_2 &= \Gamma_{TS} n_1 - (\gamma_{ST} + Q)\, n_2 \,.
\end{aligned} \tag{14.8}$$

They are similar to (8.15) except for two differences. Here we also take the PSB into account. Therefore the probability k^{g} of light absorption differs from the probability k^{e} of light-induced emission. These functions are described by (10.2). The second difference from (8.15) is to allow for the rate Q of photochemical conversion. In accordance with the model shown in Fig. 14.5, the photochemical transformation occurs in the triplet state. This assumption is rather plausible if we take into account the long life of the molecule in its triplet state. For example, this mechanism of phototransformation is found in the free-base porphyrin molecule in Fig. 14.4i. Since the quantum yield of the photochemical reaction is usually of order 10^{-3}–10^{-4}, the chain of inequalities for the rate constants can be rewritten

$$\Gamma \gg k \gg \gamma_{\mathrm{ST}} \gg Q \, , \tag{14.9}$$

i.e., the photochemical reaction is the slowest process. Only the backward processes, which are responsible for the erasure of holes, are slower than the photochemical reaction. They are not taken into account in (14.9). The temporal behavior of the populations is described by (14.8). The sum of the three equations yields

$$\dot{n}_0 + \dot{n}_1 + \dot{n}_2 = \dot{n} = -Q n_2 \, . \tag{14.10}$$

The decrease in the total number of molecules is determined by the rate constant Q. If $Q = 0$, this number is not changed. However, this does not mean that the population of each level is not changed if the sample is irradiated by laser light.

Let us analyze the population dynamics. We can use the results of Sect. 8.2, in which the temporal behavior of the full two-photon correlator was analyzed on the basis of (8.21). In accordance with Fig. 8.3 or Fig. 14.2 for the two-photon correlator, the excited singlet level is populated for a short time of order of T_1. Hence a short-lived transient hole is created. However, the hole is shallow because $k^{\mathrm{g}}/\Gamma \ll 1$. The fast growth of the photon correlator value relates to this process of the excited state population. The decrease in the two-photon correlator on the time scale of $1/k^{\mathrm{g}}$ is due to the transition of the molecule from the excited singlet level to the triplet level. Broadening of the transient hole happens at this stage. In addition, the correlator broadens at the same stage, as Fig. 8.5 shows, and the value of the correlator decreases to reach the small value $k^{\mathrm{g}}/\gamma_{\mathrm{ST}}$, as Fig. 14.2 shows. In the stationary regime, when $\mathrm{d}n_0/\mathrm{d}t = \mathrm{d}n_1/\mathrm{d}t = \mathrm{d}n_2/\mathrm{d}t = 0$, we find the following expressions for the populations:

$$n_1 = \frac{k^{\mathrm{g}}}{k^{\mathrm{e}} + \Gamma} n_0 \approx \frac{k^{\mathrm{g}}}{\Gamma} n_0 \, , \quad n_2 = \frac{\Gamma_{\mathrm{TS}}}{\gamma_{\mathrm{ST}}} \frac{k^{\mathrm{g}}}{k^{\mathrm{e}} + \Gamma} n_0 \approx \frac{\Gamma_{\mathrm{TS}}}{\Gamma} \frac{k^{\mathrm{g}}}{\gamma_{\mathrm{ST}}} n_0 \, . \tag{14.11}$$

Since $n_1 \ll n_0$ but $n_2 \gg n_0$, the majority of the molecules populate the triplet level. Therefore, a very broad hole will be burnt in the ground state population. The hole shape will repeat the homogeneous line shape only if the laser intensity is weak and it satisfies the inequality

$$\frac{k^{\mathrm{g}}}{\gamma_{\mathrm{ST}}} \ll 1 \, . \tag{14.12}$$

Hence, the chain of inequalities must be

$$\Gamma \gg \gamma_{\mathrm{ST}} \gg k^{\mathrm{g}} \gg Q \, . \tag{14.13}$$

The inequality described by (14.12) means that the majority of the molecules are in the ground electronic state, i.e., $n_0 \approx n$. Up to this time, the photochemistry channel has been switched off. Let us switch this channel on. Then taking into account the fact that

$$n_2 \approx \frac{\Gamma_{\mathrm{TS}}}{\Gamma} \frac{k^{\mathrm{g}}}{\gamma_{\mathrm{ST}}} n \, , \tag{14.14}$$

we can transform (14.10) to the form

$$\dot{n} = -\lambda \left(\omega_{\mathrm{b}} - \omega_0\right) n \, , \tag{14.15}$$

where the rate constant is given by

$$\lambda(\omega_{\mathrm{b}} - \omega_0) = \frac{\Gamma_{\mathrm{TS}}}{\Gamma} \frac{Q}{\gamma_{\mathrm{ST}}} k^{\mathrm{g}}(\omega_{\mathrm{b}} - \omega_0) \, . \tag{14.16}$$

Here the first ratio describes the quantum yield of intersystem crossing and the second ratio determines the quantum yield of photochemical conversion.

14.3 Kinetics of Persistent Hole Burning. Hole Shape in the Short Time Limit

Persistent holes are used to study the shape of the ZPL. The excited molecule can either be converted to another molecular form with photochemical rate Q or go into the ground electronic state in the basic molecular form. The probability of the second possibility is 10^4 times larger than the probability Q of the photochemical reaction. However, the molecules which left the basic molecular form for another molecular form will create a hole in the optical density. This hole can be read with the help of weak probing laser light. The hole shape and depth depend on the time t_{b} of the burning process. The solution

$$\Delta n(\omega_{\mathrm{b}}, \omega_0, t_{\mathrm{b}}) = \{1 - \exp\left[-\lambda \left(\omega_{\mathrm{b}} - \omega_0\right) t_{\mathrm{b}}\right]\} \, n(\omega_0) \tag{14.17}$$

to (14.15) determines the hole shape in the ground state population at t_{b}. By inserting this equation into (14.3), we arrive at the equation for the hole measured with the help of the probing laser:

$$h(\omega_{\mathrm{r}}, \omega_{\mathrm{b}}, t_{\mathrm{b}}) = \int\limits_0^\infty J^{\mathrm{g}}(\omega_{\mathrm{r}} - \omega_0) \{1 - \exp\left[-\lambda \left(\omega_{\mathrm{b}} - \omega_0\right) t_{\mathrm{b}}\right]\} \, n(\omega_0) \, \mathrm{d}\omega_0 \, , \tag{14.18}$$

where ω_r is the frequency of the reading laser. Persistent holes are used to study the shape of the ZPL. However, if we burn holes for a long time, we obtain spectral holes which are much broader than the ZPL. In the long time burning limit, we can neglect the exponential function in the integrand and then the hole shape is described by

$$h(\omega_r) = \int_0^\infty J^g(\omega_r - \omega_0)\, n(\omega_0)\, d\omega_0 . \tag{14.19}$$

This equation describes a very broad hole which does not reflect any features of the homogeneous band shape. Hence the burning time should be short enough to obtain a narrow hole. In the short burning time limit, we can expand the exponential function in the integrand as a power series in the burning time. In the first nonvanishing approximation, we arrive at the following expression:

$$h(\omega_r, \omega_b, t_b) = t_b \int_0^\infty J^g(\omega_r - \omega_0)\, \lambda(\omega_b - \omega_0)\, n(\omega_0)\, d\omega_0 . \tag{14.20}$$

This equation can be used to study the homogeneous ZPL only when the intensity of the burning laser is weak and satisfies the inequality (14.12). If this inequality is satisfied, we can write the relation $\lambda(\omega_b - \omega_0) \propto k^g(\omega_b - \omega_0)$ in accordance with (14.16). Therefore, for persistent hole shape in the short burning time limit, (14.20) takes the form

$$h(\omega_r, \omega_b, t_b) = t_b \frac{Q}{\gamma_{ST}} \frac{\Gamma_{TS}}{\Gamma_{TS} + 1/T_1} \int_0^\infty J^g(\omega_r - \omega_0) k^g(\omega_b - \omega_0) n(\omega_0)\, d\omega_0 . \tag{14.21}$$

Here the first ratio in front of the integral describes the quantum yield of photochemical conversion and the second ratio determines the quantum yield of intersystem crossing. Using (10.2) and (10.9), we find

$$k^g(\omega_b - \omega_0) = I \frac{8\pi^2 \omega_b}{\hbar c} J^g(\omega_b - \omega_0) , \tag{14.22}$$

where I is the laser light intensity, i.e., the number of photons per cm^2 per second. In the Condon approximation, we can take $J^g = d^2 S^g$. By inserting (14.22) into (14.21), we arrive at the following expression for the hole shape function:

$$h(\omega_r, \omega_b, t_b) \propto t_b \int_0^\infty J^g(\omega_r - \omega_0) J^g(\omega_b - \omega_0) n(\omega_0)\, d\omega_0 . \tag{14.23}$$

This function has been derived under the two assumptions

$$\frac{k^{\mathrm{g}}}{\gamma_{\mathrm{ST}}} \ll 1\,, \tag{14.24}$$

$$\frac{k^{\mathrm{g}}}{\gamma_{\mathrm{ST}}} Q t_{\mathrm{b}} \ll 1\,, \tag{14.25}$$

which are the defining conditions for low intensity and low dose of burning laser, respectively.

The hole shape function differs from the FLN function described by (13.7) in that the function J^{e} in the FLN function is replaced by the function J^{g} here. Therefore the analysis carried out in connection with the FLN spectrum can be repeated here.

The homogeneous band shape function consisting of ZPL and PSB is given by

$$J^{\mathrm{g}}(\omega_{\mathrm{b}} - \omega_0) = \alpha \frac{(\gamma/2\pi)}{(\omega_{\mathrm{b}} - \omega_0)^2 + (\gamma/2)^2} + (1-\alpha)\Phi(\omega_{\mathrm{b}} - \omega_0)\,. \tag{14.26}$$

Here we set $d^2 = 1$. After integrating both functions with respect to the frequency, we obtain unity. Inserting the last equation into (14.23), we derive the following expression for the hole shape function:

$$h(\omega_{\mathrm{r}}, \omega_{\mathrm{b}}) \propto t_{\mathrm{b}} \Bigg[\alpha^2 \frac{\gamma}{(\omega_{\mathrm{r}} - \omega_{\mathrm{b}})^2 + \gamma^2} n(\omega_{\mathrm{b}}) + \alpha(1-\alpha)\Phi(\omega_{\mathrm{b}} - \omega_{\mathrm{r}}) n(\omega_{\mathrm{b}})$$
$$+ \alpha(1-\alpha)\Phi(\omega_{\mathrm{r}} - \omega_{\mathrm{b}}) n(\omega_{\mathrm{r}}) + (1-\alpha)^2 \int_0^\infty \Phi(\omega_{\mathrm{b}} - \omega_0)\Phi(\omega_{\mathrm{r}} - \omega_0) n(\omega_0)\, \mathrm{d}\omega_0 \Bigg]\,. \tag{14.27}$$

This equation is similar to (13.10), derived for the FLN spectrum. The difference between the two equations results from the fact that the FLN spectrum is a convolution of the absorption and fluorescence homogeneous bands, whilst the persistent spectral hole function is a convolution of two absorption bands. The FLN function and persistent spectral hole function are shown in the second row of Fig. 14.6. The first row of the same figure shows the homogeneous absorption and fluorescence bands taken for calculation of the FLN and persistent spectral hole functions. The FLN spectrum resembles the homogeneous fluorescence band. However, the persistent spectral hole function has a PSB which extends in both directions from the ZPL. If the conditions of weak intensity and small dose given by (14.24) and (14.15) are not fulfilled, the spectral picture for the spectral hole function will differ from Fig. 14.6.

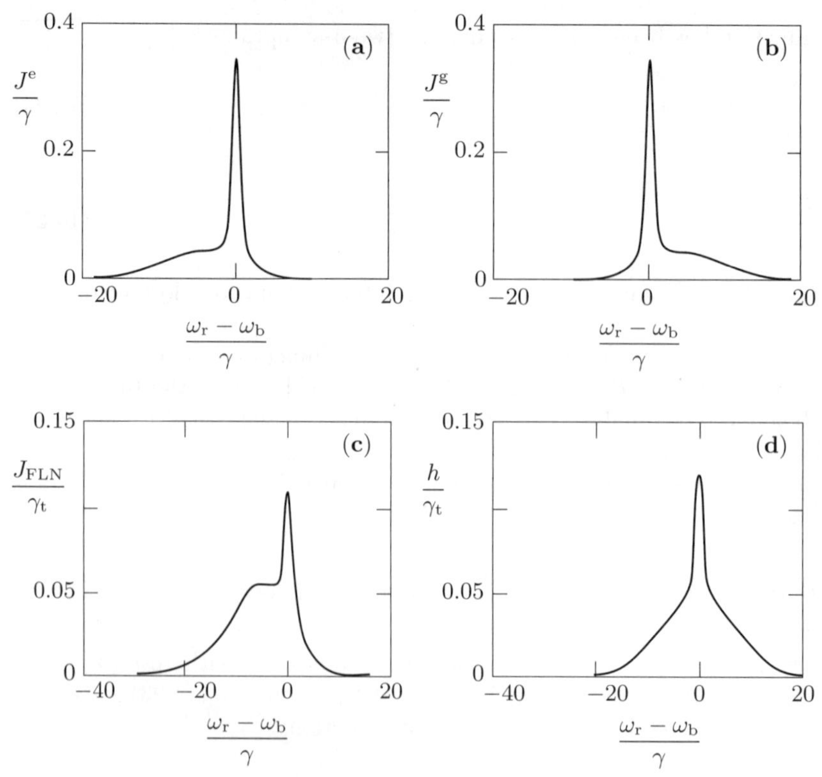

Fig. 14.6. **(a–d)** Fluorescence and absorption bands (*top*) and FLN and hole-burning spectra (*bottom*)

14.4 Hole Burning with Photoactive Photoproduct. Antiholes

Persistent spectral holes can be burnt if guest molecules can be transformed to another molecular form called a photoproduct. The theoretical considerations in the last section were carried out under the assumption that the photoproduct cannot absorb light. This assumption concerning the photoproduct is not mandatory. Persistent spectral holes will exist even when the photoproduct can absorb light. The only requirement for hole burning is that a shift in the optical resonant frequency should occur in the process of molecule–photoproduct conversion. It is known that the photoproduct can be photoactive and can absorb light in different spectral regions. However, the absorption band of the photoproduct may overlap that of the guest molecules and in this case the theory should be generalized.

When studying the kinetics of hole burning, we used a three-level energy diagram and the three equations of (14.8) for a guest molecule. The photoproduct is a slightly changed form of the guest molecule and therefore

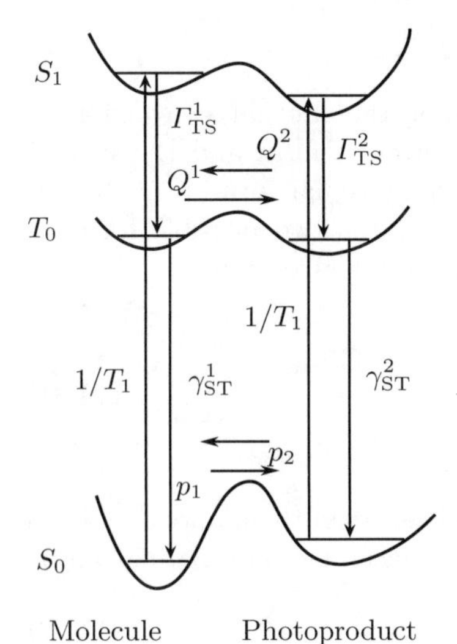

Fig. 14.7. Energy diagram of a molecule and a photoproduct

another three-level diagram can be ascribed to the photoproduct as shown in Fig. 14.7. Consequently, the guest molecule in the photoproduct form can be described by the other three rate equations. The rate constants of the molecule and of the photoproduct are marked by indices 1 and 2, respectively, as shown in Fig. 14.7. In all known cases the rates of hole-burning processes are smaller than the rates of relaxation from triplet levels. This means that the equilibrium inside the molecular and photoproduct forms is established much faster than the rates $Q^{1,2}$ and $p_{1,2}$ of a photochemical reaction. Therefore the forward and reverse photochemical reactions can be described by equations similar to (14.15), i.e., the number N_1 of molecules in the initial form and the number N_2 of molecules in a photoproduct form satisfy

$$\dot{N}_{1,2} = -P_{1,2}N_{1,2} \ , \tag{14.28}$$

where the effective rate constants are given by

$$P_{1,2} = \frac{\Gamma^{1,2}_{\rm TS}}{\Gamma^{1,2}}\frac{Q^{1,2}}{\gamma^{1,2}_{\rm ST}}k^{\rm g}_{1,2}\left(\omega_{\rm b} - \omega_{1,2}\right) + p_{1,2} \ . \tag{14.29}$$

The tunneling rates $p_{1,2}$ of the processes in the ground electronic state are included in the full effective photochemical rate constants, in contrast to (14.16). This means that both light-induced and spontaneous tunneling is taken into account.

However, (14.28) takes into account only forward processes for conversion of the molecular form. It is obvious that inverse processes for conversion of one molecular form to another must be taken into account in realistic systems. Therefore, the equations for the number N_1 of molecules in the initial form and the number N_2 of molecules in the photoproduct form will be

$$\dot{N}_1 = -P_1 N_1 + P_2 N_2 \,, \qquad \dot{N}_2 = P_1 N_1 - P_2 N_2 \,. \tag{14.30}$$

These rate equations describe the kinetics of the mutual conversion of molecules from the initial form to the photoproduct form and the inverse process after equilibrium is reached inside each molecular form. It is obvious that the number of molecules $N = N_1 + N_2$ in the two possible forms is constant because $\mathrm{d}N/\mathrm{d}t = 0$. A solution of (14.30) is given by

$$\begin{aligned} N_1(t) &= N \left[\frac{P_2}{P} + \left(\frac{p_2}{p} - \frac{P_2}{P} \right) \mathrm{e}^{-Pt} \right] , \\ N_2(t) &= N \left[\frac{P_1}{P} + \left(\frac{p_1}{p} - \frac{P_1}{P} \right) \mathrm{e}^{-Pt} \right] , \end{aligned} \tag{14.31}$$

where $P = P_1 + P_2$ and $p = p_1 + p_2$. The initial values of the numbers of both molecular forms in thermal equilibrium without laser excitation are given by

$$N_1(0) = N \frac{p_2}{p} \,, \qquad N_2(0) = N \frac{p_1}{p} \,. \tag{14.32}$$

These formulas can be derived from (14.30) for the stationary case when the laser light is switched off.

Later we shall examine the simplest case when only the resonant frequency is changed at the molecule–photoproduct conversion. Let us introduce a function $w(\omega_1, \omega_2)$ which determines the probability of finding a molecule with ZPL frequency ω_1 when the ZPL frequency of its photoproduct form is ω_2. We shall call a molecule which can be in an initial and a photoproduct form a conformant. Each conformant is characterized by the frequencies of two ZPLs. Hence $w(\omega_1, \omega_2)$ is the probability of finding a definite conformant in a solvent. Multiplying this probability by the total number N of guest molecules, one can find the total number $N(\omega_1, \omega_2) = w(\omega_1, \omega_2)N$ of definite conformants. The molecular and photoproduct forms of the conformant can be considered as the persistent and metastable forms of the conformant. It is obvious that $N(\omega_1, \omega_2) = N_1(\omega_1, \omega_2) + N_2(\omega_1, \omega_2)$, where the first and second terms on the right hand side describe the number of conformants in the molecular and photoproduct forms, respectively. The kinetics of $N_{1,2}(\omega_1, \omega_2)$ is determined by (14.31).

Allowing for the existence of the molecular and photoproduct forms in a solvent, (14.1) can be generalized to

$$\begin{aligned} J_{\chi=0}(\omega_\mathrm{r}) = \int_0^\infty \mathrm{d}\omega_1 \int_0^\infty \mathrm{d}\omega_2 \, [J_1^\mathrm{g}(\omega_\mathrm{r} - \omega_1) N_1(\omega_1, \omega_2) \\ + J_2^\mathrm{g}(\omega_\mathrm{r} - \omega_2) N_2(\omega_1, \omega_2)] \,. \end{aligned} \tag{14.33}$$

This function describes the broad band even in a case when the homogeneous optical bands of the molecule and photoproduct consist only of the ZPL. Indeed, inserting Lorentzians instead of the functions $J_{1,2}^{\mathrm{g,e}}$, we arrive at the formula

$$J_{\chi=0}(\omega_{\mathrm{r}}) = n_1(\omega_{\mathrm{r}}) + n_2(\omega_{\mathrm{r}}) \,, \tag{14.34}$$

where

$$n_1(\omega_1) = \int_0^\infty N_1(\omega_1, \omega_2)\,\mathrm{d}\omega_2 \,, \qquad n_2(\omega_2) = \int_0^\infty N_2(\omega_1, \omega_2)\,\mathrm{d}\omega_1 \tag{14.35}$$

determine the number of molecules in the persistent and metastable forms with certain frequencies of their ZPLs. These distribution functions are broad and their widths in polymers and glasses are roughly a few hundred cm^{-1}. We see that if we neglect the ground state depopulation produced by laser light, the inhomogeneous optical band of the solvent consisting of molecules and photoproduct is very broad, even when the homogeneous bands are ZPL.

Let us take into account the ground state depopulation produced by laser light. In this case the change in the functions $N_{1,2}(\omega_1, \omega_2)$ in accordance with (14.31) should be taken into account. The related changes in the inhomogeneous optical band are given by

$$\Delta J(\omega_{\mathrm{r}}, \omega_{\mathrm{b}}, t_{\mathrm{b}}) \propto \int_0^\infty \mathrm{d}\omega_1 \int_0^\infty \mathrm{d}\omega_2 \Big[J_1^{\mathrm{g}}(\omega_{\mathrm{r}} - \omega_1)\,[N_1(\omega_1, \omega_2, 0) - N_1(\omega_1, \omega_2, t_{\mathrm{b}})]$$
$$+ J_2^{\mathrm{g}}(\omega_{\mathrm{r}} - \omega_2)\,[N_2(\omega_1, \omega_2, 0) - N_2(\omega_1, \omega_2, t_{\mathrm{b}})] \Big] \,. \tag{14.36}$$

Using (14.31), we find the following differences:

$$N_1(0) - N_1(t) = \frac{P_1 p_2 - P_2 p_1}{P p} \left(1 - \mathrm{e}^{-Pt}\right) N = -\left[N_2(0) - N_2(t)\right] \,. \tag{14.37}$$

By inserting this formula into the integrand of (14.36) and assuming that the irradiation dose is small, we arrive at the following expression for the change in the optical density of the sample:

$$\Delta J \propto t_{\mathrm{b}} \left[h_+(\omega_{\mathrm{r}}, \omega_{\mathrm{b}}, t_{\mathrm{b}}) + h_-(\omega_{\mathrm{r}}, \omega_{\mathrm{b}}, t_{\mathrm{b}})\right] \,, \tag{14.38}$$

where

$$h_+(\omega_{\mathrm{r}}, \omega_{\mathrm{b}}) = \int_0^\infty \mathrm{d}\omega_1 \int_0^\infty \mathrm{d}\omega_2 \left[J_1^{\mathrm{g}} P_1 N_1(\omega_1, \omega_2, 0) + J_2^{\mathrm{g}} P_2 N_2(\omega_1, \omega_2, 0)\right] \,, \tag{14.39}$$

$$h_-(\omega_{\mathrm{r}}, \omega_{\mathrm{b}}) = -\int_0^\infty \mathrm{d}\omega_1 \int_0^\infty \mathrm{d}\omega_2 \left[J_2^{\mathrm{g}} P_1 N_1(\omega_1, \omega_2, 0) + J_1^{\mathrm{g}} P_2 N_2(\omega_1, \omega_2, 0)\right] \,. \tag{14.40}$$

The positive function h_+ describes the decrease in the absorption in the inhomogeneous optical band due to hole burning. The negative function h_- describes the increase in the absorption due to the photoactive photoproduct. Indeed, if the photoproduct does not absorb light, we have to set $J_2^{\mathrm{g}} = P_2 = 0$ and therefore $h_- = 0$. The function h_- determines the shape of the so-called antihole.

Let us turn back to the hole. The integrand in (14.39) is a sum of two terms. The first term depends on the indices of the molecule. The second term depends on the indices of the photoproduct. If the values of the probabilities p_1 and p_2 of spontaneous transitions between both forms are comparable, both persistent and metastable forms exist in a sample prior to the burning process. Both forms therefore contribute to the inhomogeneous optical band. A hole is thus burnt both in the persistent and in the metastable molecular form. In this case the spectral hole shape function can be derived in the same way as (14.27) and we arrive at the expression

$$\begin{aligned} h_+(\omega_{\mathrm{r}}, \omega_{\mathrm{b}}) \propto \sum_{j=1,2} \Bigg[& \alpha_j^2 \frac{\gamma_j}{(\omega_{\mathrm{r}} - \omega_{\mathrm{b}})^2} n_j(\omega_{\mathrm{b}}) \\ & + \alpha_j(1-\alpha_j)\Phi_j(\omega_{\mathrm{b}} - \omega_{\mathrm{r}}) n_j(\omega_{\mathrm{b}}) + \alpha_j(1-\alpha_j)\Phi_j(\omega_{\mathrm{r}} - \omega_{\mathrm{b}}) n_j(\omega_{\mathrm{r}}) \\ & + (1-\alpha_j)^2 \int_0^\infty \Phi_j(\omega_{\mathrm{b}} - \omega_j)\Phi(\omega_{\mathrm{r}} - \omega_j) n_j(\omega_j) \mathrm{d}\omega_j \Bigg] \end{aligned} \tag{14.41}$$

for the hole shape with photoactive photoproduct. The spectral hole is in fact a sum of two holes. One of them is burnt in the persistent molecular form of the conformant, the another is burnt in the unstable molecular form of the conformant, which was earlier called the photoproduct.

Let us now inspect the antiholes. The main peculiarities of antiholes can be examined for the simple case when optical bands of both forms of the conformant consist of the ZPL alone. In this case both the functions $J^{\mathrm{g,e}}$ and the rate constants

$$P_{1,2} \propto 2\chi^2 \frac{\Gamma_{\mathrm{TS}}^{1,2}}{\Gamma^{1,2}} \frac{Q^{1,2}}{\gamma_{\mathrm{ST}}^{1,2}} \frac{1/T_2\pi}{(\omega_{\mathrm{b}} - \omega_{1,2}) + 1/T_2^2} \tag{14.42}$$

are proportional to Lorentzians. Given the sharpness of the Lorentzians, we may carry out the integration with respect to frequencies in (14.40). The result is

$$h_-(\omega_{\mathrm{r}}, \omega_{\mathrm{b}}) \propto N_1(\omega_{\mathrm{b}}, \omega_{\mathrm{r}}) + N_2(\omega_{\mathrm{r}}, \omega_{\mathrm{b}}) \ . \tag{14.43}$$

One guest molecule converts to a photoproduct and changes the ZPL frequency ω_1 to another value $\omega_1 + \Delta$. Another molecule changes the ZPL frequency ω_2 to $\omega_2 + \Delta'$, and so on. If the distribution over frequency shifts Δ is considerable, then there is no correlation between resonant frequencies of the molecular and photoproduct forms. In this case the antihole will be broad, despite the fact that homogeneous bands consist of the ZPL alone. This shows (14.43). If a correlation between frequency shifts exists, the antihole will be described by a sharp function. For instance, if the degree of correlation is very high, i.e., all values Δ are the same, we find

$$N_1(\omega_1,\omega_2) \propto \delta\left(\omega_1 - \omega_2 - \Delta\right) . \tag{14.44}$$

By assuming the absence of the photoproduct before the burning process, i.e., assuming $N_2(0) = 0$, we find the following expression for the antihole:

$$h_-\left(\omega_{\mathrm{r}},\omega_{\mathrm{b}}\right) \propto -\frac{1/T_2\pi}{\left(\omega_{\mathrm{b}} - \omega_{\mathrm{r}} - \Delta\right)^2 + 1/T_2^2} , \tag{14.45}$$

i.e., an antihole of Lorentzian type will emerge after burning. The antihole is shifted by a value Δ from the hole. This example shows that antiholes do not supply us with information concerning the homogeneous optical band, as holes do. They supply us with information concerning a correlation between the frequencies of the persistent and metastable molecular forms. As a rule, antiholes are considerably broadened although narrow antiholes have been detected as well.

14.5 Polarization Aspects in Spectral Holes and Antiholes

In accordance with (10.2) and (10.3), the coefficients of absorption and light emission are proportional to an orientation factor

$$F = \frac{\left(\boldsymbol{d}\cdot\boldsymbol{E}\right)^2}{d^2E^2} = \left(\boldsymbol{n}\cdot\boldsymbol{u}\right)^2 = \cos^2\Theta , \tag{14.46}$$

which determines the orientation of the unit vector $\boldsymbol{n}$ of the dipolar moment with respect to the unit vector $\boldsymbol{u}$ of the light polarization. Therefore the use of polarized light in hole-burning spectroscopy can supply us with additional information concerning the orientation of the dipolar moments of a molecule and photoproduct.

The impurity centers in crystals are ordered in orientations. Therefore, the use of polarized light allows us to find the orientation of a guest molecule with respect to the crystallographic axes.

The guest molecules in polymers and glasses are disordered in their orientations. Therefore, the use of polarized light in the absorption experiments does not yield any advantage, because we measure the absorption coefficient averaged over various guest molecule orientations. It is impossible to gain advantage using polarized light in one-quantum spectroscopy of amorphous solids, such as absorption spectroscopy. Polarization methods can be successfully used only in two-quantum spectroscopy.

Spontaneous fluorescence is an example of the simplest two-quantum experiment. Indeed the absorbed quanta of light prepare an anisotropy in the isotropic molecular ensemble since the ensemble of excited molecules will contain an excess of molecules whose dipolar moment is parallel to the light polarization vector. Polarization of the emitted light will therefore depend on this anisotropy of the excited molecules.

Hole-burning spectroscopy is also a two-quantum spectroscopy, because one quantum of the burning light prepares an anisotropy in the solvent and the second quantum of the reading light reads this anisotropy.

Let us inspect the simplest case when the photoproduct is not optically active. The hole shape function is described by (14.23), which includes a convolution of two absorption coefficients. Each coefficient includes an orientation factor described by (14.46). If we neglect the HT interaction, all optical lines of the absorption band have the same orientation factor, i.e., this factor does not depend on the frequency and therefore the factor can be dropped from the integral. Hence, instead of (14.46) we arrive at the following orientation factors:

$$\langle h(\omega_\mathrm{r}, \omega_\mathrm{b}) \rangle = \langle F_\mathrm{r} F_\mathrm{b} \rangle h(\omega_\mathrm{r}, \omega_\mathrm{b}) \,, \tag{14.47}$$

where

$$F_\mathrm{r} = (\boldsymbol{u}_\mathrm{r} \cdot \boldsymbol{n})^2 \,, \qquad F_\mathrm{b} = (\boldsymbol{u}_\mathrm{b} \cdot \boldsymbol{n})^2 \tag{14.48}$$

are the orientation factors of the burning and reading processes. The polarizations of the burning and reading lasers differ. The guest molecules differ from one another not only by their resonant frequencies, but also by their orientations in space. The angular brackets describe an average over various guest molecule orientations.

The unit vector $\boldsymbol{n}$ of the dipolar moment of a guest molecule is connected with the molecular axes of symmetry. As usual its projections on the axes of the molecular (M) system of coordinates are known. The unit polarization vector $\boldsymbol{u}$ of the electric vector of the light is determined in the laboratory (L)

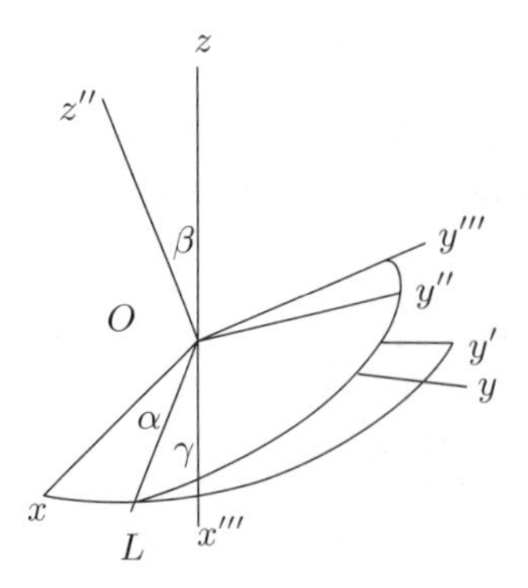

Fig. 14.8. Euler angles

system of coordinates. In order to calculate an orientation factor, we need to know the relation between the M and L systems of coordinates.

It is known from mechanics that the orientation in space of any 3D system can be described with the help of the three Euler angles. The coordinates x_M, y_M, z_M of the M system and x, y, z of the L system are shown in Fig. 14.8, together with the three Euler angles α, β, γ. The latter can be determined as follows. Let the axes of the M and L systems coincide, i.e., $x_M = x$, $y_M = y$, $z_M = z$. Let us rotate the M system through an angle α around the Oz axis. The Ox' axis lies along the OL axis. Then we rotate the M system through an angle β around the OL axis. Finally, we rotate the M system through an angle γ around the Oz'' axis. These three rotations are described by the matrices

$$S_\alpha = \begin{pmatrix} \cos\alpha & -\sin\alpha & 0 \\ \sin\alpha & \cos\alpha & 0 \\ 0 & 0 & 1 \end{pmatrix}, \qquad S_\beta = \begin{pmatrix} 1 & 0 & 0 \\ 0 & \cos\beta & -\sin\beta \\ 0 & \sin\beta & \cos\beta \end{pmatrix}, \tag{14.49}$$

$$S_\gamma = \begin{pmatrix} \cos\gamma & -\sin\gamma & 0 \\ \sin\gamma & \cos\gamma & 0 \\ 0 & 0 & 1 \end{pmatrix}.$$

If we carry out an arbitrary rotation through the three Euler angles, the components of the vector in the M system transform as follows:

$$\boldsymbol{n} = S\boldsymbol{n}', \tag{14.50}$$

where

$$S = S_\gamma S_\beta S_\alpha = \begin{pmatrix} S_{xx} & S_{xy} & S_{xz} \\ S_{yx} & S_{yy} & S_{yz} \\ S_{zx} & S_{zy} & S_{zz} \end{pmatrix} \tag{14.51}$$

$$= \begin{pmatrix} \cos\gamma\cos\alpha & -\cos\gamma\sin\alpha & \sin\gamma\sin\beta \\ -\sin\gamma\cos\beta\sin\alpha & -\sin\gamma\cos\beta\cos\alpha & \\ & & \\ \sin\gamma\cos\alpha & -\sin\gamma\sin\alpha & -\cos\gamma\sin\beta \\ +\cos\gamma\cos\beta\sin\alpha & +\cos\gamma\cos\beta\cos\alpha & \\ & & \\ \sin\beta\sin\alpha & \sin\beta\cos\alpha & \cos\beta \end{pmatrix}.$$

Using (14.50) we can express the orientation factor F in terms of the components of the vector $\boldsymbol{n}'$ in the M system and in terms of the components of the polarization vector $\boldsymbol{u}$ in the L system as

$$F = (\boldsymbol{u}S(\alpha, \beta, \gamma)\boldsymbol{n}')^2 . \tag{14.52}$$

The average over orientations is an integration over the Euler angles in accordance with the rule

$$\langle F \rangle = \frac{1}{8\pi^2} \int_0^{2\pi} \mathrm{d}\gamma \int_{-1}^{1} \mathrm{d}\cos\beta \int_0^{2\pi} \mathrm{d}\alpha F(\alpha, \beta, \gamma) . \tag{14.53}$$

In practice the average involves calculating the integrals of various powers of the sine and cosine functions.

Let us consider an average of the simplest orientation factor described by (14.52). The orientation of the M system with respect to the molecular axes can be arbitrary. It is convenient to choose the OZ' axis along the dipolar moment. Then we can write

$$S\boldsymbol{n}' = S \begin{pmatrix} 0 \\ 0 \\ 1 \end{pmatrix} = \begin{pmatrix} \sin\gamma\sin\beta \\ -\cos\gamma\sin\beta \\ \cos\beta \end{pmatrix} . \tag{14.54}$$

The third component of this vector is the simplest. Therefore the OZ axis of the L system should be chosen along the polarization vector $\boldsymbol{u}$, i.e., $\mathbf{u} = (0, 0, 1)$. Then we arrive at the following formula for the orientation factor:

$$(\boldsymbol{u}S\boldsymbol{n}') = (0\;0\;1) \begin{pmatrix} \sin\gamma\sin\beta \\ -\cos\gamma\sin\beta \\ \cos\beta \end{pmatrix} = \cos\beta . \tag{14.55}$$

Inserting $F = \cos^2\beta$ into (14.53), we find $\langle F \rangle = 1/3$.

Let us examine the orientation factor in (14.47) for the hole. The OY axis is chosen along the beams of the burning and reading light. The M and L systems are chosen as follows:

$$\boldsymbol{u}_{\mathrm{b}} = (0,\ 0,\ 1), \quad \boldsymbol{u}_{\mathrm{r}} = (\sin\Theta,\ 0,\ \cos\Theta), \quad \boldsymbol{n}' = \begin{pmatrix} 0 \\ 0 \\ 1 \end{pmatrix} . \tag{14.56}$$

Then we find the following expressions for the orientation factors:

$$F_{\mathrm{b}} = \cos^2\beta, \quad F_{\mathrm{r}} = (\sin\Theta\sin\gamma\sin\beta + \cos\Theta\cos\beta)^2 . \tag{14.57}$$

Replacing the orientation factors in (14.47) by these expressions, we find the following formula for the hole shape function:

$$\langle h\left(\omega_{\mathrm{r}}, \omega_{\mathrm{b}}\right) \rangle \propto \left(2\cos^2\Theta + 1\right) h(\omega_{\mathrm{r}}, \omega_{\mathrm{b}}) , \tag{14.58}$$

i.e., the depth of the hole depends on the angle Θ between the polarization vectors of the burning and probing light.

More interesting information can be obtained from polarization measurements in the optical band of a photoproduct if this band does not overlap with the molecular optical band. We can find an angle φ between dipolar moments of the molecular and photoproduct forms of the conformant. As shown above, if the photoproduct is able to absorb light, an antihole emerges when we burn a spectral hole in the molecular band. The antihole shape function is described by (14.40). The probabilities $P_{1,2} \propto J^{\mathrm{g}}_{1,2}$ and each of the antihole shape functions include an orientation factor (14.46). If the photoproduct was absent before burning, i.e., $N_2(0) = 0$, (14.40) for the antihole can be written

$$\langle h_-(\omega_{\mathrm{r}}, \omega_{\mathrm{b}})\rangle \propto \langle F_{2\mathrm{r}} F_{1\mathrm{b}}\rangle \, h_-(\omega_{\mathrm{r}}, \omega_{\mathrm{b}}) \;, \tag{14.59}$$

where

$$F_{2\mathrm{r}} = (\boldsymbol{u}_{\mathrm{r}} S \boldsymbol{n}'_2)^2 \;, \quad F_{1\mathrm{b}} = (\boldsymbol{u}_{\mathrm{b}} S \boldsymbol{n'}_1)^2 \;. \tag{14.60}$$

Let us chose the L and M systems such that the polarization vectors of the burning and reading lasers and the unit vector $\boldsymbol{n}'_1$ of the molecular dipolar moment are described by (14.56) and the unit vector $\boldsymbol{n}'_2$ of the photoproduct dipolar moment is given by

$$\boldsymbol{n}'_2 = \begin{pmatrix} \sin\varphi \\ 0 \\ \cos\varphi \end{pmatrix} \;. \tag{14.61}$$

Inserting (14.56) and (14.61) into the expressions for the orientation factors $F_{2\mathrm{r}}$ and $F_{1\mathrm{b}}$ and carrying out an average over the orientations, we arrive at the following equation for the antihole function:

$$\langle h_-(\omega_{\mathrm{r}}, \omega_{\mathrm{b}})\rangle \propto \left[\left(2\cos^2\Theta + 1\right)\cos^2\varphi + \left(\sin^2\Theta + 1\right)\sin^2\varphi\right] h_-(\omega_{\mathrm{r}}, \omega_{\mathrm{b}}) \;. \tag{14.62}$$

Here Θ is the angle between the polarization vectors of the burning and reading lasers and φ is an angle between dipolar moments of the molecular and photoproduct form of the conformant. Two equations can be derived for two values 0 and $\pi/2$ for the angle Θ between polarization vectors of the burning and reading lasers:

$$\begin{aligned} \langle h_-\rangle_{\uparrow\uparrow} &\propto \left(3\cos^2\varphi + \sin^2\varphi\right) h_- \;, \\ \langle h_-\rangle_{\perp} &\propto \left(\cos^2\varphi + 2\sin^2\varphi\right) h_- \;. \end{aligned} \tag{14.63}$$

By solving these equations, we find the following equation for the angle between the dipolar moments of the molecular and photoproduct forms of the conformant:

$$\tan^2\varphi = \frac{3\langle h_-\rangle_{\perp} - \langle h_-\rangle_{\uparrow\uparrow}}{2\langle h_-\rangle_{\uparrow\uparrow} - \langle h_-\rangle_{\perp}} \;. \tag{14.64}$$

The regime of low intensities and small doses of the burning laser light should be used in polarization experiments. If this regime is not fulfilled, the burning process does not prepare anisotropy in the molecular ensemble, because almost all molecules whose ZPLs coincide with the laser frequency are burnt independently of their orientations.

14.6 Photon-Gated Persistent Spectral Hole Burning

The ability of many guest–host systems to undergo persistent spectral hole burning can be used to store information. Indeed, each hole burnt in the inhomogeneous optical band can serve as a bit of information. Since a few thousand holes can be burnt in an inhomogeneous optical band, we can increase the optical storage memory thousands of times compared to non-optical devices. The gain in memory volume arises due to the existence of an additional spectral coordinate if hole burning is used. Spectral holes are of interest for a technical device if they can be burnt and erased many times. Every time we probe a hole, we irradiate the hole by weak light from the reading laser. As already mentioned, not only can persistent holes be burnt by light but they can also be erased by light. If the reading procedure is repeated many times, light from the reading laser may damage the hole and the hole may be partially erased. This is a serious shortcoming of spectral holes in their capacity as optical memory devices. It is possible to avoid this lack if photon-gated spectral hole burning is used.

In photon-gated hole burning, holes are burned after the absorption of two light quanta. Indeed, suppose that a highly excited electronic state, which is a doorway for a photochemical reaction, is inaccessible by absorption of one quantum of visible light. The doorway state can be excited either by two singlet–singlet transitions made consecutively in one guest molecule or by one singlet–singlet and one triplet–triplet transition in one molecule, as shown in Fig. 14.9. Two photons are absorbed consecutively in such an excitation process. These photons can be of red and green color when the absorber is an Sm^{2+} ion doping a BClF crystalline lattice [86]. The photon-gated hole-burning mechanism in this inorganic system is described by the energy diagram in Fig. 14.9a.

Another photon-gated mechanism has been found in the organic Shpolskii matrix doped with tetraphenylporphyrin molecules [87]. Light quanta of red and blue laser light excite the molecule in accordance with the scheme shown in Fig. 14.9b. In both cases, holes are burnt in the red part of the inhomogeneous optical band. If each of these systems is irradiated only by red light, there is no hole burning. Holes do not emerge in optical bands if the samples are irradiated only by green or by blue light. Holes are burnt only if light of two colors is used. Blue and green quanta help the system to reach doorway states. This hole burning was thus called photon-gated hole burning. Persistent holes burnt in the red optical band cannot be erased by quanta of red reading light because the energy of the red light quantum fails for the backward transition from the photoproduct.

Photochemical reactions in high electronic states are more efficient than those in low electronic states. The quantum yield of photochemical reactions in low electronic states is no more than 10^{-3}. This is two orders of magnitude less than the quantum yield of photochemical reactions in two-quantum

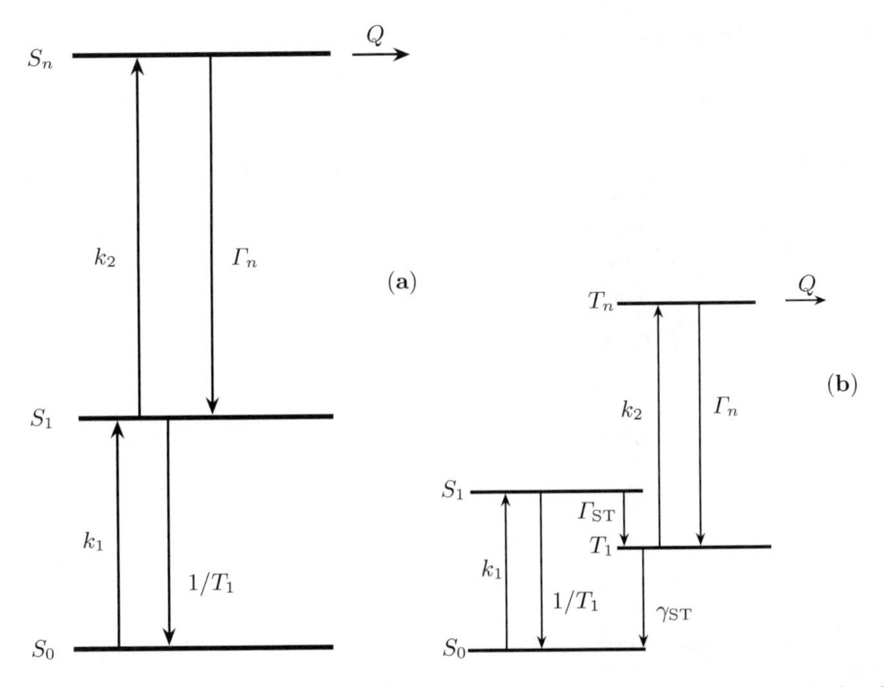

Fig. 14.9. Energy diagrams providing photon-gated hole burning via a singlet level (**a**) and a triplet level (**b**)

hole-burning experiments. Two-quantum hole burning can serve as a tool for studying physical characteristics of high electronic states.

For example, we consider the way the polarization measurements carried out in the first singlet optical band allow us to find an angle ψ between the dipolar moments $\boldsymbol{d}_1$ and $\boldsymbol{d}_2$ of two electronic transitions taking part in two-quantum hole burning, in accordance with the scheme shown in Fig. 14.9.

For concreteness, we consider the scheme shown in Fig. 14.9a. We shall call optical transitions from the ground and from the first excited singlet level the first and second transitions. The frequencies ω_1 and ω_2 relate to the ZPL of the first and second transitions. Let the function $N(\omega_1, \omega_2)$ determine the number of molecules with ZPL having frequencies ω_1 and ω_2. Then the functions $N_0(\omega_1, \omega_2)$ and $N_1(\omega_1, \omega_2)$ determine the number of these molecules in the ground and the first excited electronic states. It is obvious that $N(\omega_1, \omega_2) = N_0(\omega_1, \omega_2) + N_1(\omega_1, \omega_2)$. The values N_0 and N_1 depend on k_1, which is proportional to the intensity of the red laser. This laser burns very broad transient holes in the ground state population. This means that many of the molecules populate the first excited level and so we can neglect the orientation factor relating to this transition.

When we switch on the second (green) laser, molecular transitions from the first singlet level to the higher singlet levels begin. Therefore the function

$N_1(\omega_1, \omega_2)$ changes, and this change is given by

$$\Delta N(\omega_0, \omega_1, t_b) = \left[1 - e^{-\lambda_1(\omega_b - \omega_1)t_b}\right] N(\omega_0, \omega_1, 0) \ . \tag{14.65}$$

When both lasers are switched off, all molecules from the excited electronic states go the ground state. However, the molecules which have already been converted to photoproduct do not return to the ground state. Therefore a deficit of molecules in the ground state emerges. The deficit is given by

$$\Delta n(\omega_0, t_b) = \int_0^\infty d\omega_1 \Delta N(\omega_0, \omega_1, t_b) \ . \tag{14.66}$$

If the red reading laser is switched on, the persistent hole described by the following expression

$$\langle h(\omega_r, \omega_b)\rangle \propto \int_0^\infty \langle J_0^g(\omega_r - \omega_0)\, \Delta n(\omega_0, t_b)\rangle\, d\omega_0 \ . \tag{14.67}$$

can be measured. The polarization measurements in the hole burnt in the red band can be effective only if burning was carried out at weak intensity and small dose of the green laser, i.e., $\lambda_1 t_b \ll 1$ and $\lambda_1(\omega_b - \omega_1) \propto J_1^g(\omega_b - \omega_1)$. Taking (14.65) and (14.66) into account, we arrive at the following expression for the persistent hole burnt by red and green lasers:

$$\begin{aligned} \langle h(\omega_r, \omega_b)\rangle &\propto t_b \int_0^\infty \left\langle J_0^g(\omega_r - \omega_0) J_1^g(\omega_b - \omega_1 N(\omega_0, \omega_1, 0))\right\rangle d\omega_0 d\omega_1 \\ &\propto \langle F_0^g F_1^g \rangle\, h(\omega_r, \omega_b) \ , \end{aligned} \tag{14.68}$$

where the orientation factors are determined from

$$F_0^g = (\boldsymbol{u}_r S \boldsymbol{n}_1')^2 \ , \quad F_1^g = (\boldsymbol{u}_b S \boldsymbol{n}_2')^2 \ . \tag{14.69}$$

These equations are similar to (14.60), considered in connection with the determination of the angle φ between the dipolar moments of the molecule and photoproduct forms. However, the vectors $\boldsymbol{n}_1'$ and $\boldsymbol{n}_2'$ now determine the orientations of the dipolar moments of the first and second electronic transitions in the molecule. If the beams of the burning (green) and reading (red) laser are directed along the OY axis, and their polarization vectors are chosen in accordance with (14.56) and (14.61), where the angle φ is replaced by the angle ψ, we find the following expression after averaging over molecular orientations:

$$\langle h(\omega_r, \omega_b)\rangle \propto \left[\left(2\cos^2\Theta + 1\right)\cos^2\psi + \left(\sin^2\Theta + 1\right)\sin^2\psi\right] h(\omega_r, \omega_b) \ . \tag{14.70}$$

This formula for the persistent hole burnt in the red optical band is similar to (14.62), which relates to the antihole. Therefore, instead of (14.64), we arrive at the similar expression

$$\tan^2 \psi = \frac{3\,\langle h \rangle_{\perp} - \langle h \rangle_{\uparrow\uparrow}}{2\,\langle h \rangle_{\uparrow\uparrow} - \langle h \rangle_{\perp}} \,. \tag{14.71}$$

This equation enables one to determine the value of the angle ψ between two dipolar moments, using measurements at two polarizations of the red reading laser with respect to the polarization of the green burning laser.

Part VI

Transient Coherent Phenomena in Solids

15. Coherent Radiation of Molecular Ensembles

All equations for the absorption and fluorescence bands and spectral holes are based on the diagonal elements of the density matrix and their dependence on time and frequency. However, the density matrix of the electron–phonon–tunnelon system also has off-diagonal elements.

Off-diagonal elements of the density matrix are responsible for effects which feel a phase of the electronic excitation, i.e., for coherent optical effects. This was shown for the first time in spin echo experiments [88] and [89]. The intensity of the echo signal is determined by temporal decay of off-diagonal elements of the density matrix.

For a long time the study of coherent effects in the optical frequency domain was hampered because of the absence of sources of monochromatic and coherent light. Therefore, when lasers were developed, the photon echo [90], which is an analogue of spin echo, optical free induction decay [91] and other coherent phenomena in the optical domain [92–94] were soon observed. All these phenomena depend on phase dumping of off-diagonal elements of the density matrix. This will be shown later. All these coherent effects were named transient effects because they can be observed only for a short time after a short laser pulse is applied to a sample. Among transient effects, the photon echo is a topic of great interest. The photon echo has become a major source of information about phase and energy relaxation in solids and gases. Therefore, photon echo theory is discussed in detail in the next three chapters.

15.1 Dephasing and Energy Relaxation. Coherent Spontaneous Emission

An electronic wave function is characterized by a phase. Disappearance of phase correlation is called dephasing. Dephasing and energy relaxation are topics of major interest in the spectroscopy of non-equilibrated systems studied in this and the next few chapters. Two types of relaxation are taken into account in the optical Bloch equations (7.48), where there are two relaxation times T_2 and T_1. Their reciprocals determine the rate of relaxation of off-diagonal and diagonal elements of the density matrix. Since off-diagonal elements include the detuning Δ which determines the phase $t\Delta$ of the oscillations, these elements hold information concerning the phase of the electronic function. The constant $1/T_2$ determines the rate of exponential decay of off-diagonal elements. Therefore T_2 is called the optical dephasing time. The diagonal elements determine the probability of finding a molecule either in the ground or excited electronic states. Relaxation of the excited electronic state means relaxation of the electronic energy. Therefore T_1 is called the energy relaxation time. Dephasing and energy relaxation can be detected with the help of light emitted by a molecular ensemble.

Irradiation of a molecular ensemble is not described by a single exponential function because it has two stages: fast and slow. This fact does not result from the possible interaction between molecules of the ensemble, because it takes place even at extremely low guest molecule concentrations. Let us examine a single molecule irradiated by CW laser light which is switched off at time t_0. The energy of the system consisting of the molecule and electromagnetic field at time t_0 is given by

$$E_{\mathrm{mol+ph}}(\infty) = \mathrm{Tr}\{\hat{\rho}(\infty) H\} , \tag{15.1}$$

where the Hamiltonian of the system includes the molecule, n laser photons and the electron–photon interaction. Here $\rho(t_0 = \infty)$ is a stationary solution of the optical Bloch equations. The solution describes the initial state for the optical Bloch equation without external electromagnetic field. By carrying out the calculation of the trace in (15.1), we find

$$\begin{aligned} E_{\mathrm{mol+ph}}(\infty) = {} & (E_0 + n\hbar\omega_0)\,\rho_0(\infty) \\ & + [E_1 + (n-1)\hbar\omega_0]\,\rho_1(\infty) + \mathrm{i}\hbar\chi\,[\rho_{10}(\infty) - \rho_{01}(\infty)] , \end{aligned} \tag{15.2}$$

where E_0 and E_1 are energies of the molecule in the ground and excited electronic states. The stationary solution of the density matrix upon CW laser pumping can be found from (7.48), if all derivatives with respect to time are set to zero. This solution is given by

$$\begin{aligned} \rho_0(\infty) &= \frac{1 + kT_1}{1 + 2kT_1} , & \rho_{10}(\infty) &= \frac{\mathrm{i}\chi}{\Delta - \mathrm{i}/T_2}\frac{1}{1 + 2kT_1} , \\ \rho_1(\infty) &= \frac{kT_1}{1 + 2kT_1} , & \rho_{01}(\infty) &= \frac{-\mathrm{i}\chi}{\Delta + \mathrm{i}/T_2}\frac{1}{1 + 2kT_1} , \end{aligned} \tag{15.3}$$

where

$$k = 2\chi^2 \frac{1/T_2}{\Delta^2 + (1/T_2)^2} . \tag{15.4}$$

The first and the second terms in (15.2) describe an average molecular energy without polarization produced by the external electromagnetic field. The third term is proportional to the Rabi frequency χ. It defines an additional energy due to polarization of the molecule by the laser field. Multiplying both sides of (15.2) by the total number N of molecules in the sample we find the electronic energy of the ensemble. The third term defines the polarization energy induced by laser light in the sample. After the laser light has been switched off, the density matrix changes in accordance with the optical Bloch equations, where $\chi = \omega_0 = 0$, i.e.,

$$\begin{aligned} \rho_0(t) &= \rho_0(\infty) + \left(1 - e^{-t/T_1}\right)\rho_1(\infty) , \qquad \rho_{01}(t) = \rho_{10}^*(t) , \\ \rho_{10}(t) &= e^{-i(\Omega - t/T_2)t}\rho_{10}(\infty) , \qquad \rho_1(t) = e^{-t/T_1}\rho_1(\infty) . \end{aligned} \tag{15.5}$$

The dipolar moment of the molecule induced by laser light cannot disappear instantaneously. The dipolar moment of the molecule which has no dipolar moment in the stationary states, i.e., $d_{00} = d_{11} = 0$, is described by

$$d(t) = \mathrm{Tr}\left\{\hat{\rho}(t)\hat{d}\right\} = \rho_{10}(t)d_{01} + \rho_{01}(t)d_{10} = d\left[\rho_{10}(t) + \rho_{01}(t)\right] . \tag{15.6}$$

Substituting (15.5) for elements of the density matrix, we arrive at the following formula for the dipolar moment of the molecule:

$$d(t) = \frac{2d\chi}{\Delta^2 + (1/T_2)^2(1 + 4\chi^2 T_1 T_2)} e^{-t/T_2}\left[\Delta \sin\Omega t - \frac{1}{T_2}\cos\Omega t\right] . \tag{15.7}$$

Since $\Omega \approx 10^{15}\ \mathrm{s}^{-1}$ and $1/T_2 \leq 10^{12}\ \mathrm{s}^{-1}$, we have

$$\ddot{d}(t) \approx -\Omega^2 d(t) . \tag{15.8}$$

A system with oscillating dipolar moment radiates light at intensity given by

$$I_{\mathrm{mol}} = \frac{2\ddot{d}^2}{3c^3} = -\frac{dE_d}{dt} , \tag{15.9}$$

where E_d is the polarization energy. By inserting (15.7) into the last equation, we find that when the laser light is switched off the emission from the sample subsides in accordance with the function $\exp(-2t/T_2)$. This radiation differs from ordinary fluorescence, which subsides as $\exp(-t/T_1)$. Ordinary fluorescence subsides over a much longer time because $T_2 \ll 2T_1$. In addition, the fast part of the spontaneous radiation is of coherent type. In order to prove this, we consider a molecular ensemble consisting of N similar dipoles. The molecular dipolar moment equals $D = Nd$. Therefore the intensity of the ensemble light emission is N^2 times larger than the intensity of the molecular radiation:

$$I = N^2 I_{\mathrm{mol}} , \tag{15.10}$$

i.e., this spontaneous emission is of coherent type. The fluorescence intensity of the same ensemble is proportional to N.

Let us inspect the decay in polarization of the molecular ensemble whose molecules are distributed over resonant frequencies according to a function $N(\Omega - \Omega_0)$. The dipolar moment of the molecular ensemble can be described by

$$D(t) = \int_0^\infty d(t)\, N(\Omega - \Omega_0)\, \mathrm{d}\Omega \, . \tag{15.11}$$

Inserting (15.7) for the dipolar moment into the last equation and taking the distribution function in Gaussian form, we find the following expression for the dipolar moment:

$$D(t) = NdF(t) \cos \Omega_0 t \, , \tag{15.12}$$

where

$$F(t) \approx \mathrm{e}^{-t/T_2} \int_{-\infty}^{\infty} \mathrm{d}\Delta \frac{\mathrm{e}^{-(\Delta/B)^2}}{B\sqrt{\pi}} \frac{2\chi}{\Delta^2 + (1/T_2)^2 (1 + 4\chi^2 T_1 T_2)} \times \left[\Delta \sin \Delta t - \frac{1}{T_2} \cos \Delta t \right] . \tag{15.13}$$

The decay of the polarization calculated using (15.13) is shown in Fig. 15.1. The temporal behavior of the polarization depends strongly on the ratio of the half-width $\Delta_{1/2} = B\sqrt{\ln 2}$ of the distribution function and the homogeneous half-width $2/T_2$ of the ZPL. If this ratio is less than unity, the decay of the polarization is described by an exponential function with rate $(2/T_2)\sqrt{1 + 4\chi^2 T_1 T_2}$, which defines the half-width of the homogeneous optical line. In the opposite case when the inhomogeneous half-width $\Delta_{1/2} = B\sqrt{\ln 2}$ exceeds the homogeneous one, the decay becomes steeper. The greater the half-width of the distribution function, the steeper the decay.

Although both the inhomogeneous and the homogeneous half-width influence polarization decay, there is nevertheless a difference between the two

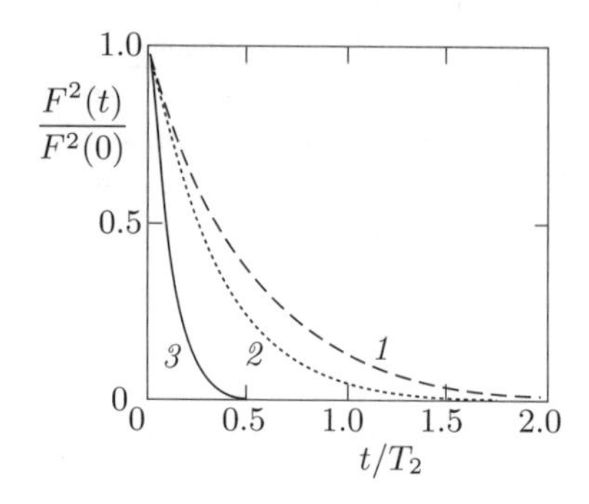

Fig. 15.1. Influence of inhomogeneous broadening on optical free induction decay (OFID) $B/T_2 = 0.3$ (curve 1), 1 (curve 2), 5 (curve 3) and $\chi T_2 = 10^{-2}$

dephasing mechanisms. For instance, the decay of the photon echo signal is determined solely by the mechanism yielding the homogeneous half-width $(2/T_2)\sqrt{1+4\chi^2 T_1 T_2}$. Therefore, dephasing resulting from this mechanism is called pure dephasing and the time T_2 is called the pure dephasing time. The distribution function does not influence echo decay.

15.2 Fast Optical Dephasing

In the last section we examined the case when pure dephasing is described by an exponential decay with rate $1/T_2$. However, nonexponential dephasing is often observed experimentally. This dephasing cannot be characterized by a single constant T_2 and therefore processes yielding nonexponential dephasing are not taken into account in the optical Bloch equations. Many attempts have been made to generalize the optical Bloch equation so that it could describe nonexponential dephasing. However, these attempts have been unsuccessful because such dephasing results from the electron–phonon interaction, which is responsible for the PSB and vibronic spectra. This interaction cannot be allowed for by the finite set of equations for the density matrix. However, the infinite set (7.35) for the density matrix includes the electron–phonon interaction yielding the PSB and vibronic spectra. Therefore (7.35) are able to describe fast nonexponential dephasing.

Fast nonexponential dephasing can be observed after an ultrashort laser pulse of duration Δt has passed through the sample. Then the first and second equations of (7.35) can be transformed to

$$\begin{aligned}\rho_{ba}(\Delta t) = \rho_{ba}(0) + \big[-\mathrm{i}\,(\omega_{ba} - \mathrm{i}/2T_1)\,\rho_{ba}(0) \\ -\chi\langle b|a\rangle\,[p_a\rho_0(0) - p_b\rho_1(0)]\,\big]\Delta t\,.\end{aligned} \tag{15.14}$$

Using the initial conditions $\rho_0(0) = 1$, $\rho_1(0) = \rho_{ab}(0) = \rho_{ba}(0) = 0$, we arrive at the following expressions for the off-diagonal elements of the density matrix:

$$\rho_{ba}(\Delta t) = \chi\Delta t\langle b|a\rangle p_a = \vartheta\langle b|a\rangle p_a = \rho_{ab}(\Delta t)\,. \tag{15.15}$$

When the laser pulse has passed, the off-diagonal elements evolve according to

$$\begin{aligned}\rho_{ba}(t) &= \mathrm{e}^{-\mathrm{i}(\Omega+\Omega_b-\Omega_a-\mathrm{i}/2T_1)t}\rho_{ba}(\Delta t)\,,\\ \rho_{ab}(t) &= \mathrm{e}^{\mathrm{i}(\Omega+\Omega_b-\Omega_a+\mathrm{i}/2T_1)t}\rho_{ab}(\Delta t)\,.\end{aligned} \tag{15.16}$$

The average value of the dipolar moment at time t is given by

$$d(t) = \mathrm{Tr}\left\{\hat{d}\,\hat{\rho}(t)\right\} = \sum_{a,b}\left[d_{ab}\rho_{ba}(t) + d_{ba}\rho_{ab}(t)\right], \tag{15.17}$$

where

$$d_{ab} = d\langle a|b\rangle = d_{ba} \; . \tag{15.18}$$

With the help of (15.15) and (15.16), we find

$$d\,(t) = 2d\vartheta\,\mathrm{Re}\left[\mathrm{e}^{-\mathrm{i}\Omega t} S^{\mathrm{g}}\,(t)\right] \; , \tag{15.19}$$

where the function $S^{\mathrm{g}}(t)$ is described by

$$S^{\mathrm{g}}(t) = \sum_{a,b} p_a \langle a|b\rangle \mathrm{e}^{-\mathrm{i}(\Omega_b - \Omega_a - \mathrm{i}/2T_1)t} \langle b|a\rangle, \tag{15.20}$$

which coincides with (10.21). The function $S^{\mathrm{g}}(t)$ has already been calculated. It was shown in Sect. 10.3 that, if we take into account only the linear FC interaction, this function is

$$S^{\mathrm{g}}(t) = \exp g(t) = \exp\left[\varphi(t,T) - \varphi(0,T)\right] \; , \tag{15.21}$$

where

$$\varphi\,(t,T) = \sum_{q=1}^{N} \frac{a_q^2}{2} \left[n_q \exp\left(\mathrm{i}\nu_q t\right) + \left(n_q + 1\right) \exp\left(-\mathrm{i}\nu_q t\right)\right] . \tag{15.22}$$

However, the quadratic FC interaction considered in Chap. 12 yields an additional term $-\mathrm{i}t\delta(T) - |t|/T_2(T)$ in the function $g(t)$, in accordance with (12.69). Therefore the function $g(t)$ is described by

$$g(t) = -\mathrm{i}t\delta(T) - \frac{|t|}{T_2} + \varphi\,(t,T) - \varphi\,(0,T) \; . \tag{15.23}$$

The function $\varphi(t)$ can be described by (15.22) if the quadratic FC interaction is smaller than the linear FC interaction. Inserting (15.23) into (15.19), we arrive at the following expression for the dipolar moment:

$$d\,(t) = 2d\vartheta \exp\left[-\varphi\,(0,T) + \mathrm{Re}\,\varphi\,(t,T) - \frac{|t|}{T_2}\right] \cos\left[\Omega t + \mathrm{Im}\,\varphi\,(t,0)\right] \; . \tag{15.24}$$

Here the shift $\delta(T)$ is included in the resonant frequency Ω.

If the linear FC interaction equals zero, the function $\varphi(t,T) = 0$ and we arrive at the formula which describes exponential decay of the polarization. If the linear FC interaction is not zero, an additional dephasing emerges because $\mathrm{Re}\varphi(t,T) \neq 0$. The additional dephasing is characterized by the function $\varphi(t,T)$, in contrast to the exponential dephasing, which is characterized by the constant T_2. The dephasing function can be transformed to

$$\mathrm{Re}\,\varphi - \mathrm{i}\,\mathrm{Im}\,\varphi = \int_0^\infty \varphi\,(\nu)\left[2n\,(\nu) + 1\right] \cos \nu t \,\mathrm{d}\nu - \mathrm{i} \int_0^\infty \varphi\,(\nu) \sin \nu t \,\mathrm{d}\nu \; , \tag{15.25}$$

where

$$\varphi(\nu) = \sum_q \frac{a_q^2}{2}\delta(\nu - \nu_q) \tag{15.26}$$

is the weighted density of phonon states. If electronic excitation interacts with acoustic phonons, this function can be taken in the following quasi-Debye form:

$$\varphi(\nu) = \varphi(0,0)9.857\frac{\nu^3}{\nu_D^{4.5}}\sqrt{\nu_D - \nu}\,, \qquad 0 \leqslant \nu \leqslant \nu_D\,. \tag{15.27}$$

The electron–phonon coupling is a linear function of the frequency. The integral of this function with respect to frequency equals the Huang–Rhys factor $\varphi(0,0)$. This factor determines the value of the electron–phonon coupling.

Figure 15.2 shows the temporal behavior of the dephasing calculated with the help of (15.25) and (15.27). At $t \approx 30/\nu_D$, the dephasing function approaches zero. The dephasing time T_2 is one or two orders of magnitude larger. Hence, nonexponential dephasing resulting from the linear FC interaction occurs 100 times faster than ordinary exponential dephasing described by the dephasing time T_2.

The expression for the dipolar moment described by (15.24) can be related to the optical band shape function $S^g(\Delta)$. Substituting $1/T_2$ for $1/2T_1$ in (10.61), we arrive at the following expression for the optical band shape function:

$$S^g(\omega_0 - \Omega) = \mathrm{Re}\int_0^\infty e^{-i(\omega_0-\Omega)t} S^g(t)\,dt \tag{15.28}$$

$$= \int_0^\infty \exp\left[-\varphi(0,T) + \mathrm{Re}\,\varphi(t,T) - \frac{|t|}{T_2}\right]\cos\left[(\omega_0 - \Omega)t + \mathrm{Im}\,\varphi(t,0)\right]dt\,.$$

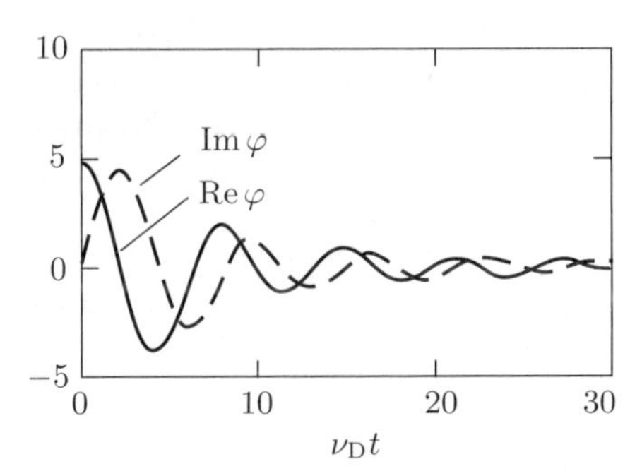

Fig. 15.2. Temporal dependence of dephasing function at $T = 0$

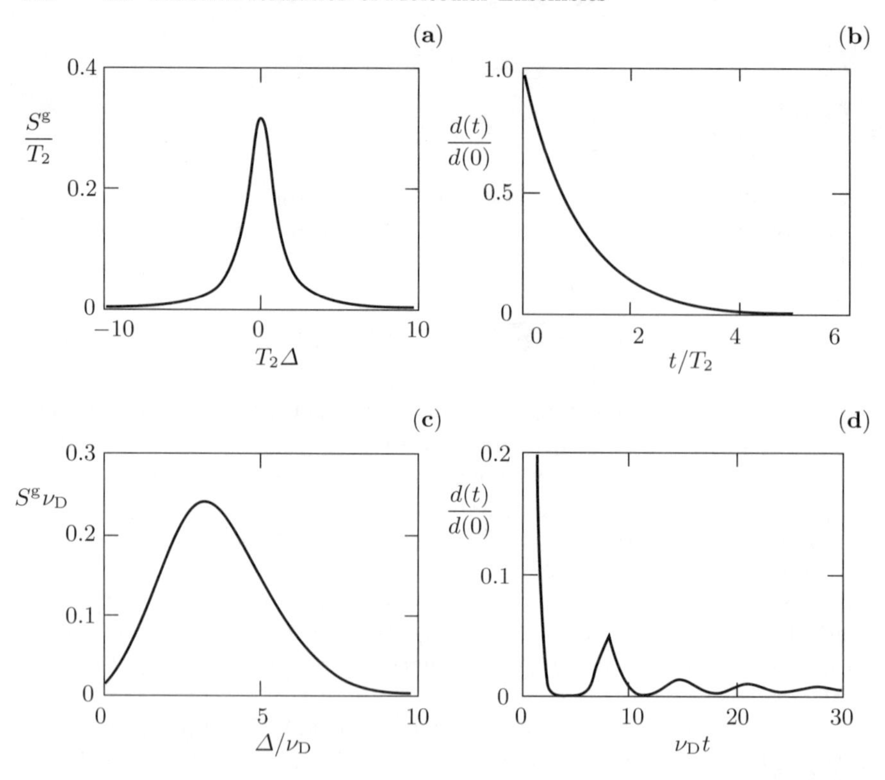

Fig. 15.3. (a–d) Optical bands consisting of ZPL (**a**) and phonon side band (**b**), and optical free induction decays relating to them

The integrand at $\omega_0 = 0$ is proportional to the dipolar moment induced by the laser light. Using (15.28) and (15.24), we can calculate the decay of the induced dipolar moment if the optical band shape function is known. Results of the calculations are shown in Fig. 15.3. In the case shown in Fig. 15.3a, the linear FC interaction equals zero and the optical band is therefore described by a ZPL of Lorentzian shape. The amplitude of the dipolar moment subsides exponentially on a time scale of T_2. The optical band and the decay of the induced dipolar moment for the case of a strong linear FC interaction is shown in Fig. 15.3b. The optical band is described by a PSB which is of Gaussian form since the ZPL is multiplied by the Debye–Waller factor $\exp[-\varphi(0,0)] = \exp(-5)$ and has small integrated intensity. In this case the decay of the induced polarization exhibits nonexponential character and occurs on a time scale of the reciprocal Debye frequency $1/\nu_D$, i.e., it is 100 times faster than the exponential decay in Fig. 15.3a. Once this fast decay is over, ordinary exponential relaxation is observed. However, the amplitude of the exponential emission is $\exp(-5)$ times weaker than the amplitude of fast emission.

16. Photon Echo

After a short laser pulse has passed through the sample, the induced polarization begins to decay. This decay takes a certain time. What happens if we irradiate the sample by another short laser pulse before the polarization decay is over? Besides an additional decay following the second laser pulse, a new coherent effect is observed. This is called photon echo. In this and the next chapter, we shall discuss the photon echo in detail.

Up to now we have assumed that the monochromatic laser field consists of photons. The quantum nature of the electromagnetic field is of great importance if we study spontaneous light emission. Such emission cannot be investigated within the framework of classical electromagnetism.

In photon echo experiments, short and ultrashort multiphoton laser pulses are used. The quantum nature of the electromagnetic field is no longer relevant if we consider photon echo. Moreover it is not convenient to use the quantum monochromatic field if we consider excitation by femtosecond laser pulses, because the Heisenberg uncertainty relation predicts that an ultrashort femtosecond pulse cannot be described by a monochromatic field. The electromagnetic field in the ultrashort laser pulse is a superposition of photons with various frequencies and there is no definite laser frequency. Therefore, femtosecond photon echo is generally considered using short time-dependent pulses of classical light.

16.1 Interaction of a Single Atom with a Classical Electromagnetic Field

The electromagnetic field in the interaction operator described by (1.40) is a quantum field. How is this interaction changed if we use a classical electromagnetic field? Let us write the interaction as

$$\hat{\Lambda} = -\frac{e}{\mu c}\hat{\boldsymbol{P}}\cdot\hat{\boldsymbol{A}} = -\frac{1}{c}\hat{\boldsymbol{j}}\cdot\hat{\boldsymbol{A}}\,, \tag{16.1}$$

where $\hat{\boldsymbol{j}}$ is the atomic current operator. The vector potential of the electromagnetic field is given by (1.33), which relates to the case of a standing wave. The standing wave is a sum of two waves running in opposite directions. Therefore, (1.33) for the vector potential can be written as

$$\hat{\boldsymbol{A}}\left(\boldsymbol{r},t\right)=\hat{\boldsymbol{A}}\left(\rightarrow\right)+\hat{\boldsymbol{A}}\left(\leftarrow\right)+\hat{\boldsymbol{A}}^{+}\left(\rightarrow\right)+\hat{\boldsymbol{A}}^{+}\left(\leftarrow\right)\,, \tag{16.2}$$

where

$$\hat{\boldsymbol{A}}(\rightarrow)=\sum_{\boldsymbol{k}}\boldsymbol{e}_{\boldsymbol{k}}\frac{c}{\omega_{\boldsymbol{k}}}\sqrt{\frac{\pi}{V}\hbar\omega_{\boldsymbol{k}}}\hat{a}_{\boldsymbol{k}}\mathrm{e}^{\mathrm{i}(\boldsymbol{k}\cdot\boldsymbol{r}-\omega_{k}t)} \tag{16.3}$$

describes a wave running from left to right. The wave $\boldsymbol{A}(\leftarrow)$ running in the opposite direction can be obtained by the substitution $\boldsymbol{k}\rightarrow-\boldsymbol{k}$. The first and second terms in (16.2) include the photon annihilation operators and the third and fourth terms include the photon creation operators. The current operator can be written as

$$\hat{\boldsymbol{j}}=\boldsymbol{j}_{10}B^{+}+\boldsymbol{j}_{01}B\,. \tag{16.4}$$

If (16.3) and (16.4) are inserted into (16.1) for the interaction operator it will include the following products of two operators: aB^{+}, $a^{+}B$, aB, $a^{+}B^{+}$. The first and second term conserve the total number of excitations in the molecule + field system. The third and forth terms do not conserve the total number of excitations. As mentioned earlier, the terms which do not conserve the total number of excitations can be omitted. Then for the wave running from left to right, we find the following interaction with a molecule:

$$\hat{\Lambda}=-\frac{1}{c}\left[\boldsymbol{j}_{10}\cdot\hat{\boldsymbol{A}}(\rightarrow)B^{+}+\boldsymbol{j}_{01}\cdot\hat{\boldsymbol{A}}^{+}(\rightarrow)B\right]\,. \tag{16.5}$$

The transition to a classical field is carried out by means of the following substitution: $a_{\boldsymbol{k}}\rightarrow\sqrt{n_{\boldsymbol{k}}}$ and $a_{\boldsymbol{k}}^{+}\rightarrow\sqrt{n_{\boldsymbol{k}}+1}\simeq\sqrt{n_{\boldsymbol{k}}}$. Taking into account the equation

$$\boldsymbol{j}_{10}=-\mathrm{i}\Omega\boldsymbol{d}\,,\qquad\boldsymbol{j}_{01}=\mathrm{i}\Omega\boldsymbol{d}\,,\qquad\boldsymbol{d}=\boldsymbol{d}_{10}=\boldsymbol{d}_{01}\,, \tag{16.6}$$

we find the following expression for the molecule–field interaction operator:

$$\hat{\Lambda}=\mathrm{i}\boldsymbol{d}\cdot\boldsymbol{E}^{*}\left(\boldsymbol{r},t\right)B-\mathrm{i}\boldsymbol{d}\cdot\boldsymbol{E}\left(\boldsymbol{r},t\right)B^{+}\,. \tag{16.7}$$

Here the classical electric field is given by

$$\boldsymbol{E}\left(\boldsymbol{r},t\right)=\sum_{\boldsymbol{k}}\boldsymbol{E}_{\boldsymbol{k}}\mathrm{e}^{\mathrm{i}(\boldsymbol{k}\cdot\boldsymbol{r}-\omega_{k}t)}\,. \tag{16.8}$$

The amplitude of such a classical wave is described by

$$\boldsymbol{E}_{\boldsymbol{k}}=\boldsymbol{e}_{\boldsymbol{k}}\sqrt{\frac{\pi}{V}\hbar\omega_{\boldsymbol{k}}n_{\boldsymbol{k}}}\,. \tag{16.9}$$

This amplitude is two times smaller than the amplitude of the standing wave used earlier.

The Hamiltonian of a two-level atom in the field with a classical wave is given by

$$H = H_{\mathrm{a}} + \Lambda(t) , \tag{16.10}$$

where the interaction operator is described by (16.7) with $\boldsymbol{r} = 0$. The eigenfunctions of the atom satisfy

$$\begin{aligned} &H_{\mathrm{a}} |0\rangle = 0 , \qquad B|0\rangle = B^{+}|1\rangle = 0 , \\ &H_{\mathrm{a}} |1\rangle = \hbar\Omega |1\rangle , \qquad B|1\rangle = |0\rangle , \qquad B^{+}|0\rangle = |1\rangle . \end{aligned} \tag{16.11}$$

The ground state energy is zero. If the laser field is absent, the temporal evolution of the atomic functions is given by

$$\mathrm{e}^{-\mathrm{i}H_{\mathrm{a}}t/\hbar}|0\rangle = |0\rangle , \qquad \mathrm{e}^{-\mathrm{i}H_{\mathrm{a}}t/\hbar}|1\rangle = \mathrm{e}^{-\mathrm{i}\Omega t}|1\rangle . \tag{16.12}$$

After the laser field is switched on, the Hamiltonian of the system is described by (16.10). The temporal evolution of an arbitrary wave function

$$|\psi\rangle = g_0(t)|0\rangle + g_1(t)\,\mathrm{e}^{-\mathrm{i}\Omega t}|1\rangle \tag{16.13}$$

of the system can therefore be determined from the Schrödinger equation using the Hamiltonian in (16.10). By substituting this function into the Schrödinger equation, we find the following equations for the unknown amplitudes:

$$\dot{g}_0 = \frac{\boldsymbol{d}\cdot\boldsymbol{E}^{*}(t)}{\hbar}\mathrm{e}^{-\mathrm{i}\Omega t}g_1 , \qquad \dot{g}_1 = -\frac{\boldsymbol{d}\cdot\boldsymbol{E}(t)}{\hbar}\mathrm{e}^{\mathrm{i}\Omega t}g_0 . \tag{16.14}$$

Let

$$\boldsymbol{d}\cdot\boldsymbol{E}(t) = \begin{cases} \hbar\chi\mathrm{e}^{-\mathrm{i}\omega_0 t} , & 0 < t < \Delta t , \\ 0 , & t > \Delta t , \end{cases} \tag{16.15}$$

where χ is the real value of the Rabi frequency and ω_0 is the frequency of the driving laser field. At the exact resonance, (16.14) take the form

$$\dot{g}_0 = \chi g_1 , \qquad \dot{g}_1 = -\chi g_0 . \tag{16.16}$$

The solution is given by the functions

$$\begin{aligned} g_0(t) &= g_0(0)\cos\vartheta + g_1(0)\sin\vartheta , \\ g_1(t) &= g_0(0)(-\sin\vartheta) + g_1(0)\cos\vartheta , \end{aligned} \tag{16.17}$$

where $\vartheta = \chi t$.

Equations (16.12), (16.13) and (16.17) define the temporal evolution of the atom without the laser field and with the driving laser field. Any relaxation of the electronic excitation is not allowed for.

16.2 Molecule Interacting with a Classical Electromagnetic Field, Phonons, and Tunnelons

The discussion will be based on the results of the last section. We have already seen that dephasing processes can be taken into consideration only if we use the density matrix approach. Indeed, all the more so if we use a classical electromagnetic field.

Let us derive equations for the density matrix of a two-level molecule interacting with a classical electromagnetic field. The elements of the density matrix emerge when we write an expression for the dipolar moment using the function described by (16.13):

$$\langle\Psi(t)|\hat{d}|\Psi(t)\rangle = d\left(g_1 g_0^* \mathrm{e}^{-\mathrm{i}\Omega t} + g_0 g_1^* \mathrm{e}^{\mathrm{i}\Omega t}\right) = d\left(G_1 G_0^* + G_0 G_1^*\right) , \quad (16.18)$$

where

$$G_0 = g_0 , \qquad G_1 = g_1 \mathrm{e}^{-\mathrm{i}\Omega t} . \quad (16.19)$$

Using

$$\dot{g}_1 \mathrm{e}^{-\mathrm{i}\Omega t} = \dot{G}_1 + \mathrm{i}\Omega G_1 , \quad (16.20)$$

the following equations for the amplitudes G_0 and G_1 can be derived from (16.14):

$$\dot{G}_0 = \chi^*\left(\boldsymbol{r}, t\right) G_1 , \qquad \dot{G}_1 = -\mathrm{i}\Omega G_1 - \chi\left(\boldsymbol{r}, t\right) G_0 , \quad (16.21)$$

where

$$\chi\left(\boldsymbol{r}, t\right) = \frac{\boldsymbol{d} \cdot \boldsymbol{E}\left(\boldsymbol{r}, t\right)}{\hbar} , \quad (16.22)$$

and the classical electromagnetic field is given by (16.8). Introducing elements of the density matrix as follows

$$\rho_{10} = G_1 G_0^* , \qquad \rho_{01} = \rho_{10}^* , \qquad \rho_0 = \left|G_0\right|^2 , \qquad \rho_1 = \left|G_1\right|^2 , \quad (16.23)$$

and using (16.21), we find the set of equations

$$\begin{aligned} \dot{\rho}_{10} &= -\mathrm{i}\Omega\rho_{10} + \chi\left(\rho_1 - \rho_0\right) , \\ \dot{\rho}_{01} &= \mathrm{i}\Omega\rho_{01} + \chi^*\left(\rho_1 - \rho_0\right) , \\ \dot{\rho}_1 &= -\chi^*\rho_{10} - \chi\rho_{01} , \\ \dot{\rho}_0 &= \chi^*\rho_{10} + \chi\rho_{01} . \end{aligned} \quad (16.24)$$

for the density matrix of the molecule. This set of equations does not take into account either spontaneous light emission or the electron–phonon interaction.

Let us compare this set of equations with (3.12), which was derived with the quantum electromagnetic field. The set (3.12) has three differences. Firstly, instead of the electronic frequency Ω, it includes a detuning $\Delta = \Omega - \omega_0$. Secondly, instead of the complex function $\chi(\boldsymbol{r}, t)$, (3.12) include the real Rabi frequency χ, and thirdly, there are additional terms in (3.12) which include the time T_1. These terms result from spontaneous light emission. Therefore, if we add such terms to (16.24), it will mean that we have taken spontaneous light emission into account. After this addition, we arrive at the following set of equations:

$$
\begin{aligned}
\dot{\rho}_{10} &= -\mathrm{i}\left(\Omega - \frac{\mathrm{i}}{2T_1}\right)\rho_{10} + \chi\left(\rho_1 - \rho_0\right) , \\
\dot{\rho}_{01} &= \mathrm{i}\left(\Omega + \frac{\mathrm{i}}{2T_1}\right)\rho_{01} + \chi^*\left(\rho_1 - \rho_0\right) , \\
\dot{\rho}_1 &= -\chi^*\rho_{10} - \chi\rho_{01} - \frac{\rho_1}{T_1} , \\
\dot{\rho}_0 &= \chi^*\rho_{10} + \chi\rho_{01} + \frac{\rho_1}{T_1} ,
\end{aligned}
\tag{16.25}
$$

which describes the two-level molecule in the classical laser driving field.

It has already been shown that it is impossible to include the true electron–phonon interaction within the framework of four equations for the electronic density matrix. The electron–phonon interaction can be taken into account by replacing the four equations of (3.12) by the infinite set of equations described by (7.29). In the course of the derivation of (7.29) from (3.12), we did not use the quantum nature of the electromagnetic field. Therefore in the case of the classical electromagnetic field, we arrive at the following infinite set of equations:

$$
\begin{aligned}
\dot{\rho}_{ba} &= -\mathrm{i}\left(\Omega_{ba} - \frac{\mathrm{i}}{2T_1}\right)\rho_{ba} + \sum_{b'}\rho_{bb'}\chi_{b'a} - \sum_{a'}\chi_{ba'}\rho_{a'a} , \\
\dot{\rho}_{ab} &= -\mathrm{i}\left(\Omega_{ab} - \frac{\mathrm{i}}{2T_1}\right)\rho_{ab} + \sum_{b'}\chi^*_{ab'}\rho_{b'b} - \sum_{a'}\rho_{aa'}\chi^*_{a'b} , \\
\dot{\rho}_{bb'} &= -\sum_{a}\left(\rho_{ba}\chi^*_{ab'} + \chi_{ba}\rho_{ab'}\right) - \mathrm{i}\left(\Omega_{bb'} - \frac{\mathrm{i}}{T_1}\right)\rho_{bb'} , \\
\dot{\rho}_{aa'} &= \sum_{b}\left(\chi^*_{ab}\rho_{ba'} + \rho_{ab}\chi_{ba'}\right) + \frac{1}{T_1}\sum_{bb'}\langle a|b\rangle\rho_{bb'}\langle b'|a'\rangle - \mathrm{i}\Omega_{aa'}\rho_{aa'} ,
\end{aligned}
\tag{16.26}
$$

where

$$
\begin{aligned}
&\Omega_{ba} = \Omega + \Omega_b - \Omega_a , \quad \Omega_{bb'} = \Omega_b - \Omega_{b'} , \quad \Omega_{aa'} = \Omega_a - \Omega_{a'} , \\
&\chi_{ab} = \frac{1}{\hbar}\langle a|\,\boldsymbol{d}\,|b\rangle \cdot \boldsymbol{E}\left(\boldsymbol{r}, t\right) = \chi_{ba} .
\end{aligned}
\tag{16.27}
$$

Here Ω_a and Ω_b are the frequencies of the phonon–tunnelon system in the ground and excited electronic states. The frequencies of the laser modes are included in the functions χ and χ^* via the classical electric field $\boldsymbol{E}(\boldsymbol{r},t)$. Therefore (16.26) can be used to examine cases with ultrashort laser excitation when the spectral width of the pulse may exceed the width of the homogeneous optical band of a guest molecule. We face just such a situation when a sample is irradiated by femtosecond laser pulses.

Although (16.26) is more complicated than the optical Bloch equations, this infinite set of equations can be used in practice. For instance, (16.26) will be used to calculate the signal of a three-pulse photon echo in Sect. 17.5. However, in some cases the simplified version of (16.26) can be used. This version appears if we neglect off-diagonal elements $\rho_{aa'}$ and $\rho_{bb'}$ in (16.26). This is an approximation described by (7.30). Therefore, we can introduce the probabilities

$$p_a = \frac{\rho_{aa}}{\rho_0}\,, \qquad p_b = \frac{\rho_{bb}}{\rho_1} \tag{16.28}$$

of finding the related tunnelon–phonon state. They have already been used when deriving the optical Bloch equations in Sect. 7.3. In this approximation, we can introduce the probabilities

$$\rho_0 = \sum_a \rho_{aa}\,, \qquad \rho_1 = \sum_b \rho_{bb} \tag{16.29}$$

of finding the system in the ground and excited electronic states, respectively. Using (7.30), we can transform (16.26) to the simpler infinite set of equations:

$$\begin{aligned}
\dot{\rho}_{ba} &= -\mathrm{i}\left(\Omega_{ba} - \frac{\mathrm{i}}{2T_1}\right)\rho_{ba} + \chi_{ba}\left(p_b\rho_1 - p_a\rho_0\right)\,, \\
\dot{\rho}_{ab} &= -\mathrm{i}\left(\Omega_{ab} - \frac{\mathrm{i}}{2T_1}\right)\rho_{ba} + \chi^*_{ab}\left(p_b\rho_1 - p_a\rho_0\right)\,, \\
\dot{\rho}_1 &= -\sum_{a,b}\left(\chi^*_{ab}\rho_{ba} + \chi_{ba}\rho_{ab}\right) - \frac{\rho_1}{T_1}\,, \\
\dot{\rho}_0 &= \sum_{a,b}\left(\chi^*_{ab}\rho_{ba} + \chi_{ba}\rho_{ab}\right) + \frac{\rho_1}{T_1}\,.
\end{aligned} \tag{16.30}$$

This simplified set of equations can be used to calculate the two- and three-pulse photon echos to the first nonvanishing order in the interaction with light.

16.3 Simplest Theory of the Two-Pulse Photon Echo

The photon echo pulse emerges after two laser pulses pass through the sample, as shown in Fig. 16.1. As a rule, the photon echo is studied using the density

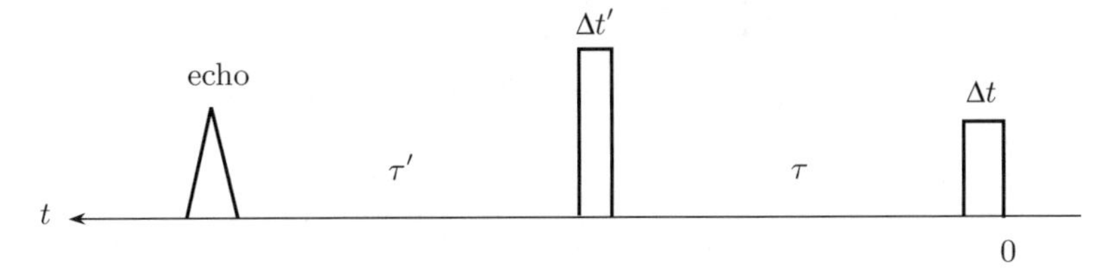

Fig. 16.1. Two laser pulses resulting in photon echo

matrix. However, it can also be understood on the basis of wave functions. Let a two-level atom be in the ground electronic state, so that it can be described by the wave function

$$|\Psi(0)\rangle = |0\rangle \ . \tag{16.31}$$

If this atom is irradiated by a laser pulse during the time interval Δt, the atom goes into a new quantum state which is a superposition of the ground and excited electronic states:

$$|\Psi(\Delta t)\rangle = \cos\vartheta\,|0\rangle - \sin\vartheta\,|1\rangle \ , \tag{16.32}$$

where $\vartheta = \chi\Delta t$. When the laser is switched off, the atomic wave function evolves according to (16.12). It therefore takes the form

$$|\Psi(\tau+\Delta t)\rangle = \cos\vartheta\,|0\rangle - \sin\vartheta\,|1\rangle\,\mathrm{e}^{-\mathrm{i}\Omega\tau} \ . \tag{16.33}$$

After the second laser pulse has passed through the sample, the atomic wave function takes the form

$$\begin{aligned}|\Psi(\Delta t'+\tau+\Delta t)\rangle &= \cos\vartheta\big(\cos\vartheta'\,|0\rangle - \sin\vartheta'\,|1\rangle\big)\\ &\qquad - \sin\vartheta\big(\sin\vartheta'\,|0\rangle + \cos\vartheta'\,|1\rangle\big)\mathrm{e}^{-\mathrm{i}\Omega\tau}\\ &= A\,|0\rangle + B\,|1\rangle \ ,\end{aligned} \tag{16.34}$$

where

$$\begin{aligned}A &= \cos\vartheta\cos\vartheta' - \sin\vartheta\sin\vartheta'\mathrm{e}^{-\mathrm{i}\Omega\tau},\\ B &= -(\cos\vartheta\sin\vartheta' + \sin\vartheta\cos\vartheta'\mathrm{e}^{-\mathrm{i}\Omega\tau}) \ .\end{aligned} \tag{16.35}$$

After the laser is switched off again, the atomic wave function evolves according to (16.12). It is therefore given by the expression

$$|\Psi(t)\rangle = |\Psi(\tau'+\Delta t'+\tau+\Delta t)\rangle = A\,|0\rangle + B\,|1\rangle\,\mathrm{e}^{-\mathrm{i}\Omega\tau'} \ . \tag{16.36}$$

The dipolar moment after two laser pulses have passed thorough the sample is described by

$$
\begin{aligned}
d(t) = \langle \Psi(t) | d | \Psi(t) \rangle &= d\Big(A^* B \mathrm{e}^{-\mathrm{i}\Omega\tau'} + \mathrm{c.c.} \Big) \\
&= d\Big[-\cos 2\vartheta \sin 2\vartheta' \cos \Omega\tau' - \cos^2\vartheta' \sin 2\vartheta \cos \Omega\,(\tau+\tau') \\
&\qquad + \sin^2\vartheta' \sin 2\vartheta \cos \Omega(\tau-\tau') \Big] .
\end{aligned}
\tag{16.37}
$$

In order to clarify the physical meaning of each term, we consider an atomic ensemble whose atoms have various resonant frequencies. The function $N(\Omega)$ describes a distribution in the resonant frequencies. The dipolar moment of this atomic ensemble is given by

$$
\begin{aligned}
D(t) &= \int_0^\infty N\,(\Omega)\, d(t) \mathrm{d}\Omega \\
&= d\Big[-\cos 2\vartheta \sin 2\vartheta' N\,(\tau') - \cos^2\vartheta' \sin 2\vartheta N\,(\tau+\tau') \\
&\qquad + \sin^2\vartheta' \sin 2\vartheta N(\tau-\tau') \Big] ,
\end{aligned}
\tag{16.38}
$$

where

$$
N(t) = \int_0^\infty N(\Omega) \cos \Omega t \,\mathrm{d}\Omega . \tag{16.39}
$$

In accordance with the general property of the Fourier transformation, the broader the function $N(\Omega)$ in the frequency scale, the narrower the function $N(t)$ in the time scale. If the width of the function $N(\Omega)$ is about 100 cm^{-1}, the width of the function $N(t)$ in the time scale should be of order 10^{-13} s. Hence, the right hand side of (16.38) is a sum of three short pulses. The first term describes optical free induction decay induced by the second laser pulse. The second term describes optical free induction decay induced by the first laser pulse. Finally, the third term describes a pulse of coherent light which emerges at $\tau' = \tau$. This pulse emerges at time $t = \tau' + \Delta t' + \tau + \Delta t' \simeq 2\tau$. This is the photon echo, illustrated in Fig. 16.1.

The photon echo is a consequence of the nonlinear dependence of the induced polarization on the intensity of laser light. In order to prove this statement, we examine the case of weak laser light when $\vartheta \ll 1$ and $\vartheta' \ll 1$. Then (16.38) takes the form

$$
D(t) = 2d \left[-\vartheta' N\,(\tau') - \vartheta N\,(\tau+\tau') + \vartheta'^2 \vartheta N(\tau-\tau') \right] . \tag{16.40}
$$

If the polarization is a linear function of the electromagnetic field, the third term describing the photon echo disappears. The photon echo is a third order effect in the atom–field interaction. It is obvious that the photon echo exists even in the case of a single atom. However, the effect is more pronounced in the atomic ensemble: the echo peak becomes narrower. If $\vartheta = \pi/4$ and $\vartheta' = \pi/2$, the echo amplitude reaches a maximum. However, the photon echo can also be observed at small values of these angles.

16.4 Bloch Vector and Its Temporal Evolution

According to (16.40), the echo signal can emerge at an arbitrary delay between the two laser pulses. This means that phase memory stored by a sample does not decay. This is a consequence of the fact that our model does not include dephasing processes, which exist in any realistic system. As we have already seen in Part III, dephasing processes can be taken into account within the scope of the density matrix approach to a molecule interacting with an electromagnetic field. Using (16.30) for the density matrix with a classic electromagnetic field, we can derive the optical Bloch equations in the same way as for the density matrix with a quantum electromagnetic field in Sect. 7.3:

$$\begin{aligned}
\dot{\rho}_{10} &= -\mathrm{i}\left(\Omega - \frac{\mathrm{i}}{T_2}\right)\rho_{10} - \chi(\boldsymbol{r},t)(\rho_0 - \rho_1) \ , \\
\dot{\rho}_{01} &= \mathrm{i}\left(\Omega + \frac{\mathrm{i}}{T_2}\right)\rho_{01} - \chi^*(\boldsymbol{r},t)(\rho_0 - \rho_1) \ , \\
\dot{\rho}_1 &= -\chi^*(\boldsymbol{r},t)\rho_{10} - \chi(\boldsymbol{r},t)\rho_{01} - \frac{\rho_1}{T_1} \ , \\
\dot{\rho}_0 &= \chi^*(\boldsymbol{r},t)\rho_{10} + \chi(\boldsymbol{r},t)\rho_{01} + \frac{\rho_1}{T_1} \ ,
\end{aligned} \tag{16.41}$$

where $\chi(\boldsymbol{r},t)$ is defined by (16.22). This set of equations can be written in a concise vector form:

$$\dot{\boldsymbol{\rho}} = \left[\hat{E} + \hat{\Lambda}(t)\right]\boldsymbol{\rho} \ , \tag{16.42}$$

where the vector and the matrices are given by

$$\boldsymbol{\rho} = \begin{pmatrix} \rho_{10} \\ \rho_{01} \\ \rho_1 \\ \rho_0 \end{pmatrix} , \quad \hat{E} = \begin{pmatrix} \varepsilon & 0 & 0 & 0 \\ 0 & \varepsilon^* & 0 & 0 \\ 0 & 0 & -1/T_1 & 0 \\ 0 & 0 & 1/T_1 & 0 \end{pmatrix} , \quad \hat{\Lambda}(t) = \begin{pmatrix} 0 & 0 & \chi & -\chi \\ 0 & 0 & \chi^* & -\chi^* \\ -\chi^* & -\chi & 0 & 0 \\ \chi^* & \chi & 0 & 0 \end{pmatrix} . \tag{16.43}$$

Here $\varepsilon = -\mathrm{i}(\Omega - \mathrm{i}/T_2)$. The four-component vector $\boldsymbol{\rho}$ can be called the Bloch vector. Its temporal evolution is determined by (16.42). The matrix E defines the temporal evolution of the Bloch vector when the external electromagnetic field is absent. The matrix Λ defines the influence of the external electromagnetic field.

The matrix elements of the operator relaxation

$$\mathrm{e}^{\hat{E}t} = \begin{pmatrix} \mathrm{e}^{\varepsilon t} & 0 & 0 & 0 \\ 0 & \mathrm{e}^{\varepsilon^* t} & 0 & 0 \\ 0 & 0 & \mathrm{e}^{-t/T_1} & 0 \\ 0 & 0 & 1 - \mathrm{e}^{-t/T_1} & 1 \end{pmatrix} \tag{16.44}$$

can be found if (16.41) is solved at $\Lambda = 0$.

Let us examine the influence of the laser pumping on the temporal evolution of the Bloch vector. It is convenient to write the Bloch vector in the form

$$\boldsymbol{\rho}(t) = \mathrm{e}^{\hat{E}t}\boldsymbol{r}(t) . \tag{16.45}$$

This representation can be called the interaction representation because the temporal evolution of the vector $\boldsymbol{r}(t)$ is determined solely by the molecule–field interaction operator. Indeed, taking into account the fact that $\mathrm{d}\rho/\mathrm{d}t = E\rho + \mathrm{e}^{Et}\mathrm{d}r/\mathrm{d}t$, we find the following equation for the vector $\boldsymbol{r}(t)$:

$$\dot{\boldsymbol{r}}(t) = \hat{\Lambda}(t)\boldsymbol{r}(t) , \tag{16.46}$$

where

$$\hat{\Lambda}(t) = \mathrm{e}^{-\hat{E}t}\hat{\Lambda}\mathrm{e}^{\hat{E}t} \tag{16.47}$$

is a pumping matrix in the interaction representation. If the laser pumping equals zero, the vector $\boldsymbol{r}$ is constant.

The vector $\boldsymbol{r}(t)$ can be written as

$$\boldsymbol{r}(t) = \hat{S}(t)\boldsymbol{r}(0) , \tag{16.48}$$

where the matrix $S(t)$ satisfies

$$\dot{\hat{S}}(t) = \hat{\Lambda}(t)\hat{S}(t) . \tag{16.49}$$

Integrating this equation with initial value $S(0) = 1$ yields the integral equation

$$\hat{S}(t) = 1 + \int_0^t \mathrm{d}x \hat{\Lambda}(x) \hat{S}(x) . \tag{16.50}$$

The solution of this equation can be expressed as

$$\begin{aligned}\hat{S}(t) = 1 + \int_0^t \mathrm{d}x \hat{\Lambda}(x) + \int_0^t \mathrm{d}x \hat{\Lambda}(x) \int_0^x \mathrm{d}y \hat{\Lambda}(y) \\ + \int_0^t \mathrm{d}x \hat{\Lambda}(x) \int_0^x \mathrm{d}y \hat{\Lambda}(y) \int_0^y \mathrm{d}z \hat{\Lambda}(z) + \dots .\end{aligned} \tag{16.51}$$

Such a series is convenient for examining various nonlinear effects, because each term includes a definite degree of the electron–photon interaction. The solution of the Bloch equations is given by

$$\boldsymbol{\rho}(t) = \mathrm{e}^{\hat{E}t}\hat{S}(t)\boldsymbol{\rho}(0) . \tag{16.52}$$

16.5 Exponential Two-Pulse Photon Echo

The two-pulse photon echo arises when the sample has been irradiated by two short laser pulses, as shown in Fig. 15.2. Using (16.52), one can write the following expression for the Bloch vector after irradiation by two laser pulses with durations Δt and $\Delta t'$:

$$\boldsymbol{\rho}(t) = \mathrm{e}^{\hat{E}(\tau' + \Delta t')} \hat{S}(\Delta t') \, \mathrm{e}^{\hat{E}(\tau + \Delta t)} \hat{S}(\Delta t) \, \boldsymbol{\rho}(0) \; . \tag{16.53}$$

The molecular polarization at time $t = \tau' + \Delta t' + \tau + \Delta t$ is defined by the formula

$$d(t) = \mathrm{Tr}\left\{ \hat{d} \, \boldsymbol{\rho}(t) \right\} = d \left[\rho_{10}(t) + \rho_{01}(t) \right] \; . \tag{16.54}$$

If we take into account all terms in the series describing the evolution operator $S(t)$, the last equation will include linear effects, like optical free induction decay, and all nonlinear effects like photon echo and others. All these effects, calculated with the help of the evolution operator $S(t)$, will be correct at arbitrary values of the interaction with light, like (16.37) for the photon echo signal derived in the last section. However, (16.37) can be transformed to the simpler (16.40) if the interaction with light is weak, i.e., $\vartheta = \chi \Delta t \ll 1$. In this case, the two-pulse echo amplitude is proportional to $\vartheta'^2 \vartheta$. This fact shows that the two-pulse photon echo is a third order effect in the interaction with light. The photon echo emerges after two pulses have passed through the sample. Therefore the expression for the echo amplitude at small intensity of the laser pulses can be derived if we take into account only the terms

$$S_1(\Delta t) = \int_0^{\Delta t} \mathrm{d}x \hat{\Lambda}(x) \; , \qquad S_2(\Delta t') = \int_0^{\Delta t'} \mathrm{d}x \hat{\Lambda}(x) \int_0^{x} \mathrm{d}y \hat{\Lambda}(y) \tag{16.55}$$

in the evolution operator $S(t)$. In this approximation, the Bloch vector is given by

$$\boldsymbol{\rho}(t) = \mathrm{e}^{\hat{E}(\tau' + \Delta t')} \hat{S}_2(\Delta t') \, \mathrm{e}^{\hat{E}(\tau + \Delta t)} \hat{S}_1(\Delta t) \, \boldsymbol{\rho}(0) \; . \tag{16.56}$$

It is convenient to measure the echo signal if the excitation pulses are shorter than the pure dephasing time T_2, i.e.,

$$\Delta t / T_2 \ll 1 \; . \tag{16.57}$$

For such short pulses, we can set $\Delta t / T_1 = 0$ and therefore the relaxation operator can be taken in the form

$$\mathrm{e}^{\hat{E}t} = \begin{pmatrix} \mathrm{e}^{\varepsilon t} & 0 & 0 & 0 \\ 0 & \mathrm{e}^{\varepsilon^* t} & 0 & 0 \\ 0 & 0 & 1 & 0 \\ 0 & 0 & 0 & 1 \end{pmatrix} \; . \tag{16.58}$$

Using this relaxation operator and (16.47), we find the formula

$$\mathrm{e}^{\hat{E}\Delta t}\hat{\Lambda}(x)=\begin{pmatrix}0 & 0 & \mathrm{e}^{\varepsilon(\Delta t-x)}\chi & \mathrm{e}^{\varepsilon(\Delta t-x)}\chi\\ 0 & 0 & \mathrm{e}^{\varepsilon^*(\Delta t-x)}\chi^* & -\mathrm{e}^{\varepsilon^*(\Delta t-x)}\chi^*\\ -\mathrm{e}^{\varepsilon x}\chi^* & -\mathrm{e}^{\varepsilon^* x}\chi & 0 & 0\\ \mathrm{e}^{\varepsilon x}\chi^* & \mathrm{e}^{\varepsilon^* x}\chi & 0 & 0\end{pmatrix}, \qquad (16.59)$$

Inserting this matrix into the operator S_1, we arrive at the matrix

$$\hat{S}_1(\Delta t)=\begin{pmatrix}0 & 0 & \Lambda_1 & -\Lambda_1\\ 0 & 0 & \Lambda_1^* & -\Lambda_1^*\\ -\Lambda_1^* & -\Lambda_1 & 0 & 0\\ \Lambda_1^* & \Lambda_1 & 0 & 0\end{pmatrix}, \qquad (16.60)$$

where

$$\Lambda_1=\chi\frac{\mathrm{e}^{\mathrm{i}(\omega_0-\Omega)\Delta t}-1}{\mathrm{i}(\omega_0-\Omega)}, \qquad (16.61)$$

where $\chi=\boldsymbol{d}\cdot\boldsymbol{E}/\hbar$. The function is nonvanishing at the resonance. Therefore we find

$$\Lambda_1\approx\Delta t\chi=\vartheta\,. \qquad (16.62)$$

The matrix described by (16.60) therefore takes the form

$$\hat{S}_1(\Delta t)=\int_0^{\Delta t}\hat{\Lambda}(x)\,\mathrm{d}x\approx\hat{S}\Delta t\chi\,, \qquad (16.63)$$

where

$$\hat{S}=\begin{pmatrix}0 & 0 & 1 & -1\\ 0 & 0 & 1 & -1\\ -1 & -1 & 0 & 0\\ 1 & 1 & 0 & 0\end{pmatrix}. \qquad (16.64)$$

Hence, in the short time limit, we can use the matrix Λ in the form

$$\hat{\Lambda}(x)\approx\hat{S}\chi\,. \qquad (16.65)$$

In this approximation the second order term of the evolution operator is given by

$$\hat{S}_2(\Delta t)\approx\frac{(\Delta t'\chi)^2}{2}\hat{S}^2=\hat{S}'\vartheta'^2\,, \qquad (16.66)$$

where

$$\hat{S}' = \begin{pmatrix} -1 & -1 & 0 & 0 \\ -1 & -1 & 0 & 0 \\ 0 & 0 & -1 & 1 \\ 0 & 0 & 1 & -1 \end{pmatrix} . \tag{16.67}$$

The four-component Bloch vector can be written as

$$\boldsymbol{\rho} = \begin{pmatrix} \boldsymbol{a} \\ \boldsymbol{b} \end{pmatrix} , \tag{16.68}$$

where the two-component vectors

$$\boldsymbol{a} = \begin{pmatrix} \rho_{10} \\ \rho_{01} \end{pmatrix} , \qquad \boldsymbol{b} = \begin{pmatrix} \rho_1 \\ \rho_0 \end{pmatrix} \tag{16.69}$$

include either off-diagonal or diagonal elements of the density matrix. The vectors $\boldsymbol{a}$ and $\boldsymbol{b}$ are called the transverse and longitudinal vectors, respectively. In fact, the four-dimensional space of the Bloch vector consists of a pair of two-dimensional subspaces of transverse and longitudinal type. In accordance with this fact, the four-dimensional matrices can be expressed in terms of two-dimensional matrices as follows:

$$\mathrm{e}^{\hat{E}t} = \begin{pmatrix} \mathrm{e}^{\hat{\omega}t} & 0 \\ 0 & \mathrm{e}^{\hat{\Gamma}t} \end{pmatrix} , \qquad \hat{S}_1 = \begin{pmatrix} 0 & \hat{\Lambda}_1 \\ \hat{\Lambda}_2 & 0 \end{pmatrix} , \qquad \hat{S}_2 = \begin{pmatrix} \hat{M} & 0 \\ 0 & \hat{N} \end{pmatrix} , \tag{16.70}$$

where

$$\begin{aligned} \hat{\Lambda}_1 &= \vartheta \begin{pmatrix} 1 & -1 \\ 1 & -1 \end{pmatrix} , \qquad \hat{\Lambda}_2 = \vartheta \begin{pmatrix} -1 & -1 \\ 1 & 1 \end{pmatrix} , \\ \hat{M} &= -\vartheta'^2 \begin{pmatrix} 1 & 1 \\ 1 & 1 \end{pmatrix} , \quad \hat{N} = \vartheta'^2 \begin{pmatrix} -1 & 1 \\ 1 & -1 \end{pmatrix} , \\ \mathrm{e}^{\hat{\omega}t} &= \begin{pmatrix} \mathrm{e}^{-\mathrm{i}(\Omega - \mathrm{i}/T_2)t} & 0 \\ 0 & \mathrm{e}^{\mathrm{i}(\Omega + \mathrm{i}/T_2)t} \end{pmatrix} , \qquad \mathrm{e}^{\hat{\Gamma}t} = \begin{pmatrix} \mathrm{e}^{-t/T_1} & 0 \\ 1 - \mathrm{e}^{-t/T_1} & 1 \end{pmatrix} . \end{aligned} \tag{16.71}$$

The Bloch vector described by (16.56) can be expressed in terms of two-dimensional matrices and vectors as follows:

$$\begin{pmatrix} \boldsymbol{a}(t) \\ 0 \end{pmatrix} = \begin{pmatrix} \mathrm{e}^{\hat{\omega}\tau'} & 0 \\ 0 & \mathrm{e}^{\hat{\Gamma}\tau'} \end{pmatrix} \begin{pmatrix} \hat{M} & 0 \\ 0 & \hat{N} \end{pmatrix} \begin{pmatrix} \mathrm{e}^{\hat{\omega}\tau} & 0 \\ 0 & \mathrm{e}^{\hat{\Gamma}\tau} \end{pmatrix} \begin{pmatrix} 0 & \hat{\Lambda}_1 \\ \hat{\Lambda}_2 & 0 \end{pmatrix} \begin{pmatrix} 0 \\ \boldsymbol{b}(0) \end{pmatrix} , \tag{16.72}$$

where we have taken into account the fact that $\Delta t' = \Delta t = 0$ in the relaxation operators, because the time between pulses is much longer than the duration of the pulses. By multiplying these matrices we arrive at the following expression for the two-dimensional transverse vector:

$$\boldsymbol{a}(t) = \mathrm{e}^{\hat{\omega}\tau'} \hat{M} \mathrm{e}^{\hat{\omega}\tau} \hat{\Lambda}_1 \boldsymbol{b}(0) . \tag{16.73}$$

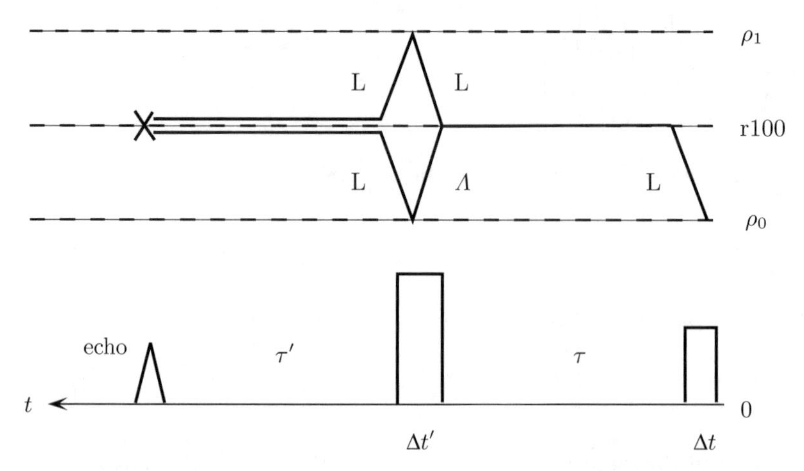

Fig. 16.2. Temporal evolution of density matrix elements (*top*) and laser pulses resulting in this evolution (*bottom*)

Figure 16.2 shows the temporal evolution of the transverse and longitudinal components of the Bloch vector in a two-pulse photon echo experiment. The dashed lines show quantum states described by transverse and longitudinal components of the Bloch vector. The solid lines show the temporal evolution of the system in the lowest approximation in the interaction Λ with light. In this approximation, the transitions between longitudinal components described by $\rho_{00} \equiv \rho_0$, $\rho_{11} \equiv \rho_1$ and transverse components described by ρ_{10}, ρ_{01} are proportional to Λ.

Multiplying the matrices in (16.73), we find the following expression for the transverse component:

$$\begin{aligned}\begin{pmatrix}\rho_{10}(t)\\ \rho_{01}(t)\end{pmatrix} &= -\vartheta'^2\vartheta\begin{pmatrix}\mathrm{e}^{-\mathrm{i}(\Omega-\mathrm{i}/T_2)\tau'} & 0\\ 0 & \mathrm{e}^{\mathrm{i}(\Omega+\mathrm{i}/T_2)\tau'}\end{pmatrix}\begin{pmatrix}1 & 1\\ 1 & 1\end{pmatrix} \\ &\quad\times\begin{pmatrix}\mathrm{e}^{-\mathrm{i}(\Omega-\mathrm{i}/T_2)\tau} & 0\\ 0 & \mathrm{e}^{\mathrm{i}(\Omega+\mathrm{i}/T_2)\tau}\end{pmatrix}\begin{pmatrix}1 & -1\\ 1 & -1\end{pmatrix}\begin{pmatrix}0\\ 1\end{pmatrix} \\ &= \vartheta'^2\vartheta 2\mathrm{e}^{-\tau/T_2}\cos\Omega\tau\begin{pmatrix}\mathrm{e}^{-\mathrm{i}(\Omega-\mathrm{i}/T_2)\tau'}\\ \mathrm{e}^{\mathrm{i}(\Omega+\mathrm{i}/T_2)\tau'}\end{pmatrix}\end{aligned} \tag{16.74}$$

Inserting this equation into (16.54), we arrive at the following equation for the induced dipolar moment of a molecule:

$$\begin{aligned}d(t) &= \vartheta'^2\vartheta 4\mathrm{e}^{-(\tau'+\tau)/T_2}\cos\Omega\tau'\cos\Omega\tau \\ &= \vartheta'^2\vartheta\, 2\mathrm{e}^{-(\tau'+\tau)/T_2}\left[\cos\Omega\left(\tau'-\tau\right)+\cos\Omega\left(\tau'+\tau\right)\right]\ .\end{aligned} \tag{16.75}$$

The sum over all molecules in the sample is

$$D = 2dN\vartheta'^2\vartheta\left[N\left(\tau'-\tau\right)+N\left(\tau'+\tau\right)\right]\mathrm{e}^{-(\tau'+\tau)/T_2}\ , \tag{16.76}$$

where $N(t)$ is given by (16.39). Here the first term defines the echo signal and the second term describes the third order contribution to the optical free induction decay. The expression for the echo amplitude in (16.40), where the electron–phonon interaction has not been allowed for, and in (16.76), where the electron–phonon interaction has been allowed for, differs by the exponential term $\exp[-(\tau+\tau')/T_2]$. This term describes the decay of the echo signal. Hence, the echo signal can be observed in realistic systems only if the pause τ between pulses is shorter than the dephasing time T_2.

By changing the delay between pulses one can find the dependence of the echo amplitude on the time delay τ, and hence measure the value of the dephasing time. This time is a characteristic of the dephasing processes in the system. Such processes are collisions if the molecules are in a gas phase, and electron–phonon interactions if the molecules are in a solid solution. It should be noted that inhomogeneous broadening does not hide pure dephasing processes as optical free induction decay does, because the integrated value of the echo signal does not depend on inhomogeneous broadening. This value characterizes only the pure dephasing time, which is a carrier of information about interactions existing in the system. The photon echo is thus a powerful method for studying these interactions.

16.6 Three-Pulse Photon Echo

After two short laser pulses have passed through the sample, a two-pulse echo (2PE) signal emerges. If we wait for a time which exceeds the dephasing time T_2 and then irradiate this sample by the third laser pulse, we observe an optical free induction decay after the third pulse. However, should we expect to observe an additional echo pulse after the third laser pulse? The experiment shows that such an echo pulse does indeed emerge. This is the so-called three-pulse photon echo (3PE) signal. The properties of the 3PE differ from those of the 2PE because the 3PE signal allows us to study both the dephasing time T_2 and the energy relaxation. Let us inspect the 3PE in detail.

If a sample is irradiated by three laser pulses as shown in Fig. 16.3, the temporal evolution of the Bloch vector is given by

$$\boldsymbol{\rho}(t)=\mathrm{e}^{\hat{E}\tau'}\hat{S}\,(\Delta t')\,\mathrm{e}^{\hat{E}t_{\mathrm{w}}}\hat{S}\,(\Delta t'')\,\mathrm{e}^{\hat{E}\tau}\hat{S}\,(\Delta t)\,\boldsymbol{\rho}(0)\,. \tag{16.77}$$

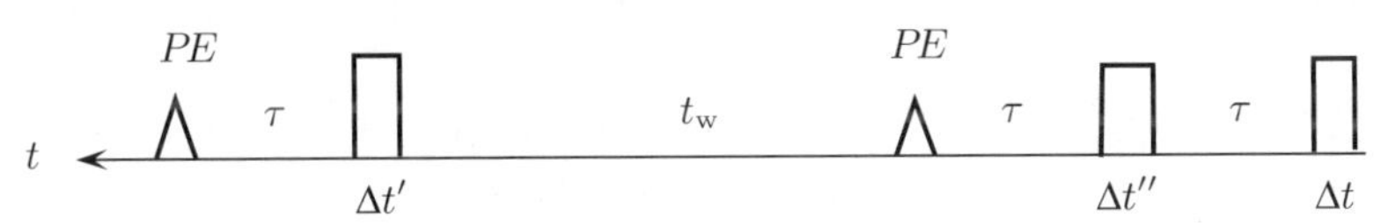

Fig. 16.3. Train of laser pulses yielding 3PE

Since the durations Δt, $\Delta t'$ and $\Delta t''$ of the pulses are smaller than each pause, these small values are omitted from the relaxation. The photon echo is a third order effect in the interaction with light. Therefore each evolution operator $S(t)$ can be taken in the first approximation, i.e., $S \approx S_1$. Then (16.77) takes the form

$$\begin{aligned}\boldsymbol{\rho}(t) &= \mathrm{e}^{\hat{E}\tau'} S_1 \mathrm{e}^{\hat{E}t_\mathrm{w}} S_1 \mathrm{e}^{\hat{E}\tau} S_1 \boldsymbol{\rho}(0) \\ &= \begin{pmatrix} \mathrm{e}^{\hat{\omega}\tau'} & 0 \\ 0 & \mathrm{e}^{\hat{\Gamma}\tau'} \end{pmatrix} \begin{pmatrix} 0 & \hat{\Lambda}_1 \\ \hat{\Lambda}_2 & 0 \end{pmatrix} \begin{pmatrix} \mathrm{e}^{\hat{\omega}t_\mathrm{w}} & 0 \\ 0 & \mathrm{e}^{\hat{\Gamma}t_\mathrm{w}} \end{pmatrix} \begin{pmatrix} 0 & \hat{\Lambda}_1 \\ \hat{\Lambda}_2 & 0 \end{pmatrix} \\ &\quad \times \begin{pmatrix} \mathrm{e}^{\hat{\omega}\tau} & 0 \\ 0 & \mathrm{e}^{\hat{\Gamma}\tau} \end{pmatrix} \begin{pmatrix} 0 & \hat{\Lambda}_1 \\ \hat{\Lambda}_2 & 0 \end{pmatrix} \begin{pmatrix} 0 \\ \boldsymbol{b}(0) \end{pmatrix} ,\end{aligned} \tag{16.78}$$

where the operators $\Lambda_{1,2}$ are defined by (16.71). Multiplying the matrices in the last equation, we find the following expression for the transverse part of the Bloch vector:

$$\boldsymbol{a}(t) = \mathrm{e}^{\omega\tau'} \hat{\Lambda}_1 \mathrm{e}^{\hat{\Gamma}t_\mathrm{w}} \hat{\Lambda}_2 \mathrm{e}^{\hat{\omega}\tau} \hat{\Lambda}_1 \boldsymbol{b}(0) \,. \tag{16.79}$$

The evolution of the Bloch vector is shown in Fig. 16.4. If the time t_w approaches zero, this diagram is transformed to the diagram shown in Fig. 16.2. Equation (16.78) for the transverse vector is transformed to (16.73). In other words the 3PE is transformed to a 2PE.

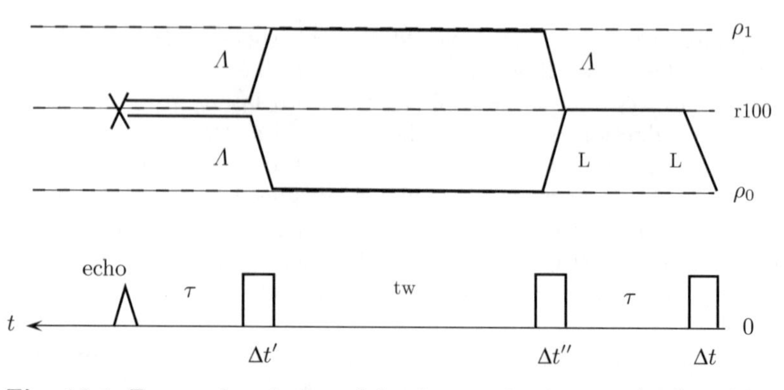

Fig. 16.4. Temporal evolution of density matrix elements (*top*) and laser pulses yielding the 3PE and resulting in this evolution (*bottom*)

Let us replace the matrices in (16.78) by their expressions from (16.71). The result is

$$\begin{aligned}\begin{pmatrix} \rho_{10}(t) \\ \rho_{01}(t) \end{pmatrix} &= \vartheta'\vartheta''\vartheta \begin{pmatrix} \mathrm{e}^{-\mathrm{i}(\Omega - \mathrm{i}/T_2)\tau'} & 0 \\ 0 & \mathrm{e}^{\mathrm{i}(\Omega + \mathrm{i}/T_2)\tau'} \end{pmatrix} \begin{pmatrix} 1 & -1 \\ 1 & -1 \end{pmatrix} \\ &\times \begin{pmatrix} \mathrm{e}^{-t_\mathrm{w}/T_1} & 0 \\ 1 - \mathrm{e}^{-t_\mathrm{w}/T_1} & 1 \end{pmatrix} \begin{pmatrix} -1 & -1 \\ 1 & 1 \end{pmatrix} \begin{pmatrix} \mathrm{e}^{-\mathrm{i}(\Omega - \mathrm{i}/T_2)\tau} & 0 \\ 0 & \mathrm{e}^{\mathrm{i}(\Omega + \mathrm{i}/T_2)\tau} \end{pmatrix} \begin{pmatrix} 1 & -1 \\ 1 & -1 \end{pmatrix} \begin{pmatrix} 0 \\ 1 \end{pmatrix} .\end{aligned} \tag{16.80}$$

The difference between the 3PE and 2PE becomes clearer if we consider this equation in parts. The action of the first, second and third matrices on the initial Bloch vector yields the following result:

$$\begin{aligned}\begin{pmatrix}\rho_1\\ \rho_0\end{pmatrix} &= \begin{pmatrix}-1 & -1\\ 1 & 1\end{pmatrix}\begin{pmatrix}\mathrm{e}^{-\mathrm{i}(\Omega-\mathrm{i}/T_2)\tau} & 0\\ 0 & \mathrm{e}^{\mathrm{i}(\Omega+\mathrm{i}/T_2)\tau}\end{pmatrix}\begin{pmatrix}1 & -1\\ 1 & -1\end{pmatrix}\begin{pmatrix}0\\ 1\end{pmatrix}\\ &= \left(\mathrm{e}^{-\mathrm{i}(\Omega-\mathrm{i}/T_2)\tau} + \mathrm{e}^{\mathrm{i}(\Omega+\mathrm{i}/T_2)\tau}\right)\begin{pmatrix}1\\ -1\end{pmatrix}\,.\end{aligned} \tag{16.81}$$

According to this equation, after applying the first and second laser pulses to the sample, the Bloch vector only has a longitudinal part, proportional to Λ^2, in the first nonvanishing approximation. However, by considering the 2PE, we have already seen that, after two pulses are applied, the Bloch vector also has a transverse component proportional to Λ^3, as shown in Fig. 16.2. Now we know that the Bloch vector will have the longitudinal component as well. This component is proportional to Λ^2, as shown in Fig. 16.4. Multiplying the last matrix by the relaxation matrix, we find that, before applying the third laser pulse, the longitudinal vector takes the form

$$\begin{pmatrix}\rho_1\left(t_{\mathrm{w}}+\Delta t''+\tau+\Delta t\right)\\ \rho_0\left(t_{\mathrm{w}}+\Delta t''+\tau+\Delta t\right)\end{pmatrix} = 2\mathrm{e}^{-t_{\mathrm{w}}/T_1}\mathrm{e}^{-\tau/T_2}\cos\Omega\tau\begin{pmatrix}1\\ -1\end{pmatrix}\,, \tag{16.82}$$

i.e., the longitudinal vector exhibits an exponential decay with characteristic relaxation time T_1. This vector describes a change in the population proportional to Λ^2 and depending on the phase $\Omega\tau$ of the electronic excitation. This frequency modulation of the population is called frequency grating. After we apply the third laser pulse to the sample, an echo signal emerges due to this grating. Because phase information is stored in the population, the information lives for a long time, comparable with T_1. This is a very important feature of the 3PE.

Inserting this longitudinal vector into (16.80), we find the following expression for the transverse vector:

$$\begin{aligned}\begin{pmatrix}\rho_{10}(t)\\ \rho_{01}(t)\end{pmatrix} &= \vartheta'\vartheta''\vartheta\begin{pmatrix}\mathrm{e}^{-\mathrm{i}(\Omega-\mathrm{i}/T_2)\tau'} & 0\\ 0 & \mathrm{e}^{\mathrm{i}(\Omega+\mathrm{i}/T_2)\tau'}\end{pmatrix}\\ &\quad\times\begin{pmatrix}1 & -1\\ 1 & -1\end{pmatrix}\begin{pmatrix}\rho_1\left(t_{\mathrm{w}}+\Delta t''+\tau+\Delta t\right)\\ \rho_0\left(t_{\mathrm{w}}+\Delta t''+\tau+\Delta t\right)\end{pmatrix}\\ &= \vartheta'\vartheta''\vartheta\,2\mathrm{e}^{-t_{\mathrm{w}}/T_1}\mathrm{e}^{-\tau/T_2}\cos\Omega\tau\begin{pmatrix}\mathrm{e}^{-\mathrm{i}(\Omega-\mathrm{i}/T_2)\tau'}\\ \mathrm{e}^{\mathrm{i}(\Omega+\mathrm{i}/T_2)\tau'}\end{pmatrix}\,.\end{aligned} \tag{16.83}$$

This expression for the three-pulse excitation is similar to (16.74) for the two-pulse excitation. Therefore, by summing up contributions from all molecules in the sample, we arrive at the following expression for the dipolar moment of the sample:

$$D = 2d\vartheta'\vartheta''\vartheta\left[N\left(\tau'-\tau\right)+N\left(\tau'+\tau\right)\right]\mathrm{e}^{-t_{\mathrm{w}}/T_1}\mathrm{e}^{-(\tau'+\tau)/T_2}\,. \tag{16.84}$$

Taking into account the fact that the function $N(t)$ describes a narrow peak with maximum at $t=0$ and using (16.76) and (16.84), we find the following expressions for the amplitudes of the 2PE and 3PE:

$$E_{\mathrm{2PE}} \propto \vartheta\vartheta'^2 \mathrm{e}^{-2\tau/T_2} , \qquad E_{\mathrm{3PE}} \propto \vartheta\vartheta'\vartheta'' \mathrm{e}^{-2\tau/T_2} \mathrm{e}^{-t_{\mathrm{w}}/T_1} . \tag{16.85}$$

By measuring the dependence of the echo amplitude on the time delay τ between the first and second laser pulses, we can find the pure dephasing time T_2, whilst measurement of the dependence of the 3PE amplitude on the time delay t_{w} between the second and third laser pulses allows us to find the lifetime T_1 of the excited electronic state. Photon echos serve as major tools for measuring the pure dephasing time and energy relaxation time in molecular ensembles with large inhomogeneous broadening.

16.7 Long-Lived 3PE

If molecules have a high quantum yield of intersystem crossing, each excited molecule goes into a long-lived triplet electronic state with high efficiency. Therefore the waiting time t_{w} can be comparable with the triplet lifetime. In this case the 3PE signal can emerge on a very long time scale after two laser pulses have been applied to the sample. Such a 3PE is called a long-lived 3PE. This very type of photon echo has been successfully used to study slow relaxations existing in polymers and glasses, and so called spectral diffusion. Therefore the long-lived 3PE should be discussed in more detail.

Let the molecule be described by the energy diagram in Fig. 8.2. Then, instead of four Bloch equations, we have five equations:

$$\begin{aligned}
\dot{\rho}_{10} &= -\mathrm{i}\left(\Omega - \frac{\mathrm{i}}{T_2}\right)\rho_{10} + \chi\left(\rho_1 - \rho_0\right) , \\
\dot{\rho}_{01} &= \mathrm{i}\left(\Omega + \frac{\mathrm{i}}{T_2}\right)\rho_{01} + \chi^*\left(\rho_1 - \rho_0\right) , \\
\dot{\rho}_1 &= -\chi^*\rho_{10} - \chi\rho_{01} - \left(\frac{1}{T_1} + \Gamma_{\mathrm{TS}}\right)\rho_1 , \\
\dot{\rho}_0 &= \chi^*\rho_{10} + \chi\rho_{01} + \frac{\rho_1}{T_1} + \gamma_{\mathrm{ST}}\rho_2 , \\
\dot{\rho}_2 &= \Gamma_{\mathrm{TS}}\rho_1 - \left(\gamma_{\mathrm{ST}} + Q\right)\rho_2 .
\end{aligned} \tag{16.86}$$

This set of equations differs from the set in (8.14). It contains the rate constant Q which allows for a phototransformation of the molecule to a photoproduct via the triplet state.

Let us consider the 3PE in such a system. Figure 16.4 and (16.78) remain correct in this case. However, the longitudinal vector has three components now:

$$\boldsymbol{b} = \begin{pmatrix} \rho_1 \\ \rho_0 \\ \rho_2 \end{pmatrix} , \tag{16.87}$$

and therefore the pumping matrices and the relaxation matrix change their dimensions to look like

$$\begin{gathered} \hat{\varLambda}_1 = \vartheta \begin{pmatrix} 1 & -1 & 0 \\ 1 & -1 & 0 \\ 0 & 0 & 0 \end{pmatrix} , \quad \hat{\varLambda}_2 = \vartheta'' \begin{pmatrix} -1 & -1 & 0 \\ 1 & 1 & 0 \\ 0 & 0 & 0 \end{pmatrix} , \\ \hat{\varGamma} = \begin{pmatrix} -\varGamma & 0 & 0 \\ 1/T_1 & 0 & \gamma_{\mathrm{ST}} \\ \varGamma_{\mathrm{TS}} & 0 & -(\gamma_{\mathrm{ST}} + Q) \end{pmatrix} , \end{gathered} \tag{16.88}$$

where $\varGamma = \varGamma_{\mathrm{TS}} + 1/T_1$ is the full rate of relaxation from the excited singlet electronic state. One part of the relaxation matrix, $\exp(\omega t)$, which describes relaxation of the transverse components, does not change. The second part, $\exp(\varGamma t)$, which describes relaxation of the longitudinal components, is given by

$$\mathrm{e}^{\hat{\varGamma} t} = \begin{pmatrix} \mathrm{e}^{-\varGamma t} & 0 & 0 \\ \begin{array}{c} \dfrac{\varGamma_{\mathrm{TS}}}{\varGamma} \dfrac{\gamma_{\mathrm{ST}}}{\gamma'_{\mathrm{ST}}} \left(1 - \mathrm{e}^{-\gamma'_{\mathrm{ST}} t}\right) \\ + \dfrac{1/T_1}{\varGamma} \left(1 - \mathrm{e}^{-\varGamma t}\right) \end{array} & 1 & \dfrac{\gamma_{\mathrm{ST}}}{\gamma'_{\mathrm{ST}}} \left(1 - \mathrm{e}^{-\gamma'_{\mathrm{ST}} t}\right) \\ \dfrac{\varGamma_{\mathrm{TS}}}{\varGamma} \left(\mathrm{e}^{-\gamma'_{\mathrm{ST}} t} - \mathrm{e}^{-\varGamma t}\right) & 0 & \mathrm{e}^{-\gamma'_{\mathrm{ST}} t} \end{pmatrix} , \tag{16.89}$$

where $\gamma'_{\mathrm{ST}} = \gamma_{\mathrm{ST}} + Q$ is the full rate of decay of the triplet state in the molecule. The elements of this matrix can be found from (16.86) at $\chi = 0$. We also take into account the inequalities

$$\varGamma_{\mathrm{TS}} \gg \gamma_{\mathrm{ST}} \gg Q . \tag{16.90}$$

It is convenient to analyze (16.79) in the form

$$\boldsymbol{a}(t) = \mathrm{e}^{\hat{\omega}\tau'} \hat{\varLambda}_1 \boldsymbol{b} \left(t_{\mathrm{w}} + \Delta t'' + \tau + \Delta t\right) , \tag{16.91}$$

where

$$\boldsymbol{b} \left(t_{\mathrm{w}} + \Delta t'' + \tau + \Delta t\right) = \mathrm{e}^{\hat{\varGamma} t_{\mathrm{w}}} \hat{\varLambda}_2 \mathrm{e}^{\hat{\omega}\tau} \hat{\varLambda}_1 \boldsymbol{b}(0) . \tag{16.92}$$

Here the transverse components of the Bloch vector are expressed in terms of the longitudinal components of the Bloch vector after two laser pulses have been applied, i.e., when frequency grating already exists in the sample. Multiplying the matrices in the last equation, we find the expression

$$\boldsymbol{b}\left(t_\mathrm{w} + \Delta t'' + \tau + \Delta t\right) = \vartheta''\vartheta 2\mathrm{e}^{-\tau/T_2}\cos\Omega\tau \begin{pmatrix} \left(\mathrm{e}^{\hat{\Gamma}t_\mathrm{w}}\right)_{11} \\ \left(\mathrm{e}^{\hat{\Gamma}t_\mathrm{w}}\right)_{01} - 1 \\ \left(\mathrm{e}^{\hat{\Gamma}t_\mathrm{w}}\right)_{21} \end{pmatrix} . \quad (16.93)$$

Inserting (16.93) into (16.91) and multiplying the matrices, we find the following expression for the transverse components of the Bloch vector:

$$\begin{pmatrix} \rho_{10}(t) \\ \rho_{01}(t) \end{pmatrix} = \vartheta'\vartheta''\vartheta\left[\left(\mathrm{e}^{\hat{\Gamma}t_\mathrm{w}}\right)_{11} - \left(\mathrm{e}^{\hat{\Gamma}t_\mathrm{w}}\right)_{01} + 1\right] \times 2\mathrm{e}^{-\tau'/T_2}\cos\Omega\tau' \begin{pmatrix} \mathrm{e}^{-\mathrm{i}(\Omega-\mathrm{i}/T_2)\tau} \\ \mathrm{e}^{\mathrm{i}(\Omega+\mathrm{i}/T_2)\tau} \end{pmatrix} . \quad (16.94)$$

Replacing the transverse components in (16.54) for the dipolar moment by (16.94) yields

$$d(t) = \vartheta'\vartheta''\vartheta\left[\left(\mathrm{e}^{\hat{\Gamma}t_\mathrm{w}}\right)_{11} - \left(\mathrm{e}^{\hat{\Gamma}t_\mathrm{w}}\right)_{01} + 1\right] \times 2\mathrm{e}^{-(\tau'+\tau)/T_2}\left[\cos\Omega(\tau'-\tau) + \cos\Omega(\tau'+\tau)\right] . \quad (16.95)$$

Using this equation, we find the following expression for the 3PE amplitude:

$$E_{3\mathrm{PE}} \propto \vartheta'\vartheta''\vartheta\left[\left(\mathrm{e}^{\hat{\Gamma}t_\mathrm{w}}\right)_{11} - \left(\mathrm{e}^{\hat{\Gamma}t_\mathrm{w}}\right)_{01} + 1\right]\mathrm{e}^{-2\tau/T_2} . \quad (16.96)$$

The terms in square brackets are the elements of the matrix described by (16.89). Using these elements, we find the final expression for the 3PE amplitude:

$$E_{3\mathrm{PE}} \propto \vartheta'\vartheta''\vartheta\left[\left(1 + \frac{1/T_1}{\Gamma}\right)\mathrm{e}^{-\Gamma t_\mathrm{w}} + \frac{\Gamma_\mathrm{TS}}{\Gamma}(1-\eta_Q)\,\mathrm{e}^{-\gamma_\mathrm{ST}t_\mathrm{w}} + \frac{\Gamma_\mathrm{TS}}{\Gamma}\eta_Q\right]\mathrm{e}^{-2\tau/T_2} , \quad (16.97)$$

where $\eta_Q = Q/(Q+\gamma_\mathrm{ST})$ is the quantum yield of phototransformation. Equation (16.97) describes the long-lived photon echo. Indeed, if $t_\mathrm{w} < T_1$, the expression in the square brackets equals 2. If $T_1 < t_\mathrm{w} < 1/\gamma_\mathrm{ST}$, the expression in these brackets is of order $\Gamma_\mathrm{TS}/\Gamma$. This is the quantum yield of intersystem crossing. It is of order 0.1–0.9 for many molecules. Hence we are able to detect the echo signal even if $t_\mathrm{w} \gg T_1$. Such an echo is said to be long-lived.

However, if $t_\mathrm{w} \gg 1/\gamma_\mathrm{ST}$, the expression in square brackets does not equal zero. It is of the same order as the photochemical quantum yield η_Q. As usual, η_Q equals 10^{-3}–10^{-4}. The echo signal of very weak intensity can in principle be detected with waiting time t_w comparable with the lifetime of the spectral hole burnt in the ground state population. An echo signal with such long delay can emerge due to frequency grating burnt into the ground state population by a photochemical reaction.

16.8 Space Anisotropy of Echo Radiation

The equations for 2PE and 3PE signals derived in the last few sections describe the echo intensity integrated over all directions in space. However, there is an anisotropy in echo radiation. In order to discuss this aspect of the photon echo, we turn back to (16.97) which describes the 3PE signal. When deriving this equation, we used the Bloch equations with the electromagnetic field taken at $\boldsymbol{r} = 0$. The field of the standing wave is described by (1.33), (1.34) and (1.35). However, we assumed that the molecule irradiated by light is located in the maximum of the electric field, i.e., $\cos \Psi_{\boldsymbol{k}} = \cos \boldsymbol{k} \cdot \boldsymbol{r} = 1$. On the other hand, the size of the sample exceeds the length of the electromagnetic wave. This means that many molecules do not get to the maximum of the electric field and therefore they will be more weakly excited. In order to take this fact into account, we must carry out the substitution

$$\Lambda \to \Lambda \cos \boldsymbol{k} \cdot \boldsymbol{r} \,, \qquad \chi \to \chi \cos \boldsymbol{k} \cdot \boldsymbol{r} \,, \qquad \vartheta \to \vartheta \cos \boldsymbol{k} \cdot \boldsymbol{r} \,. \tag{16.98}$$

After such a substitution the dipolar moment induced in the molecule by three laser pulses will be described by the expression

$$\boldsymbol{d}(\Delta, \boldsymbol{r}, t) = \boldsymbol{d}(\Delta, 0, t) \cos \boldsymbol{k}_3 \cdot \boldsymbol{r} \cos \boldsymbol{k}_2 \cdot \boldsymbol{r} \cos \boldsymbol{k}_1 \cdot \boldsymbol{r} \,. \tag{16.99}$$

Let us introduce a function $N(\Delta, \boldsymbol{r})$ which describes a distribution of the molecules in the detuning and in space. For the homogeneous distribution in space, the function takes the form $N(\Delta, \boldsymbol{r}) = N(\Delta)/V$, where V is the sample volume. Then the dipolar moment of unit volume of the sample is described by

$$\begin{aligned} D_{3\mathrm{PE}}(\boldsymbol{r}, t) &= \cos \boldsymbol{k}_3 \cdot \boldsymbol{r} \cos \boldsymbol{k}_2 \cdot \boldsymbol{r} \cos \boldsymbol{k}_1 \cdot \boldsymbol{r} \, \frac{1}{V} \int N(\Delta) \, \boldsymbol{d}(\Delta, 0, t) \, \mathrm{d}\Delta \\ &= \frac{\boldsymbol{D}(t)}{V} \cos \boldsymbol{k}_3 \cdot \boldsymbol{r} \cos \boldsymbol{k}_2 \cdot \boldsymbol{r} \cos \boldsymbol{k}_1 \cdot \boldsymbol{r} \\ &= \frac{\boldsymbol{D}(t)}{4V} \Big[\cos\left[(\boldsymbol{k}_1 + \boldsymbol{k}_2 + \boldsymbol{k}_3) \cdot \boldsymbol{r}\right] + \cos\left[(\boldsymbol{k}_1 + \boldsymbol{k}_2 - \boldsymbol{k}_3) \cdot \boldsymbol{r}\right] \\ &\quad + \cos\left[(\boldsymbol{k}_1 - \boldsymbol{k}_2 + \boldsymbol{k}_3) \cdot \boldsymbol{r}\right] + \cos\left[(\boldsymbol{k}_1 - \boldsymbol{k}_2 - \boldsymbol{k}_3) \cdot \boldsymbol{r}\right] \Big] \,. \end{aligned} \tag{16.100}$$

Hence there are four standing waves with wave vectors determined by four combinations of the wave vectors of all three laser beams. It is obvious that the echo signal will be detected effectively along these four directions. These directions do not coincide with the directions of the laser beams. Not only can the echo signal be separated from the excited light on the time scale, but a space separation can be carried out as well.

The equations for the 2PE can be derived from the equations for the 3PE by assuming that the second and third laser pulses coincide. In this case two wave vectors are the same: $\boldsymbol{k}_3 = \boldsymbol{k}_2$. If this fact is allowed for in (16.101), the expression is transformed to

$$\begin{aligned}\boldsymbol{D}_{2\mathrm{PE}}(\boldsymbol{r},t) &= \frac{\boldsymbol{D}(t)}{V}\cos^2\boldsymbol{k}_2\cdot\boldsymbol{r}\cos\boldsymbol{k}_1\cdot\boldsymbol{r} \\ &= \frac{\boldsymbol{D}(t)}{4V}\Big[\cos\left[(\boldsymbol{k}_1+2\boldsymbol{k}_2)\cdot\boldsymbol{r}\right]+2\cos\boldsymbol{k}_1\cdot\boldsymbol{r}+\cos\left[(\boldsymbol{k}_1-2\boldsymbol{k}_2)\cdot\boldsymbol{r}\right]\Big]\,.\end{aligned} \tag{16.101}$$

Hence the 3PE and 2PE echo signals propagate along a definite direction defined by

$$\boldsymbol{k}_{\mathrm{e}} = \boldsymbol{k}_1 \pm \boldsymbol{k}_2 \pm \boldsymbol{k}_3\,, \quad \boldsymbol{k}_{\mathrm{e}} = \boldsymbol{k}_1 \pm 2\boldsymbol{k}_2\,. \tag{16.102}$$

The anisotropy facilitates separation of the echo signal from the exciting light, and experimental detection of the signal.

17. Nonexponential Photon Echo

The decay of the photon echo signal is a source of information about relaxation processes in the system. Both the 2PE and 3PE exhibit exponential decay because both types of echo have been considered on the basis of the optical Bloch equations, where dephasing is characterized by a single dephasing constant T_2.

However, a nonexponential echo decay is very often observed in experiments. A theory for such photon echo cannot be developed on the basis of the optical Bloch equations. On the other hand, the optical Bloch equations are derived from the equations for the full density matrix by omitting the phonon side band, and if this PSB is taken into account, the equations for the density matrix can indeed explain nonexponential optical dephasing. Such dephasing was examined in Sect. 15.2. Therefore we should examine the nonexponential echo decay with the help of (16.26) or (16.30). These equations include the full electron–phonon interaction. Moreover, they include a classical electromagnetic field with broad spectral band and so can be used when the driving laser field is described by ultrashort laser pulses with large spectral width.

17.1 Generalized Bloch Vector

To begin with we carry out the calculation of the photon echo using (16.30). This is an infinite set of equations. Therefore if we want to use a Bloch vector approach, we have to generalize the essence of the Bloch vector. Instead of the four-dimensional Bloch vector, we introduce a multi-dimensional Bloch vector with an infinite number of components. This vector will be called the generalized Bloch vector.

Let us examine (16.30). In order to allow for a triplet level, we add the fifth equation from the set (16.85) to (16.30). Then introducing the notation

$$\begin{aligned}
&\rho_{ba} = a_n \ , \quad \rho_{ab} = a_n^* \ , \quad \chi_{ba} = \Lambda_n(t) \ , \quad \chi_{ab}^* = \Lambda_n^*(t) \ , \\
&\chi_{ba} p_b = \Lambda_n^{\mathrm{e}}(t) \ , \quad \chi_{ba} p_a = \Lambda_n^{\mathrm{g}}(t) \ , \quad \chi_{ab}^* p_b = \Lambda_n^{*\mathrm{e}}(t) \ , \quad \chi_{ab}^* p_a = \Lambda_n^{*\mathrm{g}}(t) \ , \\
&-\mathrm{i}\left(\Omega_{ba} - \frac{\mathrm{i}}{2T_1}\right) = \varepsilon_n \ , \quad -\mathrm{i}\left(\Omega_{ab} - \frac{\mathrm{i}}{2T_1}\right) = \varepsilon_n^* \ ,
\end{aligned} \tag{17.1}$$

we can write the infinite set of equations in the form

$$\begin{aligned}
\dot{a}_n &= \varepsilon_n a_n + \Lambda_n^{\mathrm{e}} \rho_1 - \Lambda_n^{\mathrm{g}} \rho_0 \,, \\
\dot{a}_n^* &= \varepsilon_n^* a_n^* + \Lambda_n^{*\mathrm{e}} \rho_1 - \Lambda_n^{*\mathrm{g}} \rho_0 \,, \\
\dot{\rho}_1 &= -\sum_n (\Lambda_n^* a_n + \Lambda_n a_n^*) - \Gamma \rho_1 \,, \\
\dot{\rho}_0 &= \sum_n (\Lambda_n^* a_n + \Lambda_n a_n^*) + \rho_1/T_1 + \gamma_{\mathrm{ST}} \rho_2 \,, \\
\dot{\rho}_2 &= \Gamma_{\mathrm{TS}} \rho_1 - \gamma_{\mathrm{ST}}' \rho_2 \,.
\end{aligned} \tag{17.2}$$

Let us introduce the generalized Bloch vector (GBV)

$$\boldsymbol{\rho} = \begin{pmatrix} \boldsymbol{a} \\ \boldsymbol{b} \end{pmatrix} , \quad \boldsymbol{a} = \begin{pmatrix} a_1 \\ a_1^* \\ a_2 \\ a_2^* \\ \vdots \end{pmatrix} , \quad \boldsymbol{b} = \begin{pmatrix} \rho_1 \\ \rho_0 \\ \rho_2 \end{pmatrix} . \tag{17.3}$$

The longitudinal part of the GBV is a three-component vector. However, the transverse part is already a multicomponent vector with an infinite number of components.

Using the GBV, we can write (17.2) as

$$\dot{\boldsymbol{\rho}} = [\hat{E} + \hat{\Lambda}(t)] \boldsymbol{\rho} \,. \tag{17.4}$$

Separating the transverse and the longitudinal parts of the GBV, we can write

$$\begin{pmatrix} \dot{\boldsymbol{a}} \\ \dot{\boldsymbol{b}} \end{pmatrix} = \begin{pmatrix} \hat{\omega} & \hat{\chi}_1 \\ \hat{\chi}_2 & \hat{\Gamma} \end{pmatrix} \begin{pmatrix} \boldsymbol{a} \\ \boldsymbol{b} \end{pmatrix} , \tag{17.5}$$

where

$$\begin{pmatrix} 0 & \hat{\chi}_1 (t) \\ \hat{\chi}_2 (t) & 0 \end{pmatrix} = \hat{\Lambda} (t) \,, \quad \begin{pmatrix} \hat{\omega} & 0 \\ 0 & \hat{\Gamma} \end{pmatrix} = \hat{E} \,. \tag{17.6}$$

This form of the equations coincides with the form of the true Bloch equations, which has already been examined. Therefore, all results derived from this representation of the true Bloch equations are correct in the more general case relating to the GBV. However, the explicit view of the matrices used in this case differs considerably. These matrices can be found if (17.5) is written in the concise form

$$\dot{\boldsymbol{a}} = \hat{\omega} \boldsymbol{a} + \hat{\chi}_1 \boldsymbol{b} \,, \qquad \dot{\boldsymbol{b}} = \hat{\chi}_2 \boldsymbol{a} + \hat{\Gamma} \boldsymbol{b} \,. \tag{17.7}$$

This set of equations has the form

$$\begin{pmatrix} \dot{a}_1 \\ \dot{a}_1^* \\ \dot{a}_2 \\ \dot{a}_2^* \\ \vdots \end{pmatrix} = \begin{pmatrix} \varepsilon_1 & 0 & 0 & 0 & \dots \\ 0 & \varepsilon_1^* & 0 & 0 & \dots \\ 0 & 0 & \varepsilon_2 & 0 & \dots \\ 0 & 0 & 0 & \varepsilon_2^* & \dots \\ . & . & . & . & \dots \\ . & . & . & . & \dots \end{pmatrix} \begin{pmatrix} a_1 \\ a_1^* \\ a_2 \\ a_2^* \\ \vdots \end{pmatrix} + \begin{pmatrix} \Lambda_1^{\mathrm{e}} & -\Lambda_1^{\mathrm{g}} & 0 \\ \Lambda_1^{*\mathrm{e}} & -\Lambda_1^{*\mathrm{g}} & 0 \\ \Lambda_2^{\mathrm{e}} & -\Lambda_2^{\mathrm{g}} & 0 \\ \Lambda_2^{*\mathrm{e}} & -\Lambda_2^{*\mathrm{g}} & 0 \\ . & . & \dots \\ . & . & \dots \end{pmatrix} \begin{pmatrix} b_1 \\ b_0 \\ b_2 \end{pmatrix} , \tag{17.8}$$

$$\begin{pmatrix} \dot{b}_1 \\ \dot{b}_0 \\ \dot{b}_2 \end{pmatrix} = \begin{pmatrix} -\Lambda_1^* & -\Lambda_1 & -\Lambda_2^* & -\Lambda_2 & \dots \\ \Lambda_1^* & \Lambda_1 & \Lambda_2^* & \Lambda_2 & \dots \\ 0 & 0 & 0 & 0 & \dots \end{pmatrix} \begin{pmatrix} a_1 \\ a_1^* \\ a_2 \\ a_2^* \\ \vdots \end{pmatrix} + \begin{pmatrix} -\Gamma & 0 & 0 \\ 1/T_1 & 0 & \gamma_{\mathrm{ST}} \\ \Gamma_{\mathrm{TS}} & 0 & -\gamma_{\mathrm{ST}}' \end{pmatrix} \begin{pmatrix} b_1 \\ b_0 \\ b_2 \end{pmatrix} . \tag{17.9}$$

It is known from linear algebra that a scalar product of two vectors can be expressed as a product of one-row and one-column matrices. This means that we have to consider row vectors and column vectors. The situation is similar to what happens in the Hilbert space of quantum mechanics, where the bra vector $\langle n|$ and ket vector $|n\rangle$ introduced by P.A.M. Dirac are row and column vectors, respectively. The scalar product of two vectors is given by $\langle n|n\rangle$.

The average dipolar moment of a molecule is given by

$$\begin{aligned} d(t) &= \mathrm{Tr}\left\{\hat{d}\hat{\rho}(t)\right\} = \sum_{a,b} \left[d_{ab}\rho_{ba}(t) + d_{ba}\rho_{ab}(t)\right] \\ &= \sum_n d_n \left(a_n + a_n^*\right) = \boldsymbol{d}^{\perp}\boldsymbol{a}(t) , \end{aligned} \tag{17.10}$$

where

$$\boldsymbol{d}^{\perp} = (d_1, d_1, d_2, d_2, \dots) , \quad d_n = d_{ab} = d\langle a|b\rangle = d_{ba} \tag{17.11}$$

is the row vector of the dipolar moment with the infinite set of electron–phonon matrix elements taken in the Condon approximation. Generally, the average value of any physical characteristic of the system is a scalar product of the row vector relating to this characteristic and the GBV given by the column vector.

17.2 Long-Lived Stimulated Photon Echo

The 3PE is frequently called the stimulated photon echo (SPE). The SPE signal can emerge with long time delay t_{w} after two laser pulses have been applied. We have already investigated the SPE on the basis of the Bloch equations. We will now examine the more general case when the system is described by the GBV.

A solution of (17.4) can be written in the form of (16.52). Replacing the operator $S(T)$ by the infinite series yields the following integral equation for the GBV:

$$\boldsymbol{\rho}(t) = \mathrm{e}^{\hat{E}t}\boldsymbol{\rho}(0) + \int_0^t \mathrm{d}t_1 \mathrm{e}^{\hat{E}(t-t_1)}\hat{\Lambda}(t_1)\boldsymbol{\rho}(t_1) , \tag{17.12}$$

where

$$\hat{\Lambda}(t) = \begin{pmatrix} 0 & \hat{\chi}_1(t) \\ \hat{\chi}_2(t) & 0 \end{pmatrix} , \tag{17.13}$$

and an explicit expression for the matrices χ_1 and χ_2 can be found from (17.8) and (17.9).

If the molecule is in the ground electronic state at $t = 0$, relaxation is impossible and the following equation holds:

$$\mathrm{e}^{\hat{E}t}\boldsymbol{\rho}(0) = \boldsymbol{\rho}(0) . \tag{17.14}$$

Solving the integral equation by an iteration procedure yields the following infinite series for the GBV:

$$\boldsymbol{\rho}(t) = \boldsymbol{\rho}(0) + \boldsymbol{\rho}_1(t) + \boldsymbol{\rho}_2(t) + \boldsymbol{\rho}_3(t) + \dots , \tag{17.15}$$

where

$$\begin{aligned}
\boldsymbol{\rho}_1(t) &= \int_0^t \mathrm{d}t_1 \mathrm{e}^{\hat{E}(t-t_1)}\hat{\Lambda}(t_1)\boldsymbol{\rho}(0) , \\
\boldsymbol{\rho}_2(t) &= \int_0^t \mathrm{d}t_1 \mathrm{e}^{\hat{E}(t-t_1)}\hat{\Lambda}(t_1) \int_0^{t_1} \mathrm{d}t_2 \mathrm{e}^{\hat{E}(t_1-t_2)}\hat{\Lambda}(t_2)\boldsymbol{\rho}(0) , \\
\boldsymbol{\rho}_3(t) &= \int_0^t \mathrm{d}t_1 \mathrm{e}^{\hat{E}(t-t_1)}\hat{\Lambda}(t_1) \int_0^{t_1} \mathrm{d}t_2 \mathrm{e}^{\hat{E}(t_1-t_2)}\hat{\Lambda}(t_2) \int_0^{t_2} \mathrm{d}t_3 \mathrm{e}^{\hat{E}(t_2-t_3)}\hat{\Lambda}(t_3)\boldsymbol{\rho}(0).
\end{aligned} \tag{17.16}$$

It is obvious that, if we know ρ_1, we can only calculate effects of first order in the interaction with light, such as the optical free induction decay. If we know ρ_2, we can calculate effects of second order in the interaction with light. Light absorption is such an effect, which is accompanied by changes in the population. The photon echo is an effect of third order in the coupling Λ with the electromagnetic field. It can thus be examined with the help of ρ_3. Hence, the photon echo is studied by examining the following expression for the induced dipolar moment:

$$d_3(t) = \mathrm{Tr}\left\{\hat{d}\hat{\rho}_3(t)\right\} = \boldsymbol{d} \cdot \boldsymbol{\rho}_3(t) = \boldsymbol{d}^{\perp}\boldsymbol{a}_3(t) \, . \tag{17.17}$$

The temporal dependence of the operators $\Lambda(t)$ in (17.16) is defined by the classical electromagnetic field. The temporal dependence of this field will be described further with the help of the delta function:

$$\boldsymbol{d} \cdot \boldsymbol{E}(\boldsymbol{r}, t) = \sum_{\boldsymbol{k}} \boldsymbol{d} \cdot \boldsymbol{E}_{\boldsymbol{k}} \mathrm{e}^{\mathrm{i}(\boldsymbol{k}\cdot\boldsymbol{r} - \omega_k t)} = \hbar\vartheta\delta\left(t - \frac{r}{c}\right) , \tag{17.18}$$

where $\boldsymbol{r}$ is a coordinate along the laser beam and the angle

$$\vartheta = \frac{1}{\hbar}\int_{-\infty}^{\infty} \boldsymbol{d} \cdot \boldsymbol{E}(\boldsymbol{r}, t)\, \mathrm{d}t \tag{17.19}$$

is proportional to the integrated value of the interaction with light. If three laser pulses are used as illustrated in Fig. 16.3, the chromophore–light interaction takes the form

$$\frac{\boldsymbol{d} \cdot \boldsymbol{E}(\boldsymbol{r}, t)}{\hbar} = \vartheta'\delta\left(t - t_{\mathrm{w}} - \tau - \frac{r}{c}\right) + \vartheta''\delta\left(t - \tau - \frac{r}{c}\right) + \vartheta\delta\left(t - \frac{r}{c}\right) . \tag{17.20}$$

Substituting this equation into (17.13) yields the following expression for the interaction operator:

$$\hat{\Lambda}(t) = \hat{\Lambda}'\delta\left(t - t_{\mathrm{w}} - \tau - \frac{r}{c}\right) + \hat{\Lambda}''\delta\left(t - \tau - \frac{r}{c}\right) + \hat{\Lambda}\delta\left(t - \frac{r}{c}\right) , \tag{17.21}$$

where the matrices in front of the delta functions do not depend on time. Therefore the expression for ρ_3 is given by

$$\begin{aligned} \boldsymbol{\rho}_3(t) &= \mathrm{e}^{\hat{E}\tau'}\hat{\Lambda}'\mathrm{e}^{\hat{E}t_{\mathrm{w}}}\hat{\Lambda}''\mathrm{e}^{\hat{E}\tau}\hat{\Lambda}\boldsymbol{\rho}(0) \\ &\times \int_0^t \mathrm{d}t_1\delta\left(t_1 - t_{\mathrm{w}} - \tau - \frac{r}{c}\right)\int_0^{t_1}\mathrm{d}t_2\delta\left(t_2 - \tau - \frac{r}{c}\right)\int_0^{t_2}\mathrm{d}t_3\delta\left(t_3 - \frac{r}{c}\right) , \end{aligned} \tag{17.22}$$

where $\tau' = t - t_{\mathrm{w}} - \tau - r/c$. The other $3^3 - 1$ products of the delta functions do not contribute to the integrals with respect to time. Carrying out the integrals, we find the expression

$$\boldsymbol{\rho}_3(t) = \theta(\tau')\,\mathrm{e}^{\hat{E}\tau'}\hat{\Lambda}'\mathrm{e}^{\hat{E}t_{\mathrm{w}}}\hat{\Lambda}''\mathrm{e}^{\hat{E}\tau}\hat{\Lambda}\boldsymbol{\rho}(0) , \tag{17.23}$$

where

$$\mathrm{e}^{\hat{E}t} = \begin{pmatrix} \mathrm{e}^{\hat{\omega}t} & 0 \\ 0 & \mathrm{e}^{\hat{\Gamma}t} \end{pmatrix} , \qquad \hat{\Lambda} = \begin{pmatrix} 0 & \hat{\chi}_1 \\ \hat{\chi}_2 & 0 \end{pmatrix} , \tag{17.24}$$

and $\theta(t)$ is a step function which equals zero for $t < 0$ and unity for $t > 0$. Equation (17.23) coincides with (16.77) derived earlier on the basis of the optical Bloch equations. However, the matrices should now be found from (17.8) and (17.9). It is obvious that the matrices χ_1 and χ_2 are described by the infinite three-column and three-row matrices, respectively. The matrix $\exp \omega t$ is of diagonal type

$$\mathrm{e}^{\hat{\omega}t} = \begin{pmatrix} \mathrm{e}^{\varepsilon_1 t} & 0 & 0 & 0 & \dots \\ 0 & \mathrm{e}^{\varepsilon_1^* t} & 0 & 0 & \dots \\ 0 & 0 & \mathrm{e}^{\varepsilon_2 t} & 0 & \dots \\ 0 & 0 & 0 & \mathrm{e}^{\varepsilon_2^* t} & \dots \\ . & . & . & . & \dots \\ . & . & . & . & \dots \end{pmatrix} , \tag{17.25}$$

and the other part $\exp \Gamma t$ of the relaxation matrix is defined by (16.89). Equation (17.23) can be written in the form

$$\begin{aligned} \boldsymbol{\rho}(t) &= \mathrm{e}^{\hat{E}\tau'} \hat{\Lambda}' \mathrm{e}^{\hat{E}t_{\mathrm{w}}} \hat{\Lambda}'' \mathrm{e}^{\hat{E}\tau} \hat{\Lambda} \boldsymbol{\rho}(0) \\ &= \begin{pmatrix} \mathrm{e}^{\hat{\omega}\tau'} & 0 \\ 0 & \mathrm{e}^{\hat{\Gamma}\tau'} \end{pmatrix} \begin{pmatrix} 0 & \hat{\chi}_1' \\ \hat{\chi}_2' & 0 \end{pmatrix} \begin{pmatrix} \mathrm{e}^{\hat{\omega}t_{\mathrm{w}}} & 0 \\ 0 & \mathrm{e}^{\hat{\Gamma}t_{\mathrm{w}}} \end{pmatrix} \begin{pmatrix} 0 & \hat{\chi}_1'' \\ \hat{\chi}_2'' & 0 \end{pmatrix} \\ &\quad \times \begin{pmatrix} \mathrm{e}^{\hat{\omega}\tau} & 0 \\ 0 & \mathrm{e}^{\hat{\Gamma}\tau} \end{pmatrix} \begin{pmatrix} 0 & \hat{\chi}_1 \\ \hat{\chi}_2 & 0 \end{pmatrix} \begin{pmatrix} 0 \\ \boldsymbol{b}(0) \end{pmatrix} . \end{aligned} \tag{17.26}$$

Multiplying out the matrices in this equation we arrive at the following expression for the induced dipolar moment:

$$d(t) = \boldsymbol{d}^{\perp} \boldsymbol{a}(t) = \boldsymbol{d}^{\perp} \mathrm{e}^{\hat{\omega}\tau'} \hat{\chi}_1' \mathrm{e}^{\hat{\Gamma}t_{\mathrm{w}}} \hat{\chi}_2'' \mathrm{e}^{\hat{\omega}\tau} \hat{\chi}_1 \boldsymbol{b}(0) . \tag{17.27}$$

The temporal evolution of the GBV is similar to the evolution of the Bloch vector. It is illustrated in Fig. 16.4. Although the last formula is similar to (16.78), the multiplication of the matrices is more complicated. It is thus convenient to analyze the temporal evolution into four stages.

Stage One. Referring to the diagram in Fig. 16.4, after two laser pulses have passed through the sample the longitudinal vector takes the form

$$\begin{aligned} \boldsymbol{b}(\tau) = \begin{pmatrix} \rho_1 \\ \rho_0 \\ \rho_2 \end{pmatrix} &= \begin{pmatrix} -\Lambda_1^{*''} & -\Lambda_1'' & -\Lambda_2^{*''} & -\Lambda_2'' & \dots \\ \Lambda_1^{*''} & \Lambda_1'' & \Lambda_2^{*''} & \Lambda_2'' & \dots \\ 0 & 0 & 0 & 0 & \dots \end{pmatrix} \\ &\quad \times \begin{pmatrix} \mathrm{e}^{\varepsilon_1 \tau} & 0 & 0 & 0 & \dots \\ 0 & \mathrm{e}^{\varepsilon_1^* \tau} & 0 & 0 & \dots \\ 0 & 0 & \mathrm{e}^{\varepsilon_2 \tau} & 0 & \dots \\ 0 & 0 & 0 & \mathrm{e}^{\varepsilon_2^* \tau} & \dots \\ . & . & . & . & \dots \\ . & . & . & . & \dots \end{pmatrix} \begin{pmatrix} \Lambda_1^{\mathrm{e}} & -\Lambda_1^{\mathrm{g}} & 0 \\ \Lambda_1^{*\mathrm{e}} & -\Lambda_1^{*\mathrm{g}} & 0 \\ \Lambda_2^{\mathrm{e}} & -\Lambda_2^{\mathrm{g}} & 0 \\ \Lambda_2^{*\mathrm{e}} & -\Lambda_2^{*\mathrm{g}} & 0 \\ . & . & . \\ . & . & . \end{pmatrix} \begin{pmatrix} 0 \\ 1 \\ 0 \end{pmatrix} . \end{aligned} \tag{17.28}$$

Multiplying the matrices yields

$$\boldsymbol{b}(\tau) = 2\vartheta''\vartheta \begin{pmatrix} 1 \\ -1 \\ 0 \end{pmatrix} I^{\mathrm{g}}(\Omega, \tau) \ , \tag{17.29}$$

where

$$I^{\mathrm{g}}(\Omega, \tau)\,\vartheta''\vartheta = \mathrm{Re}\left(\sum_n \Lambda_n^{*''} \mathrm{e}^{\varepsilon_n \tau} \Lambda_n^{\mathrm{g}}\right) = \vartheta''\vartheta\,\mathrm{Re}\left(\mathrm{e}^{-\mathrm{i}\Omega\tau} S^{\mathrm{g}}(\tau)\right) \ . \tag{17.30}$$

Equations (17.1) and (15.20) have been allowed for here. The formula for I^{g} almost coincides with (15.20) for the induced dipolar moment. The equation describes fast optical dephasing. The formula for the dipolar moment and hence also (17.30) can be transformed to (15.24), which is related to the absorption band. See, for example, (15.28) for the absorption band.

Stage Two. The longitudinal vector $b(\tau)$ will undergo temporal evolution as described by the matrix $\exp(\Gamma t_{\mathrm{w}})$. The elements of this matrix are defined by (16.89). After this temporal evolution, the longitudinal vector takes the form

$$\boldsymbol{b}(t_{\mathrm{w}} + \tau) = \mathrm{e}^{\hat{\Gamma} t_{\mathrm{w}}} \boldsymbol{b}(\tau) = 2\vartheta''\vartheta \begin{pmatrix} \left(\mathrm{e}^{\hat{\Gamma} t_{\mathrm{w}}}\right)_{11} \\ \left(\mathrm{e}^{\hat{\Gamma} t_{\mathrm{w}}}\right)_{01} - 1 \\ \left(\mathrm{e}^{\hat{\Gamma} t_{\mathrm{w}}}\right)_{21} \end{pmatrix} I^{\mathrm{g}}(\Omega, \tau) \ . \tag{17.31}$$

Stage Three. Let us now inspect the product of the first, second and third factors in (17.27):

$$\boldsymbol{d}\mathrm{e}^{\hat{\omega}\tau'} \hat{\chi}_1' = \left(d_1\ d_1\ d_2\ d_2\ \ldots\right) \times \begin{pmatrix} \mathrm{e}^{\varepsilon_1 \tau} & 0 & 0 & 0 & \ldots \\ 0 & \mathrm{e}^{\varepsilon_1^* \tau} & 0 & 0 & \ldots \\ 0 & 0 & \mathrm{e}^{\varepsilon_2 \tau} & 0 & \ldots \\ 0 & 0 & 0 & \mathrm{e}^{\varepsilon_2^* \tau} & \ldots \\ \cdot & \cdot & \cdot & \cdot & \ldots \\ \cdot & \cdot & \cdot & \cdot & \ldots \end{pmatrix} \begin{pmatrix} \Lambda_1^{\mathrm{e}} & -\Lambda_1^{\mathrm{g}} & 0 \\ \Lambda_1^{*\mathrm{e}} & -\Lambda_1^{*\mathrm{g}} & 0 \\ \Lambda_2^{\mathrm{e}} & -\Lambda_2^{\mathrm{g}} & 0 \\ \Lambda_2^{*\mathrm{e}} & -\Lambda_2^{*\mathrm{g}} & 0 \\ \cdot & \cdot & \cdot \\ \cdot & \cdot & \cdot \end{pmatrix} . \tag{17.32}$$

Multiplying the matrices, we find

$$\begin{aligned} \boldsymbol{d}\mathrm{e}^{\hat{\omega}\tau'} \hat{\chi}_1' &= 2\left(\mathrm{Re}\sum_n d_n \mathrm{e}^{\varepsilon_n \tau'} \Lambda_n^{\mathrm{e}},\ -\mathrm{Re}\sum_n d_n \mathrm{e}^{\varepsilon_n \tau'} \Lambda_n^{\mathrm{g}},\ 0\right) \\ &= 2d\vartheta'\left(I^{\mathrm{e}}(\Omega, \tau'),\ -I^{\mathrm{g}}(\Omega, \tau'),\ 0\right) \ , \end{aligned} \tag{17.33}$$

where the function of frequency and temperature is given by

$$\begin{aligned} I^{\mathrm{e}}(\Omega, \tau') &= \mathrm{Re}\left[\mathrm{e}^{-\mathrm{i}\Omega\tau'} \sum_{a,b} p_b \langle a|b\rangle^2 \mathrm{e}^{-\mathrm{i}(\Omega_b - \Omega_a - \mathrm{i}/2T_1)\tau'}\right] \\ &= \mathrm{Re}\left[\mathrm{e}^{-\mathrm{i}\Omega\tau'} S^{\mathrm{e}}(\tau')\right] \ . \end{aligned} \tag{17.34}$$

The function I^{e} is related to the fluorescence band.

Stage Four. Multiplying the row vector in (17.33) by the column vector in (17.31), we find the following expression for the dipolar moment:

$$d(t) = 4d\vartheta'\vartheta''\vartheta\left\{(\mathrm{e}^{\hat{\Gamma}t_\mathrm{w}})_{11} I^\mathrm{e}(\Omega,\tau') + \left[1 - (\mathrm{e}^{\hat{\Gamma}t_\mathrm{w}})_{01}\right] I^\mathrm{g}(\Omega,\tau')\right\} I^\mathrm{g}(\Omega,\tau)\,. \tag{17.35}$$

Substituting in the elements of the relaxation matrix, we arrive at the final expression for the dipolar moment induced by three laser pulses:

$$\begin{aligned} d(t) = 4d\vartheta'\vartheta''\vartheta\Big\{&\mathrm{e}^{-\Gamma t_\mathrm{w}} I^\mathrm{e}(\Omega,\tau') \\ &+ \left[\frac{1}{T_1\Gamma}\mathrm{e}^{-\Gamma t_\mathrm{w}} + \frac{\Gamma_\mathrm{TS}}{\Gamma}(1-\eta_Q)\,\mathrm{e}^{-\gamma'_\mathrm{ST}t_\mathrm{w}} + \frac{\Gamma_\mathrm{TS}}{\Gamma}\eta_Q\right] I^\mathrm{g}(\Omega,\tau')\Big\} I^\mathrm{g}(\Omega,\tau)\,. \end{aligned} \tag{17.36}$$

This equation is similar to (16.97), derived with the help of the optical Bloch equations. However, the optical Bloch equations do not take into account the existence of the PSB in the optical band of a molecule. Since the PSBs in the absorption and fluorescence band extend in different directions from the ZPL, the functions I^g and I^e in (17.36) relating to these bands are different as well.

In order to find the echo amplitude, we have to sum over all dipolar moments of the sample:

$$D(t) = \int_0^\infty N(\Omega)\, d(t)\mathrm{d}\Omega \propto E_\mathrm{3PE}(\tau, t_\mathrm{w}, \tau)\, N(\tau-\tau')\,. \tag{17.37}$$

Using the fact that

$$\begin{aligned} &\int_0^\infty N(\Omega)\, I^\mathrm{e}(\Omega,\tau')\, I^\mathrm{g}(\Omega,\tau)\,\mathrm{d}\Omega \\ &\qquad \approx \frac{N(\tau'-\tau)}{2}\left[\mathrm{Re}\, S^\mathrm{e}(\tau')\,\mathrm{Re}\, S^\mathrm{g}(\tau) + \mathrm{Im}\, S^\mathrm{e}(\tau')\,\mathrm{Im}\, S^\mathrm{g}(\tau)\right] \\ &\qquad = \frac{N(\tau'-\tau)}{2}\,\mathrm{Re}\left[S^\mathrm{g}(\tau)\, S^{*\mathrm{e}}(\tau')\right]\,, \end{aligned} \tag{17.38}$$

where $N(t)$ is the echo signal shape function, we arrive at

$$D(t) = E_\mathrm{3PE}(\tau, t_\mathrm{w}, \tau) N(\tau-\tau')\,, \tag{17.39}$$

where

$$\begin{aligned} E_\mathrm{3PE} = 2d\vartheta'\vartheta''\vartheta\Big\{&\mathrm{e}^{-\Gamma t_\mathrm{w}}\,\mathrm{Re}\left[S^\mathrm{g}(\tau)S^{*\mathrm{e}}(\tau)\right] \\ &+ \left[\frac{1}{T_1\Gamma}\mathrm{e}^{-\Gamma t_\mathrm{w}} + \frac{\Gamma_\mathrm{TS}}{\Gamma}(1-\eta_Q)\mathrm{e}^{-\gamma'_\mathrm{ST}t_\mathrm{w}} + \frac{\Gamma_{TS}}{\Gamma}\eta_Q\right]\mathrm{Re}\left[S^\mathrm{g}(\tau)S^{*\mathrm{g}}(\tau)\right]\Big\} \end{aligned} \tag{17.40}$$

describes the amplitude of the SPE. This amplitude is a function of the time interval τ between the first and second laser pulses. This function allows us to investigate optical dephasing resulting from the full electron–phonon interaction. Therefore, this dephasing includes both fast decay at short τ and ordinary exponential decay at long τ.

17.3 Picosecond SPE

Equation (17.40) for the SPE amplitude describes both fast dephasing and slower exponential dephasing. In Sect. 15.2 we investigated fast optical dephasing which manifests itself in optical free induction decay. It was shown that the function $S^{\mathrm{g}}(t)$ allowing for both quadratic and linear FC interaction is given by

$$S^{\mathrm{g}}(t) = \mathrm{e}^{-t/T_2+\varphi(t,T)-\varphi(0,T)} \; . \tag{17.41}$$

As a rule the influence of the quadratic FC interaction on the dephasing function $\varphi(t, T)$, which defines the PSB in the optical band, is negligible. Then we have

$$S^{\mathrm{e}}(t) = S^{*\mathrm{g}}(t) \; . \tag{17.42}$$

By inserting the last two formulas into (17.40), we arrive at the following expression for the echo amplitude:

$$E_{\mathrm{3PE}}(\tau, t_{\mathrm{w}}) \propto \left\{\mathrm{e}^{-\Gamma t_{\mathrm{w}}} \cos\left[2\,\mathrm{Im}\,\varphi(\tau)\right] + A\left(t_{\mathrm{w}}\right)\right\} \mathrm{e}^{2\,\mathrm{Re}\,\varphi(\tau,T)-2\varphi(0,T)-2\tau/T_2} \; , \tag{17.43}$$

where the function $A(t_{\mathrm{w}})$ is defined as the sum of the three terms in square brackets in (17.40).

Let us examine the last expression for the case when the PSB emerges due to creation of acoustic phonons. For the Debye phonon model, the function $\varphi(t, T)$ is nonvanishing only on the picosecond time scale. At strong electron–phonon coupling, the ZPL is small and therefore the echo signal approaches zero on the picosecond time scale. This can be seen from Fig. 17.1. It should

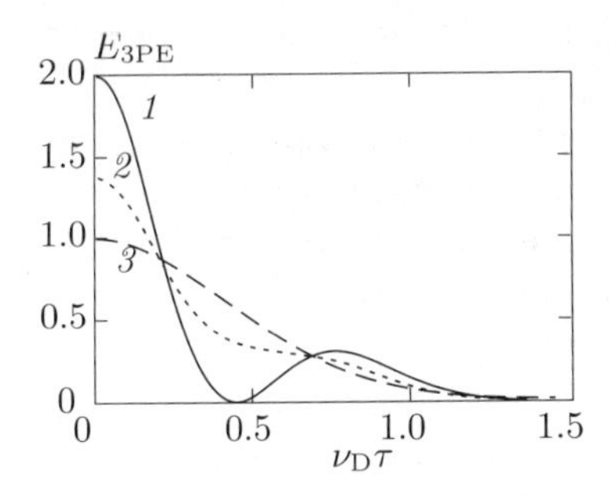

Fig. 17.1. Dependence of stimulated photon echo decay at strong FC interaction [$\varphi(0, 0) = 5$] on waiting time. $\Gamma t_{\mathrm{w}} = 0.1$ (curve 1), 1 (curve 2) and 10 (curve 3)

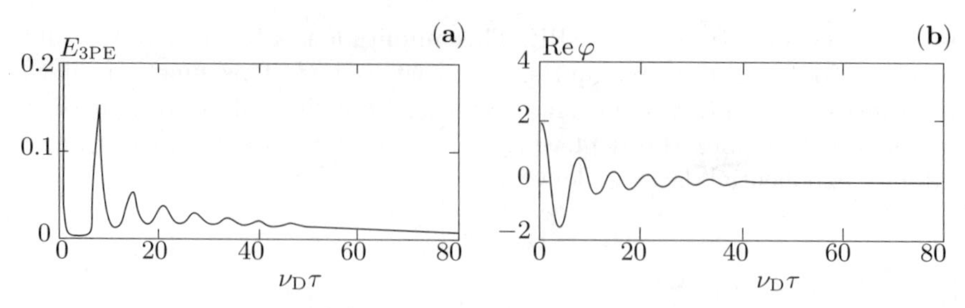

Fig. 17.2. Comparison of 3PE decay (**a**) and temporal behavior of the dephasing function (**b**) at $\Gamma t_w = 0.1$ and $\varphi(0,0) = 2$

be noted that the echo decay is of nonexponential type and that the rate of decay depends on the waiting time t_w. The dependence on t_w takes place at short waiting times up to nanoseconds when $\Gamma t_w \ll 1$. The populations of excited and ground electronic states both contribute to the echo signal if $t_w \ll 1/\Gamma$. On long time scales, when $t_w \gg 1/\Gamma$, the excited electronic state decays and the first term in (17.43) disappears. Therefore the frequency grating burnt only in the ground electronic state contributes to the echo signal. Being described by a function of nonexponential type, this signal will not depend on the waiting time t_w. This is shown by curve 3 in Fig. 17.1.

Let us now consider the case of weak electron–phonon coupling. Then the optical band consists of a ZPL and PSB with comparable intensities. The echo signal decay consists of fast oscillations on the picosecond time scale, resulting from the PSB, and slower exponential decay resulting from the ZPL. This exponential decay is not seen in Fig. 17.2a. The fast oscillations in the echo signal are due to oscillations in the dephasing function Re$\varphi(t,T)$, as shown in Fig. 17.2. The period of the oscillations equals $2\pi/\nu_D = 0.3$ ps. On a longer time scale, i.e., at $\tau > 30/\nu_D$, optical dephasing is of exponential type and is described by the function

$$E_{3PE}(\tau, t_w) \propto \left[\mathrm{e}^{-\Gamma t_w} + A(t_w)\right] \mathrm{e}^{-2\varphi(0,T)-2\tau/T_2} \,. \tag{17.44}$$

Exponential dephasing does not depend on the waiting time t_w.

17.4 Two-Pulse Femtosecond Photon Echo

Fast relaxation on picosecond and femtosecond time scales results from an electron–phonon part of the optical band. Therefore such fast relaxation can be taken into account only if we use the infinite set of equations for the density matrix, like (16.30). This set of equations was derived from the more general (16.26) with the help of the approximation

$$\rho_{aa'} = \delta_{aa'}\rho_{aa} \,, \qquad \rho_{bb'} = \delta_{bb'}\rho_{bb'} \,. \tag{17.45}$$

This approximation works well if $t_w \gg \tau$ and the photon echo signal is calculated to the first nonvanishing approximation in the interaction Λ with light. We have already seen that formulas for the 2PE can be derived from formulas for the 3PE in the limit when $t_w \to 0$. This means that we cannot use the approximation described by (17.45) when considering the 2PE. Therefore we will now use the exact (16.26).

The set of equations (16.26) can be written in the form

$$\dot{\boldsymbol{a}} = \hat{\omega}\boldsymbol{a} + \hat{\chi}_1\boldsymbol{b}\,, \qquad \dot{\boldsymbol{b}} = \hat{\chi}_2\boldsymbol{a} + \hat{\Gamma}\boldsymbol{b}\,. \tag{17.46}$$

The terms with laser pumping in these equations are given by

$$\hat{\chi}_1\boldsymbol{b}\,(\tau) = \boldsymbol{a}'\,(\tau) = \begin{pmatrix} \sum\limits_{b'} \rho_{bb'}\,(\tau)\,\chi_{b'a} - \sum\limits_{a'} \chi_{ba'}\rho_{a'a}\,(\tau) \\ \sum\limits_{b'} \chi^*_{ab'}\rho_{b'b}\,(\tau) - \sum\limits_{a'} \rho_{aa'}\,(\tau)\,\chi^*_{a'b} \end{pmatrix} = \begin{pmatrix} \rho_{ba}\,(\tau) \\ \rho_{ab}\,(\tau) \end{pmatrix}, \tag{17.47}$$

$$\hat{\chi}_2\boldsymbol{a}\,(\tau) = \boldsymbol{b}\,(\tau) = \begin{pmatrix} -\sum\limits_{a} \left(\rho_{ba}\,(\tau)\,\chi^*_{ab'} + \chi_{ba}\rho_{ab'}\,(\tau)\right) \\ \sum\limits_{b} \left(\chi^*_{ab}\rho_{ba'}\,(\tau) + \rho_{ab}\,(\tau)\,\chi_{ba'}\right) \end{pmatrix} = \begin{pmatrix} \rho_{bb'}\,(\tau) \\ \rho_{aa'}\,(\tau) \end{pmatrix}. \tag{17.48}$$

The influence of relaxation is given by

$$\mathrm{e}^{\hat{\omega}\tau}\boldsymbol{a} = \begin{pmatrix} \mathrm{e}^{\varepsilon_{ba}\tau}\rho_{ba} \\ \mathrm{e}^{\varepsilon_{ab}\tau}\rho_{ab} \end{pmatrix}, \tag{17.49}$$

where

$$\varepsilon_{ba} = -\mathrm{i}\left(\Omega_{ba} - \mathrm{i}/2T_1\right)\,, \qquad \varepsilon_{ab} = -\mathrm{i}\left(\Omega_{ab} - \mathrm{i}/2T_1\right)\,. \tag{17.50}$$

The expression for the induced dipolar moment which creates the 2PE can be derived from (17.27) for the 3PE if we set $t_w = 0$. Then we arrive at the expression

$$d(t) = \boldsymbol{d}^{\perp}\boldsymbol{a}\,(t) = \boldsymbol{d}^{\perp}\mathrm{e}^{\hat{\omega}\tau'}\hat{\chi}_1\hat{\chi}_2\mathrm{e}^{\hat{\omega}\tau}\hat{\chi}_1\boldsymbol{b}(0)\,. \tag{17.51}$$

This equation includes all terms defined by (17.47), (17.48) and (17.49). They allow us to multiply all matrices and calculate the generalized Bloch vector in (17.51).

Using the fact that

$$\boldsymbol{b}\,(0) = \begin{pmatrix} 0 \\ \rho_{aa} \end{pmatrix} = \begin{pmatrix} 0 \\ p_a \end{pmatrix}, \tag{17.52}$$

together with (17.47) and (17.49), we find the following expression for the GBV after excitation by the first laser pulse and after relaxation during the first pause:

$$e^{\hat{\omega}\tau}\hat{\chi}_1 \boldsymbol{b}(0) = \boldsymbol{a}(\tau) = \begin{pmatrix} -e^{\varepsilon_{ba}\tau}\chi_{ba}p_a \\ -p_a\chi^*_{ab}e^{\varepsilon_{ab}} \end{pmatrix} = \begin{pmatrix} \rho_{ba}(\tau) \\ \rho_{ab}(\tau) \end{pmatrix} . \tag{17.53}$$

Equation (17.51) for the dipolar moment can be written

$$d(t) = \boldsymbol{d}^{\perp} e^{\hat{\omega}\tau'} \hat{\chi}_1 \hat{\chi}_2 \boldsymbol{a}(\tau) . \tag{17.54}$$

Using (17.47), (17.48) and (17.49), the transverse part of the GBV after excitation by two laser pulses and relaxation is given by

$$\begin{aligned} \boldsymbol{a}(\tau' + \tau) &= \begin{pmatrix} \rho_{ba}(\tau' + \tau) \\ \rho_{ab}(\tau' + \tau) \end{pmatrix} = e^{\hat{\omega}\tau'}\hat{\chi}_1\hat{\chi}_2\boldsymbol{a}(\tau) = e^{\hat{\omega}\tau'}\hat{\chi}_1\boldsymbol{b}(\tau) \\ &= \begin{pmatrix} e^{\varepsilon_{ba}\tau'}\left[\sum\limits_{b'}\rho_{bb'}(\tau)\chi_{b'a} - \sum\limits_{a'}\chi_{ba'}\rho_{a'a}(\tau)\right] \\ e^{\varepsilon_{ab}\tau'}\left[\sum\limits_{b'}\chi^*_{ab'}\rho_{b'b}(\tau) - \sum\limits_{a'}\rho_{aa'}(\tau)\chi^*_{a'b}\right] \end{pmatrix} . \end{aligned} \tag{17.55}$$

Since $\rho_{ab} = \rho^*_{ba}$, we find that

$$d(t) = \sum_{ab}\left[d_{ab}\rho_{ba}(\tau' + \tau) + d_{ba}\rho_{ab}(\tau' + \tau)\right] = 2\,\mathrm{Re}\sum_{ab} d_{ab}\rho_{ba}(\tau' + \tau) . \tag{17.56}$$

Inserting the expression for $\rho_{ba}(\tau+\tau')$ and taking into account (17.50) for the complex energies, we arrive at the following formula for the dipolar moment:

$$\begin{aligned} d(t) &= e^{-\tau'/2T_1} \\ &\times 2\,\mathrm{Re}\left\{e^{-i\Omega\tau'}\sum_{ab} d_{ab}e^{-i\Omega_{ba}\tau'}\left[\sum_{b'}\rho_{bb'}(\tau)\chi_{b'a} - \sum_{a'}\chi_{ba'}\rho_{a'a}(\tau)\right]\right\} . \end{aligned} \tag{17.57}$$

Here the elements $\rho_{bb'}(\tau)$ and $\rho_{a'a}(\tau)$ are defined by (17.48), in which the elements $\rho_{ab}(\tau)$ and $\rho_{ba}(\tau)$ defined by (17.47) should by inserted. After such substitutions, we arrive at the following expressions for the matrix elements $\rho_{bb'}(\tau)$ and $\rho_{a'a}(\tau)$:

$$\begin{aligned} \rho_{bb'}(\tau) = e^{-\tau/(2T_1)}\Big\{ & e^{-i\Omega\tau}\sum_{a''}\left[e^{-i\Omega_{ba''}\tau}\chi_{ba''}p_{a''}\chi^*_{a''b'}\right] \\ & + e^{i\Omega\tau}\sum_{a''}\left[\chi_{ba''}p_{a''}\chi^*_{a''b'}e^{-i\Omega_{a''b'}\tau}\right]\Big\} , \end{aligned} \tag{17.58}$$

$$\begin{aligned} \rho_{a'a}(\tau) = -e^{-\tau/(2T_1)}\Big\{ & e^{-i\Omega\tau}\sum_{b''}\left[\chi^*_{a'b''}e^{-i\Omega_{b''a}\tau}\chi_{b''a}p_a\right] \\ & + e^{i\Omega\tau}\sum_{b''}\left[p_{a'}\chi^*_{a'b''}e^{-i\Omega_{a'b''}\tau}\chi_{b''a}\right]\Big\} . \end{aligned} \tag{17.59}$$

We are now ready to calculate the dipolar moment. By inserting (17.58) and (17.59) into (17.57), we find four terms. Summing pairs of terms with similar exponential factors, we can write (17.57) as follows:

$$d(t) = \mathrm{Re}\left[A\left(\tau', \tau\right) \mathrm{e}^{-\mathrm{i}\Omega(\tau'+\tau)} + B\left(\tau', \tau\right) \mathrm{e}^{-\mathrm{i}\Omega(\tau'-\tau)}\right] . \tag{17.60}$$

The two terms included in $A(\tau', \tau)$ result from the first terms in braces in (17.58) and (17.59). They describe the third order approximation to optical free induction decay. The two terms included in $B(\tau' , \tau)$ result from the second terms in braces in (17.58) and (17.59). They determine the echo signal. Since we are interested in the echo signal, we may set $A(\tau', \tau) = 0$ and then (17.57) for the dipolar moment takes the form

$$\begin{aligned} d_{\mathrm{2PE}}(t) &= 4\mathrm{e}^{-(\tau'+\tau)/(2T_1)} \\ &\times \mathrm{Re}\left\{\mathrm{e}^{-\mathrm{i}\Omega(\tau'-\tau)} \sum_{ab}\sum_{b'a''} d_{ab}\mathrm{e}^{-\mathrm{i}\Omega_{ba}\tau'} \chi_{ba''} p_{a''} \chi^*_{a''b'} \mathrm{e}^{-\mathrm{i}\Omega_{a''b'}\tau} \chi_{b'a}\right\} . \end{aligned} \tag{17.61}$$

Using the formulas $d_{ab} = d\langle a|b\rangle$, $\chi_{ab} = \vartheta\langle a|b\rangle$ and $\Omega_{ab} = \Omega_a - \Omega_b$, we can transform the equation for the dipolar moment to

$$\begin{aligned} d_{\mathrm{2PE}}(t) &= 4d\vartheta^3\mathrm{e}^{-(\tau'+\tau)/(2T_1)} \\ &\times \mathrm{Re}\left[\mathrm{e}^{-\mathrm{i}\Omega(\tau'-\tau)} \sum_{abb'a''} \langle a|b\rangle \mathrm{e}^{-\mathrm{i}\Omega_b\tau'} \langle b|a''\rangle \mathrm{e}^{\mathrm{i}\Omega_{a''}\tau'} p_{a''}\right. \\ &\left. \times \langle a''|b'\rangle \mathrm{e}^{\mathrm{i}\Omega_{b'}\tau} \langle b'|a\rangle \mathrm{e}^{-\mathrm{i}\Omega_a\tau}\right] . \end{aligned} \tag{17.62}$$

The phonon–tunnelon frequencies Ω_a and Ω_b are eigenvalues of the adiabatic Hamiltonians H^{g} and H^{e}. Substituting the Hamiltonians for the frequencies, we can transform the last equation to

$$\begin{aligned} d_{\mathrm{2PE}}(t) &= 4d\vartheta^3\mathrm{e}^{-(\tau'+\tau)/(2T_1)} \\ &\times \mathrm{Re}\left[\mathrm{e}^{-\mathrm{i}\Omega(\tau'-\tau)} \left\langle \mathrm{e}^{-\mathrm{i}\tau H^{\mathrm{g}}/\hbar} \mathrm{e}^{\mathrm{i}\tau H^{\mathrm{e}}/\hbar} \mathrm{e}^{\mathrm{i}\tau' H^{\mathrm{g}}/\hbar} \mathrm{e}^{-\mathrm{i}\tau' H^{\mathrm{e}}/\hbar} \right\rangle_{\mathrm{g}}\right] \\ &= 4d\vartheta^3\mathrm{e}^{-(\tau'+\tau)/(2T_1)}\, \mathrm{Re}\left[\mathrm{e}^{-\mathrm{i}\Omega(\tau'-\tau)} \left\langle \hat{S}\left(-\tau\right) \hat{S}(\tau') \right\rangle_{\mathrm{g}}\right] , \end{aligned} \tag{17.63}$$

where angle brackets denote the quantum statistical average.

Let us compare this equation with (17.36) which describes the induced dipolar moment creating the 3PE signal. The latter is expressed as the product of two averages $\langle S(\tau)\rangle$, whereas the expression for d_{2PE} is proportional to $\langle S(-\tau)S(\tau)\rangle$. Therefore the expression for d_{3PE} described by (17.36) is not converted to (17.63) for d_{2PE} as $t_w \to 0$. At first sight this result seems strange. However, it can be explained as follows. The equations for d_{3PE} and

d_{2PE} were derived using the approximate (16.30) and exact (16.26), respectively. It has already been mentioned that the approximate (16.30) can only be used to calculate the 3PE if $t_w > \tau$.

Fast optical dephasing is due to the PSB. However, the linear FC interaction yields the main contribution to the PSB. We shall thus calculate the quantum statistical average of the four exponential operators taking into account only the linear FC interaction. The calculation can be done using the technique of Sect. 10.3.

Let us consider the quantum statistical average in (17.63). Using (10.34), we can transform the adiabatic Hamiltonian $H^e(R) = H^g(R-a)$ to the adiabatic Hamiltonian $H^g(R)$ of the ground electronic state. Then the average in (17.63) takes the form

$$\begin{aligned}
\left\langle e^{-i\frac{H^g}{\hbar}\tau} e^{i\frac{H^e}{\hbar}\tau} e^{i\frac{H^g}{\hbar}\tau'} e^{-i\frac{H^e}{\hbar}\tau'} \right\rangle &= \left\langle e^{-\hat{L}} e^{i\frac{H^g}{\hbar}\tau} e^{\hat{L}} e^{i\frac{H^g}{\hbar}\tau'} e^{-\hat{L}} e^{-i\frac{H^g}{\hbar}\tau'} e^{\hat{L}} e^{-i\frac{H^g}{\hbar}\tau} \right\rangle \\
&= \left\langle e^{-\hat{L}} e^{i\frac{H^g}{\hbar}\tau} e^{\hat{L}} e^{-i\frac{H^g}{\hbar}\tau} e^{i\frac{H^g}{\hbar}(\tau+\tau')} e^{-\hat{L}} e^{-i\frac{H^g}{\hbar}(\tau+\tau')} e^{i\frac{H^g}{\hbar}\tau} e^{\hat{L}} e^{-i\frac{H^g}{\hbar}\tau} \right\rangle \\
&= \left\langle e^{-\hat{L}(0)} e^{\hat{L}(\tau)} e^{-\hat{L}(\tau+\tau')} e^{\hat{L}(\tau)} \right\rangle . \qquad (17.64)
\end{aligned}$$

The operator $\exp L(t)$ is defined by (10.30) for the case of a single phonon mode. The method of calculating the average of a product of exponential operators can be found in Appendix L. Equation (L.10) gives the result of the calculation for (17.64). By inserting this result into (17.63), we arrive at the following expression for the dipolar moment:

$$\begin{aligned}
d_{2PE}(t) &= 4d\vartheta^3 e^{-(\tau'+\tau)/2T_1} \\
&\quad \times \mathrm{Re}\left\{\exp\left[i\Omega\left(\tau'-\tau\right)\right]\exp\left[2g^*\left(\tau\right) + 2\,\mathrm{Re}\,g\left(\tau'\right) - g^*\left(\tau'+\tau\right)\right]\right\} , \qquad (17.65)
\end{aligned}$$

where the function

$$g(t) = \varphi(t,T) - \varphi(0,T) , \qquad (17.66)$$

is the difference between the dephasing function and the Huang–Rhys factor which defines the strength of the linear FC interaction. The induced dipolar moment of the sample is given by

$$D(t) = \int_0^\infty N(\Omega) d_{2PE}(t) d\Omega = E_{2PE}(\tau) N\left(\tau'-\tau\right) , \qquad (17.67)$$

where

$$\begin{aligned}
E_{2PE}(\tau) &= 4d\vartheta^3 e^{-\tau/T_1}\,\mathrm{Re}\exp\left[2g^*(\tau) + 2\,\mathrm{Re}\,g\left(\tau\right) - g^*\left(2\tau\right)\right] \qquad (17.68) \\
&= 4d\vartheta^3 e^{-\tau/T_1} e^{4\,\mathrm{Re}\,g(\tau) - \mathrm{Re}\,g(2\tau)} \cos\left[2\,\mathrm{Im}\,g(\tau) - \mathrm{Im}\,g(2\tau)\right]
\end{aligned}$$

is the 2PE amplitude.

Using (15.25) for the dephasing function together with (17.68), we can calculate the dependence of the 2PE amplitude on the delay τ. The result is shown in Fig. 17.3. The curves presented here and in Figs. 17.1 and 17.2 are calculated for the same set of parameters. Therefore comparison of these curves shows differences in the decay of the 2PE and 3PE.

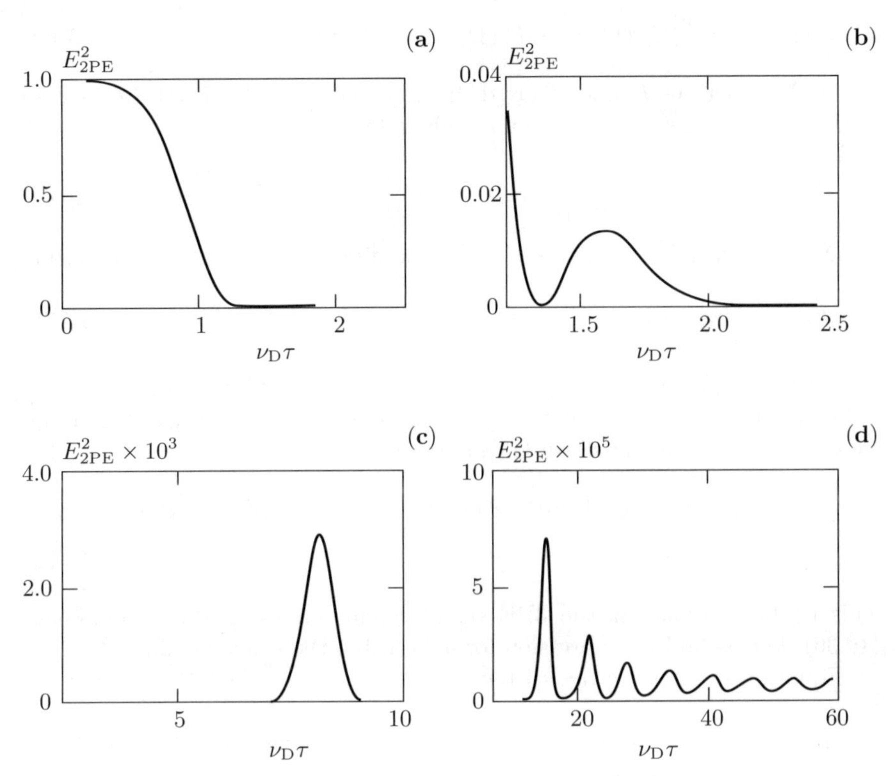

Fig. 17.3. (a–d) 2PE decay on various time scales at $\varphi(0,0) = 2$

Both types of echo signal drop off to the value of $\exp(-2\varphi(0))$ and $\exp(-3\varphi(0))$ for the 3PE and 2PE, respectively. This sharp temporal behavior is called a spike. The spike is shown in Figs. 17.2 and 17.3a. It lasts about 50 fs. Then the echo signal is absent. This pause lasts about 250 fs in the 3PE. After this pause, the oscillating echo signal emerges. It repeats the temporal behavior of the dephasing function $\varphi(\tau)$. In the case of the 2PE, the isolated peak emerges after the spike has vanished and then the oscillating signal emerges as in the case of the 3PE. However, the intensity of this oscillating signal in the 2PE is weaker than in the 3PE.

17.5 3PE at Arbitrary Waiting Time t_w

The 3PE amplitude has already been calculated in Sect. 17.3 using the approximate (16.30). When $\Gamma_{TS} = 0$, i.e., when the triplet level cannot be populated, (17.36) simplifies to

$$d(t) \propto e^{-t_w/T_1} \left[I^e(\Omega, \tau') + I^g(\Omega, \tau')\right] I^g(\Omega, \tau) \ , \tag{17.69}$$

where the functions I^g and I^e are defined by (17.30) and (17.34). In the case of the linear FC interaction, they are given by

$$I^g(\Omega, t) = \mathrm{Re}\left[e^{-i\Omega t + g(t)}\right] \ , \quad I^e(\Omega, t) = \mathrm{Re}\left[e^{-i\Omega t + g^*(t)}\right] . \tag{17.70}$$

Using these formulas the expression for the dipolar moment takes the form

$$d(t) \propto e^{-t_w/T_1} \mathrm{Re}\left[e^{-i\Omega\tau'}\left(e^{g^*(\tau')} + e^{g(\tau')}\right)\right] \mathrm{Re}\left[e^{-i\Omega\tau + g(\tau)}\right] . \tag{17.71}$$

The terms proportional to $\exp[\pm i\Omega(\tau' + \tau)]$ give a contribution of third order in the interaction with light to the optical free induction decay. Neglecting these terms, we can transform (17.71) to

$$d_{3PE}(t) \propto e^{-t_w/T_1} \mathrm{Re}\left\{e^{-i\Omega(\tau'-\tau)}\left[e^{g^*(\tau') + g^*(\tau)} + e^{g(\tau') + g^*(\tau)}\right]\right\} . \tag{17.72}$$

This is the formula for the 3PE signal calculated using the approximate (16.30). Let us find an expression for d_{3PE} using the exact (16.26).

Equation (17.27) can be written as

$$d(t) = \boldsymbol{d}^{\perp}\boldsymbol{a}(t) = \boldsymbol{d}^{\perp} e^{\hat{\omega}\tau'} \hat{\chi}_1' e^{\hat{\Gamma} t_w} \hat{\chi}_2'' e^{\hat{\omega}\tau} \hat{\chi}_1 \boldsymbol{b}(0) = d^{\perp} e^{\hat{\omega}\tau'} \hat{\chi}_1 e^{\hat{\Gamma} t_w} \boldsymbol{b}(\tau) \ , \tag{17.73}$$

where the longitudinal part $b(\tau)$ of the GBV is defined by (17.48). For simplicity, the calculations will be carried out when the waiting time satisfies the inequality

$$T_1 \gg t_w \ . \tag{17.74}$$

This inequality can be satisfied for both $t_w \ll \tau$ and $t_w \gg \tau$. The inequality described by (17.74) allows us to neglect the energy relaxation. It yields a relaxation matrix of diagonal type:

$$\exp(\hat{\Gamma} t_w) = \begin{pmatrix} e^{\varepsilon_{bb'} t_w} & 0 \\ 0 & e^{\varepsilon_{a'a} t_w} \end{pmatrix} . \tag{17.75}$$

Using this operator we arrive at

$$\mathrm{e}^{\hat{\Gamma} t_{\mathrm{w}}} \boldsymbol{b}(\tau) = \begin{pmatrix} \mathrm{e}^{\varepsilon_{bb'} t_{\mathrm{w}}} \rho_{bb'}(\tau) \\ \mathrm{e}^{\varepsilon_{a'a} t_{\mathrm{w}}} \rho_{a'a}(\tau) \end{pmatrix} = \begin{pmatrix} \rho_{bb'}(t_{\mathrm{w}} + \tau) \\ \rho_{a'a}(t_{\mathrm{w}} + \tau) \end{pmatrix} . \quad (17.76)$$

Inserting the elements $\rho_{bb'}(t_{\mathrm{w}} + \tau)$ and $\rho_{aa'}(t_{\mathrm{w}} + \tau)$ into (17.55) instead of the elements $\rho_{bb'}(\tau)$ and $\rho_{aa'}(\tau)$, we take into account the temporal evolution of the GBV during the long pause t_{w}. This evolution is described by two exponential functions. Further, we can repeat all transformations and considerations yielding (17.61). The result is given by

$$\begin{aligned} d_{\mathrm{3PE}}(t) \propto\ & 2\,\mathrm{Re}\Big[\mathrm{e}^{-\mathrm{i}\Omega(\tau'-\tau)} \sum_{ab}\sum_{b'a''} d_{ab} \mathrm{e}^{-\mathrm{i}\Omega_{ba}\tau'} \mathrm{e}^{-\mathrm{i}\Omega_{bb'} t_{\mathrm{w}}} \chi_{ba''} p_{a''} \\ & \times \chi^*_{a''b'} \mathrm{e}^{-\mathrm{i}\Omega_{a''b'}\tau} \chi_{b'a}\Big] \\ & + 2\,\mathrm{Re}\Big[\mathrm{e}^{-\mathrm{i}\Omega(\tau'-\tau)} \sum_{ab}\sum_{a'b''} d_{ab} \mathrm{e}^{-\mathrm{i}\Omega_{ba}\tau'} \chi_{ba'} \mathrm{e}^{-\mathrm{i}\Omega_{a'a} t_{\mathrm{w}}} p_{a'} \\ & \times \chi^*_{a'b''} \mathrm{e}^{-\mathrm{i}\Omega_{a'b''}\tau} \chi_{b''a}\Big] . \end{aligned} \quad (17.77)$$

When $t_{\mathrm{w}} = 0$, the formula for d_{3PE} is transformed to the formula for d_{2PE}. The last equation is indeed transformed to (17.61).

The right hand side of (17.77) can be transformed in the same way as the right hand side of (17.61). The result is given by

$$\begin{aligned} d_{\mathrm{3PE}}(t) \propto\ & 2\,\mathrm{Re}\Big\{\mathrm{e}^{-\mathrm{i}\Omega(\tau'-\tau)} \Big[\Big\langle \mathrm{e}^{-L(0)} \mathrm{e}^{L(\tau+t_{\mathrm{w}})} \mathrm{e}^{-L(\tau+t_{\mathrm{w}}+\tau')} \mathrm{e}^{L(\tau)} \Big\rangle \\ & + \Big\langle \mathrm{e}^{-L(0)} \mathrm{e}^{L(\tau)} \mathrm{e}^{-L(\tau+t_{\mathrm{w}}+\tau')} \mathrm{e}^{L(\tau+t_{\mathrm{w}})} \Big\rangle\Big]\Big\} . \end{aligned} \quad (17.78)$$

The averages in these equations are calculated using the formulas in Appendix L. The final expression for the dipolar moment yielding the 3PE signal is given by

$$d_{\mathrm{3PE}}(t) \propto 2\,\mathrm{Re}\left\{\mathrm{e}^{-\mathrm{i}\Omega(\tau'-\tau)} \left[\mathrm{e}^{g^*(\tau')+g^*(\tau)+g_1(t_{\mathrm{w}})} + \mathrm{e}^{g(\tau')+g^*(\tau)+g_2(t_{\mathrm{w}})}\right]\right\} , \quad (17.79)$$

where

$$\begin{aligned} g_1(t_{\mathrm{w}}) &= g^*(\tau + t_{\mathrm{w}}) + g(\tau' + t_{\mathrm{w}}) - g(t_{\mathrm{w}}) - g^*(\tau + t_{\mathrm{w}} + \tau') , \\ g_2(t_{\mathrm{w}}) &= g^*(\tau + t_{\mathrm{w}}) + g^*(\tau' + t_{\mathrm{w}}) - g^*(t_{\mathrm{w}}) - g^*(\tau + t_{\mathrm{w}} + \tau') . \end{aligned} \quad (17.80)$$

These equations are true for short and long waiting times t_{w}. When $t_{\mathrm{w}} = 0$, this equation for d_{3PE} can be transformed to (17.65) for the 2PE signal derived with the help of the exact (16.26). The dephasing function $\varphi(t,T)$ approaches zero on a femtosecond time scale. As shown in Fig. 15.2, the dephasing function equals zero for $t > 1$ ps. Since $g(t) = \varphi(t,T) - \varphi(0,T)$, we arrive at the conclusion that the functions $g_1(t_{\mathrm{w}})$ and $g_2(t_{\mathrm{w}})$ equal zero when $t_{\mathrm{w}} > 1$ ps. However, if these functions equal zero, (17.79) derived using the exact set (16.26) coincides with (17.72) derived using the approximate set (16.30). In real experiments, the waiting time t_w exceeds 1 ps by a large factor. Therefore, experimental data concerning the SPE can be treated with the help of the simple (17.40) or (17.43).

Part VII

Low Temperature Spectral Diffusion in Polymers and Glasses

18. Theory of Electron–Tunnelon Optical Band

The interaction of the electronic excitation of a guest molecule with phonons, and the manifestation of this interaction in electron–phonon bands have been examined in previous chapters. However, in polymers and glasses there is an additional type of low energy excitation which accompanies tunneling transitions of atoms or molecules from one potential well to another. These excitations are described within the framework of the two-level system (TLS) model. Some details of the TLS model have been discussed in Sect. 6.5. Since the excitation in the TLS is realized via a tunneling transition through a potential barrier, the excitation quantum in the TLS was called a tunnelon. Tunnelons are low frequency excitations of a few cm^{-1}. Because they exist in all polymers and glasses, they influence low temperature properties of these materials. Due to the electron–tunnelon interaction, tunnelons can be created and annihilated upon electronic excitation of the chromophore of a guest molecule. These electron–tunnelon optical bands are the topic to be considered in last part of this book.

When dealing with electron–tunnelon bands, two circumstances have to be taken into account. First of all tunnelons are excitations of Fermi type, in contrast to phonons, which are Bosons. Therefore the quantum statistical average must allow for this. The second circumstance is of a more fundamental character: tunnelon lifetimes range over more than ten decades on the time scale, from nanoseconds to days. Many tunnelons cannot reach thermal equilibrium in the course of an experiment and experimental data depend on the time scale of the experiment. In its most general sense, spectral diffusion relates to the temporal dependence of experimental spectroscopic data on the time scale of the experiment. In a narrower sense, spectral diffusion is the temporal line broadening of a single molecule. In this and the last few chapters, various temporal effects in optical spectra will be considered. However, the term spectral diffusion will be used only for the temporal line broadening of a single molecule.

Some TLSs existing in amorphous solids have lifetimes shorter than the fluorescence lifetime. These TLSs reach thermal equilibrium in the course of the experiment and can therefore be considered on an equal footing with phonons. The influence of such tunnelons on the electron–tunnelon optical band will be examined in this chapter.

18.1 General Expression for the Cumulant Function of the Electron–Tunnelon System

Let us take the equations and transformations examined in Sect. 12.2 as the basis for our considerations. The shape of the electron–tunnelon band is described by

$$J(\omega) = \frac{1}{2\pi} \int_{-\infty}^{\infty} \mathrm{e}^{\mathrm{i}(\omega - \Omega)t - t/2T_1} J(t) \mathrm{d}t \;, \tag{18.1}$$

where the dipolar correlator $J(t)$ of the chromophore takes into account its interaction with tunnelons. The correlator can be written in the following form used earlier:

$$J(t) = d^2 \operatorname{Tr} \left\{ \hat{\rho}^{\mathrm{g}}(T) \exp\left(-\mathrm{i}t \frac{H^{\mathrm{e}}}{\hbar} \right) \exp\left(\mathrm{i}t \frac{H^{\mathrm{g}}}{\hbar} \right) \right\} = d^2 \langle \hat{S}(t) \rangle_{\mathrm{g}} \;, \tag{18.2}$$

where the adiabatic Hamiltonian H^{e} of the chromophore in the excited electronic state differs from that in the ground electronic state by an operator W, viz.,

$$H^{\mathrm{e}} = H^{\mathrm{g}} + W \;. \tag{18.3}$$

The letter W denotes the quadratic electron–tunnelon interaction. It should be noted that this letter W has been used in earlier chapters to denote the quadratic electron–phonon interaction of FC type. Because the electron–phonon interaction will not be considered in this chapter, any misunderstanding is excluded.

It was shown in Sect. 12.2 that a representation of the evolution operator $\hat{S}(t) = \exp(\mathrm{i}tH^{\mathrm{g}}/\hbar) \exp(-\mathrm{i}tH^{\mathrm{e}}/\hbar)$ in terms of the infinite series

$$\hat{S}(t) = 1 + \sum_{m=1}^{\infty} \left(-\frac{\mathrm{i}}{\hbar} \right)^m \int_0^t \mathrm{d}t_1 \int_0^{t_1} \mathrm{d}t_2 \ldots \int_0^{t_{m-1}} \mathrm{d}t_m W(t_1) W(t_2) \ldots W(t_m) \tag{18.4}$$

is the most convenient representation for treating the optical band without perturbation theory. The quantum statistical average of this operator can be found using the cumulant expansion proven in Appendix I. The result is given by

$$\langle \hat{S}(t) \rangle = \exp g_{\mathrm{TLS}}(t) \;, \tag{18.5}$$

where the cumulant function is defined by

$$g_{\mathrm{TLS}}(t) = \sum_{m=1}^{\infty} \left(-\frac{\mathrm{i}}{\hbar}\right)^m \int_0^t \mathrm{d}t_1 \int_0^{t_1} \mathrm{d}t_2$$

$$\ldots \int_0^{t_{m-1}} \mathrm{d}t_m \langle W(t_1) W(t_2) \ldots W(t_m)\rangle_{\mathrm{c}} = \langle \hat{S}(t) - 1\rangle_{\mathrm{c}} \,. \tag{18.6}$$

Here the subscript c means that only the so-called coupled pairing of the interaction operators W with each other should be taken into account.

Let us begin by investigating the interaction of the chromophore with one TLS. The electron–tunnelon interaction operator is given by

$$W = \hbar \Delta c^{+} c \,, \tag{18.7}$$

where tunnelon creation and annihilation operators obey the Fermi commutation relations. The interaction operator allows for the change Δ in the TLS splitting upon electronic excitation in the chromophore. It is obvious that this electron–tunnelon interaction is the quadratic interaction of FC type. The average in (18.6) can be carried out with the help of the Wick–Bloch–Domenicis theorem for Fermi-type operators, proven in Appendix M. However, the integration is simplified when the time-ordering operator is introduced as in Sect. 12.2. After introducing this operator, the cumulant function takes the following form, similar to (12.23):

$$g_{\mathrm{TLS}}(t) = \left\langle \hat{T} \exp\left[-\frac{\mathrm{i}}{\hbar} \int_0^t W(t_1)\, \mathrm{d}t_1\right] - 1\right\rangle_{\mathrm{c}} \tag{18.8}$$

$$= \sum_{m=1}^{\infty} \frac{1}{m!} \left(-\frac{\mathrm{i}}{\hbar}\right)^m \int_0^t \mathrm{d}t_1 \int_0^t \mathrm{d}t_2 \ldots \int_0^t \mathrm{d}t_m \left\langle \hat{T}\left\{W(t_1) W(t_2) \ldots W(t_m)\right\}\right\rangle_{\mathrm{c}} \,.$$

By carrying out the average, we obtain the sum of the products of the causal tunnelon Green functions defined by

$$G(t) = -\mathrm{i}\langle \hat{T} c(t) c^{+}(0)\rangle = -\mathrm{i}\left[\langle c(t)\, c^{+}(0)\rangle\, \theta(t) - \langle c^{+} c(t)\rangle\, \theta(-t)\right] \,. \tag{18.9}$$

The causal tunnelon Green function differs from the causal phonon Green function by the sign in the square brackets. This very definition of the causal tunnelon Green function enables us to express the average in (18.8) solely in terms of these functions.

The cumulant function is a power series in the coupling constants W:

$$g_{\mathrm{TLS}}(t) = \sum_{m=1}^{\infty} g_m(t) \,. \tag{18.10}$$

The first term in this series is given by

$$g_1(t) = -\mathrm{i}\Delta \int_0^t \mathrm{d}1 \left\langle c_1^+ c_1 \right\rangle = -\Delta \int_0^t \mathrm{d}1 G(-0) \,. \tag{18.11}$$

The second term contains the average

$$\begin{aligned}\left\langle \hat{T} c_1^+ c_1 c_2^+ c_2 \right\rangle_{\mathrm{c}} &= \left\langle c_1^+ c_1 c_2^+ c_2 \right\rangle_{\mathrm{c}} \theta(1-2) + \left\langle c_2^+ c_2 c_1^+ c_1 \right\rangle_{\mathrm{c}} \theta(2-1) \\ &= \left\langle c_1^+ c_2 \right\rangle \left\langle c_1 c_2^+ \right\rangle \theta(1-2) + \left\langle c_2^+ c_1 \right\rangle \left\langle c_2 c_1^+ \right\rangle \theta(2-1) \,.\end{aligned} \tag{18.12}$$

Since

$$\begin{aligned}\left\langle c_1^+ c_2 \right\rangle \theta(1-2) &= -\mathrm{i}G(2-1)\theta(1-2) \,, \\ \left\langle c_1 c_2^+ \right\rangle \theta(1-2) &= \mathrm{i}G(1-2)\theta(1-2) \,,\end{aligned} \tag{18.13}$$

and $\theta(1-2) + \theta(2-1) \equiv 1$, we can write the second term of the cumulant function in the form

$$\begin{aligned}g_2(t) &= \frac{(-\mathrm{i}\Delta)^2}{2!} \int_0^t \mathrm{d}1 \int_0^t \mathrm{d}2 \left\langle \hat{T} c_1^+ c_1 c_2^+ c_2 \right\rangle_{\mathrm{c}} \\ &= -\frac{\Delta^2}{2} \int_0^t \mathrm{d}1 \int_0^t \mathrm{d}2 G(1-2) G(2-1) \,.\end{aligned} \tag{18.14}$$

For the third term we find

$$g_3(t) = -\frac{\Delta^3}{3} \int_0^t \mathrm{d}1 \int_0^t \mathrm{d}2 \int_0^t \mathrm{d}3 G(1-2) G(2-3) G(3-1) \,. \tag{18.15}$$

In the mth term there are $(m-1)!$ types of pairing. After integrating over all time variables, each type of pairing gives a similar contribution to the integral. Therefore the mth term is given by

$$g_m(t) = -\frac{\Delta^m}{m} \int_0^t \mathrm{d}1 \int_0^t \mathrm{d}2 \ldots \int_0^t \mathrm{d}m\, G(1-2)G(2-3)\ldots G(m-1) \,. \tag{18.16}$$

Summing over all terms yields the following expression for the cumulant function:

$$\begin{aligned}g_{\mathrm{TLS}}(t) = -\Delta \int_0^t \mathrm{d}1 \Bigg[G(-0) + \sum_{m=2}^{\infty} \frac{\Delta^{m-1}}{m} \int_0^t \mathrm{d}2 \int_0^t \mathrm{d}3 \\ \ldots \int_0^t \mathrm{d}m\, G(1-2)G(2-3)\ldots G(m-1) \Bigg] \,.\end{aligned} \tag{18.17}$$

Our task is to express this infinite series as a solution of an integral equation. In this case we could find the sum of this infinite series by solving the integral equation. Differentiating this series with respect to time, we obtain

$$\dot{g}_{\mathrm{TLS}}(t) = -\Delta\Bigg[G(-0) + \sum_{m=2}^{\infty} \Delta^{m-1} \int_0^t \mathrm{d}2 \int_0^t \mathrm{d}3 \ldots \int_0^t \mathrm{d}m\, G(t-2)G(2-3)\ldots G(m-t)\Bigg] . \tag{18.18}$$

This infinite series resembles the solution of an integral equation found by an iteration procedure. Indeed, let us introduce the following integral equation:

$$S_{\mathrm{TLS}}(x,y,t) = G(x-y) + \Delta \int_0^t \mathrm{d}z\, G(x-z)\, S_{\mathrm{TLS}}(z,y,t) . \tag{18.19}$$

This has the iterative solution

$$S_{\mathrm{TLS}}(x,y,t) = G(x-y) + \sum_{m=2}^{\infty} \Delta^{m-1} \int_0^t \mathrm{d}t_2 \int_0^t \mathrm{d}t_3 \ldots \int_0^t \mathrm{d}t_m G(x-t_2)\, G(t_2-t_3)\ldots G(t_m-y) . \tag{18.20}$$

Comparing this equation with (18.18) reveals that the derivative of the cumulant function can be expressed in terms of the solution of this integral equation as follows

$$\dot{g}_{\mathrm{TLS}}(t) = -\Delta\big[G(-0) - G(+0) + S_{\mathrm{TLS}}(t,t,t)\big] . \tag{18.21}$$

Changing the variable, we have

$$S_{\mathrm{TLS}}(t-x, t-y, t) \equiv S_{\mathrm{TLS}}(y,x,t) , \tag{18.22}$$

i.e., both functions satisfy (18.19). Therefore $S_{\mathrm{TLS}}(t,t,t) \equiv S_{\mathrm{TLS}}(0,0,t)$. Using this relation and integrating with respect to time, we find the final expression for the cumulant function:

$$g_{\mathrm{TLS}}(t) = -\Delta \int_0^t \big[G(-0) - G(+0) + S_{\mathrm{TLS}}(0,0,t)\big]\mathrm{d}t . \tag{18.23}$$

In order to calculate the cumulant function, we should find a solution of the integral equation for the function $S(x,y,t)$ at $x = y = 0$. This problem will be solved in Sect. 18.4.

18.2 Tunnelon Green Function

Equation (18.23) shows that the electron–tunnelon optical band shape function is expressed in terms of a tunnelon Green function. This function can be calculated from (18.9).

The Hamiltonian of the system consisting of one TLS interacting with phonons is given by

$$H = [\varepsilon + H_1(R) - H_0(R)]\, c^+ c + H_0(R) + U\left(c + c^+\right) . \tag{18.24}$$

Here the Hamiltonians H_1 and H_0 describe oscillators in the excited and ground states of the TLS. The last term describes tunneling in the TLS. The difference between these adiabatic Hamiltonians is the tunnelon–phonon interaction of FC type. This interaction includes both linear and quadratic terms in the phonon coordinates R.

If the tunneling operator is omitted, the tunnelon Green function is easily calculated. The result is

$$G(t) = -\mathrm{i}\mathrm{e}^{-\mathrm{i}\varepsilon t/\hbar}\Big[(1-f)\,\theta(t) - f\theta\left(-t\right)\Big] , \tag{18.25}$$

where $f = [\exp(\varepsilon/kT) + 1]^{-1}$, i.e., the Green function does not decay.

Allowing for the tunneling operator, we find the decay of the tunnelon Green function with lifetime T_1'. Therefore the tunnelon Green function takes the form

$$G(t) \approx -\mathrm{i}\exp\left(-\mathrm{i}\frac{\varepsilon}{\hbar}t - \frac{|t|}{T_1'}\right)\Big[(1-f)\,\theta(t) - f\theta\left(-t\right)\Big] . \tag{18.26}$$

More careful calculations yield a more complicated expression for the tunnelon Green function:

$$G(t) = \int_{-\infty}^{\infty} \frac{\mathrm{d}\nu}{\mathrm{i}\pi} \Gamma_{\mathrm{TLS}}\left(\nu\right) \mathrm{e}^{-\mathrm{i}t\nu}\Big\{[1 - f\left(\nu\right)]\,\theta(t) - f\left(\nu\right)\theta\left(-t\right)\Big\} . \tag{18.27}$$

This function almost coincides with the approximate (18.26) if the spectral function is a Lorentzian:

$$\Gamma_{\mathrm{TLS}}(\nu) = \frac{1/T_1'}{\left(\nu - \varepsilon/\hbar\right)^2 + \left(1/T_1'\right)^2} . \tag{18.28}$$

This can be seen from the Laplace transforms of the two functions, which are given by

$$G\left(\omega\right) = G_+\left(\omega\right) + G_-\left(\omega\right) = \frac{1-f}{\omega - \varepsilon + \mathrm{i}/T_1'} + \frac{f}{\omega - \varepsilon - \mathrm{i}/T_1'} , \tag{18.29}$$

$$G(\omega) = G_+(\omega) + G_-(\omega)$$
$$= \int_{-\infty}^{\infty} \frac{d\nu}{\pi} \Gamma_{\mathrm{TLS}}(\nu) \left(\frac{1 - f(\nu)}{\omega - \nu + i0} + \frac{f(\nu)}{\omega - \nu - i0} \right) . \quad (18.30)$$

Here the functions $G_+(\omega)$ and $G_-(\omega)$ have no poles in the upper and lower half plane of the complex variable ω.

18.3 Temperature Broadening of the Zero Tunnelon Optical Line

Let us turn back to the optical line. If a chromophore interacts with a single TLS, its optical band consists of two lines relating to $1 - 1'$ and $2 - 2'$ transitions, as shown in Fig. 9.3. Upon the $2 - 2'$ transition, a tunnelon in the ground electronic state disappears and a tunnelon in the excited electronic state is created. Therefore this transition is of electron–tunnelon type. It contrasts with the $1 - 1'$ transition, where tunnelons are not created or annihilated. This line can thus be called the zero-tunnelon optical line (ZTL). The ZTL is similar to the zero-phonon optical line (ZPL) in electron–phonon systems. The shape of the ZPL is determined by the temporal behavior of the cumulant function in the long time limit.

The first term on the right hand side of (18.17) can be written in the form

$$-\Delta G(-0)\, t = -\mathrm{i} f \Delta t = -\Delta \left(\frac{\mathrm{i}}{2} + \int_{-\infty}^{\infty} \frac{d\omega}{2\pi} G(\omega) \right) t . \quad (18.31)$$

The second term in the same equation can be transformed to

$$-\frac{\Delta^2}{2} \int_0^t d1 \int_0^t d2 G(1-2) G(2-1)$$
$$= -\frac{\Delta^2}{2} \int_{-\infty}^{\infty} \frac{d\omega_1}{2\pi} G(\omega_1) \int_{-\infty}^{\infty} \frac{d\omega_2}{2\pi} G(\omega_2) \Delta_t(1-2) \Delta_t(2-1) , \quad (18.32)$$

where the function $\Delta_t(x)$ is defined by (12.65). Allowing for (12.66), each term with $m > 1$ in the cumulant function transforms to

$$g_m(t) = -|t| \int_{-\infty}^{\infty} \frac{d\omega}{2\pi} \frac{1}{m} \left[\Delta G(\omega) \right]^m . \quad (18.33)$$

Summing over all terms yields the following expression for the cumulant function in the long time limit:

$$g^{\infty}_{\mathrm{TLS}}(t) = t\left[\frac{-\mathrm{i}\Delta}{2} + \int_{-\infty}^{\infty} \frac{\mathrm{d}\omega}{2\pi} \ln\left[1 - \Delta G(\omega, T)\right]\right] = t\left[-\mathrm{i}\delta_{\mathrm{TLS}} - \frac{\gamma_{\mathrm{TLS}}}{2}\right], \tag{18.34}$$

where the Laplace transform of the tunnelon Green function is described by (18.30) or (18.29). Equation (18.34) is similar to (12.68) for the shift and half-width of the ZPL. It is obvious that the ZTL is described by a Lorentzian. The imaginary part of (18.34) describes the ZTL shift and the real part describes the ZTL half-width.

At $T = 0$, the tunnelon Green function can be written as

$$G(\omega, 0) = \Delta_{\mathrm{TLS}}(\omega) - \mathrm{i}\Gamma_{\mathrm{TLS}}(\omega) . \tag{18.35}$$

The imaginary part is described by a spectral function and the real part is given by

$$\Delta_{\mathrm{TLS}}(\omega) = \mathrm{P} \int_{-\infty}^{\infty} \frac{\mathrm{d}\nu}{\pi} \frac{\Gamma_{\mathrm{TLS}}(\nu)}{\omega - \nu} . \tag{18.36}$$

In Sect. 12.5, the half-width and shift of the ZPL were expressed in terms of phonon Green functions. Similar transformations can be carried out for the ZTL half-width and shift. This is done in Appendix N. The result is given by

$$\delta_{\mathrm{TLS}}(T) = \int_{0}^{\infty} \frac{\mathrm{d}\omega}{2\pi} \arctan \frac{2f(\omega)\,\Delta\Gamma^{\mathrm{e}}_{\mathrm{TLS}}(\omega)\left[1 - \Delta\Delta_{\mathrm{TLS}}(\omega)\right]}{1 - 2f(\omega)\,\Delta^2\Gamma^{\mathrm{e}}_{\mathrm{TLS}}(\omega)\,\Gamma_{\mathrm{TLS}}(\omega)} , \tag{18.37}$$

$$\gamma_{\mathrm{TLS}}(t) = -\int_{0}^{\infty} \frac{\mathrm{d}\omega}{2\pi} \ln\left[1 - 4\left[1 - f(\omega)\right] f(\omega)\,\Delta^2\Gamma^{\mathrm{e}}_{\mathrm{TLS}}(\omega)\,\Gamma_{\mathrm{TLS}}(\omega)\right] . \tag{18.38}$$

Here,

$$\Gamma^{\mathrm{e}}_{\mathrm{TLS}}(\omega) = \frac{\Gamma_{\mathrm{TLS}}(\omega)}{\left[1 - \Delta\Delta_{\mathrm{TLS}}(\omega)\right]^2 + \left[\Delta\Gamma_{\mathrm{TLS}}(\omega)\right]^2} \tag{18.39}$$

describes the spectral function of the tunnelon in the excited electronic state. It is expressed in terms of the spectral function $\Gamma_{\mathrm{TLS}}(\omega)$ of the tunnelon in the ground electronic state. These formulas for the ZTL are similar to (12.84), (12.85) and (12.86) for the ZPL.

From (18.39), we can easily find the inequality

$$\Delta^2\Gamma^{\mathrm{e}}_{\mathrm{TLS}}(\omega)\Gamma_{\mathrm{TLS}}(\omega) \le 1 ,$$

which is correct at arbitrary values of the coupling constant Δ. Hence, the function $\ln(1-x)$ in (18.38) can be replaced by $-x$. We then arrive at a simpler formula for the half-width:

$$\gamma_{\mathrm{TLS}}(t) = \int_0^\infty \frac{\mathrm{d}\omega}{2\pi} \frac{\Delta^2 \Gamma^{\mathrm{e}}_{\mathrm{TLS}}(\omega)\, \Gamma_{\mathrm{TLS}}(\omega)}{\cosh^2(\hbar\omega/2kT)}\,, \tag{18.40}$$

which is similar to (12.87) for the half-width of the ZPL. Here we used the fact that $4f(\omega)[1-f(\omega)] = \cosh^{-2}(\hbar\omega/2kT)$.

The substitution of $\cosh^{-2}$ for $\sinh^{-2}$ in (12.87) yields considerable changes in the high temperature behavior of the half-width. Indeed it approaches a limit. It should be recalled that the half-width resulting from the interaction with phonons is a quadratic function of temperature in the high temperature limit. In the low temperature limit, the half-width due to electron–tunnelon coupling yields a temperature dependence which is similar to that due to the electron–phonon interaction, because $\cosh^{-2}(\hbar\omega/2kT) \approx \sinh^{-2}(\hbar\omega/2kT) \approx 4\exp(-\hbar\omega/kT)$, i.e., the difference between (18.39) and (12.87) disappears.

The expression for the tunnelon Green function described by (18.27) is more general than (18.26). In the case when the tunnelon spectral function can be approximated by a Lorentzian, the expression for the half-width can be simplified. Indeed (18.39) takes the form

$$\begin{aligned}\Gamma^{\mathrm{e}}_{\mathrm{TLS}}(\omega) &= \frac{\left[(\omega-\varepsilon)^2 + (\gamma/2)^2\right](\gamma/2)}{\left[(\omega-\varepsilon)^2 + (\gamma/2)^2 - \Delta(\omega-\varepsilon)\right]^2 + (\Delta\gamma/2)^2} \\ &\approx \frac{\gamma/2}{(\omega-\varepsilon-\Delta)^2 + (\gamma/2)^2}\,,\end{aligned} \tag{18.41}$$

where $1/T_1' = \gamma/2$ and $\hbar = 1$. By inserting the last equation into (18.42), we arrive at the simple equation

$$\frac{\gamma_{\mathrm{TLS}}}{2} \approx \frac{\Delta^2}{\cosh^2(\varepsilon/kT)} \int_0^\infty \frac{\mathrm{d}\omega}{4\pi} \Gamma^{\mathrm{e}}_{\mathrm{TLS}}(\omega)\, \Gamma(\omega) \approx \frac{1}{4} \frac{\Delta^2\gamma}{\Delta^2+\gamma^2} \frac{1}{\cosh^2(\varepsilon/kT)}\,. \tag{18.42}$$

This formula for the half-width is more convenient in practice.

18.4 Dipolar Correlator for a Chromophore–TLS System. Solution of the Integral Equation

The dipolar correlator for an electron–tunnelon system is given by

$$J(t) = d^2 \exp g_{\mathrm{TLS}}(t) . \tag{18.43}$$

The cumulant function $g_{\mathrm{TLS}}(t)$ describes the influence of the TLS on the oscillations of the electronic dipolar moment. The cumulant function is given by (18.23). Therefore, to find this function we should solve the integral equation

$$S(x,0,t) = G(x) + \Delta \int_0^t \mathrm{d}y \, G(x-y) \, S(y,0,t) \tag{18.44}$$

and evaluate it at $x = 0$.

A solution will be sought in the form

$$S(x,0,t) = -\mathrm{i} \left[S_1(t) \exp(-\mathrm{i}\Omega_1 x) + S_2(t) \exp(-\mathrm{i}\Omega_2 x) \right] , \tag{18.45}$$

where S_1, S_2, Ω_1 and Ω_2 are unknown functions of time and unknown complex frequencies, respectively. The tunnelon Green function is chosen in the form described by (18.29), with $\hbar = 1$ and $1/T_1' = \gamma/2$. Inserting (18.26) and (18.45) into the integral equation (18.44) and carrying out the integration with respect to time, we arrive at the algebraic equation

$$\begin{aligned} -\mathrm{i} \exp(-\mathrm{i}\Omega_1 x) \, S_1(t) \left(1 + \Delta \frac{1-f}{\Delta_1} + \Delta \frac{f}{\bar{\Delta}_1} \right) \\ -\mathrm{i} \exp(-\mathrm{i}\Omega_2 x) \, S_2(t) \left(1 + \Delta \frac{1-f}{\Delta_2} + \Delta \frac{f}{\bar{\Delta}_2} \right) \\ = -\mathrm{i}(1-f) \exp\left[-\mathrm{i}\left(\varepsilon - \mathrm{i}\frac{\gamma}{2} \right) x \right] \left[1 + \Delta \left(\frac{S_1}{\Delta_1} + \frac{S_2}{\Delta_2} \right) \right] \\ -\mathrm{i} f \Delta \exp\left[-\mathrm{i}\left(\varepsilon + \mathrm{i}\frac{\gamma}{2} \right) x \right] \left[\frac{S_1}{\bar{\Delta}_1} \exp(\mathrm{i}\bar{\Delta}_1 t) + \frac{S_2}{\bar{\Delta}_2} \exp(\mathrm{i}\bar{\Delta}_2 t) \right] , \end{aligned} \tag{18.46}$$

where

$$\begin{aligned} \Delta_1 &= \varepsilon - \Omega_1 - \mathrm{i}\frac{\gamma}{2} , \quad \Delta_2 = \varepsilon - \Omega_2 - \mathrm{i}\frac{\gamma}{2} , \\ \bar{\Delta}_1 &= \varepsilon - \Omega_1 + \mathrm{i}\frac{\gamma}{2} , \quad \bar{\Delta}_2 = \varepsilon - \Omega_2 + \mathrm{i}\frac{\gamma}{2} . \end{aligned} \tag{18.47}$$

If the function $S(x,0,t)$ described by (18.45) is a solution of the integral equation, the algebraic equation (18.46) must be the identity. This is possible if the coefficient multiplying each of the four exponential functions equals zero. This requirement leads to the equation

$$1 - \Delta \left(\frac{1-f}{\Omega - \varepsilon + \mathrm{i}\gamma/2} + \frac{f}{\Omega - \varepsilon - \mathrm{i}\gamma/2} \right) = 0 \tag{18.48}$$

for the two unknown complex frequencies Ω_1 and Ω_2, and two equations

$$\frac{S_1}{\Delta_1} + \frac{S_2}{\Delta_2} = -\frac{1}{\Delta} \, , \quad \frac{S_1}{\bar{\Delta}_1} \exp\left(\mathrm{i}\bar{\Delta}_1 t\right) + \frac{S_2}{\bar{\Delta}_2} \exp\left(\mathrm{i}\bar{\Delta}_2 t\right) = 0 \tag{18.49}$$

for the two unknown functions $S_1(t)$ and $S_2(t)$. Two roots of (18.48) are

$$\Omega_1 = \varepsilon + \frac{\Delta}{2} + \Omega_0 - \mathrm{i}\gamma_0 \, , \quad \Omega_2 = \varepsilon + \frac{\Delta}{2} - \Omega_0 + \mathrm{i}\gamma_0 \, , \tag{18.50}$$

where

$$\begin{aligned} \Omega_0 &= \pm \frac{1}{2\sqrt{2}} \left\{ \left[\left(\Delta^2 - \gamma^2\right)^2 + 4\Delta^2\gamma^2 (1-2f)^2 \right]^{1/2} + \Delta^2 - \gamma^2 \right\}^{1/2} , \\ \gamma_0 &= \frac{1}{2\sqrt{2}} \left\{ \left[\left(\Delta^2 - \gamma^2\right)^2 + 4\Delta^2\gamma^2 (1-2f)^2 \right]^{1/2} - \Delta^2 + \gamma^2 \right\}^{1/2} . \end{aligned} \tag{18.51}$$

The sign of the function Ω_0 must coincide with the sign of the coupling parameter Δ.

Since $G(-0) - G(+0) = \mathrm{i}$, we can express the cumulant function in terms of the functions S_1 and S_2:

$$g_{\mathrm{TLS}}(t) = -\mathrm{i}\Delta \int_0^t (1 - S_1 - S_2) \, \mathrm{d}t \, . \tag{18.52}$$

It is shown in Appendix O that

$$1 - S_1 - S_2 = \frac{\mathrm{i}}{\Delta} \frac{\mathrm{d}}{\mathrm{d}t} \ln \mathrm{Det}(t) \, , \tag{18.53}$$

where

$$\mathrm{Det} = \frac{\exp\left(\mathrm{i}\bar{\Delta}_2 t\right)}{\Delta_1 \bar{\Delta}_2} - \frac{\exp\left(\mathrm{i}\bar{\Delta}_1 t\right)}{\Delta_2 \bar{\Delta}_1} \tag{18.54}$$

is the determinant of the set of equations (18.49). Inserting (18.53) into (18.52) and carrying out the integration with respect to time, we arrive at the final expression for the cumulant function:

$$g_{\mathrm{TLS}}(t) = \ln \left[\frac{\exp\left(\mathrm{i}\bar{\Delta}_2 t\right)}{\Delta_1 \bar{\Delta}_2} - \frac{\exp\left(\mathrm{i}\bar{\Delta}_1 t\right)}{\Delta_2 \bar{\Delta}_1} \right] - \ln \left[\frac{1}{\Delta_1 \bar{\Delta}_2} - \frac{1}{\Delta_2 \bar{\Delta}_1} \right] . \tag{18.55}$$

Substituting this expression for the cumulant function into (18.43), we arrive at the following expression for the dipolar moment of a chromophore interacting with a single TLS:

$$J(t) = d^2 \left[\frac{1}{1-\alpha} \exp\left(\mathrm{i}\bar{\Delta}_2 t\right) - \frac{\alpha}{1-\alpha} \exp\left(\mathrm{i}\bar{\Delta}_1 t\right) \right] , \tag{18.56}$$

where

$$\alpha = \frac{\Delta_1 \bar{\Delta}_2}{\Delta_2 \bar{\Delta}_1} . \tag{18.57}$$

18.5 Electron–Tunnelon Optical Band Shape Function

When multiplied by the exponential factor $\exp(-\mathrm{i}\omega_0 t)$, where ω_0 is the electronic frequency, (18.56) describes the temporal behavior of the dipolar correlator. The last two formulas can be simplified using the relation $\Delta_1 + \overline{\Delta}_2 = \Delta_2 + \overline{\Delta}_1 = -\Delta$. This can be derived from (18.47), (18.50) and the equation

$$\frac{1-f}{\Delta_1} + \frac{f}{\bar{\Delta}_1} = \frac{1-f}{\Delta_2} + \frac{f}{\bar{\Delta}_2} , \tag{18.58}$$

which is in turn derived from (18.48). Then we can transform α to

$$\alpha = \frac{\bar{\Delta}_2 + f\Delta}{\bar{\Delta}_1 + f\Delta} . \tag{18.59}$$

It is shown in Appendix P that (18.51) can be simplified as follows:

$$\Omega_0 \approx \frac{\Delta}{2} - \delta_{\mathrm{TLS}} , \qquad \gamma_0 \approx \frac{\gamma}{2} - \frac{\gamma_{\mathrm{TLS}}}{2} , \tag{18.60}$$

where

$$\delta_{\mathrm{TLS}} = \frac{\gamma^2 \Delta}{\Delta^2 + \gamma^2} f(1-f) \ll \frac{\Delta}{2} , \quad \frac{\gamma_{\mathrm{TLS}}}{2} = \frac{\Delta^2 \gamma}{\Delta^2 + \gamma^2} f(1-f) \ll \frac{\gamma}{2} . \tag{18.61}$$

The last two equations lead to

$$\bar{\Delta}_2 \approx -\delta_{\mathrm{TLS}} + \mathrm{i}\frac{\gamma_{\mathrm{TLS}}}{2} , \quad \bar{\Delta}_1 \approx -\Delta + \mathrm{i}\gamma . \tag{18.62}$$

It should be noted that the equation for γ_{TLS} in (18.61) coincides with (18.42), derived from the more general (18.38). Hence we may suppose that the general (18.37) for the line shift can be transformed to the equation for δ_{TLS} found here. Inserting the last equations into (18.59) and neglecting small values in accordance with the inequalities in (18.61), we find the following simplified expression for α:

$$\alpha \approx \frac{f\Delta}{(f-1)\,\Delta + \mathrm{i}\gamma}\,. \tag{18.63}$$

The general expression for the dipolar correlator in (18.56) can be transformed to the equation

$$\begin{aligned} I(t) &= \mathrm{e}^{-\mathrm{i}\omega_0 t} J(t) \\ &= d^2\,\mathrm{e}^{-\mathrm{i}\omega_0 t}\left[\left(1 - \frac{\Delta}{\Delta - \mathrm{i}\gamma} f\right)\mathrm{e}^{-\mathrm{i}\delta_{\mathrm{TLS}} t - \gamma_{\mathrm{TLS}} t/2} + \frac{\Delta}{\Delta - \mathrm{i}\gamma} f \mathrm{e}^{-\mathrm{i}\Delta t - \gamma t}\right], \end{aligned} \tag{18.64}$$

with the help of (18.62) and (18.63). It is interesting to note that the last expression is almost the same as the expression for the dipolar correlator derived using the Anderson theory in Sect. 9.4. Indeed from (9.35), we find the following expression for the dipolar correlator:

$$I_{\mathrm{A}}(t) = d^2\,\mathrm{e}^{-\mathrm{i}\omega_0 t}\left[\left(1 - \frac{\Delta}{\Delta - \mathrm{i}R}\rho_2\right)\mathrm{e}^{-\mathrm{i}\delta t - \gamma t/2} + \frac{\Delta}{\Delta - \mathrm{i}R}\rho_2 \mathrm{e}^{-\mathrm{i}\Delta t - Rt}\right]. \tag{18.65}$$

Comparing the last two equations reveals their similarity. There are two differences, however. Firstly, the expression for γ_{TLS} is given by (18.61), whereas the expression for γ is given by (9.33). The temperature dependence manifests itself via the factor $f(1-f)$ in our theory. This fact relates to the two-tunnelon character of the dephasing processes. The temperature dependence in the Anderson theory is introduced by means of the ratio $p/(p+P)$ of the probabilities of forward and reverse transitions in the TLS. This means that we cannot generalize the Anderson theory to take into account a TLS with time dependent populations, i.e., we cannot allow for spectral diffusion. The theory presented here, however, can be generalized along these lines and this generalization will be carried out in Chap. 20.

Equation (18.64) describes the temporal behavior of the dipolar correlator for $t > 0$. The expression for $t < 0$ can be found from (18.2), with the condition

$$I(-t) = I^*(t)\,. \tag{18.66}$$

The Laplace transform of (18.64) for the dipolar correlator takes the form

$$\begin{aligned} I(\omega) = &-\mathrm{Im}\left[\left(1 - \frac{\Delta}{\Delta - \mathrm{i}\gamma} f\right)\frac{1}{\omega - \omega_0 - \delta_{\mathrm{TLS}} + \mathrm{i}\gamma_{\mathrm{TLS}}/2}\right] \\ &-\mathrm{Im}\left[\frac{\Delta}{\Delta - \mathrm{i}\gamma} f \frac{1}{\omega - \omega_0 - \Delta + \mathrm{i}\gamma/2}\right]. \end{aligned} \tag{18.67}$$

This expression is similar to (9.35) from the Anderson theory.

Our theory for a chromophore interacting with one TLS can easily be generalized to the case of an interaction with many TLSs. This very case relates to a polymer or glass matrix. Because of the low concentration of TLSs, each TLS will contribute independently to the chromophore optical band. Therefore the expression for the dipolar correlator of the system consisting of a chromophore and many TLSs is given by

$$I(t) = \prod_j^{N_0} I_j(t) = \exp\left[\mathrm{i}(\omega - \omega_0 - \delta_{\mathrm{TLS}})t - \frac{|t|\gamma_{\mathrm{TLS}}}{2}\right] \times \prod_j^{N_0} \{1 - C_j\left[1 - \exp\left(-\mathrm{i}\Delta_j t - |t|\gamma_j\right)\right]\} , \tag{18.68}$$

where the subscript j indices the TLSs and

$$C_j = \frac{\Delta_j}{\Delta_j - i\gamma_j} f_j , \tag{18.69}$$

$$\delta_{\mathrm{TLS}} = \sum_{j=1}^{N_0} \frac{\gamma_j^2 \Delta_j}{\Delta_j^2 + \gamma_j^2} f_j(1 - f_j) , \quad \frac{\gamma_{\mathrm{TLS}}}{2} = \sum_{j=1}^{N_0} \frac{\gamma_j \Delta_j^2}{\Delta_j^2 + \gamma_j^2} f_j(1 - f_j) . \tag{18.70}$$

Here N_0 is the total number of TLSs. The correlator is a sum of terms oscillating with various frequencies which are sums of various Δ_j. After Laplace transformation, each term yields a Lorentzian with the resonant frequency which relates to this sum of Δ_j. It is obvious that TLSs with $\Delta_j/\gamma_j \ll 1$ will not contribute to the optical band. In fact only TLSs with $\Delta_j/\gamma_j \geq 1$ will contribute. These TLSs are situated in the vicinity of the guest molecule, because Δ_j decreases when the distance between the guest molecule and the TLS increases. The last equations will be used in Chap. 20, where a dynamical theory for spectral diffusion is built up.

19. Chromophore Interacting with Phonons and TLSs Which Are not in Thermal Equilibrium

The equations for the density matrix derived in Part III take into account the interaction with phonons and those TLSs which have already reached thermal equilibrium. These equations were used in the last part of the book, when studying the photon echo. Let us generalize them to the density matrix and take into account interactions with TLSs whose relaxation rate is slower than the electronic relaxation rate. Such TLSs will approach thermal equilibrium during the course of optical measurements. From the mathematical point of view, this means that tunneling processes must be taken into account in the equation for the density matrix. The relevant theory was derived by the author in [95] and [96] and is discussed in this chapter.

19.1 Hamiltonian of the Electron–Phonon–Tunnelon System

Let us inspect the Hamiltonian of the system consisting of a two-level chromophore, phonons, tunneling excitations (tunnelons) and a transverse electromagnetic field. All parts of the system are shown in Fig. 19.1.

The Hamiltonian of the system is given by

$$H = H_0 + H_\perp + \hat{\Lambda} + \hat{\lambda} \,, \tag{19.1}$$

where $H_\perp$ is the Hamiltonian of the transverse electromagnetic field and the Hamiltonian of the electron–phonon–tunnelon system is

$$H_0 = \left[\hbar\Omega + \Delta H(\xi)\right] B^+ B + H(\xi) \ . \tag{19.2}$$

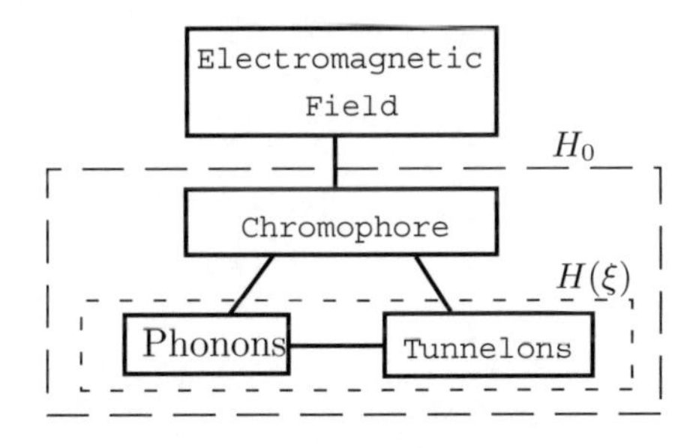

Fig. 19.1. Elements of the system considered

Here Ω defines the chromophore electronic frequency and the operators B^+ and B create and annihilate electronic excitations. The operator $H(\xi)$ is the Hamiltonian of phonons and TLSs, the operator $\Delta H(\xi)$ defines a change in the phonon–tunnelon subsystem upon electronic excitation in the chromophore of a guest molecule. This interaction of FC type results from the change in the adiabatic potentials upon electronic excitation. The operator $\Lambda = \boldsymbol{d} \cdot \boldsymbol{E}(B + B^+)$ describes the electron–photon interaction and the operator $\lambda = \lambda(c + c^+)$ defines tunneling in the TLS. It depends on the tunnelon creation and annihilation operators.

The Hamiltonian of the phonon–tunnelon subsystem is given by

$$H(\xi) = [\hbar\varepsilon + V(R)]\, c^+ c + H(R) \ . \tag{19.3}$$

Here $\hbar\varepsilon$ is the tunnelon energy and the operators c^+ and c create and annihilate tunnelons. $H(R)$ is the phonon Hamiltonian and $V(R)$ is a change in the phonon Hamiltonian upon excitation in the TLS, i.e., it defines the tunnelon–phonon interaction of FC type in the ground electronic state.

The adiabatic Hamiltonian of the whole system is changed by electronic excitation. Hence the tunnelon–phonon Hamiltonian $H(\xi)$ is changed. This change is given by

$$\Delta H(\xi) = [\hbar\Delta + \Delta V(R)]\, c^+ c + \Delta H(R) \ . \tag{19.4}$$

Here $\hbar\Delta$ defines a change in the TLS splitting, i.e., the change in the tunnelon energy upon electronic excitation. This parameter defines the strength of the electron–tunnelon coupling. The operator $\Delta H(R)$ defines the electron–phonon interaction of FC type. $\Delta V(R)$ describes a change in the tunnelon–phonon coupling upon electronic excitation. This operator has to be introduced because tunneling processes in the TLS depend on the electronic state of the chromophore.

The operators Λ and λ define the interaction of the chromophore with the electromagnetic field and the tunneling operator. If $\Lambda = \lambda = 0$, transitions between two electronic states of the chromophore and transitions between the two states of the TLS disappear. If $\lambda = 0$, only tunneling transitions disappear. The equations for the density matrix in this case have already been studied in earlier chapters. Tunneling in the TLS will be allowed for in the next section.

19.2 Equations for the Density Matrix of the Electron–Phonon–Tunnelon System

The equation for the density operator is

$$\mathrm{i}\hbar\dot{\hat{\rho}} = [H, \hat{\rho}] \ , \tag{19.5}$$

where the Hamiltonian H is given by (19.1). We need to make a choice for the set of eigenfunctions in order to find the equations for the matrix elements. For this purpose, we shall use eigenfunctions of the operator $H_0 + H_\perp$. They are a product of the electronic functions, phonon functions, tunnelon functions and functions of harmonic oscillators which describe photons.

The electronic functions of the two-level chromophore are defined by

$$\begin{aligned} B^+|0) &= |1) \ , \quad & B^+|1) &= 0 \ , \\ B|0) &= 0 \ , & B|1) &= |0) \ . \end{aligned} \tag{19.6}$$

The phonon functions in the ground electronic state can be found from

$$\begin{aligned} &H(R)|a\rangle = \hbar\Omega_a|a\rangle \ , \\ &[H(R) + V(R)]\,|\alpha\rangle = \hbar\Omega_\alpha|\alpha\rangle \ , \end{aligned} \tag{19.7}$$

and the phonon functions in the excited electronic state can be found from

$$\begin{aligned} &[H(R) + \Delta H(R)]\,|b\rangle = \hbar\Omega_b|b\rangle \ , \\ &[H(R) + V(R) + \Delta V(R) + \Delta H(R)]\,|\beta\rangle = \hbar\Omega_\beta|\beta\rangle \ . \end{aligned} \tag{19.8}$$

It is obvious that the harmonic oscillator functions which describe phonons depend on the quantum state of the TLS.

The tunnelon functions in the ground and excited electronic state satisfy the following equations:

$$\begin{aligned} \varepsilon c^+c|0\rangle &= 0 \ , \quad & (\varepsilon + \varDelta)\,c^+c|1\rangle &= 0 \ , \\ \varepsilon c^+c|2\rangle &= \varepsilon|2\rangle \ , & (\varepsilon + \varDelta)\,c^+c|3\rangle &= (\varepsilon + \varDelta)\,|3\rangle \ . \end{aligned} \tag{19.9}$$

The eigenfunctions $|\boldsymbol{n}\rangle$ of the operator $H_\perp$ are the harmonic oscillator functions which describe photons.

It is obvious that the cigenfunctions of the Hamiltonian $H_0 + H_\perp$ are given by

$$\begin{aligned} |A) &= |\boldsymbol{n}\rangle|0)|A\rangle = |\boldsymbol{n}\rangle|0) \begin{cases} |0\rangle|a\rangle \ , \\ |2\rangle|\alpha\rangle \ , \end{cases} \\ |B) &= |\boldsymbol{n}-1\rangle|1)|B\rangle = |\boldsymbol{n}-1\rangle|1) \begin{cases} |1\rangle|b\rangle \ , \\ |3\rangle|\beta\rangle \ . \end{cases} \end{aligned} \tag{19.10}$$

Here the functions of the system consisting of chromophore, phonons, and tunnelons satisfy the equations

$$\begin{aligned} H_0|0)|A\rangle &= \hbar\Omega_A|0)|A\rangle\,, \\ H_0|1)|B\rangle &= (E+\hbar\Omega_B)|1)|B\rangle\,, \end{aligned} \tag{19.11}$$

where

$$\Omega_A = \begin{cases} \Omega_a\,, \\ \Omega_\alpha + \varepsilon\,, \end{cases} \qquad \Omega_B = \begin{cases} \Omega_b\,, \\ \Omega_\beta + \Delta + \varepsilon\,. \end{cases} \tag{19.12}$$

This set of functions and the energy diagram relating to them is shown in Fig. 19.2.

Repeating the derivation presented in Part III, when we derived the equations for the density matrix which take into account spontaneous light emission, and using (19.10) and the operator equation (19.5), we can derive the following set of equations for the elements of the density matrix:

$$\begin{aligned}
\dot\rho_{BA} &= -\mathrm{i}\left(\Delta_0 + \Omega_{BA} - \frac{\mathrm{i}}{2T_1}\right)\rho_{BA} - \mathrm{i}\sum_{A'}\Lambda_{BA'}\rho_{A'A} \\
&\quad + \mathrm{i}\sum_{B'}\rho_{BB'}\Lambda_{B'A} - \mathrm{i}\underline{\sum_{B'}\lambda_{BB'}\rho_{B'A} + \mathrm{i}\sum_{A'}\rho_{BA'}\lambda_{A'A}}\,, \\
\dot\rho_{AB} &= -\mathrm{i}\left(-\Delta_0 + \Omega_{AB} - \frac{\mathrm{i}}{2T_1}\right)\rho_{AB} - \mathrm{i}\sum_{B'}\Lambda_{AB'}\rho_{B'B} \\
&\quad + \mathrm{i}\sum_{A'}\rho_{AA'}\Lambda_{A'B} - \mathrm{i}\underline{\sum_{A'}\lambda_{AA'}\rho_{A'B} + \mathrm{i}\sum_{B'}\rho_{AB'}\lambda_{B'B}}\,, \\
\dot\rho_{BB'} &= -\mathrm{i}\left(\Omega_{BB'} - \frac{\mathrm{i}}{T_1}\right)\rho_{BB'} - \mathrm{i}\sum_{A}(\Lambda_{BA}\rho_{AB'} - \rho_{BA}\Lambda_{AB'}) \\
&\quad \underline{- \mathrm{i}\sum_{B''}(\lambda_{BB''}\rho_{B''B'} - \rho_{BB''}\lambda_{B''B'})}\,, \\
\dot\rho_{AA'} &= -\mathrm{i}\Omega_{AA'}\rho_{AA'} + \frac{1}{T_1}\sum_{BB'}\langle A|B\rangle\rho_{BB'}\langle B'|A'\rangle \\
&\quad - \mathrm{i}\sum_{B}(\Lambda_{AB}\rho_{BA'} - \rho_{AB}\Lambda_{BA'}) \underline{- \mathrm{i}\sum_{A''}(\lambda_{AA''}\rho_{A''A'} - \rho_{AA''}\lambda_{A''A'})}\,.
\end{aligned} \tag{19.13}$$

Here $\Delta_0 = \Omega - \omega_0$ is the difference between the electronic and photon frequencies. The matrix elements and frequencies are defined by

$$\Lambda_{BA} = -\mathrm{i}\langle B|A\rangle\chi = \begin{cases} -\mathrm{i}\langle b|a\rangle\langle 1|0\rangle\chi = \Lambda_{ba}\,, \\ -\mathrm{i}\langle\beta|\alpha\rangle\langle 3|2\rangle\chi = \Lambda_{\beta\alpha}\,, \end{cases} \qquad \Lambda_{AB} = \Lambda^*_{BA}\,, \tag{19.14}$$

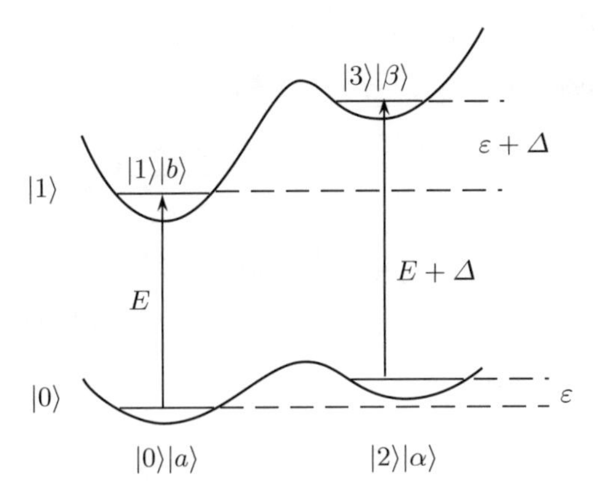

Fig. 19.2. Electron–phonon–tunnelon functions and energies

$$\lambda_{BB'} = \begin{cases} \langle b|\beta\rangle\langle 1|3\rangle\lambda = \lambda_{b\beta}\ , \\ \langle\beta|b\rangle\langle 3|1\rangle\lambda = \lambda_{\beta b}\ , \end{cases} \qquad \lambda_{AA'} = \begin{cases} \langle a|\alpha\rangle\langle 0|2\rangle\lambda = \lambda_{a\alpha}\ , \\ \langle\alpha|a\rangle\langle 2|0\rangle\lambda = \lambda_{\alpha a}\ , \end{cases} \tag{19.15}$$

$$\Omega_{BA} = \Omega_B - \Omega_A\ , \quad \Omega_{BB'} = \Omega_B - \Omega_{B'}\ , \quad \Omega_{AA'} = \Omega_A - \Omega_{A'}\ . \tag{19.16}$$

Here $\chi = \boldsymbol{d}\cdot\boldsymbol{E}/\hbar$ is the Rabi frequency. The underlined terms include the matrix elements of the tunneling operator. If we omit these terms, the set (19.13) is transformed to the set (7.29).

Let us write (19.13) in more detailed form. For this purpose we introduce the following simplified notation: $\rho_{b1,a0} = \rho_{ba}$, $\rho_{\beta 3,\alpha 2} = \rho_{\beta\alpha}$, $\rho_{b1,\beta 3} = \rho_{\beta b}$, $\rho_{a0,\alpha 2} = \rho_{a\alpha}$, etc. Optical transitions with tunneling will be neglected. This means we can neglect the following matrix elements: $\Lambda_{a\beta} = \Lambda_{b\alpha} = \rho_{a\beta} = \rho_{b\alpha} = 0$. We also neglect the off-diagonal elements $\rho_{aa'}, \rho_{bb'}, \rho_{\alpha\alpha'}$ and $\rho_{\beta\beta'}$, which can be omitted if the influence of the operators Λ and λ is studied to the first nonvanishing approximation in these operators. The validity of this approximation has already been discussed in connection with Fig. 8.1 and in Sect. 17.5. If all these approximations are taken into account, the first and second equations in the set (19.13) can be written in the form

$$\begin{aligned}
\dot\rho_{ba} &= -\mathrm{i}\left(\Delta_0 + \Omega_{ba} - \frac{\mathrm{i}}{2T_1}\right)\rho_{ba} - \mathrm{i}\Lambda_{ba}\left(\rho_{aa} - \rho_{bb}\right)\ , \\
\dot\rho_{\beta\alpha} &= -\mathrm{i}\left(\Delta_0 + \Delta + \Omega_{\beta\alpha} - \frac{\mathrm{i}}{2T_1}\right)\rho_{\beta\alpha} - \mathrm{i}\Lambda_{\beta\alpha}\left(\rho_{\alpha\alpha} - \rho_{\beta\beta}\right)\ , \\
\dot\rho_{ab} &= -\mathrm{i}\left(-\Delta_0 + \Omega_{ab} - \frac{\mathrm{i}}{2T_1}\right)\rho_{ab} - \mathrm{i}\Lambda_{ab}\left(\rho_{bb} - \rho_{aa}\right)\ , \\
\dot\rho_{\alpha\beta} &= -\mathrm{i}\left(-\Delta_0 - \Delta + \Omega_{\alpha\beta} - \frac{\mathrm{i}}{2T_1}\right)\rho_{\alpha\beta} - \mathrm{i}\Lambda_{\alpha\beta}\left(\rho_{\beta\beta} - \rho_{\alpha\alpha}\right)\ .
\end{aligned} \tag{19.17}$$

The third equation is transformed to the four equations

$$
\begin{aligned}
&\underline{\dot{\rho}_{b\beta} = -\mathrm{i}\left(-\varepsilon - \Delta + \Omega_{b\beta} - \mathrm{i}0\right)\rho_{b\beta} - \mathrm{i}\lambda_{b\beta}\left(\rho_{\beta\beta} - \rho_{bb}\right)}\,, \\
&\underline{\dot{\rho}_{\beta b} = -\mathrm{i}\left(\varepsilon + \Delta + \Omega_{\beta b} - \mathrm{i}0\right)\rho_{\beta b} - \mathrm{i}\lambda_{\beta b}\left(\rho_{bb} - \rho_{\beta\beta}\right)}\,, \\
&\dot{\rho}_{bb} = -\frac{\rho_{bb}}{T_1} - \mathrm{i}\sum_a \left(\Lambda_{ba}\rho_{ab} - \rho_{ba}\Lambda_{ab}\right) - \mathrm{i}\sum_\beta \underline{\left(\lambda_{b\beta}\rho_{\beta b} - \rho_{b\beta}\lambda_{\beta b}\right)}\,, \\
&\dot{\rho}_{\beta\beta} = -\frac{\rho_{\beta\beta}}{T_1} - \mathrm{i}\sum_\alpha \left(\Lambda_{\beta\alpha}\rho_{\alpha\beta} - \rho_{\beta\alpha}\Lambda_{\alpha\beta}\right) - \mathrm{i}\sum_b \underline{\left(\lambda_{\beta b}\rho_{b\beta} - \rho_{\beta b}\lambda_{b\beta}\right)}\,.
\end{aligned}
\tag{19.18}
$$

Finally, the fourth equation takes the form of the four equations

$$
\begin{aligned}
&\underline{\dot{\rho}_{a\alpha} = -\mathrm{i}\left(-\varepsilon + \Omega_{a\alpha} - \mathrm{i}0\right)\rho_{a\alpha} - \mathrm{i}\lambda_{a\alpha}\left(\rho_{\alpha\alpha} - \rho_{aa}\right)}\,, \\
&\underline{\dot{\rho}_{\alpha a} = -\mathrm{i}\left(\varepsilon + \Omega_{\alpha a} - \mathrm{i}0\right)\rho_{\alpha a} - \mathrm{i}\lambda_{\alpha a}\left(\rho_{aa} - \rho_{\alpha\alpha}\right)}\,, \\
&\dot{\rho}_{aa} = \frac{1}{T_1}\sum_b \langle a|b\rangle\langle 0|1\rangle\rho_{bb}\langle 1|0\rangle\langle b|a\rangle - \mathrm{i}\sum_b \left(\Lambda_{ab}\rho_{ba} - \rho_{ab}\Lambda_{ba}\right) \\
&\qquad\qquad - \mathrm{i}\sum_\alpha \underline{\left(\lambda_{a\alpha}\rho_{\alpha a} - \rho_{a\alpha}\lambda_{\alpha a}\right)}\,, \\
&\dot{\rho}_{\alpha\alpha} = \frac{1}{T_1}\sum_\beta \langle \alpha|\beta\rangle\langle 2|3\rangle\rho_{\beta\beta}\langle 3|2\rangle\langle \beta|\alpha\rangle - \mathrm{i}\sum_\beta \left(\Lambda_{\alpha\beta}\rho_{\beta\alpha} - \rho_{\alpha\beta}\Lambda_{\beta\alpha}\right) \\
&\qquad\qquad - \mathrm{i}\sum_a \underline{\left(\lambda_{\alpha a}\rho_{a\alpha} - \rho_{\alpha a}\lambda_{a\alpha}\right)}\,.
\end{aligned}
\tag{19.19}
$$

The underlined terms are due to the tunneling operator.

Figure 8.1 shows that the temporal evolution of the diagonal elements can be studied by neglecting coherent effects if the pumping is not large. In the case considered here, both the laser pumping Λ and tunneling λ are rather weak. Therefore we can neglect all derivatives with respect to time in (19.17), and in the underlined equations in the sets (19.18) and (19.19). After that we can express all off-diagonal elements of the density matrix in terms of the diagonal elements. By inserting these expressions for the off-diagonal elements into the third and fourth equations of (19.18) and (19.19), we arrive at the following rate equations:

$$
\begin{aligned}
&\dot{\rho}_{bb} = -\left(\frac{1}{T_1} + \sum_a k_{ba}\right)\rho_{bb} + \sum_a k_{ba}\rho_{aa} - \sum_\beta r_{b\beta}\rho_{bb} + \sum_\beta r_{b\beta}\rho_{\beta\beta}\,, \\
&\dot{\rho}_{aa} = \frac{1}{T_1}\sum_b \langle a|b\rangle\langle 0|1\rangle\rho_{bb}\langle 1|0\rangle\langle b|a\rangle + \sum_b k_{ba}\rho_{bb} - \sum_b k_{ba}\rho_{aa} \\
&\qquad\qquad - \sum_\alpha r_{a\alpha}\rho_{aa} + \sum_\alpha r_{a\alpha}\rho_{\alpha\alpha}\,,
\end{aligned}
$$

$$
\begin{aligned}
\dot{\rho}_{\beta\beta} &= -\left(\frac{1}{T_1} + \sum_{\alpha} k_{\beta\alpha}\right)\rho_{\beta\beta} + \sum_{\alpha} k_{\beta\alpha}\rho_{\alpha\alpha} + \sum_{b} r_{b\beta}\rho_{bb} - \sum_{b} r_{b\beta}\rho_{\beta\beta}\,, \\
\dot{\rho}_{\alpha\alpha} &= \frac{1}{T_1}\sum_{\beta} \langle\alpha|\beta\rangle\langle 2|3\rangle\rho_{\beta\beta}\langle 3|2\rangle\langle\beta|\alpha\rangle + \sum_{\beta} k_{\beta\alpha}\rho_{\beta\beta} - \sum_{\beta} k_{\beta\alpha}\rho_{\alpha\alpha} \\
&\quad + \sum_{a} r_{a\alpha}\rho_{aa} - \sum_{a} r_{a\alpha}\rho_{\alpha\alpha}\,.
\end{aligned}
\tag{19.20}
$$

Here the equations

$$
\begin{aligned}
k_{ba} &= \Lambda_{ba}\Lambda_{ab}\frac{1/T_1}{(\Delta_0 + \Omega_{ba})^2 + (1/2T_1)^2}\,, \\
k_{\beta\alpha} &= \Lambda_{\beta\alpha}\Lambda_{\alpha\beta}\frac{1/T_1}{(\Delta_0 + \Delta + \Omega_{\beta\alpha})^2 + (1/2T_1)^2}
\end{aligned}
\tag{19.21}
$$

describe the rates of the optical transitions and the equations

$$
\begin{aligned}
r_{b\beta} &= 2\pi\lambda_{b\beta}\lambda_{\beta b}\delta\left(\Omega_{b\beta} - \varepsilon - \Delta\right)\,, \\
r_{a\alpha} &= 2\pi\lambda_{a\alpha}\lambda_{\alpha a}\delta\left(\Omega_{a\alpha} - \varepsilon\right)
\end{aligned}
\tag{19.22}
$$

describe the rates of tunneling transitions in the excited and ground electronic states of the chromophore.

It is obvious that the diagonal elements of the density matrix can be written in the form

$$
\rho_{aa} = \rho_a\rho_0\,, \quad \rho_{bb} = \rho_b\rho_1\,, \quad \rho_{\alpha\alpha} = \rho_\alpha\rho_2\,, \quad \rho_{\beta\beta} = \rho_\beta\rho_3\,,
\tag{19.23}
$$

where ρ_a, ρ_b, ρ_α, and ρ_β are the probabilities of finding the system in the relevant quantum state. They satisfy

$$
\sum_{a}\rho_a = \sum_{b}\rho_b = \sum_{\alpha}\rho_\alpha = \sum_{\beta}\rho_\beta = 1\,.
\tag{19.24}
$$

Inserting (19.23) into (19.20) and summing over phonon indices, we arrive at the following rate equations for the probabilities of finding the system in one of the four states shown in Fig. 19.3:

$$
\begin{aligned}
\dot{\rho}_1 &= -\Gamma_1\rho_1 + L_0\rho_0 - b\rho_1 + B\rho_3\,, \\
\dot{\rho}_0 &= \Gamma_1\rho_1 - L_0\rho_0 \quad a\rho_0 + A\rho_2\,, \\
\dot{\rho}_3 &= -\Gamma_3\rho_3 + L_2\rho_2 + b\rho_1 - B\rho_3\,, \\
\dot{\rho}_2 &= \Gamma_3\rho_3 - L_2\rho_2 + a\rho_0 - A\rho_2\,,
\end{aligned}
\tag{19.25}
$$

where

$$
\begin{aligned}
&\Gamma_1 = \frac{1}{T_1} + L_1 = \frac{1}{T_1} + k_{10}^{\mathrm{e}}\,, \qquad \Gamma_3 = \frac{1}{T_1} + L_3 = \frac{1}{T_1} + k_{32}^{\mathrm{e}}\,, \\
&L_0 = k_{10}^{\mathrm{g}}\,, \qquad L_2 = k_{32}^{\mathrm{g}}\,,
\end{aligned}
\tag{19.26}
$$

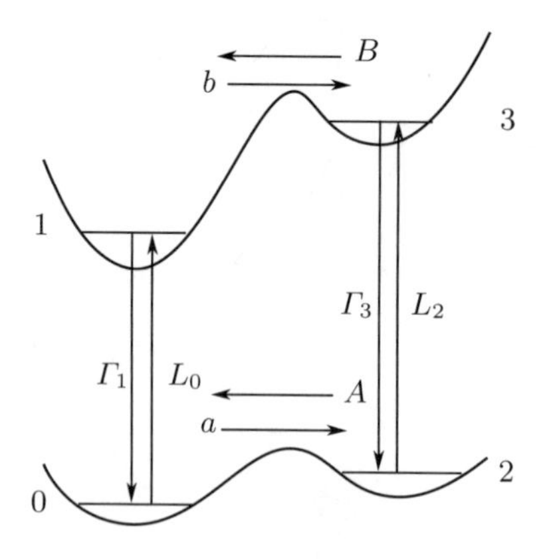

Fig. 19.3. Optical and tunneling transitions allowing for (19.25)

and the coefficients

$$
\begin{aligned}
k_{10}^{\mathrm{e}} &= \sum_{b,a} \rho_b k_{ba} = \langle 1|0\rangle^2 \chi^2 \sum_{b,a} \rho_b \langle b|a\rangle^2 \frac{1/T_1}{(\Delta_0 + \Omega_{ba})^2 + (1/2T_1)^2} , \\
k_{10}^{\mathrm{g}} &= \sum_{ba} \rho_a k_{ba} , \\
k_{32}^{\mathrm{e}} &= \sum_{\beta,\alpha} \rho_\beta k_{\beta\alpha} = \langle 3|2\rangle^2 \chi^2 \sum_{\beta,\alpha} \rho_\beta \langle \beta|\alpha\rangle^2 \frac{1/T_1}{(\Delta_0 + \Delta + \Omega_{\beta\alpha})^2 + (1/2T_1)^2} , \\
k_{32}^{\mathrm{g}} &= \sum_{\beta\alpha} \rho_\alpha k_{\beta\alpha}
\end{aligned}
\tag{19.27}
$$

define rates of light-induced transitions between levels 1, 0 and 3, 2 shown in Fig. 19.3.

The coefficients described by

$$
\begin{aligned}
b &= \sum_{\beta b} \rho_b r_{b\beta} , \quad & B &= \sum_{\beta b} \rho_\beta r_{b\beta} , \\
a &= \sum_{\alpha a} \rho_a r_{a\alpha} , \quad & A &= \sum_{\alpha a} \rho_\alpha r_{a\alpha}
\end{aligned}
\tag{19.28}
$$

are the microscopic expressions for the rates of tunneling transitions between coupled wells in the excited and ground electronic states of the chromophore. The rates b and B describe tunneling in the TLS when the chromophore is excited, whilst rates a and A describe tunneling in the TLS when the chromophore is in the ground electronic state. The probabilities of the transitions defined by (19.26), (19.27) and (19.28) are shown in Fig. 19.3.

19.3 Spontaneous and Light-Induced Transitions in TLSs

Let us consider the case when the system is excited solely via the $0-1$ transition, i.e., $L_2 = 0$ and level 3 shown in Fig. 19.3 is situated lower than level 1. Then the substitution $b \to B$ and $B \to b$ should be made in (19.25). This set of equations thus takes the form

$$\begin{aligned}
\dot{\rho}_1 &= -\left(\Gamma_1 + B\right)\rho_1 + L_0\rho_0 + b\rho_3 \, , \\
\dot{\rho}_3 &= B\rho_1 - \left(\frac{1}{T_1} + b\right)\rho_3 \, , \\
\dot{\rho}_0 &= \Gamma_1\rho_1 - \left(L_0 + a\right)\rho_0 + A\rho_2 \, , \\
\dot{\rho}_2 &= a\rho_0 + \frac{\rho_3}{T_1} - A\rho_2 \, .
\end{aligned} \tag{19.29}$$

Here and in the following, we shall assume the hierarchy of rate constants

$$\Gamma \gg L_0 \gg A, B > a, b \, . \tag{19.30}$$

With this hierarchy, the temporal evolution of the probabilities ρ_j has two stages: fast and slow relaxation. Fast relaxation occurs on a time scale of order T_1. Once this stage is over, a quasi-equilibrium is established between the probabilities ρ_1 and ρ_3 on the one hand and the probability ρ_0 on the other hand. This relation can be found if we set $\mathrm{d}\rho_1/\mathrm{d}t = \mathrm{d}\rho_3/\mathrm{d}t = 0$. Then taking into account the inequalities described by (19.30), we find the relations

$$\begin{aligned}
\rho_1 &= \frac{\left(1/T_1 + b\right)L_0\rho_0}{\left(1/T_1 + b\right)\Gamma_1 + B/T_1} \approx \frac{L_0}{\Gamma_1}\rho_0 \, , \\
\rho_3 &= \frac{BL_0\rho_0}{\left(1/T_1 + b\right)\Gamma_1 + B/T_1} \approx \frac{L_0}{\Gamma_1}T_1 B\rho_0 \, .
\end{aligned} \tag{19.31}$$

Inserting these into the third and fourth equations of the set (19.29), we obtain the two equations

$$\begin{aligned}
\dot{\rho}_0 &= -\left(\tilde{B} + a\right)\rho_0 + A\rho_2 \, , \\
\dot{\rho}_2 &= \left(\tilde{B} + a\right)\rho_0 - A\rho_2 \, .
\end{aligned} \tag{19.32}$$

The solution of these equations is

$$\begin{aligned}
\rho_0(t) &= \frac{A}{\tilde{B} + R} + \left(\rho_0(0) - \frac{A}{\tilde{B} + R}\right)\exp\left[-(\tilde{B} + R)t\right] \, , \\
\rho_2(t) &= 1 - \rho_0(t) \, ,
\end{aligned} \tag{19.33}$$

where

$$\tilde{B} = \frac{L_0}{\Gamma_1} B \,, \qquad R = A + a \,. \tag{19.34}$$

According to (19.31) and (19.33), slow relaxation is determined by tunneling between quantum states 0 and 2 in the ground electronic state and between states 1 and 3 relating to the excited electronic state of the chromophore. Tunneling in the ground and excited electronic states is called spontaneous and light-induced tunneling, respectively. The rate $\tilde{B}$ of light-induced tunneling is proportional to the pumping L_0.

Let us consider another case when the system is excited solely via the $2-3$ transition. Then (19.25) takes the form

$$\begin{aligned}
\dot{\rho}_1 &= -\left(\frac{1}{T_1} + B\right)\rho_1 + b\rho_3 \,, \\
\dot{\rho}_3 &= B\rho_1 - (\Gamma_3 + b)\,\rho_3 + L_2\rho_2 \,, \\
\dot{\rho}_0 &= \frac{\rho_1}{T_1} - a\rho_0 + A\rho_2 \,, \\
\dot{\rho}_2 &= \Gamma_3\rho_3 + a\rho_0 - (L_2 + A)\,\rho_2 \,.
\end{aligned} \tag{19.35}$$

It can be solved in the same way as (19.29). Setting $\mathrm{d}\rho_1/\mathrm{d}t = \mathrm{d}\rho_3/\mathrm{d}t = 0$, we obtain

$$\begin{aligned}
\rho_1 &= \frac{bL_2\rho_2}{(1/T_1 + B)\,\Gamma_3 + b/T_1} \approx \frac{L_2}{\Gamma_3} T_1 b\rho_2 \,, \\
\rho_3 &= \frac{(1/T_1 + B)\,L_2\rho_2}{(1/T_1 + B)\,\Gamma_3 + b/T_1} \approx \frac{L_2}{\Gamma_3}\rho_2 \,.
\end{aligned} \tag{19.36}$$

Using these relations, we find the following equations for slow relaxation:

$$\begin{aligned}
\dot{\rho}_0 &= -a\rho_0 + (\tilde{b} + A)\rho_2 \,, \\
\dot{\rho}_2 &= a\rho_0 - (\tilde{b} + A)\rho_2 \,,
\end{aligned} \tag{19.37}$$

where

$$\tilde{b} = \frac{L_2}{\Gamma_3} b \tag{19.38}$$

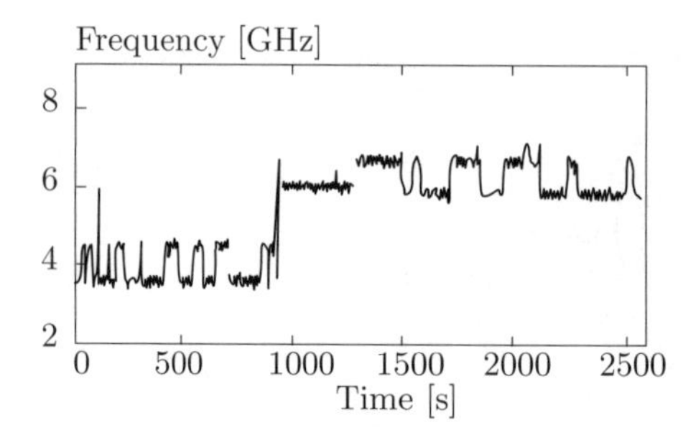

Fig. 19.4. Jumps of the 580.77 nm spectral line of a terylene molecule in polyethelene [28]

describes the rate of light-induced tunneling in the case of the excitation via the $2 - 3$ optical transition.

The solution of these equations is given by

$$\begin{aligned} \rho_2(t+t_0) &= \frac{a}{\tilde{b}+R} + \left(\rho_2(t_0) - \frac{a}{\tilde{b}+R}\right)\exp\left[-(\tilde{b}+R)t\right] , \\ \rho_0(t+t_0) &= 1 - \rho_2(t+t_0) . \end{aligned} \quad (19.39)$$

It will be shown later how we can explain jumps of optical lines like those shown in Fig. 19.4 with the help of the equations derived here.

The probability of light-induced tunneling can exceed the probability of spontaneous tunneling if the TLS describes quantum states of the chromophore itself. Such a TLS is called an extrinsic TLS because it arises due to a guest molecule embedded in the solvent. The number of extrinsic TLSs matches the total number of guest molecules. Transient and persistent spectral hole burning result from these extrinsic TLSs.

However, there is another type of TLS in polymers and glasses. These exist in pure polymers and glasses without guest molecules. They are thus called intrinsic TLSs. The number of intrinsic TLSs can considerably exceed the number of guest molecules. Rate constants b and B of tunneling in the excited electronic state in intrinsic TLSs are of the order of the rate constants a and A of spontaneous tunneling. Therefore we can neglect light-induced tunneling in an intrinsic TLS if the laser pumping is weak and satisfies the inequality (19.30), i.e., $L/\Gamma \ll 1$. It will be shown later that the interaction of the guest molecule with the huge number of intrinsic TLSs yields temporal line broadening.

19.4 Interaction with One TLS. Approximate Solution

If we want to take into account the interaction of the guest molecule with a huge number of TLSs, we will be forced to solve the infinite set of rate equations. It is impossible to find an exact solution for such a complicated system. However, this task can be considerably simplified if we make use of the large difference in value of the rates of electronic and tunneling relaxation. First, we carry out a separation of the electronic and tunneling relaxations in order to find an approximate solution, by considering the simplest model when the chromophore interacts with one TLS. It will be shown in the next section that this approximate solution can easily be used in the case when the chromophore interacts with a huge number of TLSs. This is a significant advantage of the approximate solution.

By neglecting light-induced tunneling, we can set $b = B = 0$. Therefore (19.29) takes the form

$$\begin{aligned}\dot{\rho}_1 &= -\Gamma_1\rho_1 + L_0\rho_0 \,,\\ \dot{\rho}_0 &= \Gamma_1\rho_1 - L_0\rho_0 - a\rho_0 + A\rho_2 \,,\\ \dot{\rho}_2 &= a\rho_0 - A\rho_2 \,.\end{aligned} \tag{19.40}$$

An exact solution with the initial condition $\rho_0(0) = 1$ is given by

$$\begin{aligned}\rho_1(t) &= L_0\left[\frac{A}{z_1 z_2} + \left(1 - \frac{A}{z_1}\right)\frac{e^{-z_1 t}}{z_2 - z_1} - \left(1 - \frac{A}{z_2}\right)\frac{e^{-z_2 t}}{z_2 - z_1}\right],\\ \rho_2(t) &= a\left[\frac{\Gamma_1}{z_1 z_2} + \left(1 - \frac{\Gamma_1}{z_1}\right)\frac{e^{-z_1 t}}{z_2 - z_1} - \left(1 - \frac{\Gamma_1}{z_2}\right)\frac{e^{-z_2 t}}{z_2 - z_1}\right],\\ \rho_0(t) &= 1 - \rho_1(t) - \rho_2(t)\,,\end{aligned} \tag{19.41}$$

where

$$z_{1,2} = \frac{\Gamma_1 + L_0 + R}{2} \mp \sqrt{\left(\frac{\Gamma_1 + L_0 - R}{2}\right)^2 + L_0 a}\,. \tag{19.42}$$

The rates of the electronic and tunneling relaxations are involved in these expressions. However, we can separate fast electronic relaxation and slow tunneling relaxation by finding an approximate solution.

To zeroth order in the very small ratio A/Γ_1 and to first nonvanishing order in the small ratio L_0/Γ_1, we find the approximate relations

$$\begin{aligned}&z_1 \approx R\,, \quad z_2 \approx \Gamma_1\,, \quad \frac{L_0 A}{z_1 z_2} = \frac{L_0 A}{L_0 A + \Gamma_1 R} \approx \frac{L_0}{\Gamma_1}\left(1 - \frac{a}{R}\right),\\ &\frac{A}{z_1} \approx \left(1 - \frac{a}{R}\right), \quad \frac{a}{z_2 - z_1}\frac{\Gamma_1}{z_1} \approx \frac{a}{R}\,,\\ &\frac{A}{z_2} \approx \frac{a}{z_2 - z_1}\frac{\Gamma_1}{z_2} \approx \frac{a}{z_2 - z_1} = 0\,.\end{aligned} \tag{19.43}$$

The exact (19.41) can be transformed with the help of these relations to give

$$\begin{aligned}\rho_1(t) &= \frac{L_0}{\Gamma_1}\left[\left(1 - e^{-\Gamma_1 t}\right) - \frac{a}{R}\left(1 - e^{-Rt}\right)\right],\\ \rho_2(t) &= \frac{a}{R}\left(1 - e^{-Rt}\right),\\ \rho_0(t) &= 1 - \rho_1(t) - \rho_2(t)\,.\end{aligned} \tag{19.44}$$

These approximate formulas can be written in the form

$$\begin{aligned}\rho_1(t) &\approx n_1(t) - n_1(\infty)\,p_2(t)\,,\\ \rho_2(t) &\approx n_0(\infty)\,p_2(t)\,,\\ \rho_0(t) &= 1 - \rho_1(t) - \rho_2(t) \approx n_0(t) - n_0(\infty)\,p_2(t) + n_1(\infty)\,p_2(t)\,,\end{aligned} \tag{19.45}$$

where the functions

$$n_1(t) = \frac{L_0}{\Gamma_1 + L_0}\left[1 - \mathrm{e}^{-(\Gamma_1 + L_0)t}\right], \quad n_0(t) = 1 - n_1(t) \tag{19.46}$$

describe the temporal evolution of the probabilities due to laser pumping, but without tunneling relaxation. However, the functions

$$p_2(t) = \frac{a}{R}\left(1 - \mathrm{e}^{-Rt}\right), \quad p_0(t) = 1 - p_2(t) \tag{19.47}$$

describe the temporal evolution of the probabilities when laser pumping is switched off. Since $p_2(0) = 0$ and it only reaches a very small value of order R/Γ when $t \sim \Gamma^{-1}$, we can carry out the following substitution in (19.45), within the accuracy used:

$$n_0(\infty)\, p_2(t) \to n_0(t) p_2(t), \quad n_1(\infty)\, p_2(t) \to n_1(t) p_2(t). \tag{19.48}$$

After this substitution (19.45) takes the form

$$\begin{aligned} \rho_1(t) &= n_1(t) p_0(t), \quad \rho_2(t) = n_0(t) p_2(t), \\ \rho_0(t) &= n_1(t)\left[1 - p_0(t)\right] + n_0(t) p_0(t). \end{aligned} \tag{19.49}$$

These equations coincide with the exact solution to an accuracy of order R/Γ. The approximate solution consists of the functions $n_{1,0}(t)$ and $p_{0,2}(t)$ found under the assumption that laser pumping does not influence tunneling, and tunneling does not influence the temporal dependence in the functions $n_{1,0}(t)$. In the next section, this simple property of the functions $n_{1,0}(t)$ and $p_{0,2}(t)$ allows us to build up an approximate solution for the case when the chromophore interacts with a huge number of TLSs.

The prescription for finding the approximate solution has three stages.

- In the first stage we neglect tunneling in (19.40) and solve the equations

  $$\dot{\rho}_1 = -\Gamma_1 \rho_1 + L_0 \rho_0, \qquad \dot{\rho}_0 = \Gamma_1 \rho_1 - L_0 \rho_0. \tag{19.50}$$

  The solution of these equations is denoted $n_{1,0}(t)$. This solution describes the true temporal behavior on a time scale of order of Γ^{-1}.
- In the second stage we neglect the laser pumping and electronic relaxation in (19.40) and solve the equations

  $$\dot{\rho}_0 = -a\rho_0 + A\rho_2, \qquad \dot{\rho}_2 = a\rho_0 - A\rho_2. \tag{19.51}$$

  The solution of these equations is denoted $p_{0,2}(t)$. This solution describes the true temporal behavior of the system on a time scale of order R^{-1}.
- Combining the solutions of (19.50) and (19.51), we can build up the approximate solution described by (19.49), whose temporal behavior over the whole time scale coincides with the temporal behavior of the exact solution.

19.5 Interaction with Many TLSs Undergoing Spontaneous Tunneling

First of all we use the prescription found in the last section in order to consider a chromophore interacting with two TLSs. Since one TLS has two quantum states, two TLSs have four quantum states. Figure 19.5 shows these four states of two TLSs in the ground and excited electronic states of the chromophore.

In Sect. 18.2 we transformed the infinite set (19.13) which describes a chromophore interacting with one TLS to four rate equations described by (19.25), and we found mathematical expressions for the rate constants. A similar simplification procedure can be carried out in the case of a chromophore interacting with two TLSs. We arrive at the set of eight rate equations for eight quantum states shown in Fig. 19.5. We shall only take into account optical transitions without tunneling. All these transitions are shown in Fig. 19.5. Since the tunneling operator is a linear function of the tunnelon operators c and c^+, it can connect quantum states differing by one tunnelon, i.e., the following states

$$\underleftrightarrow{\rho_0 \longleftrightarrow \rho_2 \longleftrightarrow \rho_6 \longleftrightarrow \rho_4} \qquad \underleftrightarrow{\rho_1 \longleftrightarrow \rho_3 \longleftrightarrow \rho_7 \longleftrightarrow \rho_5} \,.$$

These comments explain how the infinite set of equations for the density matrix is transformed to the following eight rate equations:

$$\begin{aligned}
\dot\rho_1 &= -\left(\Gamma_1 + b + b'\right)\rho_1 + L_0\rho_0 + B\rho_3 + B'\rho_5 \,, \\
\dot\rho_0 &= \Gamma_1\rho_1 - \left(L_0 + a + a'\right)\rho_0 + A\rho_2 + A'\rho_4 \,, \\
\dot\rho_3 &= -\left(\Gamma_3 + B + b'\right)\rho_3 + L_2\rho_2 + b\rho_1 + B'\rho_7 \,, \\
\dot\rho_2 &= \Gamma_3\rho_3 - \left(L_2 + A + a'\right)\rho_2 + a\rho_0 + A'\rho_6 \,, \\
\dot\rho_5 &= -\left(\Gamma_5 + B' + b\right)\rho_5 + L_4\rho_4 + b'\rho_1 + B\rho_7 \,, \\
\dot\rho_4 &= \Gamma_5\rho_5 - \left(L_4 + A' + a\right)\rho_4 + a'\rho_0 + A\rho_6 \,, \\
\dot\rho_7 &= -\left(\Gamma_7 + B' + B\right)\rho_7 + L_6\rho_6 + b'\rho_3 + b\rho_5 \,, \\
\dot\rho_6 &= \Gamma_7\rho_7 - \left(L_6 + A' + A\right)\rho_6 + a'\rho_2 + a\rho_4 \,,
\end{aligned} \tag{19.52}$$

where the rate constants b, B, a, A of one TLS are defined by (19.28) and the rate constants b', B', a', A' of another TLS are defined by similar expressions. The rates of laser pumping

$$\begin{aligned}
&L_0 = k^{\mathrm{g}}_{10} \,, \quad L_1 = k^{\mathrm{e}}_{10} \,, \quad L_2 = k^{\mathrm{g}}_{32} \,, \quad L_3 = k^{\mathrm{e}}_{32} \,, \\
&L_4 = k^{\mathrm{g}}_{54} \,, \quad L_5 = k^{\mathrm{e}}_{54} \,, \quad L_6 = k^{\mathrm{g}}_{76} \,, \quad L_7 = k^{\mathrm{e}}_{76}
\end{aligned} \tag{19.53}$$

define the probabilities of induced optical transitions without tunneling.

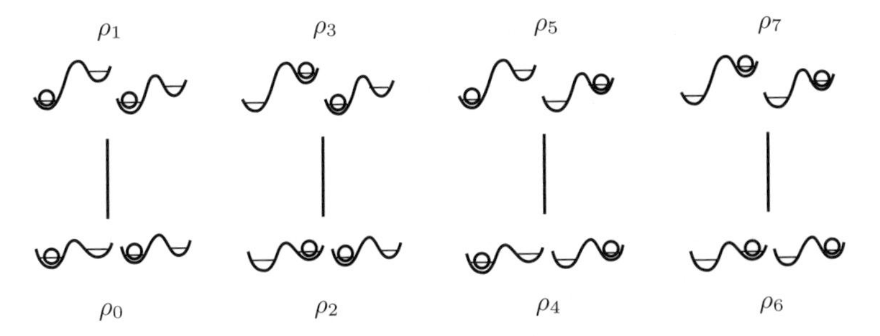

Fig. 19.5. Quantum states of a system consisting of a chromophore and two TLSs and direct optical transitions without tunneling

Let us turn back to the case when laser excitation is realized via the $0-1$ optical transition. By neglecting light-induced tunneling, we assume $b = B = b' = B' = 0$. Then (19.51) takes the following block form:

$$\begin{aligned}
\dot{\rho}_1 &= -\Gamma_1\rho_1 + L_0\rho_0 \,, \\
\dot{\rho}_0 &= \Gamma_1\rho_1 - L_0\rho_0 - (a+a')\,\rho_0 + A\rho_2 + A'\rho_4 \,, \\
\dot{\rho}_2 &= a\rho_0 - (A+a')\,\rho_2 + A'\rho_6 \,, \\
\dot{\rho}_4 &= a'\rho_0 - (A'+a)\,\rho_4 + A\rho_6 \,, \\
\dot{\rho}_6 &= a'\rho_2 + a\rho_4 - (A'+A)\,\rho_6 \,.
\end{aligned} \tag{19.54}$$

By applying the prescription derived in the last section, we can write the following approximate solution:

$$\begin{aligned}
&\rho_1(t) = n_1(t)P_0(t)\,, \quad \rho_2(t) = n_0(t)P_2(t)\,, \quad \rho_4(t) = n_0(t)P_4(t)\,, \\
&\rho_6(t) = n_0(t)P_6(t)\,, \quad \rho_0(t) = n_1(t)[1-P_0(t)] + n_0(t)P_0(t)\,,
\end{aligned} \tag{19.55}$$

where the probabilities P can be found from

$$\begin{aligned}
\dot{P}_0 &= -\,(a+a')\,P_0 + AP_2 + A'P_4 \,, \\
\dot{P}_2 &= -\,(A+a')\,P_2 + aP_0 + A'P_6 \,, \\
\dot{P}_4 &= -\,(a+A')\,P_4 + a'P_0 + AP_6 \,, \\
\dot{P}_6 &= -\,(A+A')\,P_6 + a'P_2 + aP_4 \,.
\end{aligned} \tag{19.56}$$

Their solution is given by

$$P_0 = p_0p_0'\,, \quad P_2 = p_2p_0'\,, \quad P_4 = p_0p_2'\,, \quad P_6 = p_2p_2'\,, \tag{19.57}$$

where the factors are the probabilities relating to each TLS. They can be found from

$$\begin{aligned}
&p_0 = -ap_0 + Ap_2\,, \quad p_0' = -a'p_0' + A'p_2'\,, \\
&p_2 = ap_0 - Ap_2\,, \quad\;\; p_2' = a'p_0' - A'p_2'\,.
\end{aligned} \tag{19.58}$$

The probabilities satisfy the condition

$$P_0 + P_2 + P_4 + P_6 = (p_0 + p_2)(p'_0 + p'_2) = 1 \,. \tag{19.59}$$

Equation (19.54) is an approximate solution when laser excitation is realized via the $1 \leftarrow 0$ optical transition. The case when excitation is realized via the $3 \leftarrow 2$ optical transition can be considered in similar fashion. Then, instead of (19.55), we arrive at the formulas

$$\begin{aligned} &\rho_3(t) = n_3(t)P_2(t) \,, \quad \rho_0(t) = n_2(t)P_0(t) \,, \quad \rho_4(t) = n_2(t)P_4(t) \,, \\ &\rho_6(t) = n_2(t)P_6(t) \,, \quad \rho_2(t) = n_3(t)\left[1 - P_2(t)\right] + n_2(t)P_2(t) \,, \end{aligned} \tag{19.60}$$

where

$$n_3(t) = \frac{L_2}{\Gamma_3 + L_2}\left[1 - \mathrm{e}^{-(\Gamma_3 + L_2)t}\right] \,, \qquad n_2(t) = 1 - n_3(t) \,, \tag{19.61}$$

These formulas can also be derived from (19.55) by the following substitution of indices: $0 \to 2$, $1 \to 3$ and $2 \to 0$.

The formulas for the case when laser excitation is realized via the $N \leftarrow M$ transition are

$$\begin{aligned} &\rho_N(t) = n_N(t)P_M(t) \,, \quad \rho_{M'}(t) = n_M(t)P_{M'}(t) \\ &\rho_M(t) = n_N(t)\left[1 - P_M(t)\right] + n_M(t)P_M(t) \,, \quad (M' \neq M) \,. \end{aligned} \tag{19.62}$$

These formulas can be used in the case when a guest molecule interacts with N_0 TLSs. However, the probability $P_M(t)$ is a product of N_0 probabilities $p_j(t)$ relating to each TLS.

20. Dynamical Theory of Spectral Diffusion

The phenomenon of spectral diffusion (SD) was discovered in spin systems. It manifests itself in the dependence of dephasing time T_2 on the time delay τ between two laser pulses in experiments with microwave echo [97]. Klauder and Anderson [98] put forward a theory of stochastic type to explain the SD effect in spin systems. In their theory, a dipolar interaction between resonant spin and a huge number of off-resonant spins is a source of temporal line broadening. This approach underlies all theories developed later both for spin systems [99–101] and for optical spectra of chromophores [102–104]. In optical spectra, SD manifests itself through a temporal dependence of the optical line width or optical dephasing time T_2.

The aim of this chapter is to present the dynamical theory for SD developed by the author in [105]. In this theory the half-width of a single-molecule optical line is derived purely on the basis of quantum mechanics, i.e., using the Hamiltonian of the system. The theory gives some new predictions compared to theories developed earlier. It has been successfully applied to the interpretation of experimental data in hole-burning spectroscopy and single-molecule spectroscopy.

20.1 Absorption Coefficient of a Single Guest Molecule Interacting with Nonequilibrium TLSs

All theories for SD agree that a decisive contribution to this effect emerges from spontaneous relaxations of the huge number of TLSs existing in polymers and glasses. Tunneling transitions in TLSs are realized as jumps at random times from one quantum state to another. Due to the interaction between chromophore and TLS, each jump yields a shift in the frequency of the chromophore optical line. The instant at which the jump occurs is purely random and cannot be calculated. However, the probability of this jump has a definite value. The purely stochastic theory models the probability of the frequency jump, whereas the dynamical theory offers a prescription for calculating this probability with the help of the Hamiltonian of the system and equations for the density matrix. Although a stochastic theory for line broadening is, as a rule, simpler than a dynamic theory, the latter can explain a wider range of experimental facts and provide more detailed predictions.

For instance, the dynamical theory for the optical band developed in Part IV could explain in detail the temperature behavior of the ZPL and PSB, the intensity distribution in conjugate vibronic absorption and fluorescence spectra, the breakdown of mirror symmetry in the conjugate PSB, and so on. No stochastic theory can explain all these facts. Therefore the stochastic approach is used only to study the line broadening problem. However, even in this field of application common to both approaches, the dynamical approach enables one to make more comprehensive predictions and to identify the shortcomings of existing stochastic theories. This will be discussed later.

It has been shown in previous sections how (19.62) for the density matrix of the chromophore interacting with phonons and all types of TLS can be derived from the Hamiltonian of the system. Equation (19.62) relates to the case when laser excitation is realized via the $N \leftarrow M$ optical transition. Since the Lorentzians $L_0, L_2, L_4, \ldots$ have different resonant frequencies, we can write an expression for the population of the quantum state excited by light as

$$\sum_N \rho_N(t) = \sum_M \frac{L_M P_M(t)}{\Gamma_N + L_M} \{1 - \exp\left[-\left(\Gamma_N + L_M\right) t\right]\} . \tag{20.1}$$

This expression is valid for laser excitation via an arbitrary optical transition, because only the Lorentzian in resonance with the laser line gives a nonvanishing contribution to this sum. At least, its contribution will be predominant. At $t \gg \Gamma_N^{-1}$, the population of the excited state will be given by

$$\sum_N \rho_N(t) \approx \sum_M \frac{L_M P_M(t)}{\Gamma_N} . \tag{20.2}$$

It is obvious that the function

$$k(t,T) = \sum_M L_M P_M(t) \tag{20.3}$$

determines the rate of light absorption by a chromophore interacting with phonons and with TLSs which have already reached thermal equilibrium and those which have not reached it. Therefore this absorption coefficient depends on temperature and time. By inserting the expression for L_M into the last equation, we can transform it to

$$
\begin{aligned}
k^{\mathrm{g}}(t,T) &= \sum_{N,M} P_M(t) k_{NM} \\
&= \chi^2 \sum_{N,M} P_M(t) \langle N|M\rangle^2 \sum_{b,a} \rho_a \langle b|a\rangle^2 \frac{1/T_1}{(\Delta_0 + \Omega_{ba} + \Omega_{NM})^2 + (1/2T_1)^2} \\
&\quad + \chi^2 \sum_{N,M} P_M(t) \langle N|M\rangle^2 \sum_{\beta,\alpha} \rho_\alpha \langle \beta|\alpha\rangle^2 \frac{1/T_1}{(\Delta_0 + \Omega_{\beta\alpha} + \Omega_{NM})^2 + (1/2T_1)^2} \\
&= \chi^2 \sum_{B,A} w_A(t,T) \langle B|A\rangle^2 \frac{1/T_1}{(\Delta_0 + \Omega_{BA})^2 + (1/2T_1)^2} .
\end{aligned}
\tag{20.4}
$$

Here N and M are quantum indices of the TLS in the excited and ground states of the chromophore. In the case of the interaction with a single TLS, we can write $M = 0,2$ and $N = 1,3$. This formula differs from the similar expressions considered earlier by the fact that the probability

$$
w_A(t,T) = P_M(t)\rho_A(T) = P_M(t) \begin{Bmatrix} \rho_a \\ \rho_\alpha \end{Bmatrix} \tag{20.5}
$$

of finding a definite state of the tunnelon–phonon system depends on time. Time is measured from the moment when the photon is emitted and the molecule has gone into the ground electronic state.

Using (10.19), we can transform the Lorentzian to the integral over time and then (20.4) takes the form

$$
k^{\mathrm{g}}(t) = \sum_{B,A} w_A(t,T) \Lambda_{BA} \Lambda_{AB} \int_{-\infty}^{\infty} \mathrm{e}^{-\mathrm{i}(\Delta_0 + \Omega_{BA})x - |x|/2T_1} \mathrm{d}x . \tag{20.6}
$$

The exponential function depends on the difference between the eigenvalues of the adiabatic Hamiltonian H_0 defined by (19.2). Using the equations

$$
\Omega_B |B\rangle = \left[\frac{1}{\hbar} H^{\mathrm{e}}(\xi) + \Omega \right] |B\rangle , \quad \Omega_A |A\rangle = \frac{1}{\hbar} H(\xi) |A\rangle , \tag{20.7}
$$

where $H(\xi)$ describes the Hamiltonian of the tunnelon system in the ground state and

$$
H^{\mathrm{e}}(\xi) = H(\xi) + \Delta H(\xi) \tag{20.8}
$$

in the excited electronic state of the chromophore, we can replace the frequencies Ω_A and Ω_B by the relevant Hamiltonians. After such a substitution, (20.6) takes the form

$$
k^{\mathrm{g}}(t) = \chi^2 \int_{-\infty}^{\infty} \mathrm{e}^{\mathrm{i}\Delta_0 x - |x|/2T_1} S^{\mathrm{g}}(t,x) \mathrm{d}x , \tag{20.9}
$$

where χ is the Rabi frequency and the function $S^{g}(t)$ is given by

$$S^{g}(t,x)=\mathrm{Tr}\left\{w(t,T)\exp\left(-\mathrm{i}x\frac{H^{e}(\xi)}{\hbar}\right)\exp\left(\mathrm{i}x\frac{H(\xi)}{\hbar}\right)\right\}. \quad (20.10)$$

This function describes the temporal behavior of the dipolar correlator of the chromophore interacting with phonons and tunnelons. It looks like the function defined by (10.24), which was used to calculate the electron–phonon bands. However, here the density matrix $w(t,T)$ depends on the time variable t. The existence of two time variables x and t in the last equation deserves comment.

The integration in (20.9) is carried out with respect to the time variable x. In the long time limit the temporal behavior of the dipolar correlator is defined by the optical dephasing time T_2. The function $S^{g}(t,x)$ thus approaches zero as $\exp(-|x|/T_2)$ at large x. This function of the time variable x is therefore nonvanishing over such a short time interval x that the density matrix P^{g} of the tunnelon–phonon system is in fact constant, because the temporal dependence of P^{g} is determined by slow tunneling processes. Therefore the time t which determines the time scale of tunneling dynamics can be considered as a temporal parameter which is independent of the time variable x.

The question arises: what time is zero on the tunneling time scale t? The answer can be found if we remember that a single molecule irradiated by laser light permanently emits photons at random times. It is obvious that $\sum_N \rho_N$ in (20.1) describes the temporal behavior of the full two-photon correlator. Hence t is a time interval measured from the moment of photon emission when the guest molecule is in the ground electron-tunnelon state with probability equal to unity. The physical meaning of the time variable t will be discussed in more detail in Chap. 22, where we examine single-molecule spectroscopy of systems with SD.

Let us find an expression for the density matrix $P(t)$ of the tunnelon subsystem in the ground electronic state of the chromophore. In the case of one TLS, its elements are $p_0(t)=1-p(t)$ and $p_2(t)=p(t)$. In operator form, this matrix is given by

$$\hat{P}^{g}(t)=[1-p(t)]\,cc^{+}+p(t)c^{+}c\,, \quad (20.11)$$

where c and c^{+} are the tunnelon annihilation and creation operators. In the case of many TLSs which are not interacting together, the density operator is given by

$$\hat{P}^{g}(t)=\prod_{j=1}^{N}\left\{[1-p_j(t,T)]\,c_jc_j^{+}+p_j(t,T)\,c_j^{+}c_j\right\}, \quad (20.12)$$

where $p_j(t,T) = f_j(T) + [p_j(0,T) - f_j(T)]\exp(-R_jt)]$ and $f_j(T) = [\exp(\hbar\varepsilon_j/kT)+1]^{-1}$. In the long time limit this operator takes the form

$$\hat{P}^{g}(\infty) = \prod_{j=1}^{N} \left\{ [1 - f_j(t)]\, c_j c_j^{+} + f_j(T) c_j^{+} c_j \right\} . \tag{20.13}$$

It is obvious that the matrix elements of this operator and the operator described by (M.2) coincide. Therefore all the results of Sect. 18 for the chromophore interacting with TLSs which have already reached thermal equilibrium can be derived with the density operator of (20.13).

We see that the density matrix of the tunnelons which have not reached thermal equilibrium can be obtained from the equilibrium density operator by simple substitution $f(T) \to p(t,T)$. Therefore, by carrying out this substitution in (18.68), (18.69) and (18.70), which relate to the case of thermal equilibrium, we find the following equations for the dipolar correlator, shift and half-width of the zero tunnelon line:

$$\begin{aligned} S^{g}(t,x) = \exp\left[\mathrm{i}\left[\omega - \omega_0 - \delta(t,T)\right] x - \frac{\gamma(t,T)}{2}|x| \right] \\ \times \prod_{j}^{N_0} \left\{ 1 - C_j(t,T)\left[1 - \exp\left(-\mathrm{i}\Delta_j x - |x|\,\gamma_j\right)\right] \right\} , \end{aligned} \tag{20.14}$$

where

$$C_j(t) = \frac{\Delta_j}{\Delta_j - \mathrm{i}\gamma_j} p_j(t,T) , \tag{20.15}$$

$$\frac{\gamma(t,T)}{2} = \sum_{j=1}^{N_0} \frac{\gamma_j \Delta_j^2}{\Delta_j^2 + \gamma_j^2} p_j(t,T)\left[1 - p_j(t,T)\right] , \tag{20.16}$$

$$\delta(t,T) = \sum_{j=1}^{N_0} \frac{\gamma_j^2 \Delta_j}{\Delta_j^2 + \gamma_j^2} p_j(t,T)\left[1 - p_j(t,T)\right] , \tag{20.17}$$

which relate to the case when the TLSs have not reached thermal equilibrium. According to (20.16) and (20.17), the half-width and shift of the zero tunnelon line depend on time. This dependence is called spectral diffusion. At low temperatures, we have

$$\gamma_j \approx R_j , \tag{20.18}$$

where R is the relaxation rate in the TLS.

20.2 Dependence of the Optical Dephasing Time T_2 on the Time Scale of the Experiment

Let us analyze the expression for the dipolar correlator derived in the last section. First of all we consider the product in (20.14). The coefficients $C_j(t,T)$

determine the amplitudes of the optical lines relating to optical transitions with creation and annihilation of tunnelons. In the dynamical theory, these amplitudes depend on the coupling constants Δ_j.

Equation (Q.8) for the dipolar correlator, derived in Appendix Q within the framework of the pure stochastic approach, coincides with (20.14) only if we take $\delta(t,T) = \gamma(t,T) = 0$ and $C_j(t,T) = p_j(t,T)$. This is a serious shortcoming of the stochastic theory. From the standpoint of the stochastic theory, the line broadening which emerges due to interaction with TLSs can be explained as follows. The remote TLSs interact weakly with the chromophore since their Δ_j are small. Therefore electron–tunnelon lines relating to these TLSs are shifted by very small values Δ_j from the zero tunnelon line. Although each electron–tunnelon line has the zeroth half-width, the coalescence of a great number of lines with small shifts Δ_j looks like a broadening of the zero tunnelon line.

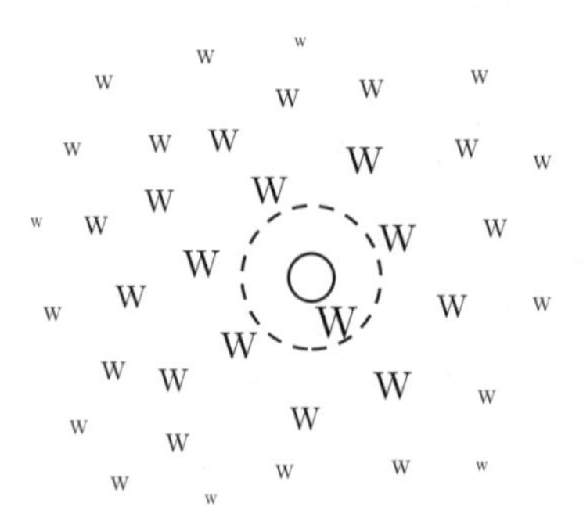

Fig. 20.1. Chromophore with TLS in a polymer. The size of the letter W describing the TLS is proportional to the interaction of the chromophore with this TLS

The line-broadening problem looks different from the point of view of (20.14). First of all, each optical line including the zero tunnelon line will already be broadened due to the factor $\exp[-x\gamma(t,T)/2]$ in front of the product in (20.14). Secondly, in contrast to the stochastic theory where all remote TLSs contribute to the optical band on an equal footing, the dynamic theory predicts that remote TLSs with $\Delta_j/\gamma_j \ll 1$ will not contribute to the optical band. Therefore all TLSs can be divided into two groups. The first group includes TLSs in the immediate vicinity of the chromophore. They are strongly coupled with the chromophore and therefore their amplitudes $C_j(t,T)$ are of order unity. The number of these TLSs is rather small. It is approximately equal to the number of well resolved lines in the optical band of the guest molecule. This local group in Fig. 20.1 consists of one TLS situated inside the circle. The optical band of this chromophore–TLS system consists of two lines. The number of TLSs from the local group and the individual characteristics of these TLSs can vary considerably for various individual guest molecules. The remaining TLSs which are situated outside the circle belong to another group. The properties of this group are similar for each guest molecule. They reflect the properties of the solvent.

Let us consider now the first exponential function in (20.14). As there are two groups of TLSs, we can write

$$\sum_j = \sum_{l=1}^{n} + \sum_{j=1+n}^{N_0} , \tag{20.19}$$

where n is the number of TLSs in the local group. Therefore (20.16) and (20.17) can be written

$$\begin{aligned} \delta\left(t_{\mathrm{w}}, T\right) &= \delta_{\mathrm{L}}(t) + \delta_{\mathrm{hom}}\left(t_{\mathrm{w}}, T\right) , \\ \gamma\left(t_{\mathrm{w}}, T\right) &= \gamma_{\mathrm{L}}(t) + \gamma_{\mathrm{hom}}\left(t_{\mathrm{w}}, T\right) , \end{aligned} \tag{20.20}$$

where δ_{L} and γ_{L} are given by the sum over l and the terms δ_{hom} and γ_{hom} are described by the sum over $s = j - n$. It is obvious that δ_{L} and γ_{L} define a contribution to the line shift and width resulting from TLSs in the local group. They are therefore quite different for different guest molecules. The distribution with respect to line width found in [29] is probably due to the different values of γ_{L} for various individual guest molecules.

It is obvious that TLSs with slow relaxation, for which $\gamma_l \to 0$, do not contribute to δ_{L} and γ_{L} because $\delta_{\mathrm{L}} \propto \gamma_l^2/\Delta_l$ and $\gamma_{\mathrm{L}} \propto \gamma_l$. Hence δ_{L} and γ_{L} can only include contributions from short-lived TLSs. However, such TLSs reach thermal equilibrium quickly. Therefore the expressions for δ_{L} and γ_{L} are given by

$$\delta_{\mathrm{L}} = \sum_l^n \frac{\gamma_l^2 \Delta_l}{\Delta_l^2 + \gamma_l^2} f_l \left(1 - f_l\right) , \quad \frac{\gamma_{\mathrm{L}}}{2} = \sum_l^n \frac{\gamma_l \Delta_l^2}{\Delta_l^2 + \gamma_l^2} f_l \left(1 - f_l\right) , \tag{20.21}$$

i.e., they depend only on temperature and cannot depend on time. These expressions for the line shift and width have already been derived at the end of Chap. 18.

The expressions for δ_{hom} and γ_{hom} include contributions from a huge number of TLSs and therefore these expressions do not reflect any peculiarities of the local environment of the chromophore. Although the contribution to the line width from one TLS with slow relaxation is rather small, the number of such TLSs is great. Therefore the contribution to the line width from all remote TLSs is conspicuous and can depend on time.

The summation over TLS in the expressions for δ_{hom} and γ_{hom} can be replaced by an integration over space. Using a distribution function $N(\varepsilon, \gamma)$, we obtain

$$\delta_{\mathrm{hom}} = \int \mathrm{d}\varepsilon \, \mathrm{d}\gamma N\left(\varepsilon, \gamma\right) p\left(1 - p\right) I_1(\gamma) , \tag{20.22}$$

$$\gamma_{\mathrm{hom}} = 2 \int \mathrm{d}\varepsilon \, \mathrm{d}\gamma N\left(\varepsilon, \gamma\right) p\left(1 - p\right) I_2(\gamma) , \tag{20.23}$$

where

$$I_1(\gamma) = \gamma^2 \sum_{s=1}^{N_0} \frac{\Delta_s}{\Delta_s^2 + \gamma^2} = \gamma^2 \frac{N_0}{V} \int_V \frac{\Delta(\boldsymbol{r})}{\Delta(\boldsymbol{r})^2 + \gamma^2} \mathrm{d}V \,,$$
$$I_2(\gamma) = \gamma \sum_{s=1}^{N_0} \frac{\Delta_s^2}{\Delta_s^2 + \gamma^2} = \gamma \frac{N_0}{V} \int_V \frac{\Delta(\boldsymbol{r})^2}{\Delta(\boldsymbol{r})^2 + \gamma^2} \mathrm{d}V \,. \tag{20.24}$$

In contrast to the similar expressions derived using the stochastic approach in Appendix Q, (20.22) and (20.23) include additional integrals I_1 and I_2. The temporal dependence in the formulas of the dynamical theory is determined by the function $p(1-p)$, rather than the function p which occurs in the formulas of the stochastic theory. The dependence of I_1 and I_2 on the rate constant γ is defined from the character of the electrostatic interaction between the chromophore and TLS. For the dipole–dipole interaction described by

$$\Delta(r) = \frac{F(\vartheta, \psi)\, \mu}{r^3} \,, \tag{20.25}$$

we arrive at the following expressions for the integrals:

$$I_1(\gamma) = \frac{n_0 \mu}{3} \left\langle F \int_{\Delta_{\min} F}^{\Delta_{\max} F} \frac{\mathrm{d}y}{y} \frac{\gamma^2}{y^2 + \gamma^2} \right\rangle \approx \frac{n_0 \mu}{3} \left\langle F \ln \frac{\gamma}{|\Delta_{\min} F|} \right\rangle \,,$$
$$I_2(\gamma) = \frac{n_0 \mu}{3} \left\langle |F| \int_{\Delta_{\min} F}^{\Delta_{\max} F} \mathrm{d}y \frac{\gamma}{y^2 + \gamma^2} \right\rangle \approx \frac{n_0 \mu}{3} \langle |F| \rangle \frac{\pi}{2} \,, \tag{20.26}$$

where the angular brackets denote integration over angles. The integral I_2 is only independent of γ for the dipole–dipole interaction. If the interaction falls off as $1/r^n$, we find that

$$I_2(\gamma) = \text{Const.} \times \gamma^{1-3/n} \,. \tag{20.27}$$

At low temperatures, the rate R_j of the TLS relaxation equals γ_j. If the distributions with respect to the TLS rate R and TLS splitting ε are independent from one another, we can write

$$N(\varepsilon, \gamma) = N(\varepsilon, R) = N_1(R)\, N_2(\varepsilon) \,. \tag{20.28}$$

Low temperature properties of glasses are well explained if the distribution with respect to the TLS splitting is assumed to have a rectangular shape, as shown in Fig. 6.6. Choosing this distribution function, we arrive at the following expression for the line width:

$$\frac{\gamma_{\text{hom}}(t,T)}{2}$$
$$= \frac{1}{\varepsilon_{\text{m}}} \int_{0}^{\varepsilon_{\text{m}}} \mathrm{d}\varepsilon \int \mathrm{d}R \, N_1(R) \, f(\varepsilon) \left(1 - \mathrm{e}^{-Rt}\right) \left[1 - f(\varepsilon)\left(1 - \mathrm{e}^{-Rt}\right)\right] I_2(R)$$
$$\approx \frac{kT}{\varepsilon_{\text{m}}} \int_{0}^{\varepsilon_{\text{m}}/kT} \mathrm{d}x f(x) \int \mathrm{d}R \, N_1(R) \left(1 - \mathrm{e}^{-Rt}\right) I_2(R) \,. \tag{20.29}$$

Here ε_{m} is the largest TLS splitting. This contribution to the line width is a linear function of temperature in the low temperature limit, that is, when $kT/\varepsilon_{\text{m}} \ll 1$. The dependence of $\gamma_{\text{hom}}(t,T)$ on time is called spectral diffusion. The optical dephasing rate is given by

$$\frac{1}{T_2(t,T)} = \frac{\gamma_{\text{L}}(T)}{2} + \frac{\gamma_{\text{hom}}(t,T)}{2} + \frac{1}{2T_1} \,. \tag{20.30}$$

It depends on temperature and time.

20.3 Logarithmic Temporal Line Broadening. Deviation from the Logarithmic Temporal Law

Comparison of the line widths described by (20.29) and (Q.22) shows that there is an additional factor $I_2(R)$ in the integrand in the formula from the dynamical theory. If the interaction between the chromophore and the TLS is of dipole–dipole type, this factor does not depend on R. By now this case is in common use and so deserves detailed consideration.

Anderson, Halperin, and Varma [50] assumed that, because of the tunneling character of the transition in the TLS, the rate constant R has the form $R \propto \exp(-\lambda)$. They assumed that the distribution function for λ is of rectangular form:

$$N(\lambda) = \frac{1}{\lambda_1 - \lambda_2} \,, \quad \lambda_2 < \lambda < \lambda_1 \,. \tag{20.31}$$

Then the distribution over R is described by a hyperbolic function with minimal and maximal values of R:

$$N_1(R) = N(\lambda)\frac{\mathrm{d}\lambda}{\mathrm{d}R} = \left[\ln\left(\frac{R_2}{R_1}\right)\right]^{-1} \frac{1}{R} \,, \quad R_1 < R < R_2 \,. \tag{20.32}$$

Inserting this function into (20.29) yields the expression

$$\frac{\gamma_{\text{hom}}(t,T)}{2} = 0.693\frac{kT}{\varepsilon_{\text{m}}}\frac{I_2}{\ln(R_2/R_1)}\int_{R_1}^{R_2}\frac{\mathrm{d}R}{R}\left(1-\mathrm{e}^{-Rt}\right)$$

$$\approx 0.693\frac{kT}{\varepsilon_{\text{m}}}\frac{I_2}{\ln(R_2/R_1)}\int_{1/t}^{R_2}\frac{\mathrm{d}R}{R}$$

$$= 0.693\,\frac{kT}{\varepsilon_{\text{m}}}\,\frac{\frac{n_0\mu}{3}\langle|F|\rangle\frac{\pi}{2}}{\ln(R_2/R_1)}\ln R_2 t\,. \tag{20.33}$$

Here we have used the fact that $\int_0^\infty \mathrm{d}x/(1+\mathrm{e}^x) = \ln 2 = 0.693$ and the integral I_2 does not depend on R for interactions of dipole–dipole type. It has been found experimentally [107] that polyethylene TLSs have a dipolar moment of order 0.4 Debye. The logarithmic temporal law has been observed in several studies [106, 108, 109].

Let us examine the temporal law of line broadening if the TLSs of a polymer have no dipole moment, but rather a quadrupole moment q. In this case $\Delta(r) = Dq/r^4$, where D is the dipole moment of the guest molecule. The integral $I_2(R)$ in (20.24) with this type of interaction is

$$I_{2q}(R) = \left(\frac{Dq}{R}\right)^{3/4}\frac{R}{4}\int_0^\infty \frac{\mathrm{d}x}{x^{1/4}}\frac{1}{1+x^2} \approx 0.4\,(Dq)^{3/4}R^{1/4}\,, \tag{20.34}$$

i.e., it depends on R. Inserting the last expression for I_2 into (20.29), we obtain the following expression for the line width:

$$\gamma_{\text{hom}}(t,T) = \text{Const.}\,\frac{kT}{\varepsilon_{\text{m}}}\left(1-\frac{1}{\sqrt[4]{R_2 t}}\right). \tag{20.35}$$

For the monopole–dipole type of chromophore–TLS interaction, we should use $\Delta(r) = De/r^2$. Inserting this function into the integral I_2, we find

$$I_{2e}(R) = \left(\frac{De}{R}\right)^{3/2}\frac{R}{2}\int_0^\infty \frac{\sqrt{x}\mathrm{d}x}{1+x^2} \approx (De)^{3/2}R^{-1/2}\,. \tag{20.36}$$

Inserting this expression into (20.29), we find the temporal behavior of the line width:

$$\gamma_{\text{hom}}(t,T) = \text{Const.}\,\frac{kT}{\varepsilon_{\text{m}}}(\sqrt{R_2 t}-1)\,. \tag{20.37}$$

We see that character of the chromophore–TLS interaction influences the temporal behavior of the line width resulting from the interaction with a huge number of TLS situated beyond the circle in Fig. 20.1. The temporal dependence of the dephasing rate according to the law $1 - t^{-1/4}$ has been observed in 3PE experiments with waiting time t_{w} on the submicrosecond time scale [110]. The temporal behavior of spectral holes according to the temporal law $t^{1/2}$ has been observed in persistent hole-burning experiments on a time scale of hours and days [109].

21. Theory of Tunneling Transitions in TLSs

21.1 General Formulas for the Tunneling Probability

Formulas for the tunneling transition were derived in Sect. 19.2. We did not discuss any details of the expressions for the rates a, A, b and B in (19.28). In this chapter we shall calculate these rates.

We consider tunneling in the ground electronic state of a chromophore. By neglecting the optical transitions in (19.19), i.e., setting $1/T_1 = \Lambda_{ab} = \Lambda_{\alpha\beta} = 0$, we arrive at the set of equations

$$
\begin{aligned}
\dot{\rho}_{a\alpha} &= -\mathrm{i}\left(\Delta_{a\alpha} - \mathrm{i}0\right)\rho_{a\alpha} - \mathrm{i}\lambda_{a\alpha}\left(\rho_{\alpha\alpha} - \rho_{aa}\right) , \\
\dot{\rho}_{\alpha a} &= -\mathrm{i}\left(\Delta_{\alpha a} - \mathrm{i}0\right)\rho_{\alpha a} - \mathrm{i}\lambda_{\alpha a}\left(\rho_{aa} - \rho_{\alpha\alpha}\right) , \\
\dot{\rho}_{aa} &= \mathrm{i}\sum_{\alpha}\left(\lambda_{\alpha a}\rho_{a\alpha} - \rho_{\alpha a}\lambda_{a\alpha}\right) , \\
\dot{\rho}_{\alpha\alpha} &= -\mathrm{i}\sum_{a}\left(\lambda_{\alpha a}\rho_{a\alpha} - \rho_{\alpha a}\lambda_{a\alpha}\right) ,
\end{aligned}
\qquad (21.1)
$$

where $\Delta_{\alpha a} = \varepsilon + \Omega_\alpha - \Omega_a = -\Delta_{a\alpha}$, and the additional term $-\mathrm{i}0$ is introduced to provide the true analytic properties of the Laplace transforms of the density matrix elements. These matrix elements will be retarded functions of time, i.e., they are zero at $t < 0$, only if their Laplace transforms are analytic functions in the upper half plane of the complex frequency. A system consisting of a TLS which interacts with phonons is described by the Hamiltonian (19.3), where $H(R)$ and $H^1(R) = H(R) + V(R)$ are phonon Hamiltonians in the ground and excited states of the TLS, respectively. The function $V(R)$ defines a change in the adiabatic potentials due to the excitation in the TLS. This function is similar to the FC interaction in electron–phonon systems. Therefore it can be written as a sum of linear and quadratic terms in the phonon coordinates R:

$$
\hat{V}(R) = (R + a)\frac{U^1}{2}(R + a) - R\frac{U^0}{2}R = a\frac{U^1}{2}a + aU^1R + R\frac{W}{2}R , \qquad (21.2)
$$

where a defines a change in the equilibrium positions of phonons and $W = U^1 - U^0$ defines a change in the force matrix upon excitation in the TLS.

Further, we can carry out transformations similar to those made in Sect. 7.3, where we derived the optical Bloch equations.

The matrix elements of the tunneling operator are very small. Therefore this case is similar to the case of weak laser pumping in the optical Bloch equations. We may thus neglect coherent effects and set

$$\dot{\rho}_{\alpha a} = \dot{\rho}_{a\alpha} = 0 \; . \tag{21.3}$$

Then, from the first and the second equations of the set described by (21.1), we find

$$\rho_{a\alpha} = -\frac{\lambda_{a\alpha}}{\Delta_{a\alpha} - \mathrm{i}0}\left(\rho_{\alpha\alpha} - \rho_{aa}\right) \, , \quad \rho_{\alpha a} = \frac{\lambda_{\alpha a}}{\Delta_{\alpha a} - \mathrm{i}0}\left(\rho_{\alpha\alpha} - \rho_{aa}\right) \, . \tag{21.4}$$

Inserting these equations into the third and fourth equations, we arrive at the following equations for the diagonal elements of the density matrix of the TLS–phonon system:

$$\begin{aligned} \dot{\rho}_{aa} &= -\sum_{\alpha} r_{a\alpha}\rho_{aa} + \sum_{\alpha} r_{a\alpha}\rho_{\alpha\alpha} \, , \\ \dot{\rho}_{\alpha\alpha} &= \sum_{a} r_{a\alpha}\rho_{aa} - \sum_{a} r_{a\alpha}\rho_{\alpha\alpha} \, , \end{aligned} \tag{21.5}$$

where

$$r_{a\alpha} = 2\pi\lambda_{a\alpha}\lambda_{\alpha a}\delta\left(\Delta_{a\alpha}\right) \tag{21.6}$$

describe the probabilities of the transitions in the TLS.

The diagonal elements of the TLS–phonon density matrix can be written in the form

$$\rho_{aa} = \rho_a \rho_0 \, , \qquad \rho_{\alpha\alpha} = \rho_\alpha \rho_2 \, , \tag{21.7}$$

where ρ_a and ρ_α are the probabilities of finding the TLS–phonon system in a definite phonon state. They satisfy the conditions

$$\sum_a \rho_a = \sum_\alpha \rho_\alpha = 1 \; . \tag{21.8}$$

Inserting (21.7) into (21.5), and carrying out the summation with respect to phonon indices, we obtain

$$\dot{\rho}_0 = -a\rho_0 + A\rho_2 \, , \qquad \dot{\rho}_2 = a\rho_0 - A\rho_2 \, , \tag{21.9}$$

where

$$\rho_2 = \sum_\alpha \rho_{\alpha\alpha} \, , \qquad \rho_0 = \sum_a \rho_{aa} \tag{21.10}$$

are the excited and ground state populations of the TLS. The rate constants are given by

$$A = 2\pi \sum_{\alpha a} \rho_\alpha \lambda_{a\alpha} \lambda_{\alpha a} \delta\left(\Delta_{a\alpha}\right), \quad a = 2\pi \sum_{\alpha a} \rho_a \lambda_{a\alpha} \lambda_{\alpha a} \delta\left(\Delta_{a\alpha}\right), \tag{21.11}$$

where

$$\rho_a = \frac{\rho_{aa}}{\rho_0}, \qquad \rho_\alpha = \frac{\rho_{\alpha\alpha}}{\rho_2} \tag{21.12}$$

are the probabilities of finding the phonon state. These probabilities are of Boltzmann type.

The equations for A and a differ from (6.20), derived earlier from the expressions for the matrix elements. The matrix elements U in (6.20) take into account only the TLS–phonon interaction of Herzberg–Teller type, whereas the matrix elements λ in (21.11) take into account the TLS–phonon interaction of Franck–Condon type. It should be emphasized that the strength of this TLS–phonon coupling can be of arbitrary value.

Equations (21.11) for the tunneling rates A and a are similar to the expressions for the absorption and fluorescence optical bands. Therefore we may use the results derived in Part IV and write down immediately the result of the calculation, which is similar to (10.50):

$$\begin{aligned} A(\varepsilon) &= \frac{\lambda^2(0)}{2\pi} \int_{-\infty}^{\infty} \mathrm{e}^{-\mathrm{i}\varepsilon(t)t - \gamma_{\mathrm{ph}}(T)|t|/2} \mathrm{e}^{\varphi(t,T) - \varphi(0,T)} \mathrm{d}t, \\ a(\varepsilon) &= \exp\left(-\varepsilon/kT\right) A(\varepsilon), \end{aligned} \tag{21.13}$$

where

$$\varepsilon(t) = \varepsilon + \delta_{\mathrm{ph}}(T), \tag{21.14}$$

and expressions for $\delta_{\mathrm{ph}}(T)$ and $\gamma_{\mathrm{ph}}(T)$ are defined by the general (12.84) and (12.85), or simpler (12.87) and (12.106). In Sect. 10.3 the expression for the electron–phonon band (10.50) has been transformed to (10.51) with (10.53). Therefore (21.13) for the tunneling rate can be transformed in the same way to give

$$A = \lambda^2(0) \left[\mathrm{e}^{-\varphi(0,T)} \frac{\gamma_{\mathrm{ph}}(t)/2\pi}{\varepsilon(T)^2 + \left(\gamma_{\mathrm{ph}}(t)/2\right)^2} + \Psi^1\left(\varepsilon(t)\right) \right], \tag{21.15}$$

i.e., the tunneling rate is the sum of two terms which are similar to the ZPL and PSB in the fluorescence band. Here the first term describes elastic tunneling, when the number of phonons is unchanged, and the second term describes phonon-assisted inelastic tunneling.

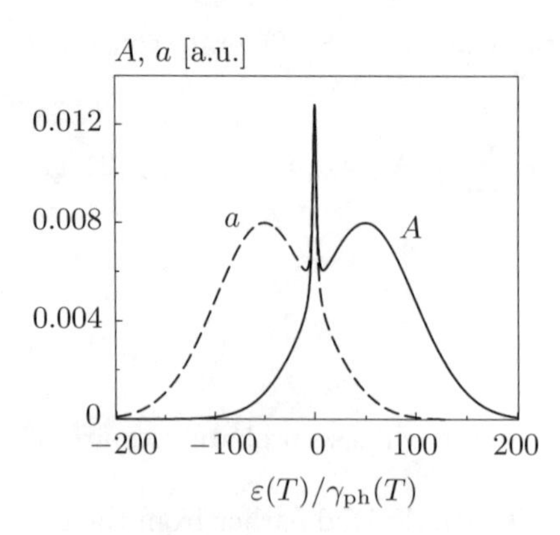

Fig. 21.1. Dependence of tunneling probabilities on tunneling splitting at $T \neq 0$, resembling optical bands

The tunneling rates A and a depend on the tunneling splitting ε in the same way as the fluorescence band depends on the photon energy. Therefore these tunneling rates look as shown in Fig. 21.1. Since level 2 in the TLS shown in Fig. 19.3 is the upper level, by definition, the TLS splitting $\varepsilon(T)$ is positive. If this splitting is of the order of the half-width $\gamma_{\text{ph}}(T)$, the main contribution to tunneling emerges from elastic phononless tunneling. There is an analogue between such resonant elastic tunneling and the resonant elastic scattering of a particle. Equation (21.15) enables one to examine in detail both elastic and inelastic tunneling.

21.2 Inelastic Tunneling Assisted by Acoustic and Localized Phonon Modes

If the TLS splitting $\varepsilon(T)$ exceeds $\gamma_{\text{ph}}(T)$, tunneling is realized via phonon creation and annihilation. This is the so-called phonon-assisted tunneling. This type of tunneling is similar to the scattering of a particle when the particle loses or acquires energy in the scattering process.

The rates of inelastic tunneling are given by

$$A\big(\varepsilon(t)\big) = \lambda^2(0)\Psi^1\big(\varepsilon(t)\big)\,, \quad a\big(\varepsilon(t)\big) = \mathrm{e}^{-\varepsilon(T)/kT}A\big(\varepsilon(T)\big)\,. \tag{21.16}$$

The function Ψ^1 is an analogue of the PSB in the optical band. By neglecting the HT interaction and quadratic FC interaction, we obtain the following expression for this function:

$$\Psi^1(\varepsilon(t)) = \sum_{m=1}^{\infty} \Psi_m^1(\varepsilon(t)) \tag{21.17}$$

$$= \mathrm{e}^{-\varphi(0,T)} \sum_{m=1}^{\infty} \frac{1}{m!} \int_{-\infty}^{\infty} \mathrm{d}\omega_1 \, \varphi(\omega_1, T)$$

$$\ldots \int_{-\infty}^{\infty} \mathrm{d}\omega_m \, \varphi(\omega_m, T) \frac{\gamma_{\mathrm{ph}}(t)/2\pi}{\left[\varepsilon(T) - \omega_1 \ldots - \omega_m\right]^2 + \left[\gamma_{\mathrm{ph}}(t)/2\right]^2} .$$

This equation can be derived in the same way as the expression for the PSB. Here

$$\varphi(\omega, T) = \int_{-\infty} \varphi(t, T) \, \mathrm{e}^{\mathrm{i}\omega t} \, \mathrm{d}t = \varphi(\omega) \left[n(\omega) + 1\right] + \varphi(-\omega) \, n(-\omega) ,$$

$$\varphi(\omega) = \sum_{q=1}^{N} \frac{a_q^2}{2} \delta(\omega - \omega_q) ,$$

$$\varphi(0, T) = \int_{-\infty}^{\infty} \varphi(\omega, T) \, \mathrm{d}\omega = \sum_q \frac{a_q^2}{2} (2n_q + 1) . \tag{21.18}$$

The function Ψ_m^1 defines inelastic phonon-assisted tunneling with creation or annihilation of m phonons.

One-Phonon Function $\varphi(\omega)$. This function plays the same role as the function of the weighted phonon density of states in the optical band. We shall use the results derived in Sects. 5.1 and 5.2, where the acoustic and localized modes were studied. Using (5.25) and (5.8), we can find a relation between the shift in the equilibrium position a_n in the node representation and the shift a_q in the natural wave representation:

$$a_n = \sum_q \sqrt{\frac{\hbar}{M\omega_q}} u(n, q) \, a_q , \quad a_q = \sum_n u(q, n) \sqrt{\frac{M\omega_q}{\hbar}} a_n , \tag{21.19}$$

where $n = \boldsymbol{n}i$. With the help of these formulas, we find the following expression for the one-phonon function:

$$\varphi(\omega) = \frac{M\omega}{2\hbar} \sum_{n,m} a_n a_m \sum_q u(n, q) \, u(m, q) \, \delta(\omega - \omega_q) . \tag{21.20}$$

Let us consider a case when, after tunneling, the distance between the chromophore and one molecule of the solvent has changed, i.e., $a_0 = a$ and

$a_1 = -a$. Here we have taken into account the fact that the sum of all displacements must be zero. In this case the last equation takes the form

$$\varphi(\omega) = a^2 \frac{M\omega}{\hbar} \frac{1}{2} \sum_q [u(0,q) - u(1,q)]^2 \, \delta(\omega - \omega_q) = a^2 \frac{M\omega}{\hbar} g(\omega) \; . \tag{21.21}$$

It was shown in Sect. 5.2 that the function $g(\omega)$ takes the form

$$g(\omega) = \frac{g_0(\omega)}{[1 - \Delta U \varDelta_0(\omega)]^2 + [\pi \Delta U g_0(\omega)/\omega]^2} \; . \tag{21.22}$$

For phonons described by the Debye model, we find

$$\begin{aligned} g_0(\omega) &= 5\frac{\omega^4}{\omega_D^5} \; , \qquad 0 < \omega < \omega_D \; , \\ \varDelta_0(\omega) &= \frac{10}{\omega_D^2}\left[-\left(\frac{1}{3} + y^2\right) + \frac{y^3}{2} \ln \frac{y+1}{|y-1|}\right] \; , \qquad y = \frac{\omega}{\omega_D} \; . \end{aligned} \tag{21.23}$$

These formulas can be used to investigate how phonon localization influences the tunneling rate. Figure 5.1 clearly demonstrates the changes in the function $g(\omega)/\omega$ resulting from the influence of the guest molecule.

Let us turn back to the calculation of the tunneling rate. If the TLS–phonon coupling is weak, we can take into account the only one-phonon term with $m = 1$ in the sum describing the function Ψ^1. This term determines the tunneling rate with creation of one phonon, i.e., so-called one-phonon tunneling. Since the Lorentzian is much narrower than the one-phonon function φ and using (21.17), we can derive the following expression for the rate of one-phonon tunneling:

$$\begin{aligned} P^1 &= \lambda^2(0) \mathrm{e}^{-\varphi(0,T)} \varphi(\omega, T) \\ &= \lambda^2(0) \, \varphi(\omega, T) \exp\left[-\int_0^\infty \varphi(\omega) \coth \frac{\hbar\omega}{2kT} \, \mathrm{d}\omega\right] \; . \end{aligned} \tag{21.24}$$

The function $\varphi(\varepsilon, T)$ defines the dependence of the tunneling rate on TLS splitting. The appearance of this function depends on the phonon localization. If insertion of the guest molecule into the solvent creates a localized mode, this fact considerably influences the tunneling rate. Figure 21.2a shows the dependence of the one-phonon tunneling rate on the degree of phonon localization. Curves 1 and 2 show the rate in the cases when a localized mode exists in the solvent, and when there is no localized mode and ordinary acoustic vibrations exist in the solvent. The narrow peak on curve 1 relates to the case when TLS splitting coincides with the frequency of the localized mode. If tunneling is assisted solely by acoustic phonons, the tunneling rate is a few orders of magnitude smaller. Curve 2 demonstrates this fact.

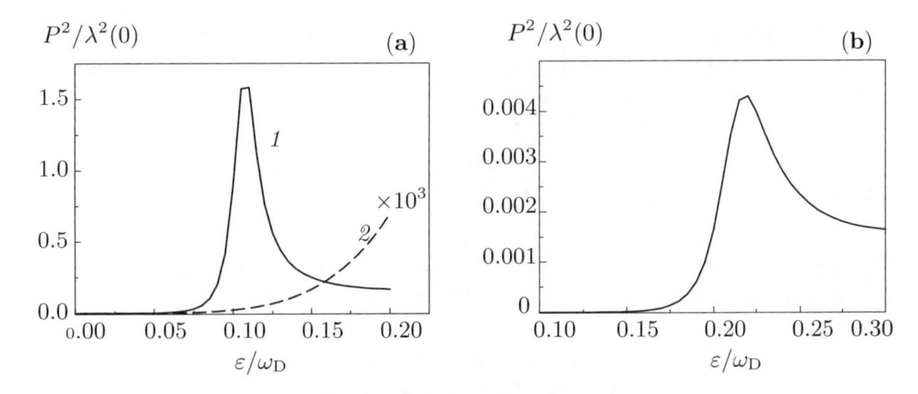

Fig. 21.2. Dependence of one-phonon (**a**) and two-phonon (**b**) tunneling probabilities on TLS splitting. Curve 1 corresponds to a pseudo-localized phonon mode and curve 2 to the Debye phonon model

Two-Phonon Tunneling. The rate of two-phonon tunneling is described by the term with $m = 2$ in (21.17). At $T = 0$, this term is given by

$$P^2(\varepsilon) = \exp\left[-\int_0^\infty \varphi(\omega)\,\mathrm{d}\omega\right] \frac{\lambda^2(0)}{2} \int_0^\infty \varphi(\omega)\,\varphi(\varepsilon - \omega)\,\mathrm{d}\omega\,. \qquad (21.25)$$

Figure 21.2b shows the result of the calculation using this formula for the case when the localized mode exists in the solvent. It is clear that, if the TLS splitting $\varepsilon(T)$ coincides with the energy of two quanta of the localized mode, the two-phonon tunneling rate has a pronounced maximum. This resonance rule also works for the m-phonon transition.

Multiphonon Tunneling. If the localized mode is absent, the tunneling rate has no resonance. Curve 2 in Fig. 21.2a confirms this conclusion. Therefore, for strong tunnelon–phonon coupling, we have to take into account the contribution from all terms of the series in (21.17). In this case the calculation can be carried out with (21.13). Figure 21.3 shows the dependence of the

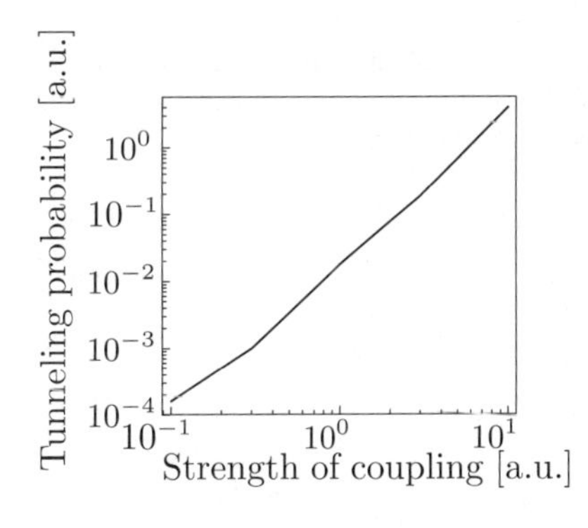

Fig. 21.3. Dependence of phonon-assisted tunneling probability on strength of TLS–phonon coupling

tunneling rate on the strength of the tunnelon–phonon coupling. An increase in the coupling by one order of magnitude yields an increase in the tunneling rate by two orders of magnitude.

21.3 Elastic Tunneling

The inelastic tunneling examined in the last section does not vanish even at $T = 0$, because the energy of TLS splitting can be transformed to the energy of created phonons. Indeed, phonon creation is possible even at $T = 0$.

However, if the TLS splitting approaches zero, the rate of inelastic tunneling approaches zero as well. In this case elastic tunneling can play an important role, as shown in Fig. 21.1, where the peak describes elastic tunneling. The rate of elastic tunneling is described by the function

$$P = p \propto \frac{\gamma_{\rm ph}(t)/2\pi}{\varepsilon(t)^2 + [\gamma_{\rm ph}(t)/2]^2} . \tag{21.26}$$

Figure 21.4a shows that $\gamma_{\rm ph}(T) \to 0$ when the temperature approaches zero. Therefore elastic tunneling is possible only at finite temperature. The temperature behavior of elastic tunneling is not monotonic, as can be seen from Fig. 21.4b.

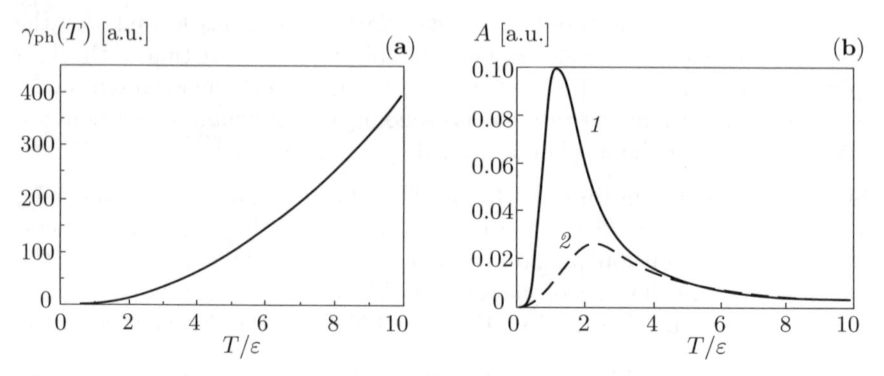

Fig. 21.4. (**a**) Typical temperature line broadening and (**b**) temperature dependence of the elastic tunneling probability calculated by means of (21.26) at $\varepsilon = 5$ (curve 1) and 20 (curve 2). ε is taken in the same units as $\gamma_{\rm ph}$

22. Investigating TLS Relaxation by Single-Molecule Spectroscopy

According to (20.14), the dipolar correlator $S^{\mathrm{g}}(t,x)$ of a guest molecule depends on two types of time variable, t and x. Using the language of a stochastic theory, we can say that the time variable x describes the time scale of fast fluctuations, whilst the time variable t denotes the time scale of slow fluctuations. In order to find an expression for the optical band, we integrate the dipolar correlator with respect to time variable x. This means that we only take into account contributions to the optical band from fast fluctuations. Since the dipolar correlator approaches zero if x exceeds $1/\gamma(t,T) = T_2(t,T)$, we can say that the time scale of fast fluctuations is of order $T_2(t,T)$. Today, single-molecule spectroscopy studies rather slow fluctuations resulting from jumps in the TLS. These occur over a longer time scale t. The optical band shape function obtained after integration over the time variable x depends on the variable t. This function is thus a good starting point for discussing slow TLS relaxations related to slow fluctuations. The manifestations of TLS relaxation in single-molecule spectroscopy are the topic of this chapter.

22.1 Two-Photon Correlator of a Molecule Interacting with TLSs

Frequency jumps of the optical line are the most typical phenomenon in single-molecule spectroscopy. Let us find a relation between these jumps and the full two-photon correlator, which has already been studied in Chap. 3. It is described by the simple expression

$$p(t) = \frac{\rho_{11}(t)}{T_1}\,, \tag{22.1}$$

where $\rho_{11}(t)$ is the probability of finding a molecule in the excited singlet electronic state. If the molecule does not interact with phonons, $\rho_{11}(t)$ is the solution of the four equations in (3.12). This set of equations differs from the optical Bloch equations in that $1/T_2$ is replaced by $1/2T_1$. If we take the electron–phonon interaction into account, the set of equations for the density matrix becomes infinite. In this case we have to use (7.29). It was shown in Sect. 7.3 that we can derive the optical Bloch equations from the more general (7.29). It was also shown that the optical Bloch equations only take into account that part of the electron–phonon interaction which yields broadening of the ZPL. This part is, in fact, the quadratic FC interaction. The half-width of the ZPL equals $2/T_2$.

Let us now calculate the full two-photon correlator for the more complex case when the guest molecule interacts with TLSs which are not in thermal equilibrium. Such a physical system has already been examined in Sects. 19.5 and 20.1 and we shall use the results of those sections. The expression for the two-photon correlator includes the probability of the excited electronic state which can be reached after absorption of the laser photon. Hence, if we excite the system via the Lorentzian which relates to the $N \leftarrow M$ electron–tunnelon transition, the expression for the two-photon correlator is

$$p(t) = \frac{\rho_{\mathrm{N}}(t)}{T_1} . \tag{22.2}$$

The Lorentzians relating to the other optical transitions are not in resonance with the laser light and so do not absorb this light. Therefore we can generalize (22.2) to

$$p(t) = \frac{1}{T_1} \sum_N \rho_N(t) = \frac{1}{T_1} \sum_M \frac{L_M P_M(t)}{\Gamma_N + L_M} \{1 - \exp\left[-\left(\Gamma_N + L_M\right) t\right]\} . \tag{22.3}$$

This expression describes the two-photon correlator upon laser excitation with the arbitrary frequency of the laser light. The main term in this sum is the one whose transition frequency coincides with the laser frequency. Using the inequality $L_M/\Gamma \ll 1$ and the approximate equation $1/\Gamma \approx 1/T_1$, we can write the expression for the two-photon correlator as

$$p(t) = k\left(t, T\right) \left(1 - \mathrm{e}^{-t/T_1}\right) , \tag{22.4}$$

where

$$k\left(t, T\right) = \sum_M L_M P_M(t) \tag{22.5}$$

is the absorption coefficient of a guest molecule interacting with phonons and TLSs. This coefficient has already been calculated in Sect. 20.1. It is given by (20.9), where the dipolar correlator $S^{\mathrm{g}}(t, x)$ is defined by (20.14), (20.15), (20.16) and (20.17).

Let us apply this formula to the dipolar correlator for the case when the chromophore interacts strongly with a single TLS in its vicinity. This means that the coupling constant Δ_j of this TLS satisfies the inequalities $\Delta_j \gg 1/T_2 < \gamma_j$. This situation is shown in Fig. 20.1. The energy diagram for such a system is shown in Fig. 19.3. In this case the product in (20.14) includes only one factor, since the other factors have $\Delta_j \ll 1/T_2$. Therefore, after integration with respect to 'fast' time x, (20.14) yields the following formula for the absorption coefficient:

$$k^{g}(t) = 2\chi^2 \left[[1 - p_2(t)] \frac{1/T_2(t)}{(\omega - \Omega)^2 + [1/T_2(t)]^2} + p_2(t) \frac{1/T_2(t)}{(\omega - \Omega - \Delta)^2 + [1/T_2(t)]^2} \right], \tag{22.6}$$

where

$$\frac{1}{T_2(t,T)} = \frac{\gamma_{\text{ph}}(T)}{2} + \frac{\gamma_{\text{hom}}(t,T)}{2} + \frac{1}{2T_1}, \tag{22.7}$$

$$p_2(t) = f(t) + [p_2(0) - f(t)] \exp(-Rt), \tag{22.8}$$

and the line shift $\delta(T)$ is included in the resonance frequency. Here $\gamma_{\text{ph}}(T)$ allows for broadening due to the electron–phonon interaction and $\gamma_{\text{hom}}(t,T)$ takes into account a contribution to the line width from the interaction with all TLSs situated outside the circle shown in Fig. 20.1.

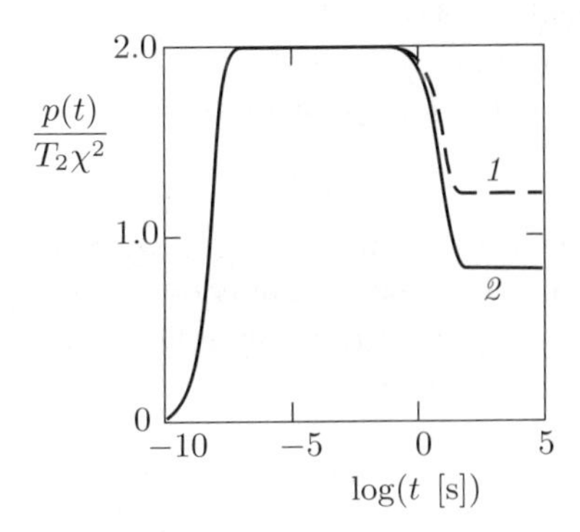

Fig. 22.1. Temporal dependence of the two-photon correlator for excitation via ω_0 (curve 1) and via $\omega_0 + \Delta$ (curve 2). $f = 0.4$, $\Delta = 8/T_2$, $T_1 = 10^{-8}$ s, $R = 10^{-1}$ s^{-1}

How does the TLS in the immediate vicinity of the chromophore manifest itself in the two-photon correlator? Figure 22.1 shows the two-photon correlator as a function of the time delay between two emitted photons, calculated using (22.4) and (22.6), upon laser excitation via each of two optical lines. Under laser excitation with light frequencies ω_0 and $\omega_0 + \Delta$, we should take $p_2(0) = 0$ and 1, respectively. The following information can be found from the two-photon correlator. Times relating to the increase and decrease of this correlator enable one to find the time constants T_1 and $1/R$, respectively. The ratio of correlators 1 and 2 occurring on the long time scale in Fig. 22.1 equals the ratio of two optical lines in thermal equilibrium when relaxation of the TLS is over. The correlator is shown in Fig. 22.2 as a function of the laser frequency. In the following sections, we shall use this figure for the two-photon correlator to discuss such interesting phenomena as jumps of the optical line of a single molecule.

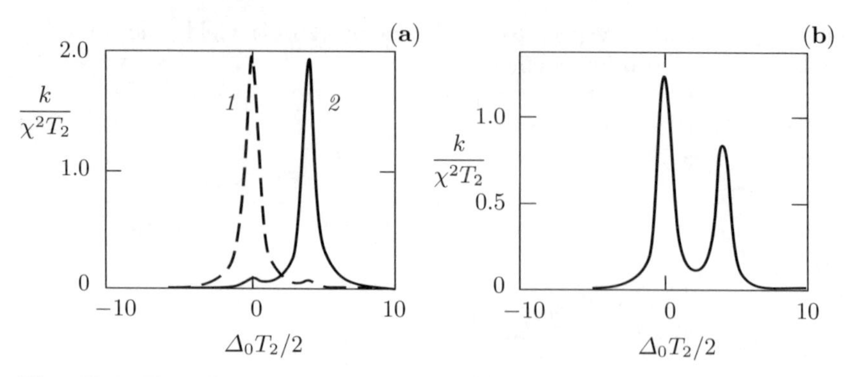

Fig. 22.2. Two-photon correlator as a function of the frequency of the exciting light. $Rt \ll 1$ (**a**) and $Rt \gg 1$ (**b**). $f = 0.4$, $\Delta = 8/T_2$, $T_1 = 10^{-8}$ s, $R = 10^{-1}$ s^{-1}

22.2 Spontaneous and Light-Induced Jumps of the Optical Line. Relation to Hole Burning

When one-photon counting methods are used, the laser frequency is scanned in a frequency range which includes the optical line. All photons emitted by the molecule are counted. The number of these emitted photons is a function of the laser frequency. Let the laser scan be carried out in the frequency range shown in Fig. 22.2. These lines relate to two vertical transitions from the 0 and 2 states shown in Fig. 19.3. If we carry out the laser scan quickly, i.e., the time t_s of the laser scan satisfies the inequality $t_s R < 1$, we observe only one optical line in each laser scan: either line 1 or line 2. This situation is reflected by the two-photon correlator at short time delay between photons, as shown in Fig. 22.2a. Here R is the rate of the tunneling transition between the left and right quantum states. Which of the two lines is observed depends on a random circumstance, i.e., in which of two possible states the chromophore absorbs the laser photon. Indeed the chromophore can absorb the laser photon when it occurs either in quantum state 0 or quantum state 2. Hence, for fast scans, the spectral trajectory exhibits jumps of the absorption resonance frequency between two possible positions ω_0 and $\omega_0 + \Delta$ at random times.

In the case of slow laser scans, when $t_s R > 1$, we simultaneously observe two optical lines in each laser scan. This means that we cannot detect jumping lines if scans are too slow. This situation is similar to that shown in Fig. 22.2b. The intensities of optical lines are proportional to the probabilities $1 - f$ and f of finding the system in the left or right well. This situation is completely analogous to that observed in ordinary spectroscopic experiments with homogeneous molecular ensembles. We may thus conclude that the two-photon correlator, as a function of the laser frequency for short time delays between photons, can be related to the case with jumping optical lines, whereas for long time delays, the two-photon correlator displays a spectral picture similar to that observed in a homogeneous molecular ensemble. This means that the full two-photon correlator as a function of the laser frequency and time delay between emitted photons can be used to find more information than the ordinary absorption coefficient. In fact, the coefficient of light absorption coincides with the full two-photon correlator only at infinite time delay.

The physical picture drawn up with the help of the full two-photon correlator can be linked with the temporal behavior of trains of emitted photons, as shown in Fig. 22.3. Both trains of photons shown in this figure exhibit photon bunching. When we excite a molecule by laser light with frequency ω_0, a pause in the photon emission emerges at a random time when this molecule has jumped to another quantum state with another resonant frequency $\omega_0 + \Delta$. Therefore, when the scanned laser frequency reaches the new value of the resonant frequency, light emission begins again. Photon bunching arises due to jumps of the molecule from one quantum state to another. This situation resembles the one already discussed in connection with Fig. 8.4, which relates to the case when the molecule jumps from optically active singlet states to a 'dark' triplet state. These have been called 'on' and 'off' states. For instance, if the molecule is in the right well but is not able to absorb the laser light, the low photon train in Fig. 22.3 disappears and we arrive at the situation shown in Fig. 8.4.

$\longrightarrow \omega_0$

$\longrightarrow \omega_0 + \Delta$

Fig. 22.3. Sequence of photons emitted by a single molecule upon excitation in lines with resonance frequencies ω_0 and $\omega_0 + \Delta$

The rate of tunneling transition from one well to another in the ground electronic state of the chromophore does not depend on the intensity of the excited laser light. However, tunneling can also happen after the chromophore absorbs a photon. The rate of such light-induced tunneling will depend on the laser intensity. Light-induced tunneling is typical for hole-burning spectroscopy.

Persistent spectral hole burning can occur if the excited electronic state 3 in Fig. 19.3 has lower energy than the excited electronic state 1, and the tunneling transitions in the ground electronic state are suppressed, i.e., $a \sim A \sim 0$. Let us consider this very case under laser excitation via the $0-1$ transition. Let $t=0$ be the time of registration of the first photon from the photon pair. Then $\rho_0(0) = 1$ and we have the following expression for the two-photon correlator:

$$p(t) = \frac{\rho_1(t) + \rho_3(t)}{T_1} \approx L_0 \exp(-\tilde{B}t) \,. \tag{22.9}$$

At times satisfying the inequality $\tilde{B}t_0 \gg 1$, light emission is over. Hence, for random times of order t_0, the spectral line jumps to another spectral position.

If we change the laser frequency and excite via the $2-3$ optical transition we start to register light emission again since, according to (19.36) and (19.39), the expression for the two-photon correlator is given by

$$p\left(t+t_0\right)=\frac{\rho_1\left(t+t_0\right)+\rho_3\left(t+t_0\right)}{T_1}\approx L_2\exp(-\tilde{b}t)\;. \tag{22.10}$$

We took $\rho_2(t_0)=1$ in (19.39).

From the last two formulas, we deduce that $1/\tilde{B}$ and $1/\tilde{b}$ are the 'lifetimes' of the optical lines relating to the $0-1$ and $2-3$ transitions, respectively. They can be found with the help of (8.32) if we sum over all on-intervals. These τ_{on} will depend on the reciprocal laser intensity. The dependence of τ_{on} on the laser intensity is a telling feature of light-induced spectral jumps.

22.3 Analysis of Complicated Spectral Trajectories Using the Two-Photon Correlator

An example of a more complicated spectral trajectory can be found in Fig. 19.4. Here the optical line jumps among four spectral positions. Therefore we are forced to examine the full two-photon correlator which describes a guest molecule interacting with two different TLSs. Indeed two TLSs have four quantum states in each electronic state of the guest molecule. Such a full two-photon correlator is given by

$$\begin{aligned}k\left(\Delta_0,t,T\right)&=(1-\rho)\left(1-\rho'\right)k\left(\Delta_0\right)+\rho\left(1-\rho'\right)k\left(\Delta_0-\Delta\right)\\&\quad+(1-\rho)\,\rho' k\left(\Delta_0-\Delta'\right)+\rho\rho' k\left(\Delta_0-\Delta-\Delta'\right)\;.\end{aligned} \tag{22.11}$$

Here $k=2\chi^2L(\omega)$, where L is a Lorentzian and ρ and ρ' describe the probabilities of finding the TLS in the excited state. They are given by equations similar to (22.8). The frequency of the laser light determines the initial value of the probabilities ρ and ρ'. For example if we excite the molecule via the transition with $\Delta_0=0$, only the first term in (22.11) is operative. In this case we should take the initial condition $\rho(0)=\rho'(0)=0$. Under this initial condition, the system occurs in the quantum state with $\Delta_0=0$ at $t=0$. If excitation is realized via the optical line with $\Delta_0-\Delta=0$, we have to use the initial condition $\rho(0)=1$ and $\rho'(0)=0$, and so on.

The two-photon correlator calculated with (22.4) and (22.11) is shown in Fig. 22.4. Here curves 1, 2, 3 and 4 relate to laser excitation via each of the four Lorentzians in (22.11), i.e., the initial conditions: $\rho(0)=\rho'(0)=0$, $\rho(0)=1$ and $\rho'(0)=0$, $\rho(0)=0$ and $\rho'(0)=1$, and $\rho(0)=\rho'(0)=1$. Figures 22.4a, b and c correspond to various time delays between photons in the photon pairs.

For $t<1/R<1/R'$, we have Fig. 22.4a. For such short time delays, neither of the two TLSs can reach thermal equilibrium and therefore the correlator as a function of the laser frequency is described by a single Lorentzian jumping among four spectral positions.

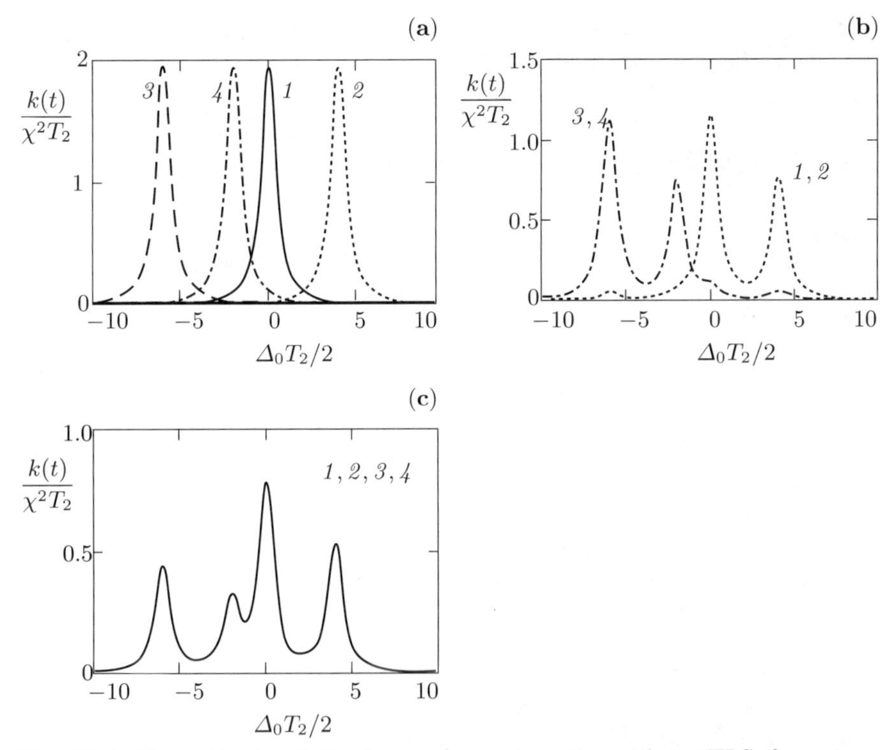

Fig. 22.4. Absorption band of a chromophore interacting with two TLSs for various durations of the laser scan. $t_{sc} = 10^{-5}$s (**a**), 1 s (**b**), 102 s (**c**). TLS parameters are $E = 0.5$ K, $R = 6 \times 10^2$ s^{-1}, $T_2\Delta = 4$ and $E' = 1$ K, $R' = 10^{-1}$ s^{-1}, $T_2\Delta' = -6$, $T = 1.7$ K

For $1/R < t < 1/R'$, we have Fig. 22.4b. For this time delay, one TLS with relaxation rate R has already reached thermal equilibrium. However, the second TLS cannot reach thermal equilibrium. Therefore the two-photon correlator as a function of the laser frequency is described by a doublet of lines and the doublet jumps among two spectral positions. For excitation via optical lines 1 or 2, the two-photon correlator is described by the left doublet. If we excite via optical lines 3 or 4, the correlator will be described by the right doublet of lines.

For $1/R < 1/R' < t$, we have Fig. 22.4c. For such long time delays, both TLSs occur in thermal equilibrium. Therefore the correlator is a sum of four optical lines. There are no jumps in this case.

The spectral pictures shown in Fig. 22.4 are related to the two-photon correlator when we count pairs of photons. However, these figures can also be related to spectral pictures observed if we use one-photon counting methods with laser scans. For instance, for fast laser scans, when the time t_s of the scan satisfies the inequality $t_s < 1/R < 1/R'$, we see only one optical line jumping among four spectral positions, as shown in Figs. 19.4 and 22.4a. If

$1/R < t_s < 1/R'$, we see jumps of coupled lines among two spectral positions. Finally, if the inequality $1/R < 1/R' < t_s$ holds, there are no jumps of lines.

22.4 Single-Molecule Chemical Reactions

Single-molecule spectroscopy can be used to study chemical reactions of individual molecules. Such chemical reactions have been observed in [111]. The molecule flavin-adenine dinucleotide (FAD) is fluorescent in its oxidized form but not in its reduced form. FAD is first reduced by a cholesterol molecule to $FADH_2$, and then oxidized by O_2 to yield H_2O_2. FAD emission peaks at 520 nm. A single FAD molecule was irradiated by He–Cd laser light with $\lambda = 442$ nm at room temperature. At high cholesterol concentrations, the chemical reaction of reduction and then oxidation can be realized. In the ensemble of FAD molecules, we detect both chemical forms of FAD. However, the molecule is oxidized and reduced at random times. Therefore, in experiments with single molecules, the authors of [111] observed blinking fluorescence. At the moment when fluorescence emerged, the FAD was oxidized and this time could be detected in the experiment with a single molecule.

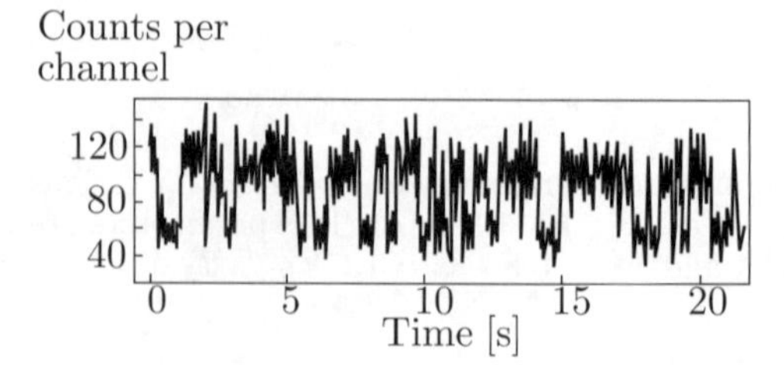

Fig. 22.5. Temporal behavior of the fluorescence of a single FAD molecule under CW irradiation [111]

The blinking fluorescence shown in Fig. 22.5 resembles the fluorescence from a molecule with triplet level discussed in connection with the photon-bunching phenomenon shown in Fig. 8.4. The authors of [111] used one-photon registration methods. In this case the quantum trajectory shown in Fig. 22.5 has to be treated statistically after measurement. The quantum trajectory of blinking fluorescence consists of on- and off-intervals. Almost 500 such intervals were treated in [111]. Figure 22.6 shows the distribution of on-intervals with respect to their duration. The curve is given by the function

$$p(t) = \frac{k_1 k_2}{k_2 - k_1} \left[\exp(-k_1 t) - \exp(-k_2 t) \right] , \tag{22.12}$$

with $k_1 = 2.9\ \mathrm{s}^{-1}$ and $k_2 = 17\ \mathrm{s}^{-1}$. This distribution resembles the two-photon correlator of the molecule with triplet level. This means that dynamical processes of various types can have similar two-photon correlators. Here, we consider the case when relaxation is described by a rate constant, and

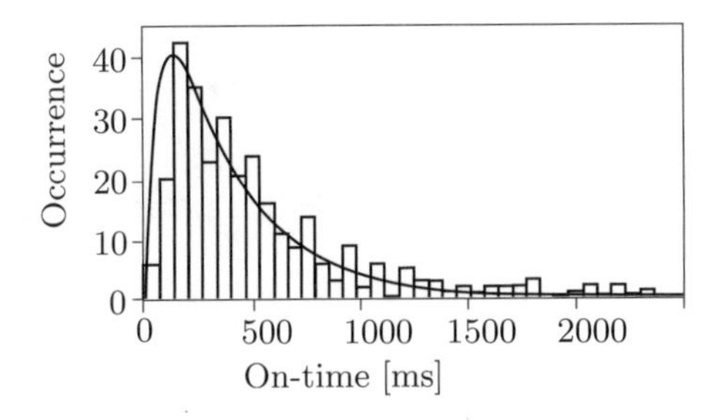

Fig. 22.6. Distribution of on-intervals [111]

these constants can be easily found by means of the one-photon counting method. However, the relaxation can be of a more complicated type.

We have already seen an example of this type when we discussed logarithmic spectral diffusion. It is very difficult to investigate such relaxation with one-photon methods. The full two-photon correlator is a more convenient tool for investigating complicated relaxation. An example of this type is examined in the next section.

22.5 Manifestation of Spectral Diffusion in the Two-Photon Correlator

According to (22.6), the temporal dependence of the two-photon correlator can be due to the temporal dependence of Lorentzian amplitudes and half-widths. The temporal dependence of the amplitudes is of exponential type. However, the half-width of the Lorentzians depends logarithmically on time, over a time scale covering several orders of magnitude.

Equations (22.6) and (22.7) determine the temporal behavior of the two-photon correlator. The optical dephasing rate has been taken in the form

$$\frac{1}{T_2(t)} = \frac{\gamma_{\mathrm{ph}}(T)}{2}\left(1 + C \ln R_2 t\right) . \tag{22.13}$$

The parameter C determines the ratio of γ_{hom} and γ_{ph}. Figure 22.7 shows the temporal dependence of optical dephasing at $\gamma_{\mathrm{ph}} \gg \gamma_{\mathrm{hom}}$, $\gamma_{\mathrm{ph}} \sim \gamma_{\mathrm{hom}}$, and $\gamma_{\mathrm{ph}} \ll \gamma_{\mathrm{hom}}$. The temporal dependence of the two-photon correlator calculated for these three cases of optical dephasing is shown in Fig. 22.8. The calculation was carried out for laser excitation via the $0-1$ optical transition. Fast decay of the correlator at $t = 10^{-6}$ s is due to exponential relaxation of the TLS in the immediate vicinity of the guest molecule. However, the slight decrease of the correlator over the whole time scale is due to spectral diffusion, i.e., the logarithmic increase in the dephasing rate.

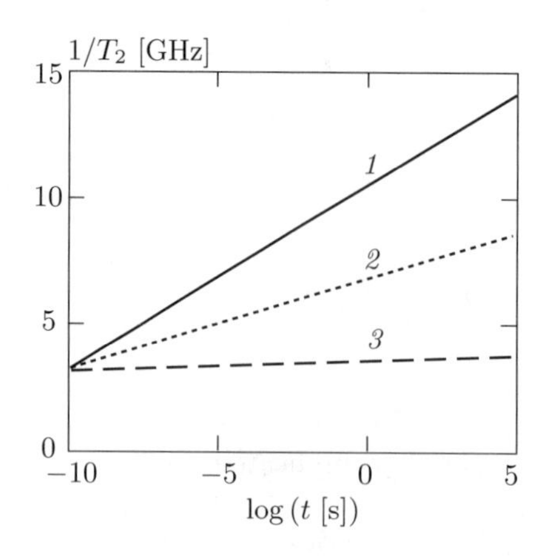

Fig. 22.7. Temporal dependence of optical dephasing calculated by means of (21.13) at $C = 10^{-1}$ (curve 1), 5×10^{-2} (curve 2), 5×10^{-3} (curve 3) and $\gamma_{\mathrm{ph}} = 10/\pi$ GHz

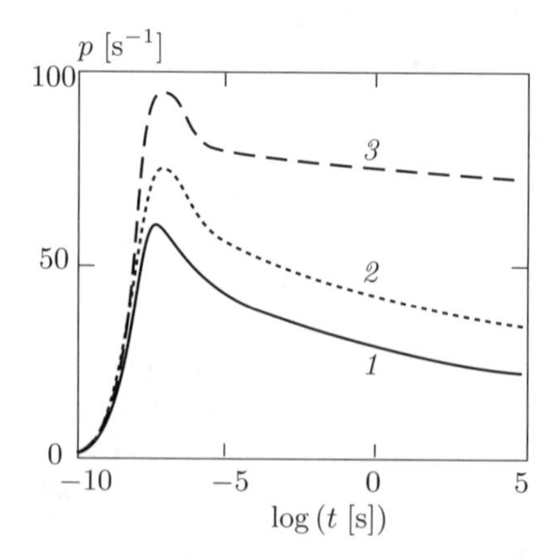

Fig. 22.8. Temporal dependence of the full two-photon correlator upon excitation in the 0 − 1 optical line at the following values of the parameters: $\chi = 10^6$ s^{-1}, $f = 0.2$, $R = 10^6$ s^{-1}, $\gamma_{\mathrm{ph}} = 10^{10}$ s^{-1}, with $C = 10^{-1}$ (curve 1), 5×10^{-2} (curve 2), 5×10^{-3} (curve 3) and $\gamma_{\mathrm{ph}} = 10/\pi$ GHz

22.6 Diversity of the Local Dynamics of Individual Molecules

The spectral trajectory shown in Fig. 19.4 is only one type of trajectory so far measured in experiments with single pentacene molecules in a *p*-terphenyl crystal. The measurements show that each guest molecule has, in fact, its own individual spectral trajectory. It should be noted that some differences in the trajectories can be related to the random character of quantum jumps predicted by quantum mechanics. These differences do not result from any difference in the quantum dynamics of individual guest molecules. Therefore differences due to dynamics can only be found if we measure some kind of probability, rather than a random characteristic like the spectral trajectory. The half-width of the optical line and the two-photon correlator are probabilities which contain information concerning the fast and slow dynamics of individual molecules.

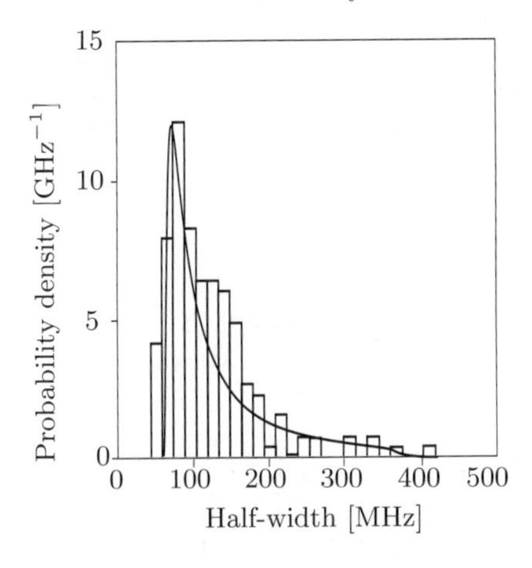

Fig. 22.9. Distribution of line widths of 176 terylene molecules doping polyethylene at 1.7 K

Line Width Distributionof Individual Molecules. The ZPL half-width equals $2/T_2$ and it exceeds the natural half-width $1/T_1$ connected with the fluorescence lifetime. The ZPL half-width is due to the interaction of the electronic excitation with phonons and TLSs. The question as to whether each individual guest molecule has the same strength of electron–phonon and electron–tunnelon interaction remained open until single molecule spectroscopy found the answer. It was shown that the line width distribution of individual guest molecules exists even in an anthracene single crystal [112]. The distribution is larger for guest molecules in polymers and glasses. Figure 22.9 shows such a distribution for the line widths of a terylene molecule doping polyethylene [29]. The half-width distributions for terylene molecules doping Shpolskii matrices and some polymers can be found in [112]. The minimal half-width of these distributions is quite different for each host matrix. In fact the width and the left boundary of the distribution depends on the kind of pair constituted by the guest–host molecules. The existence of such a distribution proves that the immediate vicinity of the guest molecule plays an important role in optical dephasing processes.

Varied Temporal Behavior of the Two-Photon Correlator. The full two-photon correlator is the probability of finding pairs of photons with a definite delay. We have already seen that this probability, as a function of the time delay, describes the dynamics of the fluorescent electronic state of an individual guest molecule.

In single-molecule experiments, an autocorrelation function (AF) $g^{(2)}(t)$ is measured, which is expressed in terms of the fluorescence intensity I. The autocorrelation function is related to the full two-photon correlator by

$$\frac{p\left(t_0\right)}{p\left(\infty\right)} = g^{(2)}\left(t_0\right) = \lim_{t\to\infty} \frac{\langle I(t)I\left(t+t_0\right)\rangle}{\langle I(t)I\left(t+\infty\right)\rangle} = \lim_{t\to\infty} \frac{\langle I(t)I\left(t+t_0\right)\rangle}{\langle I(t)\rangle^2} , \tag{22.14}$$

where t is the time of photon pair counts and t_0 is the time delay between photons in a pair. The infinite time should be interpreted physically as a delay between photons which exceeds all relaxation times in the system consisting of the guest molecule and TLSs. The AF approaches unity when the time delay t_0 approaches physical infinity.

It is more convenient to use the ratio which describes the AF than to use the full two-photon correlator, because those factors which have nothing to do with molecular dynamics are canceled. These factors cannot be controlled in real experiments. However, the AF also has one drawback. For instance, in the case when the probability $p(\infty)$ is very small, it is difficult to measure and errors can be large. For instance we, face this situation when the molecule has a large intersystem crossing quantum yield and small rate constant γ_{ST} for the triplet state. In this case $p(\infty) \approx \gamma_{ST}$, in accordance with (8.21). The situation for finding $p(\infty)$ becomes even less clear in the case when the guest molecule undergoes spectral burning. In this case fluorescence disappears at random times, when the molecule is converted to a photoproduct, and the dark interval can be very long.

Figure 22.10 shows the AF of three individual terylene molecules located at various sites in polyethylene. The curves are different. The AF can thus serve as a source of information about the local dynamics of the guest molecule. It should be recalled that exponential decay looks like a smooth step taking one order of magnitude in a logarithmic time scale. We have already seen that the relaxation of TLSs in the immediate vicinity of the guest

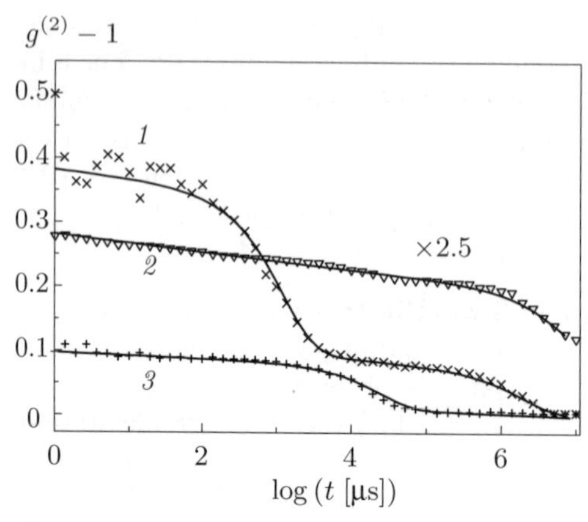

Fig. 22.10. Autocorrelation functions for three terylene molecules in PE at 1.8 K [29] and theoretical curves calculated by means of (21.4) and (21.13) at the following parameter values ($T_1 = 10^{-8}$ s, $R_2 = 10^{10}$ s^{-1}): (curve 1) $f = 0.18$, $R = 8\times10^2$ s^{-1}, $f' = 0.065$, $R' = 0.5$ s^{-1}, $p(\infty) = p(10$ s), (curve 2) $f = 0.09$, $R = 0.12$ s^{-1}, $f' = R' = 0$, $p(\infty) = p(10^9$ s), (curve 3) $f = 0.062$, $R = 47$ s^{-1}, $f' = R' = 0$, $p(\infty) = p(10$ s)

molecule can manifest itself as exponential decay in the two-photon correlator. We can thus say that there are two TLSs in the vicinity of molecule 1, with lifetimes 10^{-3} s and 1 s, whereas there is only one TLS in the vicinity of molecules 2 and 3, with relaxation times 10 s and 10^{-2} s, respectively. There is also a weak logarithmic relaxation which covers a time scale of at least six orders of magnitude. We have already seen this type of relaxation in the optical dephasing rate. Such relaxation is called spectral diffusion. Here we see how spectral diffusion can manifest itself in the full two-photon correlator of an individual guest molecule.

22.7 Contribution to the Optical Band from Different Types of TLS

Spectral hole burning and photon echo were major tools for studying the ZPL in amorphous solids before single-molecule spectroscopy was developed. It was established that spectral holes are broadened in the low temperature region according to a linear temperature law [113–115] and that spectral holes also exhibit temporal broadening, which was referred to as spectral diffusion. It was found that a logarithmic temporal law is typical for the broadening of spectral holes [106]. However, very recently deviations from logarithmic temporal broadening have been found using the stimulated photon echo technique on sub-microsecond time scales [110] and the persistent spectral hole-burning method on a time scale of hours and days [109]. A possible theoretical explanation for these facts was proposed in [116].

All these facts are anomalies from the standpoint of a theory that only takes into account the interaction with phonons. Since these observations are typical for amorphous solids like polymers and glasses, they have been explained by means of the chromophore–TLS interaction. It has been shown that both the linear temperature law and the logarithmic temporal law can be explained if the guest molecule interacts with the huge number of TLSs existing in polymers and glasses. However, such an approach cannot explain the differences in the line broadening of individual molecules detected in single-molecule spectroscopy and shown in Figs. 22.9 and 22.10. These differences are due to the fact that the local environment of individual guest molecules is different. It turns out that a few TLSs in this local environment and the sea of TLSs in the polymer give different kinds of contribution to spectroscopic effects.

Let us consider the product in (20.14). The majority of TLSs are situated far from the guest molecule and therefore the change Δ_j in the energy splitting of these TLSs upon electronic excitation is very small. However, the amplitude C_j approaches zero at small Δ_j. Therefore these remote TLSs do not contribute to the product in (20.14). Only TLSs with $\Delta_j > R_j$, i.e., TLSs coupling strongly with the guest molecule, will contribute to the product. However, such TLSs are situated in the immediate vicinity of the molecule.

In fact there are only a few TLSs with strong chromophore–TLS interaction. Only TLSs in the immediate vicinity of the molecule are able to satisfy the inequality

$$\Delta_j > 2/T_2 \,, \tag{22.15}$$

where the optical dephasing time T_2 includes all dephasing mechanisms. It is obvious that all TLSs satisfying this inequality will manifest themselves via well resolved optical lines in the chromophore–TLS optical band. Therefore the spectral resolution condition can be written as

$$\frac{2}{T_2} = \Delta(r) = \frac{Dd}{r^3} \,. \tag{22.16}$$

It can serve as an equation for finding the radius of the local neighborhood. By taking $2/T_2 = 200$ MHz and $D = d = 0.4$ Debye, we find the values $r_l = 4.7$ nm and $V_l = 471\ \mathrm{nm}^3$ for the radius and volume of this neighborhood. At a TLS concentration of order $2 \times 10^{-2}\ \mathrm{nm}^{-3}$, only one TLS can occur in this neighborhood. This evaluation agrees with experimental facts found in [117]. The authors of this work, after studying the optical bands of 80 individual molecules, found that the optical bands of more than 70% of the molecules consisted of two optical lines. This means that only one TLS is situated in their vicinity. 20% of the molecules had two TLSs in their immediate vicinity and 5% of the molecules had three or more TLSs.

22.8 Deviations from the Standard TLS Model

According to the standard TLS model, polymer and glass TLSs do not interact with each other and the rate constants R of these TLSs are distributed according to the hyperbolic law $1/R$. The simplicity of the standard TLS model and its ability to explain a wide variety of experimental facts relating to low temperature properties of polymers and glasses seemed surprising and attracted much attention. Some attempts were made to introduce the interaction between TLSs because this very point is considered by many as a weak point of the TLS model. So far, these attempts cannot be described as successful.

Single-molecule spectroscopy gives a unique opportunity to check the standard TLS model because it enables one to visualize TLSs via well-resolved optical lines. The investigation undertaken by Michel Orrit's group [117] aimed to test how far the standard TLS model was able to describe results obtained using single-molecule spectroscopy. This investigation gives many important new results.

About 70 terylene molecules in polyethylene with polycrystalline structure and 14 terylene molecules in the amorphous solid polyisobutylene were probed by CW laser light. The concentration was 10^{-6}–10^{-7} M. Their optical bands were measured at $T = 1.8$ K. If tunneling systems of polymers can be modeled by TLSs, the optical band of the guest molecule interacting with TLSs must indeed consist of 2^N optical lines, where N is the number of TLSs in the immediate vicinity of the guest molecule.

Approximately 60% of the optical bands of individual molecules can be described by the standard TLS model. However, about 40% of the optical spectra cannot be described by invoking this model. Either they consist of an odd number of lines or the temporal behavior of measured optical lines cannot be explained by the TLS model.

Indeed the standard temporal behavior of the optical band of the guest molecule interacting with one TLS is shown in Fig. 22.11. Here the result of 1 000 scans of the laser frequency is presented. The duration of one laser scan is about 1 s. We shall call the vertical black band with white intervals of random lengths a trail, following the terminology proposed in [117]. The trails on a 2D plot visualize the frequencies of absorbed photons and times of absorption. Two trails mean that the optical line is jumping between two spectral positions. All white intervals in one trail coincide exactly on the time scale with black intervals in another trail. This 2D picture of trails shown in Fig. 22.11 is typical for a guest molecule interacting with one TLS. By the way, each trail visualizes a train of photons with photon bunching like that shown in Fig. 22.3.

It becomes clear that the trails shown in Fig. 22.12 cannot be explained by the standard TLS model. About 40% of the molecules studied in [117] demonstrate various types of anomalies, including those shown in Fig. 22.12. It is difficult to find deviations from the standard TLS model in spectral hole-burning or photon echo experiments because the optical bands of about 60% of the molecules can be described by the TLS model.

Single-molecule spectroscopy serves as a unique tool for studying tunneling relaxation in polymers and glasses.

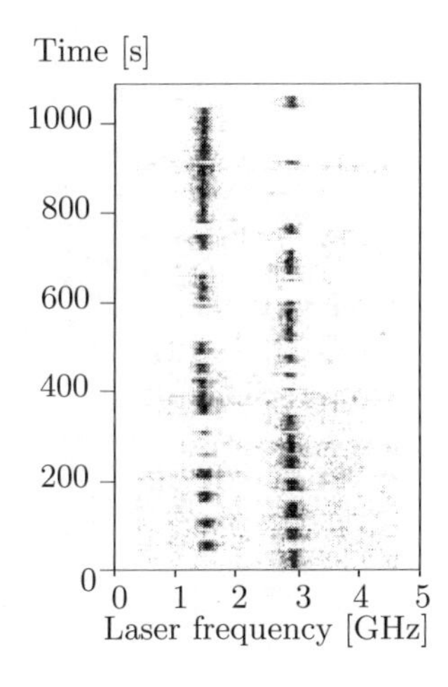

Fig. 22.11. Jumps of the spectral line according to the standard TLS model [117]

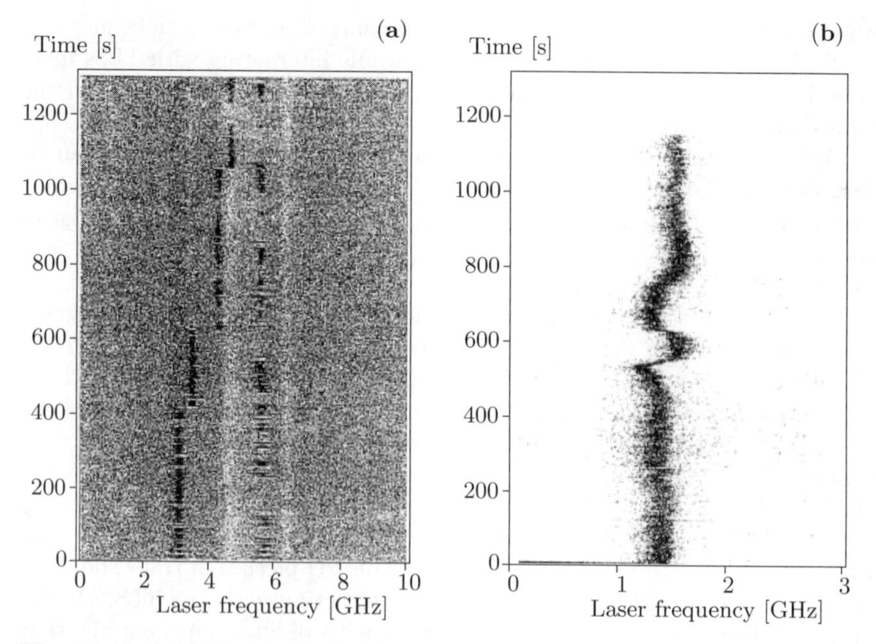

Fig. 22.12. (a,b) Two examples of trails which disagree with the TLS model [117]

A. Appendices

A. Calculation of the Function $\gamma(\omega)$

Inserting (2.6) into (2.8), we find the expression

$$\gamma(\omega) = \frac{8\pi^2 \Omega^2 d^2}{\hbar(\omega + \Omega)} N(\omega + \Omega) , \tag{A.1}$$

where

$$N(\omega + \Omega) = \frac{1}{V} \sum_{\boldsymbol{k}} \cos^2 \alpha_{\boldsymbol{k}} \delta(\omega + \Omega - \omega_{\boldsymbol{k}}) . \tag{A.2}$$

Here $\alpha_{\boldsymbol{k}}$ is the angle between the dipole moment and the polarization vector. The volume V is taken as L^3. The shape of the volume does not influence the final result if the volume approaches infinity. The length of standing waves must satisfy the condition

$$n \frac{\lambda_n}{2} = L , \quad n = 1, 2, 3, \dots . \tag{A.3}$$

Since $k = 2\pi/\lambda$, we can write all possible wave vectors in the form

$$\boldsymbol{k} = \left(\frac{\pi}{L} n_x, \frac{\pi}{L} n_y, \frac{\pi}{L} n_z \right) , \tag{A.4}$$

where n_x, n_y, n_z are positive integers. It is obvious that the minimal volume in k-space is given by

$$\Delta^3 k = \left(\frac{\pi}{L} \right)^3 = \frac{\pi^3}{V} . \tag{A.5}$$

The sum with respect to wave vectors in (A.1) can be written

$$\begin{aligned} N(\omega + \Omega) &= \frac{1}{\pi^3} \sum_{\boldsymbol{k}} \cos^2 \alpha_{\boldsymbol{k}} \delta(\omega + \Omega - \omega_{\boldsymbol{k}}) \Delta^3 k \\ &\approx \frac{1}{\pi^3} \int \cos^2 \alpha_{\boldsymbol{k}} \delta(\omega + \Omega - \omega_k) \, \mathrm{d}\boldsymbol{k} . \end{aligned} \tag{A.6}$$

Wave vectors with positive components occupy only 1/8 of the phase space. Therefore, when carrying out the integration with respect to $\boldsymbol{k}$, we have to take this fact into account:

$$\begin{aligned} N(\omega+\Omega) &= \frac{1}{8\pi^3}\int_0^\infty k^2 \mathrm{d}k \int_{-1}^{1} \mathrm{d}\cos\theta_k \int_0^{2\pi} \mathrm{d}\varphi_k \left(\cos\alpha_{\boldsymbol{k}}\right)^2 \delta\left(\omega+\Omega-\omega_{\boldsymbol{k}}\right) \\ &= \left(\frac{1}{2\pi c}\right)^3 (\omega+\Omega)^2 \int_{-1}^{1} \mathrm{d}\cos\theta_k \int_0^{2\pi} \mathrm{d}\varphi_k \left(\cos\alpha_{\boldsymbol{k}}\right)^2 . \end{aligned} \tag{A.7}$$

The vectors $\boldsymbol{k}$ and $\boldsymbol{e}_{\boldsymbol{k}}$ are mutually perpendicular. Let these vectors be taken along the x and z axes. Then we find

$$\left(\boldsymbol{d}\cdot\boldsymbol{e}_{\boldsymbol{k}}\right)^2 \equiv \left(\boldsymbol{d}_\perp\cdot\boldsymbol{e}_{\boldsymbol{k}}\right)^2 = d^2\sin^2\theta_k\cos^2\varphi_k = d^2\cos^2\alpha_{\boldsymbol{k}} . \tag{A.8}$$

Inserting this formula into (A.7), we find that

$$N(\omega+\Omega) = \frac{(\omega+\Omega)^2}{6\pi^2 c^3} . \tag{A.9}$$

Substituting this function into (A.1) yields the final result

$$\gamma(\omega) = \frac{4d^2\Omega^2(\omega+\Omega)}{3c^3\hbar} . \tag{A.10}$$

B. Probability of Spontaneous Photon Emission

Using (2.15) and well known properties of the delta function, we find

$$\begin{aligned} &\sum_k |G_k(t)|^2 \qquad \text{(B.1)} \\ &= \sum_k \frac{|\lambda_k|^2}{\Delta_k^2+(\gamma/2)^2}\left(1+\mathrm{e}^{-t\gamma}-2\mathrm{e}^{-\gamma t/2}\cos\Delta_k t\right)\int_{-\infty}^{\infty}\mathrm{d}\Delta\,\delta\left(\Delta-\Delta_k\right) \\ &= \int_{-\infty}^{\infty}\frac{\mathrm{d}\Delta}{\pi}\frac{\pi\sum_k|\lambda_k|^2\delta\left(\Delta-\Delta_k\right)}{\Delta^2+(\gamma/2)^2}\left(1+\mathrm{e}^{-t\gamma}-2\mathrm{e}^{-\gamma t/2}\cos\Delta t\right) \\ &= \int_{-\infty}^{\infty}\frac{\mathrm{d}\Delta}{\pi}\frac{\gamma/2}{\Delta^2+(\gamma/2)^2}\left(1+\mathrm{e}^{-t\gamma}-2\mathrm{e}^{-\gamma t/2}\cos\Delta t\right) . \end{aligned}$$

C. Unitarity of Amplitudes

Using (2.31) for the probability amplitudes and similar equations for the conjugate amplitudes, we can derive the following equations:

$$\frac{\mathrm{d}}{\mathrm{d}t}\left|G_n^0\right|^2 = -\mathrm{i}\Lambda^* G_{n-1}^1 G_n^{*0} + \mathrm{i}\Lambda G_{n-1}^{*1} G_n^0 ,$$

$$\frac{\mathrm{d}}{\mathrm{d}t}\left|G_{n-1}^1\right|^2 = -\mathrm{i}\Lambda G_n^0 G_{n-1}^{*1} + \mathrm{i}\Lambda^* G_n^{*0} G_{n-1}^1$$
$$-\mathrm{i}\sum_{\boldsymbol{k}}\left(\lambda_{\boldsymbol{k}} G_{n-1\boldsymbol{k}}^0 G_{n-1}^{*1} - \lambda_{\boldsymbol{k}}^* G_{n-1\boldsymbol{k}}^{*0} G_{n-1}^1\right) ,$$

$$\frac{\mathrm{d}}{\mathrm{d}t}\sum_{\boldsymbol{k}}\left|G_{n-1\boldsymbol{k}}^0\right|^2 = -\mathrm{i}\sum_{\boldsymbol{k}}\left(\lambda_{\boldsymbol{k}}^* G_{n-1\boldsymbol{k}}^{*0} G_{n-1}^1 - \lambda_{\boldsymbol{k}} G_{n-1\boldsymbol{k}}^0 G_{n-1}^{*1}\right) . \tag{C.1}$$

The sum of these equations yields (2.33).

D. Derivation of (3.8)

Using the Laplace transformation for the derivative, we can transform the fourth equation of the set described by (3.6) to give

$$G_{\boldsymbol{k}}^1 = \frac{\Lambda' G_{\boldsymbol{k}}^0}{\omega - \Delta - \Delta_{\boldsymbol{k}} + \mathrm{i}\gamma/2} . \tag{D.1}$$

Here we omit the index of the laser mode. Inserting this into the equation for the Laplace transform derived from the third equation of (3.6), we arrive at the expression

$$G_{\boldsymbol{k}}^0 = \frac{\lambda_{\boldsymbol{k}}^* G^1}{\omega - \Delta_{\boldsymbol{k}} - \dfrac{|\Lambda'|^2}{\omega - \Delta - \Delta_{\boldsymbol{k}} + \mathrm{i}\gamma/2}} = \frac{\lambda_{\boldsymbol{k}}^* G^1}{\omega - \Delta_{\boldsymbol{k}}' + \mathrm{i}\gamma'/2} . \tag{D.2}$$

Here $\Delta_{\boldsymbol{k}}'$ and γ' depend on Λ'. Strictly speaking, this equation will include a ratio with Λ'', Λ''', and so on, as occurs in (2.26). However, we can artificially cut off this continued fraction using the arguments discussed in connection with (2.26). After this cutoff procedure, we arrive at (D.2).

Using (D.1) and (D.2), we find the expression

$$\sum_{\boldsymbol{k}} \lambda_{\boldsymbol{k}} G_{\boldsymbol{k}}^1 = \Lambda' G^1 \sum_{\boldsymbol{k}} \frac{|\lambda_{\boldsymbol{k}}|^2}{\omega - \Delta - \Delta_{\boldsymbol{k}} + \mathrm{i}\gamma/2}\,\frac{1}{\omega - \Delta_{\boldsymbol{k}}' + \mathrm{i}\gamma'/2}$$
$$\approx \frac{\Lambda' G^1}{\Delta' + \mathrm{i}(\gamma' - \gamma)/2}\left(\sum_{\boldsymbol{k}} \frac{|\lambda_{\boldsymbol{k}}|^2}{\omega - \Delta - \Delta_{\boldsymbol{k}} + \mathrm{i}\gamma/2} - \sum_{\boldsymbol{k}} \frac{|\lambda_{\boldsymbol{k}}|^2}{\omega - \Delta_{\boldsymbol{k}}' + \mathrm{i}\gamma'/2}\right)$$
$$\approx \frac{\Lambda' G^1}{\Delta' + \mathrm{i}(\gamma' - \gamma)/2}\left[-\mathrm{i}\frac{\gamma(\omega - \Delta)}{2} + \mathrm{i}\frac{\gamma(\omega)}{2}\right]$$
$$= \frac{\Lambda' G^1}{\Delta' + \mathrm{i}(\gamma' - \gamma)/2}\,\frac{\mathrm{i}\gamma}{2}\,\frac{\Delta}{\Omega} \approx \frac{\mathrm{i}\gamma}{2}\,\frac{\Lambda'}{\Omega} G^1 . \tag{D.3}$$

Here we used the arguments discussed in connection with (2.8) and the equation

$$\sum_{\boldsymbol{k}} \frac{|\lambda_{\boldsymbol{k}}|^2}{\omega - \Delta - \Delta_{\boldsymbol{k}} + \mathrm{i}\gamma/2} \approx -\mathrm{i}\frac{\gamma(\omega - \Delta)}{2}, \tag{D.4}$$

where $\gamma(\omega)$ is defined by (2.9).

E. Derivation of (3.24)

Using (1.61) and (1.64) for the direct and inverse Laplace Transformation, we find

$$\begin{aligned}
\int_{-\infty}^{\infty} s(\omega)\, p(\omega)\, \mathrm{e}^{-\mathrm{i}t\omega} \frac{\mathrm{d}\omega}{2\pi} &= \int_{-\infty}^{\infty} s(\omega) \int_{0}^{\infty} p(x) \mathrm{e}^{\mathrm{i}\omega x} \mathrm{d}x\, \mathrm{e}^{-\mathrm{i}t\omega} \frac{\mathrm{d}\omega}{2\pi} \\
&= \int_{0}^{\infty} \mathrm{d}x\, p(x) \int_{-\infty}^{\infty} s(\omega) \mathrm{e}^{-\mathrm{i}(t-x)\omega} \frac{\mathrm{d}\omega}{2\pi} \\
&= \int_{0}^{\infty} p(x) s(t-x) \theta(t-x)\, \mathrm{d}x \\
&= \int_{0}^{t} s(t-x)\, p(x)\, \mathrm{d}x\,.
\end{aligned} \tag{E.1}$$

Here we used the Heaviside step function $\theta(t)$.

F. Proof of the Approximation (7.16)

With the help of the Laplace transformation of the first, second and third equations in (7.14), we find the following equations:

$$\begin{aligned}
G_0^a &= \frac{1}{\omega - \omega_0^a + \mathrm{i}0} + \frac{1}{\omega - \omega_0^a + \mathrm{i}0} \sum_b \Lambda_{ab} G_0^b\,, \\
G_0^b &= \frac{1}{\omega - \omega_0^b + \mathrm{i}0} \sum_a \left[\Lambda_{ba} G_0^a + \sum_{\boldsymbol{k}} \lambda_{\boldsymbol{k}}^{ba} G_{\boldsymbol{k}}^a \right], \\
G_{\boldsymbol{k}}^a &= \frac{1}{\omega - \omega_{\boldsymbol{k}}^a + \mathrm{i}0} \sum_b \left[\lambda_{\boldsymbol{k}}^{ab} G_0^b + \underline{\Lambda'_{ab} G_{\boldsymbol{k}}^b} \right].
\end{aligned} \tag{F.1}$$

The underlined term can be omitted for same reason that justified the cutoff procedure in the continued fraction (2.26). From the third equation,

$$\begin{aligned}\sum_{\boldsymbol{k}} \lambda_{\boldsymbol{k}}^{ba} G_{\boldsymbol{k}}^{a'} &= \sum_{\boldsymbol{k}} \frac{\lambda_{\boldsymbol{k}}^{ba}}{\omega - \omega_{\boldsymbol{k}}^{a'} + \mathrm{i}0} \sum_{b'} \lambda_{\boldsymbol{k}}^{a'b'} G_0^{b'} \\ &= \sum_{b'} \langle b|a\rangle\langle a'|b'\rangle \sum_{\boldsymbol{k}} \frac{|\lambda_{\boldsymbol{k}}|^2}{\omega - \omega_{\boldsymbol{k}}^{a'} + \mathrm{i}0} G_0^{b'} \\ &= -\frac{\mathrm{i}}{2T_1} \sum_{b'} \langle b|a\rangle\langle a'|b'\rangle G_0^{b'} \,. \end{aligned} \tag{F.2}$$

The other relations described by (7.16) can be derived in a similar way. It should be noted that a new type of term $\lambda_{\boldsymbol{k}'}^{ab} G_{\boldsymbol{k}}^{b}$ emerges in the fifth equation in (7.17). However, it can be shown that terms of such type can be neglected because of their small contribution to the sums on the left hand side of (7.16).

G. The Wick–Bloch–Dominicis Theorem for Bosons

This theorem gives a prescription for calculating quantum statistical averages of products of Bose operators. Since phonons are bosons, their commutation relations are

$$b_l b_k^+ - b_k^+ b_l = \delta_{lk} \,, \quad b_l b_k - b_k b_l = 0 \,, \quad b_l^+ b_k^+ - b_k^+ b_l^+ = 0 \,. \tag{G.1}$$

The quantum statistical average with the density operator

$$\hat{\rho}(T) = \exp\left(\frac{F - \sum_q \varepsilon_q b_q^+ b_q}{kT} \right) \tag{G.2}$$

is defined by the following formulas:

$$\operatorname{Tr}\left(\hat{\rho}(T)\, b_l^+ b_k\right) \equiv \langle b_l^+ b_k \rangle = \delta_{lk} n_l(T) = \frac{\delta_{lk}}{\exp(\varepsilon_l/kT) - 1} \,, \tag{G.3}$$

$$\operatorname{Tr}\left(\hat{\rho}(T)\, b_k b_l^+\right) \equiv \langle b_k b_l^+ \rangle = \delta_{lk} \left[n_l(T) + 1\right] = \frac{\delta_{lk}}{1 - \exp(-\varepsilon_l/kT)} \,. \tag{G.4}$$

After replacing the Kronecker symbol δ_{lk} by the relevant commutation relation, we may write these relations in the form

$$\langle a_l a_k \rangle = \frac{[a_l, a_k]}{1 - \exp(\pm \varepsilon_l/kT)} \,, \tag{G.5}$$

where the upper and lower signs are for $a_l = b_l^+$ and $a_l = b_l$, respectively. This equation is also correct in the case when both operators are phonon annihilation or creation operators.

Let us examine the relation $b(\beta) = \exp(\beta H) b \exp(-\beta H)$, where $H = E b^+ b$. Differentiating this operator with respect to β, we find

$$\frac{\mathrm{d}}{\mathrm{d}\beta} b(\beta) = \exp(\beta H)[H, b]\exp(-\beta H) = -Eb(\beta) \ . \tag{G.6}$$

The solution of this equation with initial value $b(0) = b$ is

$$b(\beta) = \exp(\beta H)\, b \exp(-\beta H) = \exp(-E\beta)\, b \ . \tag{G.7}$$

Similarly, it can be shown that

$$\exp(\beta H)\, b^+ \exp(-\beta H) = \exp(E\beta)\, b^+ \ . \tag{G.8}$$

Using these two formulas, the commutation relation for the density operator is

$$\hat{\rho}(T)\, a_l = \exp\left(\pm\frac{\varepsilon_l}{kT}\right) a_l \hat{\rho}(T) \ , \tag{G.9}$$

where the upper and lower signs are for $a_l = b_l^+$ and $a_l = b_l$, respectively.

Let us consider now the commutation relation

$$\begin{aligned}[][a_1, a_2 a_3 a_4 \ldots a_{2N}] = [a_1, a_2]\, a_3 a_4 \ldots a_{2N} + [a_1, a_3]\, a_2 a_4 \ldots a_{2N} \\ + \ldots + [a_1, a_{2N}]\, a_2 a_3 a_4 \ldots a_{2N-1} \ .\end{aligned} \tag{G.10}$$

The calculation of the quantum statistical average over all terms yields

$$\begin{aligned}\langle a_1 a_2 a_3 a_4 \ldots a_{2N}\rangle - \langle a_2 a_3 a_4 \ldots a_{2N} a_1\rangle = [a_1, a_2]\, \langle a_3 a_4 \ldots a_{2N}\rangle \\ + [a_1, a_3]\, \langle a_2 a_4 \ldots a_{2N}\rangle + \ldots + [a_1, a_{2N}]\, \langle a_2 a_3 a_4 \ldots a_{2N-1}\rangle \ .\end{aligned} \tag{G.11}$$

Using (G.9) and the fact that cyclic permutations leave the trace invariant, we arrive at

$$\begin{aligned}\langle a_2 a_3 a_4 \ldots a_{2N} a_1\rangle &= \mathrm{Tr}\left(\hat{\rho}(T)\, a_2 a_3 a_4 \ldots a_{2N} a_1\right) \\ &= \mathrm{Tr}\left(a_1 \hat{\rho}(T)\, a_2 a_3 a_4 \ldots a_{2N}\right) \\ &= \exp\left(\pm\frac{\varepsilon_1}{kT}\right) \mathrm{Tr}\left(\hat{\rho}(T)\, a_1 a_2 a_3 a_4 \ldots a_{2N}\right) \\ &= \exp\left(\pm\frac{\varepsilon_1}{kT}\right) \langle a_1 a_2 a_3 a_4 \ldots a_{2N}\rangle \ .\end{aligned} \tag{G.12}$$

Inserting the right hand side of this equation into the left hand side of (G.11), we find the following fundamental equation

$$\begin{aligned}\langle a_1, a_2 a_3 a_4 \ldots a_{2N}\rangle = \langle a_1, a_2\rangle\langle a_3 a_4 \ldots a_{2N}\rangle + \langle a_1, a_3\rangle\langle a_2 a_4 \ldots a_{2N}\rangle \\ + \ldots + \langle a_1, a_{2N}\rangle\langle a_2 a_3 a_4 \ldots a_{2N-1}\rangle \ .\end{aligned} \tag{G.13}$$

If we apply this equation to the average of a product of $2N-2$ operators and then a product of $2N-4$ operators and so on, we shall be able express the average for N operators as a sum of terms in which each term is a product of averages of pairs of Bose operators. Each term of this sum relates to a definite type of pairing of the operators a_l.

H. Influence of FC and HT Interactions on the Optical Band Shape Function

Let us examine the following equation:

$$J^{g,e}(\Delta) = \frac{1}{2\pi} \int_{-\infty}^{\infty} dt \, e^{\pm i\Delta t - |t|/2T_1} J^{g,e}(t) \, , \tag{H.1}$$

where the dipolar correlators are defined by

$$\begin{aligned} J^{g}(t) &= \mathrm{Tr}\left\{ \rho^{g} \exp\left(i\frac{H^{g}}{\hbar}t \right) d \exp\left(-i\frac{H^{e}}{\hbar}t \right) d \right\} \, , \\ J^{e}(t) &= \mathrm{Tr}\left\{ \rho^{e} \exp\left(i\frac{H^{e}}{\hbar}t \right) d \exp\left(-i\frac{H^{g}}{\hbar}t \right) d \right\} \, . \end{aligned} \tag{H.2}$$

Here $H^{e}(Q) = H(Q-a)$ and $H^{g}(Q) = H(Q)$. After changing the variable $Q - a \rightarrow Q$ in the expression for J^{e}, this quantum statistical average includes the density matrix ρ^{g}. Therefore the last equations can be written

$$\begin{aligned} J^{g}(t) &= \left\langle \exp\left[i\frac{H(Q)}{\hbar} \right] d(Q) \exp\left[-i\frac{H(Q-a)}{\hbar} \right] d(Q) \right\rangle \, , \\ J^{e}(t) &= \left\langle \exp\left[i\frac{H(Q)}{\hbar} \right] d(Q+a) \exp\left[-i\frac{H(Q+a)}{\hbar} \right] d(Q+a) \right\rangle \, . \end{aligned} \tag{H.3}$$

The results for the multimode variable Q can be derived from the simplest generalization of the results for the one-mode variable Q. We therefore begin by examining the one-mode case. The dipole moment can be written in the form

$$d(Q) = \int_{-\infty}^{\infty} dk \, d(k) e^{-ikQ} \, . \tag{H.4}$$

By using the coordinate shift operator defined by (10.30) and (10.33), we can write (H.3) as

$$J^{g,e}(t) = \int_{-\infty}^{\infty} dk \, d(k) \int_{-\infty}^{\infty} dk' \, d(k') J_{kk'}(\pm a) e^{-(k+k')a(1\mp 1)/2} \, , \tag{H.5}$$

where

$$\begin{aligned} J_{kk'}(\pm a) &= \left\langle e^{iHt/\hbar} e^{-ikQ} e^{\mp L} e^{-iHt/\hbar} e^{\pm L} e^{-ik'Q} \right\rangle \\ &= \left\langle e^{-ikQ(t)} e^{\mp L(t)} e^{\pm L} e^{-ik'Q} \right\rangle \, . \end{aligned} \tag{H.6}$$

Here the upper and lower signs are taken for absorption and emission, respectively. Once we have done the calculation for the absorption band, the result for the fluorescence band can be obtained by changing the sign of the shift a.

The last equation can be written

$$J_{kk'} = \left\langle \mathrm{e}^{F_1}\mathrm{e}^{F_2}\mathrm{e}^{F_3}\mathrm{e}^{F_4} \right\rangle . \tag{H.7}$$

Using the formulas

$$\mathrm{e}^{F_1}\mathrm{e}^{F_2} = \mathrm{e}^{F_1+F_2}\mathrm{e}^{\frac{1}{2}[F_1,F_2]} , \qquad \left\langle \mathrm{e}^{F_1+F_2} \right\rangle = \exp \frac{\left\langle (F_1+F_2)^2 \right\rangle}{2} , \tag{H.8}$$

we can transform (H.7) to the form

$$\begin{aligned} J_{kk'} &= \left\langle \mathrm{e}^{F_1}\mathrm{e}^{F_2}\mathrm{e}^{F_3}\mathrm{e}^{F_4} \right\rangle = \left\langle \mathrm{e}^{F_1+F_2+F_3+F_4} \right\rangle \exp\left[\frac{1}{2}\sum_{j=1}^{4}\sum_{s=1}^{j}[F_s,F_j]\right] \\ &= \exp\left[\frac{1}{2}\left(\sum_{j=1}^{4}\langle F_j^2\rangle + \sum_{j=1}^{4}\sum_{s=1}^{j}\Psi_{sj}\right)\right] , \end{aligned} \tag{H.9}$$

where

$$\Psi_{sj} = \langle F_s F_j \rangle + \langle F_j F_s \rangle + [F_s, F_j] . \tag{H.10}$$

Let us calculate the averages in the last equations only for the case when the upper sign is chosen in (H.6). We find the following formulas for the averages:

$$\begin{aligned} &\left\langle F_1^2 \right\rangle = -\frac{k^2}{2}(2n+1) , \quad \left\langle F_2^2 \right\rangle = -\frac{a^2}{2}(2n+1) = \left\langle F_3^2 \right\rangle , \\ &\left\langle F_4^2 \right\rangle = -\frac{k'^2}{2}(2n+1) , \\ &\Psi_{12} = -\mathrm{i}ka , \quad \Psi_{13} = \mathrm{i}ka\left[(n+1)\mathrm{e}^{-\mathrm{i}\nu t} - n\mathrm{e}^{\mathrm{i}\nu t}\right] , \\ &\Psi_{14} = -kk'\left[(n+1)\mathrm{e}^{-\mathrm{i}\nu t} + n\mathrm{e}^{\mathrm{i}\nu t}\right] , \quad \Psi_{23} = \frac{a^2}{2}\left[(n+1)\mathrm{e}^{-\mathrm{i}\nu t} + n\mathrm{e}^{\mathrm{i}\nu t}\right] , \\ &\Psi_{24} = \mathrm{i}k'a\left[(n+1)\mathrm{e}^{-\mathrm{i}\nu t} - n\mathrm{e}^{\mathrm{i}\nu t}\right] , \quad \Psi_{34} = -\mathrm{i}k'a , \end{aligned} \tag{H.11}$$

where $n = [\exp(\hbar\nu/kT) - 1]^{-1}$. Inserting these expressions into (H.9), we arrive at the expression

$$J_{kk'}(a) = \exp\left[-\frac{k^2+k'^2}{4}(2n+1) - \frac{a^2}{2}(2n+1) - \mathrm{i}\frac{k+k'}{2}a + f_{kk'}(a)\right] , \tag{H.12}$$

where

$$f_{kk'}(a) = \frac{(a+ik)}{\sqrt{2}}(n+1)\mathrm{e}^{-\mathrm{i}\nu t}\frac{(a+ik')}{\sqrt{2}} + \frac{(a-ik)}{\sqrt{2}}n\mathrm{e}^{\mathrm{i}\nu t}\frac{(a-ik')}{\sqrt{2}} \; . \tag{H.13}$$

In the case of the fluorescence band, we change the sign of a. Equation (H.5) for the dipolar correlators takes the form

$$J^{\mathrm{g,e}} = \exp\left[-\frac{a^2}{2}(2n+1)\right]\int_{-\infty}^{\infty}\mathrm{d}k\, D(k)\int_{-\infty}^{\infty}\mathrm{d}k'\, D(k')\exp f_{kk'}(\pm a) \; , \tag{H.14}$$

where

$$D(k) = d(k)\exp\left[-\frac{k^2+k'^2}{4}(2n+1) - \mathrm{i}\frac{k+k'}{2}a\right] \; . \tag{H.15}$$

The function $\exp(f_{kk'})$ includes various powers of the variable k. The integration in (H.14) is carried out using the formula

$$\int_{-\infty}^{\infty}\mathrm{d}k\, D(k)(-\mathrm{i}k)^m = \left[\left(\frac{\partial}{\partial x}\right)^m D(x)\right]_{x=0} \; , \tag{H.16}$$

where

$$D(x) = \int_{-\infty}^{\infty}\mathrm{d}k\, D(k)\mathrm{e}^{-\mathrm{i}kx} = \int_{-\infty}^{\infty}\frac{\mathrm{d}Q(T)}{\sqrt{\pi}}\, d\left(Q+\frac{a}{2}+x\right)\mathrm{e}^{-Q^2(T)} \; . \tag{H.17}$$

The dimensionless coordinate

$$Q(T) = \frac{Q}{\sqrt{\langle Q^2\rangle}} = \frac{Q}{\sqrt{n+1/2}} \tag{H.18}$$

depends on the temperature. The coordinate has a simple physical meaning. It is a ratio of the coordinate Q and the root-mean-squared value of the same coordinate. We have used the relation

$$\exp\left[-\frac{k^2}{4}(2n+1)\right] = \int_{-\infty}^{\infty}\frac{\mathrm{d}Q}{\sqrt{\pi(n+1/2)}}\exp\left(\mathrm{i}kQ - \frac{Q^2}{n+1/2}\right) \; , \tag{H.19}$$

From the last equation, we can transform (H.14) to obtain

$$J^{\mathrm{g,e}}(t) = \mathrm{e}^{-\varphi(0,T)}\left[\sum_{m=0}^{\infty}\frac{1}{m!}\hat{\varphi}^m(\pm a,t,T)D(x)D(x')\right]_{x=x'=0} \; , \tag{H.20}$$

where

$$\hat{\varphi}(\pm a,t,T)=\frac{1}{2}\left[\left(\pm a-\frac{\partial}{\partial x}\right)(n+1)\mathrm{e}^{-\mathrm{i}\nu t}\left(\pm a-\frac{\partial}{\partial x'}\right)+\left(\pm a+\frac{\partial}{\partial x}\right)n\mathrm{e}^{\mathrm{i}\nu t}\left(\pm a+\frac{\partial}{\partial x'}\right)\right] \tag{H.21}$$

is a differential operator. If the dipole moment does not depend on the coordinates, we can omit the derivatives in the last equation and we arrive at the well known result for the purely linear FC interaction.

If we carry out the substitution $a,\nu,n,x\rightarrow a_q,\nu_q,n_q,x_q$ and sum over the mode index q in (H.21), we arrive at the operator $\varphi(\pm a,t,T)$ for the multimode case:

$$\hat{\varphi}(\pm a,t,T)=\frac{1}{2}\sum_q\left[\left(\pm a_q-\frac{\partial}{\partial x_q}\right)(n_q+1)\mathrm{e}^{-\mathrm{i}\nu_q t}\left(\pm a_q-\frac{\partial}{\partial x'_q}\right)+\left(\pm a_q+\frac{\partial}{\partial x_q}\right)n_q\mathrm{e}^{\mathrm{i}\nu_q t}\left(\pm a_q+\frac{\partial}{\partial x'_q}\right)\right]. \tag{H.22}$$

After inserting (H.20) into (H.1) and integrating over time, we arrive at the following expression for the band shape functions:

$$J^{\mathrm{g,e}}(\Delta)=D^2(0)\mathrm{e}^{-\varphi(0,T)}\frac{1/2T_1\pi}{\Delta^2+(1/2T_1)^2}+\Psi^{\mathrm{g,e}}(\Delta)\,, \tag{H.23}$$

where the first term describes the ZPL and the second the electron–vibration part of the optical band. The second term is given by

$$\begin{aligned}\Psi^{\mathrm{g,e}}(\Delta)&=\sum_{m=1}^{\infty}\Psi_m^{\mathrm{g,e}}(\Delta)\\&=\mathrm{e}^{-\varphi(0,T)}\left[\sum_{m=1}^{\infty}\frac{1}{m!}\int_{-\infty}^{\infty}\mathrm{d}\nu_1\,\hat{\varphi}(\nu_1)\right.\\&\qquad\left.\ldots\int_{-\infty}^{\infty}\mathrm{d}\nu_m\,\hat{\varphi}(\nu_m)\frac{D(x)D(x')/2T_1\pi}{(\Delta\mp\nu_1\ldots\mp\nu_m)+(1/2T_1)^2}\right]_{x=x'=0}.\end{aligned} \tag{H.24}$$

After applying the operator

$$\hat{\varphi}(\nu)=\frac{1}{2}\sum_q\left[\left(\pm a_q-\frac{\partial}{\partial x_q}\right)(n_q+1)\delta(\nu-\nu_q)\left(\pm a_q-\frac{\partial}{\partial x'_q}\right)+\left(\pm a_q+\frac{\partial}{\partial x_q}\right)n_q\delta(\nu+\nu_q)\left(\pm a_q+\frac{\partial}{\partial x'_q}\right)\right] \tag{H.25}$$

to the functions of the dipole moment, we should set $x=0$ and $x'=0$. The formula (11.24) is derived from the last equations with the help of

$$\left[\left(\frac{\partial}{\partial x}\right)^m D(x)\right]_{x=0} \approx \left[\left(\frac{\partial}{\partial Q}\right)^m d(Q)\right]_{Q=a/2} . \tag{H.26}$$

I. Cumulant Expansion

According to the Wick–Bloch–Dominicis theorem, the quantum statistical average of a product of operators can be expressed as a sum of terms with various types of operator pairings. For instance, if $W = R^2$, the average of two operators is given by

$$\begin{aligned}\langle W_1 W_2\rangle &= \langle R_1 R_1 R_2 R_2\rangle \\ &= \langle R_1 R_1\rangle\langle R_2 R_2\rangle + 2\langle R_1 R_2\rangle\langle R_1 R_2\rangle \\ &= \langle W_1\rangle\langle W_2\rangle + \langle W_1 W_2\rangle_{\mathrm{c}} ,\end{aligned} \tag{I.1}$$

i.e., the average is the sum of two terms. The first and second terms are examples of uncoupled and coupled averages of two operators W. The cumulant expansion enables one to express the average of the product of the operators solely in terms of the coupled average.

The time-ordered product of the operators is a sum in which each term relates to a definite order of the time arguments. Since the Wick–Bloch–Dominicis theorem applies to each term of this sum, the theorem is also valid for the time-ordered product of the operators. That is, using (I.1), we can write

$$\langle \hat{T} W_1 W_2\rangle = \langle \hat{T} W_1\rangle\langle \hat{T} W_2\rangle + \langle \hat{T} W_1 W_2\rangle_{\mathrm{c}} . \tag{I.2}$$

Let us introduce the notation

$$(-\mathrm{i})^n \int_0^t \mathrm{d}t_1 \ldots \int_0^t \mathrm{d}t_n \langle \hat{T} W_1 W_2 \ldots W_n\rangle = \langle\langle \overbrace{W \ldots W}^{n}\rangle\rangle . \tag{I.3}$$

Then (12.22) takes the form

$$\langle S(t)\rangle = 1 + \sum_{n=1}^{\infty} \frac{1}{n!} \langle\langle \overbrace{W \ldots W}^{n}\rangle\rangle . \tag{I.4}$$

Each term of this sum is a sum of terms consisting solely of coupled averages. The general appearance of such a term is

$$S(n_1, n_2, \ldots n_k) = \langle\langle W\rangle\rangle_{\mathrm{c}}^{n_1} \langle\langle WW\rangle\rangle_{\mathrm{c}}^{n_2} \ldots \langle\langle \overbrace{W \ldots W}^{k}\rangle\rangle_{\mathrm{c}}^{n_k} , \tag{I.5}$$

where all numbers $n_1, n_2, \ldots, n_k$ must satisfy

$$n_1 + 2n_2 + 3n_3 + \ldots + kn_k = n . \tag{I.6}$$

Equation (I.4) can be written

$$\langle S(t)\rangle = \sum_{n=0}^{\infty}\frac{1}{n!}\sum_{n_1=0}^{\infty}\sum_{n_2=0}^{\infty}\dots\langle\langle W\rangle\rangle_{\mathrm{c}}^{n_1}\langle\langle WW\rangle\rangle_{\mathrm{c}}^{n_2} \\ \dots N(n_1,n_2,\dots)\delta(n-n_1-2n_2-\dots)\,, \tag{I.7}$$

where $N(n_1, n_2, \dots)$ is the number of similar terms relating to the same combination of the numbers $n_1, n_2, \dots n_k$.

In order to find this number, we must count how many ways there are of dividing kn_k elements into groups with k elements. For the case $k = 3$, we find

$$123\dots 3n_3 \to (123)(456)\dots(3n_3-2, 3n_3-1, 3n_3), \dots \,. \tag{I.8}$$

By carrying out all possible permutations, we find $(3n_3)!$ different expressions. Among these expressions, the following will be identical:

- those which relate to a permutation inside one average, e.g., (123) and (132),
- those which relate to a permutation between averages, e.g., (123)(456) and (456)(123).

The number of such permutations equals $n_3!(3!)^{n_3}$. Therefore the number of different terms equals

$$N_3 = \frac{(3n_3)!}{n_3!(3!)^{n_3}}\,. \tag{I.9}$$

For the case of kn_k operators, the number of different terms is

$$N_k = \frac{(kn_k)!}{n_k!(k!)^{n_k}}\,. \tag{I.10}$$

Let us now find the number $N(n_1, n_2, \dots)$. Clearly,

$$C_m^{kn_k} = \frac{m!}{(kn_k)!(m-kn_k)!} \tag{I.11}$$

is the number of ways of choosing m elements from kn_k elements. Therefore

$$N(n_1, n_2, \dots) = C_n^{n_1} N_1 C_{n-n_1}^{2n_2} N_2 C_{n-n_1-2n_2}^{3n_3} N_3 \dots \tag{I.12}$$

is the number of similar terms in (I.7). Since

$$N(n_1, n_2, \dots) = \frac{n!}{n_1!(1!)^{n_1} n_2!(2!)^{n_2} n_3!(3!)^{n_3}\dots}\,, \tag{I.13}$$

and

$$\sum_{n=0}^{\infty} \delta\left(n - n_1 - 2n_2 - 3n_3 - \ldots\right) = 1 \tag{I.14}$$

is correct for any set of numbers $n_1, n_2, \ldots$, we have

$$\begin{aligned}
\langle S(t) \rangle &= \sum_{n_1=0}^{\infty} \frac{1}{n_1!}\left(\frac{\langle\langle W\rangle\rangle_{\mathrm{c}}}{1!}\right)^{n_1} \\
&\quad \times \sum_{n_2=0}^{\infty} \frac{1}{n_2!}\left(\frac{\langle\langle WW\rangle\rangle_{\mathrm{c}}}{2!}\right)^{n_2} \sum_{n_3=0}^{\infty} \frac{1}{n_3!}\left(\frac{\langle\langle WWW\rangle\rangle_{\mathrm{c}}}{3!}\right)^{n_3} \ldots \\
&= \exp\left(\langle\langle W\rangle\rangle_{\mathrm{c}} + \frac{\langle\langle WW\rangle\rangle_{\mathrm{c}}}{2!} + \frac{\langle\langle WWW\rangle\rangle_{\mathrm{c}}}{3!} + \ldots\right) \\
&= \exp g(t) \,,
\end{aligned} \tag{I.15}$$

where

$$g(t) = \sum_{n=1}^{\infty} \frac{(-\mathrm{i})^n}{n!} \int_0^t \mathrm{d}t_1 \ldots \int_0^t \mathrm{d}t_n \langle \hat{T} W_1 W_2 \ldots W_n \rangle_{\mathrm{c}} = \langle S(t) - 1 \rangle_{\mathrm{c}} \tag{I.16}$$

is the cumulant function, which involves only coupled averages marked by the subscript c.

J. Relation Between Phonon Green Functions

Let us inspect the causal Green functions for the ground and excited electronic states at $T = 0$:

$$D^{\mathrm{g,e}}(t) = -\mathrm{i}\langle g, e|\hat{T}\, R^{\mathrm{g,e}}(t) R(0)|g, e\rangle \,, \tag{J.1}$$

where

$$R^{\mathrm{g,e}}(t) = \exp\left(\mathrm{i}\frac{H^{\mathrm{g,e}}}{\hbar} t\right) R(0) \exp\left(-\mathrm{i}\frac{H^{\mathrm{g,e}}}{\hbar} t\right) . \tag{J.2}$$

With the Hamiltonian

$$H^{\mathrm{g}} = \sum_q \hbar\nu_q (b_q^+ b_q + 1/2) \,,$$

we find the following expression for the coordinate:

$$R^{\mathrm{g}}(t) = \sum_q l_q (b_q \mathrm{e}^{-\mathrm{i}\nu_q t} + b_q^+ \mathrm{e}^{\mathrm{i}\nu_q t}) \,. \tag{J.3}$$

Inserting this coordinate into (J.1) yields

$$
\begin{aligned}
D^{\mathrm{g}}(t) &= -\mathrm{i}\left[\langle 0|\,R^{\mathrm{g}}(t)R^{\mathrm{g}}(0)\,|0\rangle\,\vartheta(t) + \langle 0|\,R^{\mathrm{g}}(0)R^{\mathrm{g}}(t)\,|0\rangle\,\vartheta(-t)\right] \\
&= -\mathrm{i}\sum_q l_q^2\left[\mathrm{e}^{-\mathrm{i}\nu_q t}\vartheta(t) + \mathrm{e}^{\mathrm{i}\nu_q t}\vartheta(-t)\right] \\
&= \int_{-\infty}^{\infty}\frac{\mathrm{d}\nu}{\mathrm{i}\pi}\Gamma_0(\nu)\left[\mathrm{e}^{-\mathrm{i}\nu t}\vartheta(t) + \mathrm{e}^{\mathrm{i}\nu t}\vartheta(-t)\right] ,
\end{aligned}
\tag{J.4}
$$

where

$$
\Gamma_0(\nu) = \pi\sum_q l_q^2\delta(\nu - \nu_q) . \tag{J.5}
$$

Calculating the Laplace transform of the phonon Green function, we arrive at the expression

$$
D_0^{\mathrm{g}}(\omega) = \int_{-\infty}^{\infty}\frac{\mathrm{d}\nu}{\pi}\Gamma_0(\nu)\left(\frac{1}{\omega - \nu + \mathrm{i}0} - \frac{1}{\omega + \nu - \mathrm{i}0}\right) . \tag{J.6}
$$

This equation coincides with (5.49). Since we are inspecting the case when $T = 0$, the wave functions $|g\rangle$ and $|e\rangle$ in (J.1) describe the ground state of the harmonic oscillator when the system is in the ground and excited electronic states. The following expression is derived in quantum field theory for zero temperature [118]:

$$
D^{\mathrm{e}}(t) = -\mathrm{i}\langle e|\hat{T}\,R^{\mathrm{e}}(t)R(0)|e\rangle = -\mathrm{i}\langle g|\hat{T}\,R^{\mathrm{g}}(t)R(0)\hat{S}(\infty)|g\rangle_{\mathrm{c}} , \tag{J.7}
$$

where

$$
\begin{aligned}
\hat{S}(\infty) &= \hat{T}\,\exp\left(-\frac{\mathrm{i}}{\hbar}\int_{-\infty}^{\infty}\mathrm{d}x\,W(x)\right) , \\
W(x) &= \exp\left(\mathrm{i}\frac{H^{\mathrm{g}}}{\hbar}x\right)(H^{\mathrm{g}} - H^{\mathrm{e}})\exp\left(-\mathrm{i}\frac{H^{\mathrm{g}}}{\hbar}x\right) ,
\end{aligned}
\tag{J.8}
$$

and the subscript c means that only the coupled averages should be taken into account. Equation (J.7) can be written in the form

$$
\begin{aligned}
D^{\mathrm{e}}(t) &= -\mathrm{i}\langle g|\hat{T}R^{\mathrm{g}}(t)R(0)|g\rangle \\
&-\mathrm{i}\sum_{m=1}^{\infty}\frac{1}{m!}\left(-\frac{\mathrm{i}}{\hbar}\right)^m\int_{-\infty}^{\infty}\mathrm{d}t_1\ldots\int_{-\infty}^{\infty}\mathrm{d}t_m\langle g|\hat{T}\,R^{\mathrm{g}}(t)R(0)W_1W_2\ldots W_m|g\rangle_{\mathrm{c}} ,
\end{aligned}
\tag{J.9}
$$

The calculation of the averages is carried out as for (12.58). Since all mathematical expressions that differ by time permutations are equal, we arrive at the following expression for the causal phonon Green function in the excited electronic state:

$$D^{\mathrm{e}}(t) = D^{\mathrm{g}}(t) + \sum_{m=1}^{\infty} W^m \int_{-\infty}^{\infty} \mathrm{d}t_1 \ldots \int_{-\infty}^{\infty} \mathrm{d}t_m D^{\mathrm{g}}(t-t_1) D^{\mathrm{g}}(t_1-t_2) \ldots D^{\mathrm{g}}(t_m) \,. \tag{J.10}$$

Carrying out the Laplace transformation of this equation, we arrive at the following relation between phonon Green functions for the ground and excited electronic states:

$$D^{\mathrm{e}}(\omega,0) = D^{\mathrm{g}}(\omega,0) + \sum_{m=1}^{\infty} W^m \left(D^{\mathrm{g}}(\omega,0)\right)^{m+1} = \frac{D^{\mathrm{g}}(\omega,0)}{1 - W D^{\mathrm{g}}(\omega,0)} \,. \tag{J.11}$$

K. Shift of the ZPL at $T = 0$

The line shift at $T = 0$ is given by (12.74), i.e.,

$$\delta_{\mathrm{ph}}^{\mathrm{g}}(0) = \int_0^{\infty} \frac{\mathrm{d}\omega}{2\pi} \arctan \frac{W \Gamma^{\mathrm{g}}(\omega)}{1 - W \Delta^{\mathrm{g}}(\omega)} \,. \tag{K.1}$$

Let us apply this formula to the case of one localized mode. Then we have

$$\begin{aligned} \Gamma^{\mathrm{g}}(\omega) &= \frac{\pi}{2\nu_{\mathrm{g}}} \delta(\omega - \nu_{\mathrm{g}}) \,, \\ \Delta^{\mathrm{g}}(\omega) &= \mathrm{P} \int_0^{\infty} \frac{\mathrm{d}\nu}{\pi} \Gamma^{\mathrm{g}}(\nu) \frac{2\nu}{\omega^2 - \nu^2} = \frac{1}{\omega^2 - \nu_{\mathrm{g}}^2} \,. \end{aligned} \tag{K.2}$$

Inserting these formulas into the argument of the arctangent, we find

$$\frac{W \Gamma^{\mathrm{g}}(\omega)}{1 - W \Delta^{\mathrm{g}}(\omega)} = \frac{W\pi}{2\nu_{\mathrm{g}}} \frac{\left(\omega^2 - \nu_{\mathrm{g}}^2\right) \delta(\omega - \nu_{\mathrm{g}})}{\omega^2 - \nu_{\mathrm{e}}^2} = \begin{cases} +0 \,, & 0 < \omega < \nu_{\mathrm{g}} \,, \\ -0 \,, & \nu_{\mathrm{g}} < \omega < \nu_{\mathrm{e}} \,, \\ +0 \,, & \nu_{\mathrm{e}} < \omega < \infty \,, \end{cases} \tag{K.3}$$

where

$$\nu_{\mathrm{e}}^2 = \nu_{\mathrm{g}}^2 + W \,. \tag{K.4}$$

Since $\arctan(+0) = 0$ and $\arctan(-0) = \pi$, we find that

$$\delta_{\mathrm{ph}}^{\mathrm{g}}(0) = \int_{\nu_{\mathrm{g}}}^{\nu_{\mathrm{e}}} \frac{\mathrm{d}\omega}{2\pi} \pi = \frac{\nu_{\mathrm{e}} - \nu_{\mathrm{g}}}{2} \,, \tag{K.5}$$

i.e., the line shift equals the energy difference between the zero point vibrations in the excited and ground electronic states. It is obvious that (K.1) defines this difference for the multimode case.

L. Derivation of (17.65)

Let us consider the one-mode case and introduce the following notation:

$$L(t_k) = \frac{a}{\sqrt{2}}[b(t_k) - b^+(t_k)] = L_k = a_k[b_k - b_k^+] \,. \tag{L.1}$$

Using the operator equation $e^F e^G = e^{F+G} e^{[F,G]/2}$, we can derive the following operator equation for the product of N exponential operators:

$$\prod_{k=1}^{N} \exp L_k = \exp\left(\sum_{k=1}^{N} L_k\right) \exp\left(\frac{1}{2}\sum_{k=1}^{N-1}\sum_{j=k+1}^{N}[L_k, L_j]\right) \,. \tag{L.2}$$

The calculation of the quantum statistical average yields the following result:

$$\begin{aligned}\left\langle \exp\left(\sum_{k=1}^{N} L_k\right)\right\rangle &= \exp\left\langle \frac{1}{2}\left(\sum_{k=1}^{N} L_k\right)^2\right\rangle \\ &= \exp\left(\frac{1}{2}\sum_{k=1}^{N}\langle L_k^2\rangle + \frac{1}{2}\sum_{k=1}^{N}\sum_{j=1(j\neq k)}^{N}\langle L_k L_j\rangle\right) \,.\end{aligned} \tag{L.3}$$

Therefore the quantum statistical average of the product of the exponential operators is given by

$$\left\langle \prod_{k=1}^{N} \exp L_k \right\rangle = \exp\left(\frac{1}{2}\sum_{k=1}^{N}\varphi_{kk} + \sum_{k=1}^{N-1}\sum_{j=k+1}^{N}\varphi_{kj}\right) \,, \tag{L.4}$$

where

$$\begin{aligned}\varphi_{kj} &= \frac{1}{2}\left(\langle L_j L_k\rangle + \langle L_k L_j\rangle + [L_k, L_j]\right) \\ &= -a_k a_j\left[(n+1)\,\mathrm{e}^{-\mathrm{i}\nu(k-j)} + n\mathrm{e}^{\mathrm{i}\nu(k-j)}\right] \,.\end{aligned} \tag{L.5}$$

Let us apply the last formulas to the product of four exponential operators. Using the notation

$$\begin{aligned} &a_1 = -\frac{a}{\sqrt{2}}\,, \quad t_1 = 0\,, \qquad a_2 = \frac{a}{\sqrt{2}}\,, \quad t_2 = \tau\,, \\ &a_3 = -\frac{a}{\sqrt{2}}\,, \quad t_3 = \tau + \tau'\,, \qquad a_4 = \frac{a}{\sqrt{2}}\,, \quad t_4 = \tau\,, \end{aligned} \tag{L.6}$$

we arrive at the expression

$$\begin{aligned} &\varphi_{12} = \varphi(-\tau) , \qquad \varphi_{13} = -\varphi(-\tau' - \tau) , \qquad \varphi_{14} = \varphi(-\tau) , \\ &\varphi_{23} = \varphi(-\tau') , \qquad \varphi_{24} = -\varphi(0) , \qquad \varphi_{34} = \varphi(\tau') , \\ &\varphi_{11} = \varphi_{22} = \varphi_{33} = \varphi_{44} = -\varphi(0) , \end{aligned} \tag{L.7}$$

where

$$\varphi(t) = \frac{a^2}{2}\left[(n+1)\mathrm{e}^{-\mathrm{i}\nu t} + n\mathrm{e}^{\mathrm{i}\nu t}\right] . \tag{L.8}$$

In the multimode case, we arrive at the following expression for the function $\varphi(t)$:

$$\varphi(t, T) = \sum_q \frac{a_q^2}{2}\left[(n_q+1)\mathrm{e}^{-\mathrm{i}\nu_q t} + n_q\mathrm{e}^{\mathrm{i}\nu_q t}\right] . \tag{L.9}$$

Inserting (L.7) into (L.4), we arrive at the expression

$$\begin{aligned} &\left\langle \exp[-\hat{L}(0)]\exp[\hat{L}(\tau)]\exp[-\hat{L}(\tau'+\tau)]\exp[\hat{L}(\tau)]\right\rangle \\ &\qquad = \exp\left[2g^*(\tau) + 2\,\mathrm{Re}\,g(\tau') - g^*(\tau'+\tau)\right] , \end{aligned} \tag{L.10}$$

where

$$g(t) = \varphi(t, T) - \varphi(0, T) = g^*(-t) . \tag{L.11}$$

M. The Wick–Bloch–Dominicis Theorem for Fermions

Suppose a set of operators satisfies the anticommutation relations

$$c_l c_k^+ + c_k^+ c_l = \delta_{lk} , \quad c_l c_k + c_k c_l = 0, \quad c_l^+ c_k^+ + c_k^+ c_l^+ = 0 . \tag{M.1}$$

The quantum statistical average with the density operator

$$\hat{\rho}(T) = \exp\left(\frac{F - \sum\limits_q \varepsilon_q c_q^+ c_q}{kT}\right) \tag{M.2}$$

is defined by

$$\mathrm{Tr}\left\{\hat{\rho}(T) c_l^+ c_k\right\} \equiv \langle c_l^+ c_k \rangle = \delta_{lk} f_l(T) = \frac{\delta_{lk}}{\exp(\varepsilon_l/kT) + 1} , \tag{M.3}$$

$$\mathrm{Tr}\left\{\hat{\rho}(T)c_k c_l^+\right\} \equiv \langle c_k c_l^+\rangle = \delta_{lk}\big(1 - f_l(T)\big) = \frac{\delta_{lk}}{1+\exp(-\varepsilon_l/kT)} \;. \tag{M.4}$$

The Kronecker symbol can be replaced by a commutation relation. Therefore the last equations can be written in the form

$$\langle a_l a_k\rangle = \frac{[a_l, a_k]_+}{1+\exp(\pm\varepsilon_l/kT)} \;, \tag{M.5}$$

where the upper and lower signs are for $a_l = c_l^+$ and $a_l = c_l$, respectively. Let us consider an operator $c(\beta) = \exp(\beta H)c\exp(-\beta H)$, where $H = Ec^+c$. Differentiating this operator with respect to β, we find

$$\frac{\mathrm{d}}{\mathrm{d}\beta}c\,(\beta) = \exp\left(\beta H\right)[H, c]\exp\left(-\beta H\right) = -Ec\,(\beta) \;. \tag{M.6}$$

The solution of this equation with the initial value $c(0) = c$ is given by

$$c\,(\beta) = \exp\left(\beta H\right)\, c\, \exp\left(-\beta H\right) = \exp\left(-E\beta\right)c \;. \tag{M.7}$$

Similarly,

$$\exp\left(\beta H\right)c^+ \exp\left(-\beta H\right) = \exp\left(E\beta\right)c^+ \;. \tag{M.8}$$

Using the last equations, we arrive at the following commutation relation for the density operator:

$$\hat{\rho}\,(T)\, a_l = \exp\left(\pm\frac{\varepsilon_l}{kT}\right) a_l\hat{\rho}\,(T) \;, \tag{M.9}$$

where the upper and lower signs are for $a_l = c_l^+$ and $a_l = c_l$, respectively. Let us consider the anticommutation relation

$$\begin{aligned}[a_1, a_2a_3a_4\ldots a_{2N}]_+ = {} & [a_1, a_2]_+\, a_3a_4\ldots a_{2N} - [a_1, a_3]_+\, a_2a_4\ldots a_{2N} \\ & + \ldots + [a_1, a_{2N}]_+\, a_2a_3a_4\ldots a_{2N-1} \;. \end{aligned} \tag{M.10}$$

Here the minus sign corresponds to terms with odd permutations of the Fermi operators. Carrying out the quantum statistical average of the last operator equation, we find

$$\begin{aligned}&\langle a_1a_2a_3a_4\ldots a_{2N}\rangle + \langle a_2a_3a_4\ldots a_{2N}a_1\rangle = [a_1, a_2]_+\langle a_3a_4\ldots a_{2N}\rangle \\ &- [a_1, a_3]_+\langle a_2a_4\ldots a_{2N}\rangle + \ldots + [a_1, a_{2N}]_+\langle a_2a_3a_4\ldots a_{2N-1}\rangle \;. \end{aligned} \tag{M.11}$$

Using (M.9) and the possibility of cycling permutations under the trace sign, we find

$$\begin{aligned}\langle a_2a_3a_4\ldots a_{2N}a_1\rangle &= \mathrm{Tr}\left\{\hat{\rho}\,(T)\, a_2a_3a_4\ldots a_{2N}a_1\right\} \\ &= \mathrm{Tr}\left\{a_1\hat{\rho}\,(T)\, a_2a_3a_4\ldots a_{2N}\right\} \\ &= \exp\left(\pm\frac{\varepsilon_1}{kT}\right)\mathrm{Tr}\left\{\hat{\rho}\,(T)\, a_1a_2a_3a_4\ldots a_{2N}\right\} \\ &= \exp\left(\pm\frac{\varepsilon_1}{kT}\right)\langle a_1a_2a_3a_4\ldots a_{2N}\rangle \;. \end{aligned} \tag{M.12}$$

Inserting the right hand side of the last equation into the left hand side of (M.11) yields the final result

$$\begin{aligned}\langle a_1, a_2a_3a_4 \dots a_{2N}\rangle &= \langle a_1, a_2\rangle \langle a_3a_4 \dots a_{2N}\rangle - \langle a_1, a_3\rangle \langle a_2a_4 \dots a_{2N}\rangle \\ &+ \dots + \langle a_1, a_{2N}\rangle \langle a_2a_3a_4 \dots a_{2N-1}\rangle \ .\end{aligned} \tag{M.13}$$

If we apply this equation to each of the averages from the product of $2N-2$ operators, then from the product of $2N-4$ operators, and so on, we shall be able to express the average from N operators as a sum in which each term is a product of averages of pairs of Fermi operators. Each term in this sum relates to a definite type of pairing of the operators a_l.

N. Derivation of (18.37) and (18.38)

Using the formula

$$\begin{aligned}G(\omega, T) &= \int_{-\infty}^{\infty} \frac{\mathrm{d}\nu}{\pi} \Gamma(\nu) \left[\frac{1-f(\nu)}{\omega-\nu+\mathrm{i}0} + \frac{f(\nu)}{\omega-\nu-\mathrm{i}0}\right] \\ &= G(\omega, 0) + \mathrm{i}2f(\omega)\Gamma(\omega) \ ,\end{aligned} \tag{N.1}$$

we can write

$$\int_{-\infty}^{\infty} \frac{\mathrm{d}\omega}{2\pi} \ln\left[1-\Delta G(\omega, T)\right] = \int_{-\infty}^{\infty} \frac{\mathrm{d}\omega}{2\pi} \ln\left[1-\Delta G(\omega, 0)\right] \tag{N.2}$$

$$+ \int_{-\infty}^{\infty} \frac{\mathrm{d}\omega}{2\pi} \ln\left[1-2\mathrm{i}f(\omega)\frac{\Delta\Gamma(\omega)}{1-\Delta G(\omega,0)}\right] .$$

Using the analytical properties of the Green function, we find

$$\begin{aligned}\int_{-\infty}^{\infty} \frac{\mathrm{d}\omega}{2\pi} \ln\left[1-\Delta G(\omega, 0)\right] &= -\int_{-\infty}^{\infty} \frac{\mathrm{d}\omega}{2\pi} \Delta G(\omega, 0) \\ &= \frac{\mathrm{i}\Delta}{2} \int_{-\infty}^{\infty} \frac{\mathrm{d}\omega}{\pi} \Gamma(\omega) = \frac{\mathrm{i}\Delta}{2} \ .\end{aligned} \tag{N.3}$$

Let us examine the second integral in (N.2). Using the notation $G(\omega, 0) = x - \mathrm{i}y$, we can write the argument of the logarithm as

$$Z = |Z|\,\mathrm{e}^{\mathrm{i}\varphi} = \frac{(1-\Delta x)^2 + (\Delta y)^2 - 2f(\Delta y)^2 - \mathrm{i}2f\Delta y(1-\Delta x)}{(1-\Delta x)^2 + (\Delta y)^2} \ . \tag{N.4}$$

The modulus and argument of this complex function are given by

$$|Z| = 1 - 4f(1-f)\frac{(\Delta y)^2}{(1-\Delta x)^2 + (\Delta y)^2}\,, \tag{N.5}$$

$$\varphi = \arctan\frac{2f\Delta y\,(1-\Delta x)}{(1-\Delta x)^2 + (\Delta y)^2 - 2f(\Delta y)^2}\,. \tag{N.6}$$

Returning to the notation

$$x = \Delta_{\mathrm{TLS}}(\omega)\,, \quad y = \Gamma_{\mathrm{TLS}}(\omega)\,, \quad \frac{y}{(1-\Delta x)^2 + (\Delta y)^2} = \Gamma^{\mathrm{e}}_{\mathrm{TLS}}(\omega)\,, \tag{N.7}$$

we arrive at the formula

$$\begin{aligned}\int\limits_{-\infty}^{\infty}\frac{\mathrm{d}\omega}{2\pi}\ln\left[1-\Delta G(\omega,T)\right] &\\ = \int\limits_{0}^{\infty}\frac{\mathrm{d}\omega}{4\pi}\ln\left\{1-4\left[1-f(\omega)\right]f(\omega)\,\Delta^2\Gamma^{\mathrm{e}}_{\mathrm{TLS}}(\omega)\,\Gamma_{\mathrm{TLS}}(\omega)\right\} &\\ +\frac{\mathrm{i}\Delta}{2} - \mathrm{i}\int\limits_{0}^{\infty}\frac{\mathrm{d}\omega}{2\pi}\arctan\frac{2f(\omega)\,\Delta\Gamma^{\mathrm{e}}_{\mathrm{TLS}}(\omega)\left[1-\Delta\Delta_{\mathrm{TLS}}(\omega)\right]}{1-2f(\omega)\,\Delta^2\Gamma^{\mathrm{e}}_{\mathrm{TLS}}(\omega)\,\Gamma_{\mathrm{TLS}}(\omega)}\,.\end{aligned} \tag{N.8}$$

Inserting this formula into (18.34), we arrive at (18.37) and (18.38).

O. Derivation of (18.53)

The determinant of (18.49) is given by

$$\mathrm{Det} = \frac{\exp\left(\mathrm{i}\bar{\Delta}_2 t\right)}{\Delta_1\bar{\Delta}_2} - \frac{\exp\left(\mathrm{i}\bar{\Delta}_1 t\right)}{\bar{\Delta}_1\Delta_2}\,. \tag{O.1}$$

The solution of (18.49) is given by the functions

$$S_1 = \frac{-\dfrac{1}{\Delta\bar{\Delta}_2}\exp\left(\mathrm{i}\bar{\Delta}_2 t\right)}{\mathrm{Det}}\,, \quad S_2 = \frac{\dfrac{1}{\Delta\bar{\Delta}_1}\exp\left(\mathrm{i}\bar{\Delta}_1 t\right)}{\mathrm{Det}}\,. \tag{O.2}$$

Using these formulas, we find

$$1 - S_1 - S_2 = \frac{\left(\dfrac{1}{\Delta}+\dfrac{1}{\Delta_1}\right)\dfrac{\exp\left(\mathrm{i}\bar{\Delta}_2 t\right)}{\bar{\Delta}_2} - \left(\dfrac{1}{\Delta}+\dfrac{1}{\Delta_2}\right)\dfrac{\exp\left(\mathrm{i}\bar{\Delta}_1 t\right)}{\bar{\Delta}_1}}{\dfrac{\exp\left(\mathrm{i}\bar{\Delta}_2 t\right)}{\Delta_1\bar{\Delta}_2} - \dfrac{\exp\left(\mathrm{i}\bar{\Delta}_1 t\right)}{\Delta_2\bar{\Delta}_1}}\,. \tag{O.3}$$

Allowing for the relations

$$\varDelta_1 + \bar{\varDelta}_2 = \varDelta_2 + \bar{\varDelta}_1 = -\varDelta \,, \tag{O.4}$$

we find

$$\frac{1}{\varDelta} + \frac{1}{\varDelta_1} = -\frac{\bar{\varDelta}_2}{\varDelta_1 \varDelta} \,, \qquad \frac{1}{\varDelta} + \frac{1}{\varDelta_2} = -\frac{\bar{\varDelta}_1}{\varDelta_2 \varDelta} \,. \tag{O.5}$$

Inserting this expression into (O.3) yields the result

$$1 - S_1 - S_2 = -\frac{1}{\varDelta} \frac{\dfrac{\exp(\mathrm{i}\bar{\varDelta}_2 t)}{\varDelta_1} - \dfrac{\exp(\mathrm{i}\bar{\varDelta}_1 t)}{\varDelta_2}}{\dfrac{\exp\left(\mathrm{i}\bar{\varDelta}_2 t\right)}{\varDelta_1 \bar{\varDelta}_2} - \dfrac{\exp\left(\mathrm{i}\bar{\varDelta}_1 t\right)}{\varDelta_2 \bar{\varDelta}_1}} = \frac{\mathrm{i}}{\varDelta} \frac{\mathrm{d}}{\mathrm{d}t} \ln \mathrm{Det} \,. \tag{O.6}$$

P. Derivation of (18.60)

Let us inspect the formulas

$$\begin{aligned} \varOmega_0 &= \pm \frac{1}{2\sqrt{2}} \left\{ \left[\left(\varDelta^2 - \gamma^2\right)^2 + 4\varDelta^2\gamma^2 \left(1 - 2f\right)^2 \right]^{1/2} + \varDelta^2 - \gamma^2 \right\}^{1/2} \,, \\ \gamma_0 &= \frac{1}{2\sqrt{2}} \left\{ \left[\left(\varDelta^2 - \gamma^2\right)^2 + 4\varDelta^2\gamma^2 \left(1 - 2f\right)^2 \right]^{1/2} - \varDelta^2 + \gamma^2 \right\}^{1/2} \,. \end{aligned} \tag{P.1}$$

The square root can be transformed as

$$\begin{aligned} \left[\left(\varDelta^2 - \gamma^2\right)^2 + 4\varDelta^2\gamma^2 \left(1 - 2f\right)^2 \right]^{1/2} &= \left[\left(\varDelta^2 + \gamma^2\right)^2 - 16\varDelta^2\gamma^2 f \left(1 - f\right) \right]^{1/2} \\ &= \left(\varDelta^2 + \gamma^2\right) \sqrt{1 - w} \,, \end{aligned} \tag{P.2}$$

where

$$w = \frac{16\varDelta^2\gamma^2}{\left(\varDelta^2 + \gamma^2\right)^2} f \left(1 - f\right) \leqslant 1 \,. \tag{P.3}$$

The left hand side of this inequality equals unity only in the special case when $f = 1/2$ and $\varDelta = \gamma$. In other cases the left hand side is much smaller than unity, and we may therefore carry out the expansion of the square root as a power series in this small value. To the first approximation, we arrive at the expression

$$\sqrt{1 - w} \approx 1 - w/2 = 1 - \frac{8\varDelta^2\gamma^2}{\left(\varDelta^2 + \gamma^2\right)^2} f \left(1 - f\right) \,. \tag{P.4}$$

Inserting this equation into (P.2), we find

$$\left(\Delta^2+\gamma^2\right)\sqrt{1-w}\approx\Delta^2+\gamma^2 \quad \frac{8\Delta^2\gamma^2}{\Delta^2+\gamma^2}f\,(1-f)\ . \tag{P.5}$$

Substituting this equation into the expression for Ω_0 yields the expression

$$\begin{aligned}\Omega_0 &\approx \pm\frac{1}{2\sqrt{2}}\sqrt{2\Delta^2-2\Delta^2\frac{4\gamma^2}{\Delta^2+\gamma^2}f\,(1-f)}\\ &=\frac{\Delta}{2}\sqrt{1-\frac{4\gamma^2}{\Delta^2+\gamma^2}f\,(1-f)}\approx\frac{\Delta}{2}-\frac{\Delta\gamma^2}{\Delta^2+\gamma^2}f\,(1-f)\ .\end{aligned} \tag{P.6}$$

A similar chain of transformations yields the following expression for the half-width:

$$\gamma_0\approx\frac{\gamma}{2}-\frac{\Delta^2\gamma}{\Delta^2+\gamma^2}f\,(1-f)\ . \tag{P.7}$$

Q. Stochastic Approach to Spectral Diffusion

If the electronic transition does not interact with TLSs, the temporal behavior of the dipolar correlator is described by the function

$$I(t)=\mathrm{e}^{-\mathrm{i}\Omega t}\ . \tag{Q.1}$$

The optical line shape of this system is given by the delta function,

$$I(\omega)=\int_{-\infty}^{\infty}I(t)\mathrm{e}^{\mathrm{i}\omega t}\mathrm{d}t=2\pi\delta(\omega-\Omega)\ . \tag{Q.2}$$

Suppose that the frequency of the electronic transition can jump among two positions Ω and $\Omega+\Delta$ due to the interaction with TLSs. Then the dipolar correlator can be described by

$$I(t,t_{\mathrm{w}})=\mathrm{e}^{-\mathrm{i}\Omega t}[1-\rho(t_{\mathrm{w}})]+\mathrm{e}^{-\mathrm{i}(\Omega+\Delta)t}\rho(t_{\mathrm{w}})\ , \tag{Q.3}$$

where ρ is the probability of finding the system with the resonant frequency $\Omega+\Delta$. It is obvious that the optical band of the chromophore interacting with one TLS consists of two delta functions with resonant frequencies Ω and $\Omega+\Delta$. The dipolar correlator does not approach zero on the long time scale t. Strictly speaking, this is therefore an unphysical model. However, it will be shown later that the interaction with an infinite number of TLSs corrects this shortcoming of the model.

It will be very useful to compare this expression for the dipolar correlator with that given by the dynamical theory in Sect. 20.1. The dipolar correlator is given by

$$I(t, t_{\mathrm{w}}) = \mathrm{e}^{-\mathrm{i}\delta t - \gamma t/2} \left[\mathrm{e}^{-\mathrm{i}\Omega t} [1 - C(t_{\mathrm{w}})] + \mathrm{e}^{-\mathrm{i}(\Omega + \Delta)t} C(t_{\mathrm{w}}) \right] , \tag{Q.4}$$

where

$$C(t_{\mathrm{w}}) = \frac{\Delta}{\Delta - \mathrm{i}R} \rho(t_{\mathrm{w}}) . \tag{Q.5}$$

There are two dfferences:

- the amplitude $C \to 0$ if $\Delta \to 0$,
- the correlator approaches zero at large t.

The optical band relating to this dipolar correlator consists of two Lorentzians. The dipolar correlator of the dynamical theory has no serious shortcomings, in contrast to (Q.3). Nevertheless, (Q.3) can serve as a starting point for the expression for the line width which is used in practice. The derivation goes as follows.

The probability ρ changes over a long time scale t_{w} according to the equation

$$\dot{\rho} = -P\rho + p(1 - \rho) = -R\rho + p , \tag{Q.6}$$

where P is the rate of transitions to the state with resonant frequency $\Omega + \Delta$ and p is the rate of transitions to the state with resonant frequency Ω. The solution of (Q.6) with $\rho(0) = 0$ is

$$\rho(t_{\mathrm{w}}) = \frac{p}{R} \left(1 - \mathrm{e}^{-Rt_{\mathrm{w}}} \right) = f(T) \left(1 - \mathrm{e}^{-Rt_{\mathrm{w}}} \right) , \tag{Q.7}$$

where $f(T) = [\exp(-\varepsilon/kT) + 1]^{-1}$ is the probability of finding the state with resonant frequency $\Omega + \Delta$ and energy ε of TLS splitting in thermal equilibrium.

If the electronic dipole moment interacts with many TLSs, the expression for the dipolar correlator is

$$I(t, t_{\mathrm{w}}) = \mathrm{e}^{-\mathrm{i}\Omega t} \prod_{j=1}^{N_0} \left[1 - \rho_j(t_{\mathrm{w}}) \left(1 - \mathrm{e}^{-\mathrm{i}\Delta_j t} \right) \right] . \tag{Q.8}$$

The next step in stochastic theories is to transform the last equation to

$$I(t, t_{\mathrm{w}}) \approx \exp \left[-\mathrm{i}\Omega t - \sum_{j=1}^{N_0} \rho_j(t_{\mathrm{w}}) \left(1 - \mathrm{e}^{-\mathrm{i}\Delta_j t} \right) \right] . \tag{Q.9}$$

Let us introduce a distribution function

$$N(\varepsilon, R, \Delta) = \sum_{j=1}^{N_0} \delta(\varepsilon - \varepsilon_j) \delta(R - R_j) \delta(\Delta - \Delta_j) . \tag{Q.10}$$

With the help of this distribution function, we can transform the sum in (Q.9) to obtain

$$\sum_{j=1}^{N_0} \rho_j(t_\mathrm{w}) \left(1 - \mathrm{e}^{-\mathrm{i}\Delta_j t}\right) = \int \mathrm{d}\varepsilon \int \mathrm{d}R \int \mathrm{d}\Delta N(\varepsilon, R, \Delta)\rho(\varepsilon, R) \left(1 - \mathrm{e}^{-\mathrm{i}\Delta t}\right) . \tag{Q.11}$$

As a rule, the following approximation to statistical independence is used:

$$N(\varepsilon, R, \Delta) = N_1(R) N_2(\varepsilon) N_3(\Delta) , \tag{Q.12}$$

where

$$\int \mathrm{d}\varepsilon N_1(\varepsilon) = \int \mathrm{d}R N_2(R) = 1 \tag{Q.13}$$

and

$$N_3(\Delta) = \sum_{j=1}^{N_0} \delta(\Delta - \Delta_j) . \tag{Q.14}$$

Therefore the dipolar correlator can be transformed to the expression

$$I(t, t_\mathrm{w}) \approx \exp\left[-\mathrm{i}\Omega t - \int \mathrm{d}\varepsilon N_2(\varepsilon) \int \mathrm{d}R N_1(R) \rho(\varepsilon, R, t_\mathrm{w}) \sum_{j=1}^{N_0} \left(1 - \mathrm{e}^{-\mathrm{i}\Delta_j t}\right)\right] . \tag{Q.15}$$

The sum over the TLS index j can be replaced by the sum over a lattice node index n as follows:

$$\sum_{j}^{N_0} \approx \frac{N_0}{N} \sum_{n}^{N} , \tag{Q.16}$$

where N is the number of nodes. The sum with respect to nodes can then be replaced by integration over space,

$$\begin{aligned} &I(t, t_w) \\ &\approx \exp\left[-\mathrm{i}\Omega t - \int \mathrm{d}\varepsilon N_2(\varepsilon) \int \mathrm{d}R N_1(R) \rho(\varepsilon, R, t_\mathrm{w}) \frac{N_0}{V} \int_V \left[1 - \mathrm{e}^{-\mathrm{i}\Delta(r)t}\right] \mathrm{d}V\right] . \end{aligned} \tag{Q.17}$$

The integral over space depends on time. For the dipole–dipole interaction between chromophore and TLS, the line shift can be taken in the form

$$\Delta(r) = DdF(\vartheta, \psi)/r^3 . \tag{Q.18}$$

The integral over space can thus be transformed to

$$\frac{N_0}{V}\int_V \left[1-\mathrm{e}^{-\mathrm{i}\Delta(r)t}\right]\mathrm{d}V = \frac{N_0}{V}\frac{Dd}{3}\left\langle \int_{\Delta_{\min}}^{\Delta_{\max}} \frac{\mathrm{d}\Delta}{\Delta^2}\left\{1-\exp[-\mathrm{i}F(\vartheta,\psi)\Delta t]\right\}\right\rangle , \tag{Q.19}$$

where angular brackets mean integration with respect to angles ϑ and ψ. In the long time limit, this expression takes the form

$$\begin{aligned} &\frac{N_0}{V}\frac{Dd}{3}\left\langle \int_{\Delta_{\min}}^{\Delta_{\max}} \frac{\mathrm{d}\Delta}{\Delta^2}\left\{1-\exp[-\mathrm{i}F(\vartheta,\psi)\Delta t]\right\}\right\rangle \\ &\qquad\approx \frac{N_0}{V}\frac{Dd}{3}\left[\mathrm{i}\left\langle F\ln\frac{1}{\Delta_{\min}Ft}\right\rangle + \pi\left\langle |F|\right\rangle |t|\right] . \end{aligned} \tag{Q.20}$$

According to this equation, the dipolar correlator approaches zero at large t. Indeed

$$I(t,t_{\mathrm{w}}) \propto \exp[-\gamma(t_{\mathrm{w}})\,|t|] , \tag{Q.21}$$

where

$$\gamma(t_{\mathrm{w}}) = \frac{N_0}{V}\frac{Dd}{3}\pi\left\langle |F|\right\rangle \int \mathrm{d}\varepsilon N_2(\varepsilon)\int \mathrm{d}R N_1(R) f(T)\left(1-\mathrm{e}^{-Rt_{\mathrm{w}}}\right) . \tag{Q.22}$$

This expression for the line width is used in many studies where spectral diffusion is discussed on the basis of the stochastic approach. It is interesting that, although the dipolar correlator of the chromophore interacting with one or more TLSs does not approach zero, the dipolar correlator of the chromophore interacting with an infinite number of TLSs does approach zero at large t. This new property of the dipolar correlator emerged when we replaced the sum with respect to the TLS index j by the integral over space in (Q.17).

Bibliography

1. Pohl D.W., Courjon D. (Eds.): Near Field Optics, NATO ASI Ser. E, Appl. Sci. **242** (Kluwer, Dordrecht, 1996)
2. Binning G., Roher H.: Rev. Mod. Phys. **59**, 615 (1987)
3. Moerner W.E., Kador L.: Phys. Rev. Lett. **62**, 2535 (1989)
4. Orrit M., Bernard J.: Phys. Rev. Lett. **63**, 2716 (1990)
5. Shpol'skii E.V., Il'ina A.A., Klimova L.A.: Dokl. Akad. Nauk SSSR **87**, C.935 (1952)
6. Shpol'skii E.V.: Usp. Fis. Nauk **71**, C.215 (1960)
7. Svishchev G.M.: Opt. Spektr. **16**, 96 (1964)
8. Denisov Iu.V., Kizel V.A.: Opt. Spektr. **23**, C.472 (1967)
9. Szabo A.: Phys. Rev. Lett. **25**, 924 (1970)
10. Personov R.I., Al'shits E.I., Bykovskaya L.A.: Opt. Comm. **6**, 169 (1972)
11. Kharlamov B.M., Personov R.I., Bykovskaya L.A.: Opt. Comm. **12**, 191 (1974)
12. Gorokhovski A., Kaarli R.K., Rebane L.A.: JETP Lett. **20**, 474 (1974)
13. Kimble H.J., Dagenais M., Mandel L.: Phys. Rev. A **18**, 201 (1978)
14. Dagenais M., Mandel L.: Phys. Rev. A **18**, 2217 (1978)
15. Diedrich F., Walther H.: Phys. Rev. Lett. **58**, 203 (1987)
16. Ohtani H., Wilson R.J., Chiang S., Mate C.M.: Phys. Rev. Lett. **60**, 2398 (1988)
17. Shera E.B., Seitzinger N.K., Davis L.M., Keller R.A., Soper S.A.: Chem. Phys. Lett. **174**, 553 (1990)
18. Soper S.A., Shera E.B., Martin J.S., Jett J.H., Hahn J.H., Nutter H.L., Keller R.A.: Anal. Chem. **63**, 432 (1991)
19. Hirschfeld T.: Appl. Opt.**15**, 2965, 3135 (1976)
20. Morikawa K., Yanagida M.: J. Biochem. **89**, 693 (1981)
21. Dovichi N.J., Martin J.C., Jett J.H., Trkula M., Keller R.A.: Anal. Chem. **56**, 348 (1984)
22. Itano W.M., Berquist L.C., Wineland D.J.: Science **237**, 612 (1987)
23. Wrachtrup J., von Borczyskowski C., Bernard J., Orrit M., Brown R.: Nature **363**, 244 (1993)
24. Kohler J., Disselhorst J.A.J.M., Donckers M.C.J.M., Grouenen E., Schmidt J., Moerner W.E.: Nature **363**, 242 (1993)
25. Wrachtrup J., von Borczyskowski C., Bernard J., Brown R., Orrit M.: Chem. Phys. Lett. **245**, 262 (1995)
26. Fleury L., Tamarat Ph., Lounis B., Bernard J., Orrit M.: Chem. Phys. Lett. **236**, 87 (1995)
27. Ambrose W.P., Basche Th., Moerner W.E.: J. Chem. Phys. **95**, 7150 (1991)
28. Thenio P., Myers A.B., Moerner W.E.: J. Lumin. **56**, 1 (1993)
29. Fleury L., Zumbusch A., Orrit M., Brown R., Bernard J.: J. Lumines. **56**, 15 (1993)

30. Funatsu T., Harada Y., Tokunaga M., Saito K., Yanagida T.: Nature **374**, 555 (1995)
31. Noji H., Yasuda R., Yoshida M., Kinosita K.: Nature **386**, 299 (1997)
32. Dickson R.M., Cubitt A.B., Tsien R., Moerner W.E.: Nature **388**, 355 (1997)
33. van Oijen A.M., Ketelaars M., Kohler J., Aartsma T.J., Schmidt J.: Science **285**, 400 (1999)
34. Weiss S.: Science **283**, 1676 (1999)
35. Basche Th., Moerner W.E., Orrit M., Wild U. (Eds.): Single Molecule Optical Detection, Imaging and Spectroscopy (VCH, Weinheim, New York, Basel, Cambridge, Tokyo, 1996)

Chapter 1

36. Landau L.D., Lifshits E.M.: Teoriya polya, Fizmatgiz, Moskva (1960)
37. Goldstein H.: Classical Mechanics (Addison-Wesley, Cambridge, 1950)
38. Davydov A.S.: Kvantovaya mechanika (GIFML, Moskva, 1963)

Chapter 2

39. Ahiezer A.I., Beresteckii V.B.: Kvantovaya elektrodinamika (GIPhML, Moskva, 1963)

Chapter 3

40. Osad'ko I.S.: JETP **86**, 875 (1998)
41. Bloch F.: Phys. Rev. **70**, 460 (1946)

Chapter 4

42. Born M., Oppeheimer R.: Ann. Physik **84**, 457 (1927)
43. Born M., Huang K.: Dynamical Theory of Crystal Lattices (Clarendon Press, Oxford, 1954)
44. Franck J.: Trans. Faraday Soc. **21**, 536 (1925)
45. Condon E.: Phys. Rev. **28**, 1182 (1926); **32**, 858 (1928)

Chapter 6

46. Vol'kenshtein M.V., Gribov L.A., El'iashevich M.A., Stepanov B.I.: Kolebaniia molekul (Nauka, Moskva, 1983)
47. Osad'ko I.S.: In: Spectroscopy and Excitation Dynamics of Condensed Molecular Systems, ed. by V. Agranovich, R. Hochstrasser (North-Holland, Amsterdam, 1983)
48. Kulagin S.A., Osad'ko I.S.: Phys. Stat. Sol. (b) **110**, 57–67 (1982)

49. Zeller R.C., Pohl R.O.: Phys. Rev. **B4**, 2029 (1971)
50. Anderson P.W., Halperin B.I., Varma C.M.: Philos. Mag. **25**, 1 (1972)
51. Phillips W.A.: J. Low Temp. Phys. **7**, 351 (1972)
52. Phillips W.A. (Ed.): Topics in Current Physics Vol. 24, Amorphous Solids: Low Temperature Properties (Springer, Heidelberg, 1981)

Chapter 7

53. DeVoe R.G., Brewer R.G.: Phys. Rev. Lett. **50**, 1269 (1983)

Chapter 9

54. Kubo R.: Fluctuation, Relaxation and Resonance in Magnetic Systems, ed. by D. ter Haar (Oliver Boyd, Edinburgh, 1962)
55. Anderson P.W., Weiss P.R.: Rev. Mod. Phys. **25**, 269 (1953)
56. Talon H., Fleury L., Bernard J., Orrit M.: J. Opt. Soc. Am. **B9**, 825–828 (1992)
57. Anderson P.W.: J. Phys. Soc. Japan **9**, 316 (1954)

Chapter 10

58. Lax M.: J. Chem. Phys. **20**, 1752 (1952)
59. Kubo R., Toyozawa Y.: Progr. Theor. Phys. **13**, 160 (1955)
60. Pekar S.I.: Issledovanie po elektronnoi teorii kristallov (Gostehizdat, 1951)
61. Huang K., Rhys A.: Proc. Roy. Soc. **204**, 406 (1950)
62. Personov R.I., Osad'ko I.S., Godiaev E.D., Al'shic E.I.: Sov. Phys. Solid State **13**, 2223 (1971)

Chapter 11

63. Avarmaa R.: Opt. Spektr. **32**, 959 (1972)
64. Osad'ko I.S.: Sov. Phys. Usp. **22**, 311 (1979)
65. Korotaev O.N., Kaliteevskii M.Iu.: Sov. Phys. JETP **52**, 220 (1980)
66. Nersesova G.N., Shtrokirh O.Ph.: Opt. Spektr. (USSR), **44**, 58 (1978)
67. Nersesova G.N., Chigirev A.R., Egenburg Ph., Shtrokirh O.Ph., Gliadkovskii V.I.: Opt. spektr. **54**, 635 (1983)
68. Gastilovich E.A.: Sov. Phys. Usp. **34**, 592 (1991)
69. Plakhotnic T.V., Personov R.I., Gastilovich E.A.: Chem. Phys. **150**, 429 (1991)
70. Broude V.L., Rashba E.I., Sheka E.Ph.: Spektroskopiia molekuliarnih eksitonov (Energoizdat, Moskva, 1981)

Chapter 12

71. McCumber D.E., Sturge M.D.: J. Appl. Phys. **34**, 1682 (1963)
72. Krivoglaz M.A.: Sov. Phys. Solid State **6**, 1340 (1964)
73. Osad'ko I.S.: Sov. Phys. Solid State **13**, 974 (1971); **14**, 2252 (1972); **17**, 2098 (1975)
74. Osad'ko I.S.: Sov. Phys. Solid State **15**, 1614 (1974)
75. Osad'ko I.S.: Sov. Phys. JETP **45**, 827 (1977)
76. Hsu D., Skinner J.L.: J. Chem. Phys. **81**, 1606, 5471 (1984); **83**, 2097, 2107 (1985)
77. Hsu D., Skinner J.L.: J. Lumines. **37**, 331 (1987)
78. Zaitsev N.N., Osad'ko I.S.: JETP **77**, 950 (1993)
79. Donskoi D.V., Zaitsev N.N., Osad'ko I.S.: Chem. Phys. **176**, 135 (1993)
80. Molenkamp L.W., Wiersma D.A.: J. Chem. Phys. **80**, 3054 (1984)
81. Hesselink W.E., Wiersma D.A.: J. Chem. Phys. **73**, 648 (1980)

Chapter 13

82. Personov R.I.: Spectrochimica Acta **386**, 1535 (198[illegible])
83. Bykovskaya L.A., Personov R.I., Romanovskii Yu.V.: Analytica Chimica Acta **125**, 1 (1981)

Chapter 14

84. Small G.: In: Spectroscopy and Excitation Dynamics of Condensed Molecular Systems, ed. by V. Agranovich, R. Hochstrasser (North-Holland, Amsterdam, 1983)
85. Haarer D.: In: Persistent Spectral Hole Burning: Science and Applications, ed. by W.E. Moerner (Springer, 1987)
86. Winnecker A., Shelby R.M., Macfarlane R.M.: Opt. Lett. **10**, 350 (1985)
87. Korotaev O.N., Donskoi E.I., Gliadkovkii V.I.: Opt. spektr. **59**, 492 (1985)

Chapter 15

88. Hahn E.L.: Phys. Rev. **80**, 580 (1950)
89. Mims W.B.: In: Electron Paramagnetic Resonance, ed. by S. Geschwind (Plenum, NY, 1972)
90. Kurnit N.A., Abella I.D., Hartmann S.R.: Phys. Rev. Lett. **13**, 567 (1464)
91. Brewer R.G., Shoemaker R.L.: Phys. Rev. **A6**, 2001 (1972)
92. Allen L., Eberly J.H.: Optical Resonance and Two-Level Atoms, (Wiley Interscience Publication, NY, 1975)
93. Hesselink W.H., Wiersma D.A.: In: Spectroscopy and Excitation Dynamics of Condensed Molecular Systems, ed. by V. Agranovich, R. Hochstrasser (North-Holland, Amsterdam, 1983)
94. Mukamel S.: Principles of Nonlinear Optical Spectroscopy (Oxford University Press, NY, 1995)

Chapter 19

95. Osad'ko I.S.: JETP **89**, 513 (1999)
96. Osad'ko I.S., Yershova L.B.: J. Chem. Phys. **112**, 9645 (2000)

Chapter 20

97. Mims W.B., Nassau K., McGee J.D.: Phys. Rev. **123**, 2059 (1961)
98. Klauder R., Anderson P.W.: Phys. Rev. **125**, 912 (1962)
99. Hu P., Hartmann S.B.: Phys. Rev. **B9**, 1 (1974)
100. Hu P., Walker L.R.: Phys. Rev. **B18**, 1300 (1978)
101. Black J.L., Halperin B.I.: Phys. Rev. **B16**, 2879 (1977)
102. Reinecke T.L.: Solid State Commun. **32**, 1103 (1989)
103. Bay Y.S., Fayer M.D.: Phys. Rev. **B39**, 11066 (1989)
104. Reilly P.D., Skinner J.L.: J. Chem. Phys. **101**, 959, 965 (1994)
105. Osad'ko I.S.: JETP **82**, 434 (1996)
106. Littau K., Bay Y.S., Fayer M.D.: J. Chem. Phys. **92**, 4145 (1990)
107. Maier H., Wunderlich R., Haarer D., Kharlamov B.M., Kulikov S.G.: Phys. Rev. Lett. **74**, 5252 (1995)
108. Kohler W., Meiler J., Friedrich J.: Phys. Rev. **B35**, 4031 (1987)
109. Maier H., Kharlamov B.M., Haarer D.: Phys. Rev. Lett. **76**, 2085 (1996)
110. Zilker S.J., Haarer D.: Chem. Phys. **220**, 167 (1997)

Chapter 22

111. Lu H.P., Xun L., Xie X.S.: Science **282**, 1877 (1998)
112. Kozankevich B., Bernard J., Orrit J.: J. Chem. Phys. **101**, 9377 (1994)
113. Muller K.-P., Haarer D.: Phys. Rev. Lett. **66**, 2344 (1991)
114. Thijssen H.P.H., van der Berg R., Volker S.: Chem. Phys. Lett. **97**, 295 (1983)
115. Volker S.: Ann. Rev. Phys. Chem. **40**, 499 (1989)
116. Osad'ko I.S., Yershova L.B.: J. Chem. Phys. **111**, 7652 (1999)
117. Boiron A.M., Tamarat Ph., Lounis B., Brown R., Orrit M.: Chem. Phys. **247**, 119 (1999)

Appendices

118. Abrikosov A.A., Gor'kov L.P., Dzialoshinskii I.E.: Metodi Kvantovoi teorii polia v statisticheskoi phizike (GIPhML, Moskva, 1962)

Index

Springer Series in Chemical Physics

Editors: Vitalii I. Goldanskii Fritz P. Schäfer J. Peter Toennies

1 **Atomic Spectra and Radiative Transitions**
By I. I. Sobelman
2 **Surface Crystallography by LEED**
Theory, Computation and Structural Results. By M. A. Van Hove, S. Y. Tong
3 **Advances in Laser Chemistry**
Editor: A. H. Zewail
4 **Picosecond Phenomena**
Editors: C. V. Shank, E. P. Ippen, S. L. Shapiro
5 **Laser Spectroscopy**
Basic Concepts and Instrumentation
By W. Demtröder 3rd Printing
6 **Laser-Induced Processes in Molecules** Physics and Chemistry
Editors: K. L. Kompa, S. D. Smith
7 **Excitation of Atoms and Broadening of Spectral Lines** By I. I. Sobelman, L. A. Vainshtein, E. A. Yukov
8 **Spin Exchange**
Principles and Applications in Chemistry and Biology
By Yu. N. Molin, K. M. Salikhov, K. I. Zamaraev
9 **Secondary Ion Mass Spectrometry SIMS II** Editors: A. Benninghoven, C. A. Evans, Jr., R. A. Powell, R. Shimizu, H. A. Storms
10 **Lasers and Chemical Change**
By A. Ben-Shaul, Y. Haas, K. L. Kompa, R. D. Levine
11 **Liquid Crystals of One- and Two-Dimensional Order**
Editors: W. Helfrich, G. Heppke
12 **Gasdynamic Laser** By S. A. Losev
13 **Atomic Many-Body Theory**
By I. Lindgren, J. Morrison
14 **Picosecond Phenomena II**
Editors: R. M. Hochstrasser, W. Kaiser, C. V. Shank
15 **Vibrational Spectroscopy of Adsorbates** Editor: R. F. Willis
16 **Spectroscopy of Molecular Excitions**
By V. L. Broude, E. I. Rashba, E. F. Sheka
17 **Inelastic Particle-Surface Collisions**
Editors: E. Taglauer, W. Heiland
18 **Modelling of Chemical Reaction Systems** Editors: K. H. Ebert, P. Deuflhard, W. Jäger
19 **Secondary Ion Mass Spectrometry SIMS III**
Editors: A. Benninghoven, J. Giber, J. László, M. Riedel, H. W. Werner
20 **Chemistry and Physics of Solid Surfaces IV** Editors: R. Vanselow, R. Howe
21 **Dynamics of Gas-Surface Interaction**
Editors: G. Benedek, U. Valbusa
22 **Nonlinear Laser Chemistry**
Multiple-Photon Excitation
By V. S. Letokhov
23 **Picosecond Phenomena III**
Editors: K. B. Eisenthal, R. M. Hochstrasser, W. Kaiser, A. Laubereau
24 **Desorption Induced by Electronic Transitions DIET I** Editors: N. H. Tolk, M. M. Traum, J. C. Tully, T. E. Madey
25 **Ion Formation from Organic Solids**
Editor: A. Benninghoven
26 **Semiclassical Theories of Molecular Scattering** By B. C. Eu
27 **EXAFS and Near Edge Structures**
Editors: A. Bianconi, L. Incoccia, S. Stipcich
28 **Atoms in Strong Light Fields**
By N. B. Delone, V. P. Krainov
29 **Gas Flow in Nozzles**
By U. G. Pirumov, G. S. Roslyakov
30 **Theory of Slow Atomic Collisions**
By E. E. Nikitin, S. Ya. Umanskii
31 **Reference Data on Atoms, Molecules, and Ions** By A. A. Radzig, B. M. Smirnov
32 **Adsorption Processes on Semiconductor and Dielectric Surfaces I**
By V. F. Kiselev, O. V. Krylov
33 **Surface Studies with Lasers**
Editors: F.R. Aussenegg, A. Leitner, M.E. Lippitsch
34 **Inert Gases**
Potentials, Dynamics, and Energy Transfer in Doped Crystals. Editor: M. L. Klein
35 **Chemistry and Physics of Solid Surfaces V** Editors: R. Vanselow, R. Howe
36 **Secondary Ion Mass Spectrometry, SIMS IV** Editors: A. Benninghoven, J. Okano, R. Shimizu, H. W. Werner
37 **X-Ray Spectra and Chemical Binding**
By A. Meisel, G. Leonhardt, R. Szargan
38 **Ultrafast Phenomena IV**
By D. H. Auston, K. B. Eisenthal
39 **Laser Processing and Diagnostics**
Editor: D. Bäuerle

Springer Series in Chemical Physics

Editors: Vitalii I. Goldanskii Fritz P. Schäfer J. Peter Toennies

Managing Editor: H. K. V. Lotsch

40 **High-Resolution Spectroscopy of Transient Molecules**
By E. Hirota

41 **High Resolution Spectral Atlas of Nitrogen Dioxide 559–597 nm**
By K. Uehara and H. Sasada

42 **Antennas and Reaction Centers of Photosynthetic Bacteria**
Structure, Interactions, and Dynamics
Editor: M. E. Michel-Beyerle

43 **The Atom-Atom Potential Method**
Applications to Organic Molecular Solids
By A. J. Pertsin and A. I. Kitaigorodsky

44 **Secondary Ion Mass Spectrometry SIMS V**
Editors: A. Benninghoven, R. J. Colton, D. S. Simons, and H. W. Werner

45 **Thermotropic Liquid Crystals, Fundamentals**
By G. Vertogen and W. H. de Jeu

46 **Ultrafast Phenomena V**
Editors: G. R. Fleming and A. E. Siegman

47 **Complex Chemical Reaction Systems**
Mathematical Modelling and Simulation
Editors: J. Warnatz and W. Jäger

48 **Ultrafast Phenomena VI**
Editors: T. Yajima, K. Yoshihara, C. B. Harris, and S. Shionoya

49 **Vibronic Interactions in Molecules and Crystals**
By I. B. Bersuker and V. Z. Polinger

50 **Molecular and Laser Spectroscopy**
By Zu-Geng Wang and Hui-Rong Xia

51 **Space-Time Organization in Macro molecular Fluids**
Editors: F. Tanaka, M. Doi, and T. Ohta

52 **Multiple-Photon Laser Chemistry**
By. R. V. Ambartzumian, C. D. Cantrell, and A. Puretzky

53 **Ultrafast Phenomena VII**
Editors: C. B. Harris, E. P. Ippen, G. A. Mourou, and A. H. Zewail

54 **Physics of Ion Impact Phenomena**
Editor: D. Mathur

55 **Ultrafast Phenomena VIII**
Editors: J.-L. Martin, A. Migus, G. A. Mourou, and A. H. Zewail

56 **Clusters of Atoms and Molecules**
Solvation and Chemistry of Free Clusters, and Embedded, Supported and Compressed Clusters
Editor: H. Haberland

57 **Radiationless Transitions in Polyatomic Molecules**
By E. S. Medvedev and V. I. Osherov

58 **Positron Annihilation in Chemistry**
By O. E. Mogensen

59 **Soot Formation in Combustion**
Mechanisms and Models
Editor: H. Bockhorn

60 **Ultrafast Phenomena IX**
Editors: P. Barbara, W. H. Knox, G. A. Mourou, and A. H. Zewail

61 **Gas Phase Chemical Reaction Systems**
Experiments and Models 100 Years After Max Bodenstein
Editors: J. Wolfrum, H.-R. Volpp, R. Rannacher, and J. Warnatz

62 **Ultrafast Phenomena X**
Editors: P. F. Barbara, J. G. Fujimoto, W. H. Knox, and W. Zinth

Druck: Strauss Offsetdruck, Mörlenbach
Verarbeitung: Schäffer, Grünstadt